PARTIALLY ORDERED ABELIAN GROUPS WITH INTERPOLATION

MATHEMATICAL SURVEYS AND MONOGRAPHS

NUMBER 20

PARTIALLY ORDERED ABELIAN GROUPS WITH INTERPOLATION

K. R. GOODEARL

American Mathematical Society
Providence, Rhode Island

1980 *Mathematics Subject Classification* (1985 *Revision*). Primary 06F20, 46A55; Secondary 19A49, 19K14, 16A30, 16A54, 46L05, 46L80.

Library of Congress Cataloging-in-Publication Data

Goodearl, K. R.

Partially ordered abelian groups with interpolation.

(Mathematical surveys and monographs, ISSN 0076-5376; no. 20)

Bibliography: p.

Includes index.

1. Abelian groups. 2. Interpolation. I. Title. II. Series.

QA171.G625 1986 512'.2 86-7876

ISBN 0-8218-1520-2 (alk. paper)

Contents

PREFACE ix

PROLOGUE: PARTIALLY ORDERED GROTHENDIECK GROUPS xiii

NOTATIONAL CONVENTIONS xxi

1. BASIC NOTIONS 1

• Partially ordered abelian groups • Infima and suprema • Ideals and quotient groups • Categories of partially ordered abelian groups • Pullbacks, pushouts, and coproducts • Additional concepts

2. INTERPOLATION 22

• Riesz interpolation and decomposition properties • Ideals and quotient groups • Extensions • Products, pullbacks, and pushouts • 2-unperforated interpolation groups • Relatively bounded homomorphisms

3. DIMENSION GROUPS 44

• Dimension groups • Products, pullbacks, and pushouts • Simplicial groups • Direct limits of simplicial groups

4. STATES 60

• Existence • Values of states • Uniqueness • Additional uniqueness criteria • Discrete states

5. COMPACT CONVEX SETS 73

• Basic definitions • Categorical concepts • Extreme points and faces • Separation by hyperplanes • Existence of extreme points • Probability measures • Faces of probability measures

6. State Spaces 94

• Basic structure • Some examples • Functoriality • Products and limits • Faces • Change of order-unit • Discrete states

7. Representation by Affine Continuous Functions 113

• Affine continuous function spaces • Affine representations • Order-unit norms • Bounded homomorphisms

8. General Comparability 126

• Characteristic elements • Projection bases • Comparability • Extremal states • Closures of faces • Functional representations

9. Dedekind σ-Completeness 141

• Prototypical examples • Additional examples • General comparability • Functional representations

10. Choquet Simplices 153

• Simplices • Faces • Complementary faces • Choquet simplices • Categorical properties

11. Affine Continuous Functions on Choquet Simplices 166

• Interpolation • Inverse limits • Semicontinuous functions • Compact sets of extreme points • Closed faces • Complementary faces

12. Metric Completions 188

• Completions with respect to positive homomorphisms • Dedekind completeness • Completions with respect to extremal states • Criterion for extremal states • Closed faces

13. Affine Continuous Functions on State Spaces 207

• Approximations • Compact sets of extremal states • Closed faces

14. Simple Dimension Groups 217

• Simplicity • State spaces • Classification • Finite-dimensional state spaces • Finite rank

15. Norm-Completeness 236

• Norm-completeness • Norm-completions • Quotient groups • Functional representations • Compact sets of extremal states • Closed faces • Maximal ideals

16. COUNTABLE INTERPOLATION AND MONOTONE σ-COMPLETENESS 263

• Countable interpolation • Monotone σ-completeness • Norm-completeness • Functional representations • Compact sets of extremal states • Closed faces • Quotient groups

17. EXTENSIONS OF DIMENSION GROUPS 285

• Extensions • Some examples • Extensions with order-units • Some examples • Existence of extensions with order-units

EPILOGUE: FURTHER K-THEORETIC APPLICATIONS 309

OPEN PROBLEMS 317

BIBLIOGRAPHY 325

INDEX 333

Preface

Ordered algebraic structures—particularly ordered fields, ordered groups, and ordered vector spaces—have a well-established tradition in mathematics, partly due to their intrinsic interest, and partly due to their applications in other areas. A new branch of this subject, motivated by K-theoretic applications (as sketched in the Prologue and the Epilogue), has grown out during the past decade. This branch, which is mainly concerned with partially ordered abelian groups satisfying the Riesz interpolation property ("interpolation groups", for the sake of abbreviation), is the subject of the present book. The purpose of the book is to provide a solid foundation in the theory of interpolation groups, for the use of researchers or students pursuing the applications. In particular, detailed developments are presented for that part of the subject on which a majority of the current applications rest. That a few topics related to newly developing applications are not included is due partly to the lack of a good perspective on these evolving developments, and partly to the considerable additional space that would be required to treat them coherently.

Although interpolation groups are defined as purely algebraic structures, their development has been strongly influenced by functional analysis. The interpolation property itself was introduced in F. Riesz's fundamental 1940 paper, as a key to his investigations of partially ordered real vector spaces, and the study of ordered vector spaces with the interpolation property has been continued by many functional analysts since that time. A number of the techniques developed in this area have been adapted to interpolation groups, after being stripped of scalars. On the other hand, parts of functional analysis dealing with compact convex sets have been applied to interpolation groups via certain dual objects. Any partially ordered abelian group possessing an "order-unit" (i.e., a positive element whose positive multiples provide upper bounds for all elements of the group) has a natural dual object called its "state space", consisting of all normalized positive real-valued homomorphisms on the group. This state space is a compact convex subset of the space of all real-valued functions on the group, and in the case of an interpolation group, the state space is a very special type of compact convex set known as a "Choquet simplex". A major line of development for interpolation groups has concerned the interrelationships among an interpolation group, its state space, and the partially ordered real Banach space of all affine continuous real-valued functions on the state space.

This cross-cultural development has left interpolation groups somewhat estranged from both algebraists, who may feel intimidated by compact convex sets, and functional analysts, who may feel handicapped by the lack of scalars. The intention of this book is to make the subject of interpolation groups accessible to readers from each culture, by providing sufficient details for both the algebraic and analytic aspects of the subject. Aside from a few basic concepts from the general theories of abelian groups and partially ordered sets, the algebraic theory of interpolation groups is developed directly from the definitions. A bit of the language of category theory is used in places, and the reader may need to remind himself of a few basic concepts such as products and coproducts, direct and inverse limits, and pullbacks and pushouts. On the other hand, the analytic theory of interpolation groups depends on selected nontrivial portions of the theory of compact convex sets. This is developed, as needed, from an elementary level, assuming only that the reader is (or can become) familiar with some very basic functional analysis, such as linear topological spaces, the Hahn-Banach Theorem, and the Riesz Representation Theorem. Readers with stronger backgrounds can easily skim over any familiar parts of this development.

The first four chapters of the book are purely algebraic, beginning with introductions to partially ordered abelian groups in general and to interpolation groups in particular. Of necessity, the highest density of definitions occurs in these introductory chapters, and some readers may therefore wish to skim the first two chapters, referring back to them later as needed. Chapter 3 is concerned with "dimension groups", which may be described as upward-directed interpolation groups in which positive integers can be cancelled from inequalities. (The importance of the class of dimension groups lies in the fact that in most applications, all the interpolation groups which appear are dimension groups.) A major milestone of the theory is located here, namely, a structural characterization of dimension groups as direct limits of finite products of copies of the integers. Chapter 4, another introductory chapter, is concerned with states (normalized positive real-valued homomorphisms on a partially ordered abelian group with an order-unit) as individual functions. Existence theorems for states, formulas for the ranges of values of states on group elements, and conditions for the uniqueness of states are developed here.

In Chapter 5, a purely functional-analytic interlude occurs, designed to introduce the more algebraic reader to some basic aspects of compact convex sets. In addition to presenting a number of basic concepts, the Krein-Mil′man Theorem (the fundamental existence theorem for extreme points of compact convex sets) is derived. The following two chapters introduce the connections through which functional analysis is brought to bear on interpolation groups. Chapter 6 is concerned with the state space (the collection of all states) on a partially ordered abelian group with order-unit, and with the structure of the state space as a compact convex subset of the linear topological space of all real-valued functions on the group, while Chapter 7 is concerned with the natural representation of the group (via evaluations) as affine continuous real-valued functions on the state space. Also included here is another section of functional analysis, on spaces of affine continuous real-valued functions on compact convex sets.

A long-range goal is to describe any interpolation group with order-unit as completely as possible in terms of affine continuous functions on its state space. The cases in which this goal is most readily accessible are developed in Chapters 8 and 9, namely, the cases of interpolation groups with a sufficiently large supply of direct sum decompositions, and Dedekind σ-complete lattice-ordered abelian groups. Here is a second milestone: a complete description of any Dedekind σ-complete lattice-ordered abelian group with order-unit in terms of continuous real-valued functions on a basically disconnected compact Hausdorff space (the collection of extremal states).

The remaining portions of functional analysis needed in the book, namely, some of the general properties of Choquet simplices and their affine continuous function spaces, are developed in Chapters 10 and 11. In particular, Choquet simplices are characterized as those compact convex sets on which the space of affine continuous real-valued functions is an interpolation group. The following chapter concerns completions of interpolation groups with respect to pseudometrics derived from states, a process by which certain questions and calculations can be reduced to the case of a Dedekind complete lattice-ordered abelian group.

In the remainder of the book, some of the fruits of the combined algebraic and analytic developments appear. For example, a characterization of the closure of the image of the natural affine continuous function representation of an interpolation group with order-unit is obtained in Chapter 13. This result provides a strong approximation to the desired description of the interpolation group in terms of affine continuous functions on its state space and may be regarded as another milestone of the theory, in spite of its rather technical nature. One effective application of this result is the development, in Chapter 14, of a complete description and classification of all "simple" dimension groups (i.e., dimension groups with no nontrivial ideals) in terms of Choquet simplices.

Chapter 15 is concerned with the class of those interpolation groups (with order-unit) which are complete with respect to the natural norm derived from their states. For such a group, the approximation theorems of Chapter 13 yield an exact description of the group as isomorphic to a well-described group of affine continuous functions. In addition, there is a strong duality between the algebraic structure of the group and the geometric structure of its state space. For instance, there are natural bijections between the extremal states and the maximal ideals of the group, and between the closed faces of the state space and the norm-closed ideals of the group. In Chapter 16, which may be viewed as a corollary of the preceding chapter, the results from the norm-complete case find application to groups which either satisfy the countably infinite version of the interpolation property or are monotone σ-complete.

The final chapter is a signpost indicating one of the directions in which open problems may be found. It is an introduction to the problem of classifying dimension groups arising as extensions of one given dimension group by another and also serves to present some techniques for constructing new dimension groups as extensions of old ones. A list of some open problems concerning interpolation groups is given following the Epilogue.

To use this book for an introductory course on interpolation groups, we suggest developing the basic concepts and methods as far as the classification of simple dimension groups (Chapter 14) and the structure theory of archimedean norm-complete dimension groups (Chapter 15). In such a course, the following sections may be omitted without loss of continuity: Chapter 1—Pullbacks, pushouts, and coproducts; Chapter 2—Extensions; Products, pullbacks, and pushouts; Chapter 3—Products, pullbacks, and pushouts; Chapter 4—Additional uniqueness criteria; Chapter 5—Categorical concepts; Faces of probability measures; Chapter 6—Products and limits; Change of order-unit; Chapter 9—Additional examples; Chapter 10—Complementary faces; Categorical properties; Chapter 11—Inverse limits; Complementary faces; Chapter 14—Finite-dimensional state spaces; Finite rank; Chapter 15—Norm-completions. In a few instances, an item from one of these sections is referred to in an example or a discussion in another section, but this should cause the reader little trouble.

A more drastic shortcut, for the reader who wishes to see the general structure theory of interpolation groups before delving into the supporting technical details, would be to skip over most of Chapters 8–13, referring back to the statements of the results in these chapters as needed. In such a case, the reader may find it helpful to read through the definition of a simplex and the section on Choquet simplices in Chapter 10, as well as the statements of Theorem 11.4 and Corollary 13.6, before embarking on Chapter 14.

We have resisted the temptation to coin any abbreviation for the term "partially ordered abelian group". (In particular, the ugliness of the acronym "poag" makes it easy to reject.) In discussions and in statements of theorems, we have pedantically included the phrase "as an ordered group" wherever it seemed useful to avoid potential confusion. This usage is intended to make hypotheses and conclusions more readily apparent to the browsing reader, and any reader can more easily disregard redundant usage than worry over omissions.

It is a pleasure to acknowledge the influences which led to my interest in and research on interpolation groups: George Elliott's work on the classification of ultramatricial algebras and approximately finite-dimensional C^*-algebras and David Handelman's work on the structure of $\aleph_0$-continuous regular rings and finite Rickart C^*-algebras, via the partially ordered Grothendieck group K_0. Their lead in developing other applications for partially ordered K_0 has helped to maintain the momentum of the research and to attract other mathematicians to the area. In regard to the manuscript, I would like to thank Robert Phelps for reading an early draft of the functional analysis sections and offering comments which helped improve the exposition.

Prologue: Partially Ordered Grothendieck Groups

Motivation for the study of partially ordered abelian groups has come from many different parts of mathematics, for mathematical systems with compatible order and additive (or linear) structures are quite common. This is particularly evident in functional analysis, where spaces of various kinds of real-valued functions provide impetus for investigating partially ordered real vector spaces. In the past decade, the observation that a Grothendieck group (such as K_0 of a ring or algebra) often possesses a natural partially ordered abelian group structure has led to new directions of investigation, whose goals have been to develop structure theories for certain types of partially ordered abelian groups to the point where effective application to various Grothendieck groups is possible. These recent developments in the area of partially ordered abelian groups are the subject of this book. In order to give the reader some hint of the motivations for these developments, we present here a sketch of the construction of Grothendieck groups as abelian groups equipped with pre-orderings that are often partial orderings, together with brief sketches of several situations in which the theory of partially ordered abelian groups (particularly those known as interpolation groups) can be applied, via the Grothendieck groups K_0, to the study of certain rings and C^*-algebras. This discussion is admittedly cursory, in the interest of avoiding technicalities, and for reasons of space. For the same reasons, we do not discuss the recent applications of interpolation groups to topological Markov chains [**88**, **22**, **65**, **68**] or to positive polynomials and compact group actions [**70**, **71**, **73**, **74**]. We shall continue the discussion in the Epilogue, where further applications of the theory developed in the book will be exhibited.

A Grothendieck group is an invariant attached to a collection of objects, such as the vector bundles on a given topological space, the finitely generated projective modules over a given ring, or the projection operators associated with a given C^*-algebra. The construction only requires a collection of objects equipped with a notion of isomorphism and some means of combining any two objects from the collection into a third object. While it is traditional to develop Grothendieck

---

Versions of some of the discussions in the Prologue and the Epilogue appear in [**54**, **55**].

groups using short exact sequences as the means of combining objects, in many important applications this reduces to direct sums (or direct products). Since the construction process is simpler using direct sums, we shall restrict our discussion to that case.

Thus for our basic data we take a set $\mathcal{P}$ of objects in some category, such that $\mathcal{P}$ has a zero object and every finite set of objects in $\mathcal{P}$ has a direct sum (coproduct) in $\mathcal{P}$. The easiest algebraic system to build using this data is an abelian semigroup, whose elements are the isomorphism classes of objects in $\mathcal{P}$ and whose operation is addition induced from the direct sum operation in $\mathcal{P}$. However, the semigroup obtained in this manner need not have cancellation and so cannot always be embedded in an abelian group. This problem can be circumvented either by reducing the semigroup modulo a suitable congruence relation or, more conveniently, by using an equivalence relation on $\mathcal{P}$ slightly coarser than isomorphism, as follows.

Objects $A, B \in \mathcal{P}$ are said to be *stably isomorphic* (in $\mathcal{P}$) if and only if there is an object $C \in \mathcal{P}$ such that $A \oplus C \cong B \oplus C$. Stable isomorphism is an equivalence relation on $\mathcal{P}$, and we write $[A]$ for the stable isomorphism class of an object $A \in \mathcal{P}$. Let $\text{Grot}(\mathcal{P})^+$ denote the collection of all stable isomorphism classes in $\mathcal{P}$. The direct sum operation in $\mathcal{P}$ induces an addition operation in $\text{Grot}(\mathcal{P})^+$, where $[A]+[B] = [A \oplus B]$ for all $A, B \in \mathcal{P}$, and using this operation, $\text{Grot}(\mathcal{P})^+$ becomes an abelian semigroup with cancellation. By formally adjoining additive inverses to $\text{Grot}(\mathcal{P})^+$, we obtain an abelian group $\text{Grot}(\mathcal{P})$, the *Grothendieck group of* $\mathcal{P}$. (Reminder: If short exact sequences are available in $\mathcal{P}$, the Grothendieck group constructed from $\mathcal{P}$ using short exact sequences may well be different from the group constructed here.) All elements of $\text{Grot}(\mathcal{P})$ have the form $[A] - [B]$ for $A, B \in \mathcal{P}$, and elements $[A] - [B]$ and $[C] - [D]$ in $\text{Grot}(\mathcal{P})$ are equal if and only if $A \oplus D$ is stably isomorphic to $B \oplus C$.

For example, if $\mathcal{P}$ is the collection of all real (complex) vector bundles on some compact Hausdorff space X, then $\text{Grot}(\mathcal{P})$ is the real (complex) K-group $K^0(X)$. For another example, let R be a ring with 1, and let $\mathcal{P}$ be the collection of all finitely generated projective right R-modules (i.e., all direct summands of free right R-modules of finite rank). In this case, $\text{Grot}(\mathcal{P})$ is the algebraic K-group $K_0(R)$. Alternatively, $K_0(R)$ may be constructed by taking $\mathcal{P}$ to be the category of all rectangular matrices over R. Then the objects in $\mathcal{P}$ are all idempotent square matrices over R, and the direct sum of idempotent matrices e and f is the block matrix $\left(\begin{smallmatrix} e & 0 \\ 0 & f \end{smallmatrix}\right)$. In case R is a C^*-algebra, the set of idempotent matrices over R may be reduced to the set of self-adjoint idempotent matrices (see [**52**, Chapter 19]).

In order not to lose track of the original semigroup $\text{Grot}(\mathcal{P})^+$ inside $\text{Grot}(\mathcal{P})$, we make it the "positive cone" of an order relation. Namely, for $x, y \in \text{Grot}(\mathcal{P})$, we define $x \leq y$ if and only if $y - x$ lies in $\text{Grot}(\mathcal{P})^+$. This relation is a pre-order (i.e., a reflexive, transitive relation) which is invariant under translation (that is, $x \leq y$ implies $x + z \leq y + z$). The combined structure $(\text{Grot}(\mathcal{P}), +, \leq)$ is called a *pre-ordered abelian group*. In case the relation $\leq$ is a partial order (i.e., an antisymmetric pre-order), $\text{Grot}(\mathcal{P})$ is a *partially ordered abelian group*.

Various relatively mild assumptions on $\mathcal{P}$ will force $\text{Grot}(\mathcal{P})$ to be partially ordered. A common assumption is that all the objects in $\mathcal{P}$ are "directly finite", i.e., objects $A, B \in \mathcal{P}$ can satisfy $A \oplus B \cong A$ only if B is a zero object. For example, if $\mathcal{P}$ is the collection of all finitely generated projective right modules over a unital ring R, the objects in $\mathcal{P}$ are directly finite if and only if all square matrices over R that satisfy $xy = 1$ also satisfy $yx = 1$. In particular, this holds if R is commutative, or if R is a directed union of finite-dimensional algebras, or if R is noetherian on either side. It also holds if R is a *unit-regular* ring, meaning that for any $x \in R$ there exists a unit (invertible element) $u \in R$ such that $xux = x$, for in that case the objects in $\mathcal{P}$ may be cancelled from direct sums [**63**, Theorem 2; **49**, Theorem 4.5].

In the case that $\mathcal{P}$ is the collection of all finitely generated projective right modules over a unital ring R, the module R plays a special role in $\mathcal{P}$, for every object in $\mathcal{P}$ is isomorphic to a direct summand of a finite direct sum of copies of R. As a consequence, $[R]$ plays a special role in $K_0(R)$: given any $x \in K_0(R)$ there exists a positive integer n such that $x \leq n[R]$. By virtue of this property, $[R]$ is called an *order-unit* of $K_0(R)$.

The construction of $K_0(R)$ for a ring R without 1 is not based on some class of non-unital projective modules but instead is obtained from the "unitification" of R, namely the ring R^1 based on the abelian group $\mathbb{Z} \times R$, with multiplication given by the rule $(m,r)(n,s) = (mn, ms + nr + rs)$. The original ring R may be identified with the set $\{0\} \times R$, which is an ideal of R^1, and then $R^1/R \cong \mathbb{Z}$. Then $K_0(R)$ is defined to be the subgroup of $K_0(R^1)$ consisting of those elements $[A] - [B]$ in $K_0(R^1)$ for which A/AR and B/BR are (stably) isomorphic free abelian groups, and $K_0(R)$ is equipped with the pre-ordered abelian group structure inherited from $K_0(R^1)$. In general, $K_0(R)$ need not have an order-unit. To take the place of an order-unit, we may use the subset $D(R)$ of $K_0(R)^+$ consisting of those elements $x \in K_0(R)$ for which $0 \leq x \leq [R^1]$.

Grothendieck groups having been constructed, two basic meta-questions arise: What sort of information about the data $\mathcal{P}$ is stored in $\text{Grot}(\mathcal{P})$, and how may this information be retrieved? For instance, since $\text{Grot}(\mathcal{P})$ is an invariant of $\mathcal{P}$, there can be situations in which it can be proved that data $\mathcal{P}$ and $\mathcal{P}'$ are not equivalent by showing that $\text{Grot}(\mathcal{P})$ and $\text{Grot}(\mathcal{P}')$ are not isomorphic. Also, by its construction $\text{Grot}(\mathcal{P})$ reflects the arrangement of direct sum decompositions in $\mathcal{P}$, and we may ask how much of the direct sum decomposition structure of $\mathcal{P}$, and what other structural information about $\mathcal{P}$, may be recovered from $\text{Grot}(\mathcal{P})$. By way of illustration, we shall discuss a number of situations in which these questions have been successfully answered. Due to the author's bias, these examples are K_0's of certain rings and C^*-algebras, but the patterns of these examples are to be expected in Grothendieck groups of other mathematical systems.

• **Irrational Rotation Algebras.** Let T be the unit circle in the plane, and let $C(T)$ denote the algebra of all continuous complex-valued functions on T. We may view the functions in $C(T)$ as bounded linear operators on the Hilbert space $L^2(T)$ (where a function from $C(T)$ acts on functions from $L^2(T)$ by multiplication). Given a positive real number α, let $\rho_\alpha \colon T \to T$ be counterclockwise rotation through the angle $2\pi\alpha$. The rule $\rho_\alpha^*(f) = f\rho_\alpha$ then defines a bounded linear operator ρ_α^* on $L^2(T)$, and we let A_α denote the closed self-adjoint subalgebra of bounded linear operators on $L^2(T)$ generated by $C(T)$ and ρ_α^*. This algebra is known as the "transformation group C^*-algebra of the rotation ρ_α".

It is fairly easy to distinguish among the algebras A_α for rational α, and it is also easy to distinguish the rational cases from the irrational cases. However, distinguishing among the irrational cases is a subtler problem, which was only solved when Pimsner, Voiculescu, and Rieffel calculated K_0 of these algebras. Namely, for any irrational number $\alpha \in (0,1)$, there is an ordered group isomorphism from $K_0(A_\alpha)$ onto the subgroup $\mathbb{Z} + \alpha\mathbb{Z}$ of $\mathbb{R}$, under which the canonical order-unit $[A_\alpha]$ in $K_0(A_\alpha)$ corresponds to the real number 1 [**103**, Corollary 2.6; **104**, Corollary 1; **108**, Theorem 1].

Thus if $A_\alpha \cong A_\beta$ for some irrational numbers $\alpha, \beta \in (0,1)$, there must be an ordered group isomorphism f of $\mathbb{Z} + \alpha\mathbb{Z}$ onto $\mathbb{Z} + \beta\mathbb{Z}$ such that $f(1) = 1$. After approximating elements of $\mathbb{Z} + \alpha\mathbb{Z}$ and $\mathbb{Z} + \beta\mathbb{Z}$ by rational numbers, clearing denominators, and using the relation $f(1) = 1$, it is easily seen that $\mathbb{Z} + \alpha\mathbb{Z} = \mathbb{Z} + \beta\mathbb{Z}$. From this a quick computation leads to the conclusion that either $\beta = \alpha$ or $\beta = 1 - \alpha$, whence either $\rho_\beta = \rho_\alpha$ or $\rho_\beta = \rho_\alpha^{-1}$.

• **Ultramatricial Algebras.** Fix a field F. A *matricial* F-algebra is any F-algebra that is isomorphic to a finite direct product of full matrix algebras over F. (In case F is algebraically closed, the matricial F-algebras are exactly the finite-dimensional semisimple F-algebras.) An *ultramatricial* F-algebra is an F-algebra that is a union of a countable ascending sequence of matricial subalgebras (equivalently, any F-algebra that is isomorphic to a direct limit of a countable sequence of matricial F-algebras and F-algebra homomorphisms). It is easily checked that matricial F-algebras are unit-regular. Hence, any unital ultramatricial F-algebra R is unit-regular, and so $K_0(R)$ is a partially ordered abelian group. Using F-algebra unitifications, it follows that K_0 of any ultramatricial F-algebra is partially ordered.

These partially ordered abelian groups may be used to classify ultramatricial F-algebras, following a method of Elliott. If R and S are unital ultramatricial F-algebras, then $R \cong S$ if and only if there exists an ordered group isomorphism of $K_0(R)$ onto $K_0(S)$ mapping $[R]$ to $[S]$ [**35**, Theorem 4.3; **49**, Theorem 15.26]. If R and S are non-unital ultramatricial F-algebras, then $R \cong S$ if and only if there exists an ordered group isomorphism of $K_0(R)$ onto $K_0(S)$ mapping $D(R)$ onto $D(S)$ [**35**, Theorem 4.3].

The partially ordered abelian groups which can appear as K_0 of ultramatricial

F-algebras are just those which are isomorphic (as ordered groups) to direct limits of countable sequences of finite products of copies of $\mathbb{Z}$ [**35**, Theorems 5.1, 5.5]. However, it is usually impossible to check directly whether a given partially ordered abelian group is isomorphic to such a direct limit. Some obvious properties of these direct limits are that they are countable, they are directed (upward and downward), and they are *unperforated* (any x satisfying $nx \geq 0$ for some $n \in \mathbb{N}$ also satisfies $x \geq 0$). A more fundamental property, also easily checked, is the *Riesz interpolation property*: given any x_1, x_2, y_1, y_2 such that $x_i \leq y_j$ for all i, j, there exists z such that $x_i \leq z \leq y_j$ for all i, j. The direct limits of sequences of finite products of copies of $\mathbb{Z}$ were characterized by Effros, Handelman, and Shen as exactly those countable partially ordered abelian groups which are directed and unperforated and which satisfy the Riesz interpolation property [**30**, Theorem 2.2; **52**, Corollary 21.8].

Partially ordered abelian groups with the latter three properties are now called *dimension groups*. Thus, given a partially ordered abelian group G and an order-unit $u \in G$, there exists an ordered group isomorphism of G onto K_0 of some unital ultramatricial F-algebra R, with u corresponding to $[R]$, if and only if G is a countable dimension group. For the non-unital case, replace the order-unit u by an upward-directed subset $D \subseteq G^+$ such that every element of G^+ is a sum of elements from D and such that any element of G^+ which lies below an element of D must lie in D. Then there exists an ordered group isomorphism of G onto K_0 of some ultramatricial F-algebra R, with D mapping onto $D(R)$, if and only if G is a countable dimension group. (These results are obtained by combining Elliott's results [**35**, Theorems 5.1, 5.5] with those of Effros, Handelman, and Shen [**30**, Theorem 2.2].)

For example, the subgroup $\{a/2^n \mid a \in \mathbb{Z} \text{ and } n \in \mathbb{N}\}$ of $\mathbb{Q}$ appears as K_0 of a direct limit of matrix algebras

$$M_2(F) \to M_4(F) \to M_8(F) \to \cdots$$

using block diagonal maps $x \mapsto \left(\begin{smallmatrix} x & 0 \\ 0 & x \end{smallmatrix}\right)$. The lexicographic product of $\mathbb{Z}$ with itself appears as K_0 of a direct limit

$$F \times M_2(F) \to F \times M_3(F) \to F \times M_4(F) \to \cdots$$

using maps $(x, y) \mapsto \left(x, \left(\begin{smallmatrix} x & 0 \\ 0 & y \end{smallmatrix}\right)\right)$. The subgroup $\mathbb{Z} + \mathbb{Z}\sqrt{2}$ of $\mathbb{R}$ appears as K_0 of a direct limit

$$F \times M_2(F) \to M_3(F) \times M_4(F) \to M_7(F) \times M_{10}(F) \to \cdots$$

using maps

$$(x, y) \mapsto \left(\begin{pmatrix} x & 0 \\ 0 & y \end{pmatrix}, \begin{pmatrix} x & 0 & 0 \\ 0 & x & 0 \\ 0 & 0 & y \end{pmatrix} \right).$$

An algorithm for obtaining $\mathbb{Z} + \alpha\mathbb{Z}$ (where α is a positive irrational number) as K_0 of an ultramatricial algebra, by using the continued fraction expansion of α, was developed by Effros and Shen [**32**, Theorem 3.2].

The ability to realize any countable dimension group as K_0 of an ultramatricial algebra is an aid to constructing examples, since dimension groups are easier to construct than ultramatricial algebras. For instance, we may use this method to construct algebras with various sorts of ideal-theoretic structure, for this corresponds directly to ideal-theoretic structure in K_0. An *ideal* in a partially ordered abelian group G is any directed subgroup H of G such that whenever $x_1, x_2 \in H$ and $y \in G$ with $x_1 \leq y \leq x_2$, then $y \in H$. For any unital ultramatricial algebra R (or, more generally, for any unit-regular ring), the lattice of two-sided ideals of R is isomorphic to the lattice of ideals of $K_0(R)$ [**49**, Corollary 15.21]. Given any countable ordinal α, it is easy to construct a countable dimension group, with an order-unit, whose lattice of ideals is isomorphic to the interval $[1, \alpha+1]$ (see [**50**, Proposition 1.3]). Consequently, there exists a unital ultramatricial F-algebra whose lattice of two-sided ideals is isomorphic to $[1, \alpha+1]$. Similarly, there exists a unital ultramatricial F-algebra whose lattice of ideals is anti-isomorphic to $[1, \alpha]$. (These ideal-theoretic examples are the basic ingredients in two corresponding module-theoretic examples constructed by the author [**50**, Corollaries 1.6, 1.7].)

Any family of ultramatricial algebras defined by properties which are reflected in K_0 may be classified by a corresponding family of countable dimension groups. For example, since an ultramatricial algebra R is simple if and only if $K_0(R)$ is simple (i.e., the only ideals in $K_0(R)$ are $K_0(R)$ and $\{0\}$), the family of simple unital ultramatricial F-algebras is classified by the family of countable simple dimension groups with order-unit. The task of constructing all (countable) simple dimension groups was initiated by Effros, Handelman, and Shen [**30**, Lemmas 3.1, 3.2, Theorem 3.5] and completed by the author and Handelman [**57**, Theorem 4.11].

• **Approximately Finite-Dimensional C^*-Algebras.** A complex C^*-algebra is *approximately finite-dimensional* (abbreviated *AF*) if it is isomorphic (as a C^*-algebra) to a direct limit of a countable sequence of finite-dimensional C^*-algebras. Since all finite-dimensional C^*-algebras are semisimple, any AF C^*-algebra A contains a dense ultramatricial complex *-subalgebra R. Moreover, the inclusion map $R \to A$ induces an ordered group isomorphism of $K_0(R)$ onto $K_0(A)$ (see [**52**, Corollary 19.10] for the unital case). Hence, K_0 of any AF C^*-algebra is a countable dimension group. Bratteli proved that AF C^*-algebras are determined up to isomorphism by their dense ultramatricial *-subalgebras [**15**, Theorem 2.7; **52**, Theorem 20.7]. Consequently, AF C^*-algebras are classified in terms of countable dimension groups, via K_0, in exactly the same manner as ultramatricial algebras [**35**, Theorems 4.3, 5.1, 5.5; **30**, Theorem 2.2; **52**, Theorems 20.7, 21.10].

• **Pseudo-Rank Functions.** A ring R is (*von Neumann*) *regular* provided that for each $x \in R$ there exists $y \in R$ such that $xyx = x$. (This ensures a large supply of idempotents, for whenever $xyx = x$, the elements xy and yx are idempotent.) A (normalized) *pseudo-rank function* on a regular ring R with 1 is any map N from R to the unit interval $[0, 1]$ such that (a) $N(1) = 1$; (b) $N(xy) \leq N(x)$ and $N(xy) \leq N(y)$ for all $x, y \in R$; (c) $N(e + f) = N(e) + N(f)$ for all orthogonal idempotents $e, f \in R$. For example, if R is the ring of all $n \times n$ matrices over a field F, then normalized matrix rank defines a pseudo-rank function N on R (that is, $N(x) = \text{rank}(x)/n$ for all matrices $x \in R$). For another example, if R is a ring of subsets of some set X, and $X \in R$, then a pseudo-rank function on R is just a nonnegative finitely additive measure μ on R such that $\mu(X) = 1$.

Pseudo-rank functions on R correspond to certain real-valued functions on $K_0(R)$, normalized with respect to the canonical order-unit $[R]$. Namely, a *state on* $(K_0(R), [R])$ is any group homomorphism $s\colon K_0(R) \to \mathbb{R}$ such that $s(K_0(R)^+) \subseteq \mathbb{R}^+$ and $s([R]) = 1$. There is a canonical bijection between the set of such states and the set of pseudo-rank functions on R [**56**, Proposition 2.4; **49**, Proposition 17.12]. Thus questions of existence and/or uniqueness for pseudo-rank functions may be translated into questions of existence and/or uniqueness for states.

For example, the author and Handelman showed that any nonzero partially ordered abelian group with an order-unit has at least one state [**56**, Corollary 3.3; **49**, Corollary 18.2]. Since K_0 of any nonzero unit-regular ring is nonzero and partially ordered, it follows that any nonzero unit-regular ring has at least one pseudo-rank function [**56**, Corollary 3.5; **49**, Corollary 18.5]. The author and Handelman also developed various criteria for a nonzero partially ordered abelian group G with an order unit u to have a unique state. For instance, this happens if and only if there exist integers $s > t > 0$ such that given any $x, y \in G^+$ with $x + y = u$, there is some $n \in \mathbb{N}$ for which either $ntx \leq nsy$ or $nty \leq nsx$. As a consequence, a nonzero unit-regular ring R possesses a unique pseudo-rank function if and only if there exist integers $s > t > 0$ such that given any orthogonal idempotents $e, f \in R$ with $e + f = 1$, there is some $n \in \mathbb{N}$ for which either the direct sum of nt copies of eR embeds in the direct sum of ns copies of fR or the direct sum of nt copies of fR embeds in the direct sum of ns copies of eR [**56**, Theorem 4.6; **49**, Theorem 18.16].

• **Hiatus.** Other uses of Grothendieck groups occur in cases where a structure theory for a class of partially ordered abelian groups is used, via K_0, to derive a corresponding structure theory for a class of rings or C^*-algebras. As these applications are more easily discussed with the relevant ordered group theory at hand, we postpone them until the Epilogue. In these applications, the Grothendieck groups that appear satisfy the Riesz interpolation property and in fact are usually dimension groups. Therefore our main interest is in developing structure theory for dimension groups, although in many instances the development relies only on the interpolation property.

Notational Conventions

The main purpose of this section is to provide a quick review of those basic concepts from the general theory of partially ordered sets that will be needed in the book. As the few notations and definitions from abelian group theory that we use are quite standard, we do not review them.

A *partial order* on a set X is any reflexive, antisymmetric, transitive relation on X. In most cases, partial orders are denoted $\leq$. When this notation is used, associated relations $\geq, <, >$ are defined on X so that $x \geq y$ if and only if $y \leq x$, while $x < y$ if and only if $x \leq y$ but $x \neq y$, and $x > y$ if and only if $y < x$. A *partially ordered set* (or *poset*) is a set X equipped with a specified partial order. The *dual ordering* on a partially ordered set $(X, \leq)$ is the partial order $\geq$. To avoid confusion, the dual ordering is often denoted $\leq^*$. The *dual* of $(X, \leq)$ is the partially ordered set $(X, \leq^*)$.

Let $(X, \leq)$ be a partially ordered set, and let Y be an arbitrary set. There is a natural partial order $\leq$ on the set X^Y of all functions from Y to X, where $f \leq g$ if and only if $f(y) \leq g(y)$ for all $y \in Y$. This partial order is called the *pointwise ordering* on X^Y. With respect to the pointwise ordering, $f < g$ means only that $f \leq g$ and $f(y) < g(y)$ for at least one $y \in Y$. For uniformly strict inequalities in X^Y, we use the notation $\ll$. Thus $f \ll g$ if and only if $f(y) < g(y)$ for all $y \in Y$. The corresponding partial order $\lll$ on X^Y is called the *strict ordering*. (That is, $f \lll g$ if and only if either $f \ll g$ or $f = g$.)

Intervals in a partially ordered set $(X, \leq)$ are defined in the same manner as intervals on the real line, and with the same notation. Thus for elements $a, b \in X$, the *closed interval* $[a, b]$ is the set of those $x \in X$ for which $a \leq x \leq b$, and the *open interval* (a, b) is the set of those $x \in X$ for which $a < x < b$. (Of course, if $a \not\leq b$, then $[a, b]$ and (a, b) are empty.)

A partially ordered set X is *totally ordered* if every pair of elements $x, y \in X$ are comparable, i.e., either $x \leq y$ or $y \leq x$. In this case, the partial order $\leq$ on X is called a *total order*. The set X is *upward* (*downward*) *directed* provided every finite subset of X has an upper (lower) bound in X. If every finite subset of X has a supremum (least upper bound) and an infimum (greatest lower bound) in X, then X is a *lattice*. A lattice X is *complete* provided every subset of X (including the empty subset) has a supremum and an infimum in X. (The notations $\wedge$ and $\vee$ are used for infima and suprema.) A *boolean algebra* is a lattice X which is distributive (that is,

$$x \wedge (y \vee z) = (x \wedge y) \vee (x \wedge z) \quad \text{and} \quad x \vee (y \wedge z) = (x \vee y) \wedge (x \vee z)$$

for all $x, y, z \in X$) and also complemented (that is, X contains a smallest element 0 and a largest element 1, and for each element $x \in X$ there is an element $y \in X$ such that $x \wedge y = 0$ and $x \vee y = 1$).

Let X and Y be partially ordered sets. A map $f\colon X \to Y$ is said to be *order-preserving* (or *isotone*) if f preserves all the order relations in X, that is, whenever $x, y \in X$ with $x \leq y$, then $f(x) \leq f(y)$. Dually, f is *order-reversing* (or *antitone*) provided that whenever $x, y \in X$ with $x \leq y$, then $f(x) \geq f(y)$. An *order-isomorphism* of X onto Y is a bijection $f\colon X \to Y$ such that f and f^{-1} are both order-preserving. Such a map preserves whatever infima and suprema may exist in X. Namely, if an element $a \in X$ is the infimum (supremum) of a subset $A \subseteq X$, then $f(a)$ is the infimum (supremum) of $f(A)$ in Y. An *order anti-isomorphism* of X onto Y is a bijection $f\colon X \to Y$ such that f and f^{-1} are both order-reversing. (Then f provides an order-isomorphism of X onto the dual of Y.) Such a map reverses infima and suprema; namely, if an element $a \in X$ is the infimum (supremum) of a subset $A \subseteq X$, then $f(a)$ is the supremum (infimum) of $f(A)$ in Y. An *order-automorphism* of X is any order-isomorphism of X onto itself, while an *order anti-automorphism* of X is any order anti-isomorphism of X onto itself.

Finally, let X and Y be lattices. A *lattice homomorphism* from X to Y is any map $f\colon X \to Y$ that preserves finite infima and suprema, that is,

$$f(x_1 \wedge \cdots \wedge x_n) = f(x_1) \wedge \cdots \wedge f(x_n),$$
$$f(x_1 \vee \cdots \vee x_n) = f(x_1) \vee \cdots \vee f(x_n)$$

for all $x_1, \cdots, x_n \in X$. All lattice homomorphisms are order-preserving. A *lattice isomorphism* of X onto Y is any bijection $f\colon X \to Y$ such that f and f^{-1} are both lattice homomorphisms. A map $f\colon X \to Y$ is a lattice isomorphism if and only if it is an order-isomorphism.

We follow the prevalent American usage (as distinct from, e.g., Bourbaki) in letting $\mathbb{N}$ denote the set $\{1, 2, 3, \ldots\}$ of positive integers. The symbol $\mathbb{Z}^+$ is used for the set $\{0, 1, 2, \cdots\}$ of nonnegative integers, in order to be consistent with the notation G^+ for the subset $\{x \in G \mid x \geq 0\}$ of a partially ordered abelian group G.

CHAPTER 1

Basic Notions

This chapter, like most introductory chapters, is designed to establish a basic working vocabulary and grammar for the subject at hand. Here the language to be presented is that of partially ordered abelian groups. A basic fluency with the languages of abelian groups and partially ordered sets is assumed, and attention is mainly directed toward the hybrid.

The introductory vocabulary here consists of a few dozen items which recur throughout the subject. Only a few basic concepts corresponding to nouns are introduced, particularly "partially ordered abelian group", "positive cone", "order-unit", "ideal", and "positive homomorphism", and illustrations are provided. A number of adjectival phrases are discussed, most intended as modifiers for "partially ordered abelian group", such as "totally ordered", "lattice-ordered", "Dedekind complete", "directed", "unperforated", and "archimedean". At this introductory level, those concepts which may be viewed as verb forms are mainly imperatives of the type "construct —". The constructions requested include quotients, direct products and sums, lexicographic direct products and sums, direct and inverse limits, pullbacks and pushouts. Finally, some basic grammar governing the use of this vocabulary is presented. Further vocabulary and more extensive grammar will be introduced in later chapters, as needed.

• Partially Ordered Abelian Groups.

DEFINITION. A *pre-ordered abelian group* is an abelian group G equipped with a specified pre-order $\leq$ (i.e., a reflexive, transitive relation) which is *translation-invariant*, that is, given any $x, y, z \in G$ with $x \leq y$, it follows that $x + z \leq y + z$. A *partially ordered abelian group* is an abelian group equipped with a specified translation-invariant *partial* order. A *totally ordered abelian group* is any partially ordered abelian group in which the given partial order is actually a total order. A *partially ordered vector space* over a subfield F of $\mathbb{R}$ is any F-vector space V equipped with a specified translation-invariant partial order $\leq$ such that given any $x, y \in V$ and $\alpha \in F$ with $x \leq y$ and $\alpha \geq 0$, it follows that $\alpha x \leq \alpha y$.

For example, any additive subgroup of $\mathbb{R}$, when equipped with the usual ordering of real numbers, becomes a totally ordered abelian group. We shall always assume, unless otherwise specified, that any given subgroup of $\mathbb{R}$ has been made into a totally ordered abelian group with the usual ordering. Of course, any rational linear subspace of $\mathbb{R}$ is a partially ordered rational vector space, while $\mathbb{R}$ itself is a partially ordered real vector space.

Although pre-ordered abelian groups often appear in applications, there is little need to develop a more general theory to cover them, because any pre-ordered abelian group has closely associated with it a canonical partially ordered abelian group, as in the following proposition. Consequently, we may restrict our attention to partially ordered abelian groups.

PROPOSITION 1.1. *Let G be a pre-ordered abelian group, and set*

$$H = \{x \in G \mid x \geq 0 \text{ and } x \leq 0\}.$$

Then H is a subgroup of G, and the pre-order on G induces a partial order on G/H under which $x + H \leq y + H$ if and only if $x \leq y$ (for any $x, y \in G$). Using this partial order, G/H is a partially ordered abelian group.

PROOF. Obviously $0 \in H$. If $x, y \in H$, then

$$x + y \geq x + 0 \geq 0 \quad \text{and} \quad x + y \leq x + 0 \leq 0,$$

whence $x + y \in H$. Adding $-x$ to each side of the relations $x \geq 0$ and $x \leq 0$, we obtain $0 \geq -x$ and $0 \leq -x$, so that $-x \in H$. Thus H is a subgroup of G.

Given any $x, y \in G$, define $x \sim y$ if and only if $x \leq y$ and $x \geq y$. It is clear that $\sim$ is an equivalence relation on G and that the pre-order on G induces a partial order on the set of equivalence classes of $\sim$ in G. For any $x, y \in G$, observe that $x \sim y$ if and only if $x - y \in H$, if and only if $x + H = y + H$. Thus the equivalence classes of $\sim$ in G are exactly the cosets of H in G. Hence, there is a well-defined partial order $\leq$ on G/H such that for all $x, y \in G$, we have $x + H \leq y + H$ if and only if $x \leq y$.

Since the given pre-order on G is translation-invariant, we conclude that the induced partial order on G/H must be translation-invariant. Thus G/H is a partially ordered abelian group. $\square$

DEFINITION. A *positive element* in a partially ordered abelian group G is any element $x \in G$ such that $x \geq 0$, while a *strictly positive element* of G is any element $x \in G$ such that $x > 0$.

While this terminology conflicts slightly with the standard usage for real numbers, it is convenient and widely used, since this notion of a positive element cannot be replaced by the notion of a nonnegative element (meaning an element $x \in G$ such that $x \not< 0$) except when G is totally ordered. For real numbers, however, we shall continue to use "positive" and "negative" with their traditional meanings.

Note that the partial order on G can be specified just by fixing the set of positive elements in G, since for any $x, y \in G$, we have $x \leq y$ if and only if $y - x \geq 0$. (Namely, if $x \leq y$ then

$$0 = x + (-x) \leq y + (-x),$$

while if $y - x \geq 0$, then $y = (y-x) + x \geq 0 + x$.) The set of positive elements of G is a "cone" in the following sense.

DEFINITION. A *cone* in an abelian group G is any subset C of G such that $0 \in C$ and C is closed under addition (that is, C is a submonoid of G). A cone C in G is called a *strict cone* provided that 0 is the only element $x \in G$ for which both $x \in C$ and $-x \in C$. Given a cone C in G, we may define a relation $\leq_C$ on G so that $x \leq_C y$ if and only if $y - x \in C$ (for any $x, y \in G$). The relation $\leq_C$ is a translation-invariant pre-order on G, and $\leq_C$ is a partial order if and only if C is a strict cone.

DEFINITION. The *positive cone* of a partially ordered abelian group G is the set G^+ of all positive elements of G.

Given a partially ordered abelian group G, note that its positive cone G^+ is a strict cone and that the given partial order on G coincides with $\leq_{G^+}$. Conversely, if C is a strict cone in an abelian group G, then G becomes a partially ordered abelian group when equipped with $\leq_C$, in which case $G^+ = C$. In many cases, it is easier to specify a partially ordered abelian group this way (i.e., by describing its positive cone) than by describing the partial order relation. However, one must be certain that the subset G^+ described is actually a strict cone.

One consequence of using this notation is that since we are by convention viewing $\mathbb{Z}$ as a partially ordered abelian group with the usual order, $\mathbb{Z}^+$ must stand for the set of nonnegative integers. Similarly, $\mathbb{Q}^+$ and $\mathbb{R}^+$ stand for the sets of nonnegative rational and real numbers.

DEFINITION. In any abelian group G, the relation $=$ is a translation-invariant partial order, called the *discrete ordering*. When equipped with the discrete ordering, G becomes a partially ordered abelian group with positive cone $\{0\}$.

DEFINITION. Let H be a subgroup of a partially ordered abelian group G. Unless otherwise specified, we make H into a partially ordered abelian group using the partial order inherited from G, so that $H^+ = H \cap G^+$.

Observe that in a partially ordered abelian group G, inequalities with the same sense may be added together. Namely, given $x_1, y_1, \ldots, x_n, y_n$ in G such that $x_i \leq y_i$ for all i, each of the differences $y_i - x_i$ lies in G^+ and so their sum lies in G^+, whence $\sum x_i \leq \sum y_i$. In particular, it follows that if $x, y \in G$ with $x \leq y$, then $nx \leq ny$ for all positive integers n. Observe also that the operation of taking additive inverses reverses inequalities in G. Namely, if $x, y \in G$ with $x \leq y$, then $y - x$ is in G^+ and so $(-x) - (-y)$ is in G^+, whence $-x \geq -y$.

PROPOSITION 1.2. *Let G be a partially ordered abelian group, and let $w \in G$.*

(a) *The map $x \mapsto x + w$ is an order-automorphism of G.*

(b) *The map $x \mapsto w - x$ is an order anti-automorphism of G.*

PROOF. (a) The maps $x \mapsto x + w$ and $x \mapsto x - w$ are inverse bijections of G onto itself. As these maps are order-preserving, they must be order-automorphisms.

(b) This map is a self-inverse bijection of G onto itself. As it is order-reversing, it must be an order anti-automorphism. $\square$

PROPOSITION 1.3. *Let H be a subgroup of a partially ordered abelian group G. Then the following conditions are equivalent*:

(a) *H is upward directed.*

(b) *H is downward directed.*

(c) *H is generated (as a subgroup of G) by a subset of G^+.*

(d) *All elements of H have the form $x - y$ for $x, y \in H^+$.*

PROOF. (a)⇔(b): These conditions are equivalent because the rule $x \mapsto -x$ defines an order anti-automorphism of H onto itself.

(a)⇒(d): Given $a \in H$, there exists $x \in H$ such that $a \leq x$ and $0 \leq x$. Set $y = x - a$. Then $x, y \in H^+$ and $a = x - y$.

(d)⇒(a): Given $a_1, a_2 \in H$, there exist elements x_1, y_1, x_2, y_2 in H^+ such that each $a_i = x_i - y_i$. Then each $a_i \leq x_i \leq x_1 + x_2$. Since $x_1 + x_2 \in H$, this shows that H is upward directed.

(d)⇒(c): This is clear.

(c)⇒(d): We are given that H is generated, as a group, by some set $X \subseteq G^+$. Then $X \subseteq H^+$, and so H is generated by H^+. Hence, every element of H is a sum of elements from $H^+ \cup (-H^+)$, where $-H^+$ denotes $\{-x \mid x \in H^+\}$. Since H^+ and $-H^+$ are closed under addition, we conclude that every element of H is a sum of one element from H^+ and one element from $-H^+$. $\square$

DEFINITION. A *directed subgroup* of a partially ordered abelian group G is any subgroup H of G that satisfies the equivalent conditions of Proposition 1.3. If G is a directed subgroup of itself, then we just say that G is *directed.* A *directed abelian group* is any directed partially ordered abelian group.

For example, any totally ordered abelian group is directed. On the other hand, if G is a nonzero abelian group equipped with the discrete ordering, then G is not directed.

DEFINITION. An *order-unit* in a partially ordered abelian group G is any positive element $u \in G^+$ such that given any $x \in G$, there is some positive integer n for which $x \leq nu$.

For instance, if G is a subgroup of $\mathbb{R}$, then any strictly positive element of G is an order-unit. We observe that any partially ordered abelian group G which contains an order-unit u must be directed. Namely, given $x_1, x_2 \in G$ there exist positive integers n_1, n_2 such that each $x_i \leq n_i u$, whence each $x_i \leq (n_1 + n_2)u$.

In a nonzero partially ordered abelian group G, the element 0 cannot be an order-unit. Hence, if u is an order-unit in G, then $u > 0$. In addition, $u \leq nu$ for all $n \in \mathbb{N}$, so that $nu > 0$ for all $n \in \mathbb{N}$, and consequently, $mu < 0$ for all negative integers m.

DEFINITION. Given a set X and an abelian group A, we use A^X to denote the abelian group of all functions from X to A (with pointwise addition). In case A is a partially ordered abelian group, we may define the *pointwise ordering* on

A^X, so that for $f, g \in A^X$, we have $f \le g$ if and only if $f(x) \le g(x)$ for all $x \in X$. Unless otherwise specified, we always assume that A^X has been equipped with the pointwise ordering, which makes it into a partially ordered abelian group. There is another useful ordering on A^X called the *strict ordering*, which we denote by $\lll$. Given $f, g \in A^X$, we define $f \ll g$ if and only if $f(x) < g(x)$ for all $x \in X$ (note that $f \ll g$ is a stronger condition than $f < g$, which means only that $f \le g$ and $f \ne g$). Of course, $f \lll g$ if and only if either $f \ll g$ or $f = g$. Equipped with the strict ordering, A^X again becomes a partially ordered abelian group.

DEFINITION. Given a topological space X and an additive subgroup A of $\mathbb{R}$, we use $C(X, A)$ to denote the abelian group of all continuous functions from X to A (with pointwise addition). Then $C(X, A)$ is a subgroup of A^X, so that using either the pointwise ordering or the strict ordering, $C(X, A)$ becomes a partially ordered abelian group. If unspecified, we assume that the partial order on $C(X, A)$ is the pointwise ordering.

Observe that if X is compact, then any strictly positive function in $C(X, A)$ is an order-unit in $C(X, A)$ with respect to either the pointwise ordering or the strict ordering.

• **Infima and Suprema.** In a partially ordered abelian group G, infima and suprema do not always exist, but there are useful computational relationships among those infima and suprema that do exist. Remember that lattice-theoretic equations concerning elements of G should not be blithely presented and manipulated unless it is known that the relevant infima and suprema exist in G.

PROPOSITION 1.4. *Let G be a partially ordered abelian group, let $w \in G$, and let $\{x_i \mid i \in I\}$ be a nonempty subset of G. The following identities hold in the sense that if either side exists, then the other exists, in which case the two sides agree.*

(a) $(\bigwedge x_i) + w = \bigwedge(x_i + w)$.
(b) $(\bigvee x_i) + w = \bigvee(x_i + w)$.
(c) $w - (\bigwedge x_i) = \bigvee(w - x_i)$.
(d) $w - (\bigvee x_i) = \bigwedge(w - x_i)$.

PROOF. Since the map $x \mapsto x + w$ is an order-automorphism of G, it preserves whatever infima or suprema exist in G. Since the map $x \mapsto w - x$ is an order anti-automorphism of G, it reverses whatever infima or suprema exist in G. □

DEFINITION. A *lattice-ordered abelian group* is any partially ordered abelian group which, as a partially ordered set, is a lattice.

For example, any totally ordered abelian group is lattice-ordered. Also, if X is a topological space and A is an additive subgroup of $\mathbb{R}$, then $C(X, A)$ is a lattice-ordered abelian group, with

$$(f \wedge g)(x) = \min\{f(x), g(x)\} \quad \text{and} \quad (f \vee g)(x) = \max\{f(x), g(x)\}$$

for all $f, g \in C(X, A)$ and all $x \in X$. However, if $C(X, A)$ is given the strict ordering, then it is usually not lattice-ordered. (E.g., consider the case in which X is the unit interval, $f(x) = x$, and $g(x) = 1 - x$.)

PROPOSITION 1.5. *For a partially ordered abelian group G, the following conditions are equivalent:*

(a) *G is lattice-ordered.*

(b) *G is directed, and in the partially ordered set G^+, every pair of elements has an infimum.*

(c) *G is directed, and in the partially ordered set G^+, every pair of elements has a supremum.*

PROOF. (a)⇒(b): Since any lattice is upward and downward directed, G must be directed. Any pair of elements $x_1, x_2 \in G^+$ has an infimum x in G. As $0 \le x_i$ for each i, we have $0 \le x$, and so $x \in G^+$. Then x is also the infimum of x_1 and x_2 within G^+.

(b)⇒(a): By assumption, any pair of elements $x_1, x_2 \in G^+$ has an infimum x within G^+. We claim that x is also the infimum of x_1 and x_2 in G. Hence, consider any $y \in G$ such that $y \le x_1$ and $y \le x_2$. As G is directed, there exists $a \in G$ such that $a \ge 0$ and $a \ge -y$. Now $y + a$ is an element of G^+, and $y + a \le x_i + a$ for each i. Let z be the infimum of $x_1 + a$ and $x_2 + a$ within G^+, and note that $y + a \le z$. In addition, $a \le x_i + a$ for each i, whence $a \le z$. Consequently, $z - a$ is an element of G^+ such that $y \le z - a$ and $z - a \le x_i$ for each i. Since x is the infimum of x_1 and x_2 within G^+, we obtain $z - a \le x$, and hence $y \le x$. Therefore x is the infimum of x_1 and x_2 in G, as claimed. In particular, we may now write $x = x_1 \wedge x_2$ without ambiguity.

Given any elements $u, v \in G$, there exists $w \in G$ such that $w \le u$ and $w \le v$. As $u - w$ and $v - w$ lie in G^+, they have an infimum in G, and

$$[(u - w) \wedge (v - w)] + w = u \wedge v$$

by Proposition 1.4. Thus every pair of elements in G has an infimum. Applying the order anti-automorphism $x \mapsto -x$, we conclude that every pair of elements in G also has a supremum.

(a)⇔(c): These implications are proved in the same manner, except that no extra effort is required to see that a supremum within G^+ of two elements of G^+ is also their supremum in G. □

In particular, Proposition 1.5 shows that a directed abelian group G is lattice-ordered if and only if G^+ is a lattice.

PROPOSITION 1.6. *Let G be a partially ordered abelian group, and let $x, y \in G$. Then $x \wedge y$ exists in G if and only if $x \vee y$ exists in G, in which case*

$$(x \wedge y) + (x \vee y) = x + y.$$

PROOF. Assuming that $x \wedge y$ exists, we compute using Proposition 1.4 that

$$x + y - (x \wedge y) = (x + y - x) \vee (x + y - y) = x \vee y.$$

Similarly, if $x \vee y$ exists, then

$$x + y - (x \vee y) = (x + y - x) \wedge (x + y - y) = x \wedge y. \quad \square$$

PROPOSITION 1.7. *Let G be a partially ordered abelian group, let $x \in G$, and let $\{y_i \mid i \in I\}$ be a nonempty subset of G. The following identities hold provided the relevant infima and suprema exist in G.*

(a) $x \wedge (\bigvee y_i) = \bigvee (x \wedge y_i)$.

(b) $x \vee (\bigwedge y_i) = \bigwedge (x \vee y_i)$.

PROOF. (a) Set $y = \bigvee y_i$. Certainly, $\bigvee (x \wedge y_i) \leq x \wedge y$, because $x \wedge y_i \leq x \wedge y$ for all $i \in I$. For each $i \in I$, we compute that

$$(x \wedge y) + y_i = (x + y_i) \wedge (y + y_i) \leq (x + y) \wedge (y_i + y) = (x \wedge y_i) + y.$$

As a result,

$$(x \wedge y) + y = \bigvee [(x \wedge y) + y_i] \leq \bigvee [(x \wedge y_i) + y] = \left[\bigvee (x \wedge y_i)\right] + y,$$

and consequently $x \wedge y \leq \bigvee (x \wedge y_i)$. Thus $x \wedge y = \bigvee (x \wedge y_i)$.

(b) This follows from (a) on applying the order anti-automorphism $z \mapsto -z$. $\square$

In particular, it follows from Proposition 1.7 that any lattice-ordered abelian group is distributive as a lattice.

PROPOSITION 1.8. *Let G be a partially ordered abelian group, let $x \in G$, and let $\{x_i \mid i \in I\}$ be a nonempty subset of G.*

(a) *If $x = \bigwedge x_i$ and the set $\{x_i \mid i \in I\}$ is downward directed, then $nx = \bigwedge nx_i$ for all $n \in \mathbb{N}$.*

(b) *If $x = \bigvee x_i$ and the set $\{x_i \mid i \in I\}$ is upward directed, then $nx = \bigvee nx_i$ for all $n \in \mathbb{N}$.*

PROOF. (a) In case $n = 1$, there is nothing to prove. Now assume that the identity holds for some n, and note that $(n+1)x \leq (n+1)x_i$ for all $i \in I$. Then consider any element $y \in G$ such that $y \leq (n+1)x_i$ for all $i \in I$. Given any $i, j \in I$, there exists $k \in I$ such that $x_k \leq x_i$ and $x_k \leq x_j$, whence

$$y \leq (n+1)x_k = nx_k + x_k \leq nx_i + x_j.$$

Thus $y - x_j \leq nx_i$ for all $i, j \in I$. Since $nx = \bigwedge nx_i$ by the induction hypothesis, it follows that $y - x_j \leq nx$ for all $j \in I$. Then $y - nx \leq x_j$ for all $j \in I$ and so $y - nx \leq x$, whence $y \leq (n+1)x$. Therefore $(n+1)x = \bigwedge (n+1)x_i$, and the induction step is complete.

(b) This follows from (a) on applying the order anti-automorphism $z \mapsto -z$. $\square$

DEFINITION. A partially ordered abelian group G is *Dedekind complete* (or *conditionally complete*) provided that every nonempty subset of G which is bounded above in G has a supremum in G. Equivalently, G is Dedekind complete if and only if every nonempty subset of G which is bounded below in G has an infimum in G.

For example, $\mathbb{R}$ is Dedekind complete. More generally, $\mathbb{R}^X$ is Dedekind complete, for any set X. However, if $\mathbb{R}^X$ is given the strict ordering, then as long as X has more than one element, $\mathbb{R}^X$ is not Dedekind complete.

Note that a Dedekind complete partially ordered abelian group G is not required to be a lattice, since two-element subsets of G need not be bounded above or below. For instance, if G is a nonzero abelian group equipped with the discrete ordering, then G is Dedekind complete, but elements $x, y \in G$ have a supremum in G only if $x = y$. In general, a Dedekind complete partially ordered abelian group is a lattice if and only if it is directed, in which case we refer to it as a Dedekind complete *lattice-ordered* abelian group.

DEFINITION. A partially ordered abelian group G is *Dedekind σ-complete* (or *conditionally σ-complete*) provided that every nonempty countable subset of G which is bounded above in G has a supremum in G.

For example, if X is an uncountable set and G is the subgroup of $\mathbb{R}^X$ consisting of those functions in $\mathbb{R}^X$ with countable support, then G is a Dedekind σ-complete lattice-ordered abelian group, but G is not Dedekind complete.

• Ideals and Quotient Groups.

DEFINITION. A *convex subset* of a partially ordered set G is any subset $H \subseteq G$ with the property that whenever $x, z \in H$ and $y \in G$ with $x \leq y \leq z$, then $y \in H$. (If there is any danger of confusion with the concept of a convex subset of a real vector space, a set H with the property described above may be called an *order-convex* subset of G.) A *convex subgroup* of a partially ordered abelian group G is any subgroup H of G which is also a convex subset of G.

To prove that a subgroup H of a partially ordered abelian group G is convex, it suffices to show that whenever $a \in G$ and $b \in H$ with $0 \leq a \leq b$, then $a \in H$. (Namely, if $x \leq y \leq z$ with $x, z \in H$ and $y \in G$, then $0 \leq y - x \leq z - x$ with $z - x \in H$. If $y - x \in H$, then also $y \in H$.)

For example, $\{0\}$ and G are always convex subgroups of G. If G is a subgroup of $\mathbb{R}$, then it has no other convex subgroups.

DEFINITION. In a partially ordered abelian group G, the intersection of any family of convex subgroups is again a convex subgroup. Thus given any subset X of G, there is a smallest convex subgroup of G containing X, called the *convex subgroup of G generated by X*.

For instance, the convex subgroup generated by a single positive element u in G is the subgroup

$$\{x \in G \mid -nu \leq x \leq nu \text{ for some } n \in \mathbb{N}\}.$$

Note that u is an order-unit for this subgroup.

DEFINITION. An *ideal* of a partially ordered abelian group G is any directed convex subgroup of G.

For example, $\{0\}$ is always an ideal of G. However, G is an ideal of itself only if it is directed.

If G is a partially ordered real vector space, then any ideal H of G is necessarily a linear subspace of G. Namely, given $x \in H$ and $\alpha \in \mathbb{R}$, write $x = x_1 - x_2$ for some $x_1, x_2 \in H^+$, and choose integers m, n such that $m \leq \alpha \leq n$. Then $mx_i \leq \alpha x_i \leq nx_i$ for $i = 1, 2$, whence each $\alpha x_i \in H$, and so $\alpha x \in H$.

PROPOSITION 1.9. *Let G be a partially ordered abelian group.*

(a) *If X is a nonempty subset of G^+ and K is the convex subgroup of G generated by X, then K is an ideal of G.*

(b) *If H is a directed subgroup of G and K is the convex subgroup of G generated by H^+, then K is the smallest ideal of G that contains H.*

(c) *If H is a convex subgroup of G and*

$$K = \{x - y \mid x, y \in H^+\}$$

(so that K equals the subgroup of G generated by H^+), then K is the largest ideal of G that is contained in H.

PROOF. (a) Let Y denote the subset of G^+ consisting of all sums of elements from X. Then K also equals the convex subgroup of G generated by Y, and hence there is no loss of generality in assuming that X is closed under addition. Set

$$K' = \{a \in G \mid -x \leq a \leq x \text{ for some } x \in X\}.$$

Since X is a subset of G^+ that is closed under addition, we infer that K' is a convex subgroup of G. Obviously $X \subseteq K'$, whence $K \subseteq K'$. On the other hand, the elements of K' must belong to any convex subgroup of G that contains X. Thus $K = K'$.

Now given any $a_1, a_2 \in K$, there exist $x_1, x_2 \in X$ such that $-x_i \leq a_i \leq x_i$ for each i. Then each $a_i \leq x_i \leq x_1 + x_2$. Since $x_1 + x_2$ lies in X and so in K, this proves that K is directed. Therefore K is an ideal of G.

(b) We know from (a) that K is an ideal of G. Since H is directed, it is generated by H^+ (as a group), and hence $H \subseteq K$. On the other hand, any ideal I of G that contains H clearly must contain K, because I is a convex subgroup containing H^+.

(c) It is clear that K is a directed subgroup of G. Now consider $a \in G$ and $b \in K$ with $0 \leq a \leq b$. Then $b = x - y$ for some $x, y \in H^+$, and so $0 \leq a \leq b \leq x$. Since H is convex, we obtain $a \in H^+$, whence $a \in K$. Thus K is a convex subgroup of G, and hence an ideal of G.

Obviously $K \subseteq H$. If L is an ideal of G that is contained in H, then $L^+ \subseteq H^+$. Since L is directed, it is generated (as a group) by L^+, and therefore $L \subseteq K$. $\square$

COROLLARY 1.10. *If G is a partially ordered abelian group, then the family $\mathcal{L}$ of ideals of G forms a complete lattice under inclusion. Given any nonempty collection $\{H_i \mid i \in I\}$ of ideals of G, the infimum $\bigwedge H_i$ in $\mathcal{L}$ is the subgroup of*

G generated by $\bigcap H_i^+$, *while the supremum* $\bigvee H_i$ *in* $\mathcal{L}$ *is the convex subgroup of G generated by* $\bigcup H_i^+$.

PROOF. If $H = \bigcap H_i$, then H is a convex subgroup of G, and $H^+ = \bigcap H_i^+$. The subgroup K of G generated by H^+ is, by Proposition 1.9, the largest ideal of G that is contained in H. Thus K is the largest ideal of G that is contained in all the H_i, so that K is the infimum of the H_i in $\mathcal{L}$.

Set $L = \sum H_i$, and note that L is generated (as a group) by $\bigcup H_i^+$. If M is the convex subgroup of G generated by $\bigcup H_i^+$, then $M \supseteq L$, so that M is also the convex subgroup of G generated by L^+. According to Proposition 1.9, M is the smallest ideal of G that contains L. Thus M is the smallest ideal of G that contains all the H_i, so that M is the supremum of the H_i in $\mathcal{L}$. □

LEMMA 1.11. *Let G be a partially ordered abelian group, and let H be a subgroup of G. Then the set* $(G^+ + H)/H$ *is a cone in the group* G/H. *This cone is strict if and only if H is a convex subgroup of G.*

PROOF. Set $C = (G^+ + H)/H$. Since G^+ contains 0 and is closed under addition, the same is true of C, so that C is a cone in G/H.

First assume that H is convex, and consider any $x \in G/H$ such that both x and $-x$ lie in C. Then there exist $a, b \in G^+$ such that $x = a+H$ and $-x = b+H$. Hence, $a + b + H = x - x = 0$, and so $a + b \in H$. Since $0 \le a \le a + b$, it follows from the convexity of H that $a \in H$, whence $x = 0$. Thus C is a strict cone.

Conversely, assume that C is strict, and consider any $a \in G$ and $b \in H$ such that $0 \le a \le b$. Then $a + H \in C$ and also $-a + H = b - a + H \in C$. Since C is strict, $a + H = 0$, so that $a \in H$. Therefore H is a convex subgroup of G. □

DEFINITION. Let H be a subgroup of a partially ordered abelian group G. By Lemma 1.11, the set $(G^+ + H)/H$ is a cone in G/H, and hence we may make G/H into a pre-ordered abelian group with this cone as positive cone, that is,

$$(G/H)^+ = (G^+ + H)/H.$$

The resulting pre-order on G/H is called the *quotient ordering*, and a computational description of this relation may be found in the following proposition.

Unless otherwise specified, we shall always assume that G/H has been made into a pre-ordered abelian group using the quotient ordering. Note from Lemma 1.11 that G/H is a partially ordered abelian group if and only if H is a convex subgroup of G.

PROPOSITION 1.12. *Let H be a subgroup of a partially ordered abelian group G, and let* $x, y \in G$. *Then* $x + H \le y + H$ *in* G/H *if and only if* $x \le y + a$ *for some* $a \in H$. *In case H is a directed subgroup of G, then* $x + H \le y + H$ *if and only if* $x \le y + b$ *for some* $b \in H^+$.

PROOF. If $x+H \le y+H$, then $y-x+H$ lies in $(G/H)^+$, and so $y-x+H = z+H$ for some $z \in G^+$. Hence, $z = y-x+a$ for some $a \in H$, so that $y-x+a \ge 0$ and thus $x \le y + a$.

If $x \leq y + a$ for some $a \in H$, then the element $z = y + a - x$ is an element of G^+ such that $z + H = y - x + H$. Consequently, $y - x + H$ lies in $(G/H)^+$, so that $x + H \leq y + H$.

Now assume that H is directed, and that $x \leq y + a$ for some $a \in H$. There exists $b \in H$ such that $a \leq b$ and $0 \leq b$. Then $b \in H^+$ and $x \leq y + b$. $\square$

In working with the quotient ordering on a quotient group G/H, where G is a partially ordered abelian group and H is a subgroup of G, it is obviously quite helpful to assume that H is a convex subgroup of G, so that the quotient ordering is a partial order. The further assumption that H be an ideal of G is also quite useful, not so much for the slight improvement in the description of the quotient ordering given in Proposition 1.12 as for the consequence that a finite number of inequalities in G/H can be pulled back to inequalities in G in which the same correction factor from H is used. To be more precise, consider elements $x_1, y_1, \ldots, x_n, y_n$ in G such that $x_i + H \leq y_i + H$ for all i. Then for each $i = 1, \ldots, n$, there is some $a_i \in H$ such that $x_i \leq y_i + a_i$. If H is only a convex subgroup of G, we cannot make any useful adjustments to relate these a_i to each other. However, if H is an ideal, then there is some $a \in H^+$ such that each $a_i \leq a$, and then $x_i \leq y_i + a$ for all i. As this property is needed in most arguments involving quotient orderings, we shall usually investigate quotients of partially ordered abelian groups by ideals rather than by convex subgroups.

PROPOSITION 1.13. *Let H be an ideal of a partially ordered abelian group G, and let $x, x_1, \ldots, x_n$ be elements of G.*

(a) *If $x = x_1 \wedge \cdots \wedge x_n$, then $x + H = (x_1 + H) \wedge \cdots \wedge (x_n + H)$.*

(b) *If $x = x_1 \vee \cdots \vee x_n$, then $x + H = (x_1 + H) \vee \cdots \vee (x_n + H)$.*

PROOF. (a) Obviously $x + H \leq x_i + H$ for all i. Now consider any coset $y + H$ in G/H such that $y + H \leq x_i + H$ for all i. Then there exists an element $a \in H^+$ such that $y \leq x_i + a$ for all i. Now $y - a \leq x_i$ for all i, so that $y - a \leq x$, whence $y + H \leq x + H$. Thus $x + H$ is the infimum of the $x_i + H$.

(b) This follows from (a) on applying the order anti-automorphisms $z \mapsto -z$ in G and in G/H. $\square$

COROLLARY 1.14. *If G is a lattice-ordered abelian group and H is an ideal of G, then G/H is a lattice-ordered abelian group, and the quotient map $G \to G/H$ is a lattice homomorphism.* $\square$

• **Categories of Partially Ordered Abelian Groups.** We may form a natural category by taking all partially ordered abelian groups together with all order-preserving additive maps between them. It is also convenient to have a category in which all the objects have order-units. Here we take as objects all partially ordered abelian groups with specified order-units and as morphisms all order-unit-preserving, order-preserving, additive maps between them. We establish some terminology for these two categories, and then observe some of their properties.

DEFINITION. Let G and H be partially ordered abelian groups. A *positive homomorphism* from G to H is any abelian group homomorphism $f: G \to H$ that maps positive elements to positive elements, that is, $f(G^+) \subseteq H^+$. Note that a group homomorphism from G to H is positive if and only if it is order-preserving. Now suppose that we are given order-units $u \in G$ and $v \in H$. A *normalized positive homomorphism from* (G, u) *to* (H, v) is any positive homomorphism $f: G \to H$ such that $f(u) = v$.

Note that if $f: G \to H$ is a positive homomorphism between partially ordered abelian groups, then $\ker(f)$ is a convex subgroup of G. However, even if G and H are directed, $\ker(f)$ need not be directed, and so need not be an ideal of G. For example, let $f: \mathbb{Z}^2 \to \mathbb{Z}$ be the positive homomorphism given by the rule $f(x, y) = x + y$, where $\mathbb{Z}^2$ has the product ordering (defined below).

DEFINITION. By *the category of partially ordered abelian groups* we shall mean the category whose objects are all partially ordered abelian groups and whose morphisms are all positive homomorphisms between them. By *the category of partially ordered abelian groups with order-unit* we shall mean the category whose objects are all pairs (G, u), where G is a partially ordered abelian group and u is an order-unit in G, and whose morphisms are all normalized positive homomorphisms between these objects.

In either of these categories, morphisms which are set-theoretically bijective need not be isomorphisms. For instance, if G denotes the abelian group $\mathbb{Z}$ equipped with the discrete ordering, then the set-theoretic identity map $G \to \mathbb{Z}$ is a bijective positive homomorphism but not an isomorphism of ordered groups. In general, an isomorphism $f: H \to K$ in the category of partially ordered abelian groups must obviously satisfy $f(H^+) = K^+$. Conversely, a group isomorphism $f: H \to K$ satisfying $f(H^+) = K^+$ must be an isomorphism of partially ordered abelian groups, because the homomorphisms f and f^{-1} are both positive.

If G and H are isomorphic objects in the category of partially ordered abelian groups, we usually write just that G and H are "isomorphic as ordered groups". Similarly, we usually write that isomorphic objects (G, u) and (H, v) in the category of partially ordered abelian groups with order-unit are "isomorphic as ordered groups with order-unit".

DEFINITION. Let $\{G_i \mid i \in I\}$ be a nonempty collection of partially ordered abelian groups. There is a natural partial order on the abelian group $G = \prod_{i \in I} G_i$, where for $x, y \in G$ we have $x \leq y$ if and only if $x_i \leq y_i$ for all $i \in I$. This partial order is called the *product ordering* on G, and unless otherwise specified, we assume any direct product of partially ordered abelian groups to be equipped with the product ordering. Using the product ordering, G becomes a partially ordered abelian group with positive cone $\prod_{i \in I} G_i^+$. Then G, together with the projection maps $G \to G_i$, is a product of the family $\{G_i\}$ in the category of partially ordered abelian groups. Another partial order on G which is sometimes useful is the *strict ordering*, denoted $\ll$, where for $x, y \in G$ we have $x \ll y$ if and only if $x_i < y_i$ for all $i \in I$.

Note that in the case that all the G_i equal a common partially ordered abelian group A, so that $G = A^I$, the product ordering on G coincides with the pointwise ordering, and the strict ordering as defined above coincides with the strict ordering as defined earlier.

DEFINITION. Let $\{(G_i, u_i) \mid i \in I\}$ be a nonempty collection of partially ordered abelian groups with order-unit. Let u denote the element with components u_i in the product group $\prod G_i$. If I is finite, then u is an order-unit in $\prod G_i$, but if I is infinite, then u is usually not an order-unit in $\prod G_i$. Thus we define the *restricted direct product* of the (G_i, u_i) to be the convex subgroup G of $\prod G_i$ generated by u, that is

$$G = \{x \in \prod G_i \mid -nu \leq x \leq nu \text{ for some } n \in \mathbb{N}\}.$$

Then u is an order-unit in G, and (G, u) together with the projection maps $G \to G_i$ is a product of the family $\{(G_i, u_i)\}$ in the category of partially ordered abelian groups with order-unit. Hence, we use the notation $(G, u) = \prod_{i \in I}(G_i, u_i)$.

Given a collection $\{G_i \mid i \in I\}$ of partially ordered abelian groups, by our conventions the abelian group $G = \bigoplus G_i$ is given the product ordering, unless otherwise specified. Then G together with the injection maps $G_i \to G$ is a coproduct for the family $\{G_i\}$ in the category of partially ordered abelian groups.

Coproducts also exist in the category of partially ordered abelian groups with order-unit but are messier to construct. Rather than construct them directly, we leave their existence as a consequence of the construction of pushouts (Proposition 1.19).

If a partially ordered abelian group G is, as an abelian group, the direct sum of a family $\{G_i \mid i \in I\}$ of subgroups, then the natural map from the external direct sum $\bigoplus G_i$ into G is a bijective positive homomorphism. However, this map is an isomorphism of ordered groups only if $G^+ = \sum G_i^+$. If this is the case, then we say that G, *as an ordered group*, is the internal direct sum of the subgroups G_i.

PROPOSITION 1.15. *Let $\{G_i, f_{ji}\}$ be a direct system of partially ordered abelian groups and positive homomorphisms, indexed by a directed set I. Let G be the abelian group direct limit of this system, and for each $i \in I$ let $q_i : G_i \to G$ be the natural map. Then G can be made into a partially ordered abelian group with positive cone*

$$G^+ = \bigcup_{i \in I} q_i(G_i^+),$$

and G together with the maps q_i is a direct limit for the given system in the category of partially ordered abelian groups.

PROOF. As each G_i^+ is a cone in G_i, each $q_i(G_i^+)$ is a cone in G. Note that whenever $i \leq j$ in I, we have $q_i(G_i^+) = q_j f_{ji}(G_i^+) \subseteq q_j(G_j^+)$. Since I is directed, it follows that the cones $q_i(G_i^+)$ are directed under inclusion. Consequently, the set $G^+ = \bigcup q_i(G_i^+)$ is a cone in G.

Now consider any $x \in G$ such that $x, -x \in G^+$. Then there is some $i \in I$ such that x and $-x$ are in $q_i(G_i^+)$. Choose $a, b \in G_i^+$ such that $x = q_i(a)$ and $-x = q_i(b)$. Now $q_i(a+b) = 0$, and hence there is an index $j \geq i$ for which $f_{ji}(a+b) = 0$. Since

$$0 \leq f_{ji}(a) = -f_{ji}(b) \leq 0,$$

it follows that $f_{ji}(a) = 0$, whence $x = q_j f_{ji}(a) = 0$.

Thus G^+ is a strict cone, so that G does become a partially ordered abelian group with this positive cone.

Given a partially ordered abelian group H and positive homomorphisms $g_i \colon G_i \to H$ for all $i \in I$ such that $g_j f_{ji} = g_i$ whenever $i \leq j$ in I, there at least exists a unique group homomorphism $g \colon G \to H$ such that $gq_i = g_i$ for all $i \in I$. Since $gq_i(G_i^+) = g_i(G_i^+) \subseteq H^+$ for all $i \in I$, we also have $g(G^+) \subseteq H^+$, so that g is a positive homomorphism.

Therefore G is a direct limit of the G_i in the category of partially ordered abelian groups. □

In the situation of Proposition 1.15, we write $G = \varinjlim G_i$ by way of abbreviation, and we observe from the description of G how elements can be manipulated within it. For instance, given $x_1, x_2 \in G$ with $x_1 \leq x_2$, we can choose an index $j \in I$ and elements $y_1, y_2 \in G_j$ such that each $x_i = q_j(y_i)$, and by increasing j if necessary, we may assume in addition that $y_1 \leq y_2$.

PROPOSITION 1.16. *Let $\{(G_i, u_i), f_{ji}\}$ be a direct system of partially ordered abelian groups with order-unit and normalized positive homomorphisms, indexed by a directed set I. Let G be a direct limit of the system $\{G_i, f_{ji}\}$ in the category of partially ordered abelian groups, and for each $i \in I$ let $q_i \colon G_i \to G$ be the natural map. Then there exists a unique order-unit $u \in G$ such that $q_i(u_i) = u$ for all $i \in I$, and (G, u) together with the maps q_i is a direct limit for the given system in the category of partially ordered abelian groups with order-unit.*

PROOF. Since $f_{ji}(u_i) = u_j$ whenever $i \leq j$ in I, there exists a unique element $u \in G$ such that $q_i(u_i) = u$ for all $i \in I$. As each u_i is an order-unit in G_i, it follows that u is an order-unit in G. By definition of u, each of the maps q_i is a normalized positive homomorphism from (G_i, u_i) to (G, u). That (G, u) is a direct limit of the (G_i, u_i) is clear. □

In the situation of Proposition 1.16, we write $(G, u) = \varinjlim (G_i, u_i)$.

Let $\{G_i, f_{ij}\}$ be an inverse system of partially ordered abelian groups and positive homomorphisms, indexed by a directed set I. If

$$G = \{x \in \prod G_i \mid f_{ij}(x_j) = x_i \text{ whenever } i \leq j \text{ in } I\},$$

then G together with the projection maps $p_i \colon G \to G_i$ is an inverse limit of the system $\{G_i, f_{ij}\}$ in the category of partially ordered abelian groups, and we write $G = \varprojlim G_i$.

Let $\{(G_i, u_i), f_{ij}\}$ be an inverse system of partially ordered abelian groups with order-unit and normalized positive homomorphisms, indexed by a directed set I. If $(H, u) = \prod (G_i, u_i)$ and

$$G = \{x \in H \mid f_{ij}(x_j) = x_i \text{ whenever } i \leq j \text{ in } I\},$$

then (G, u) together with the projection maps $p_i\colon G \to G_i$ is an inverse limit of the system $\{(G_i, u_i), f_{ij}\}$ in the category of partially ordered abelian groups with order-unit, and we write $(G, u) = \varprojlim (G_i, u_i)$.

Note that G can be identified with the convex subgroup of $\varprojlim G_i$ generated by u. In particular, G is an ideal of $\varprojlim G_i$.

• Pullbacks, Pushouts, and Coproducts.

DEFINITION. Recall that a *pullback* of a nonempty collection

$$\{f_i\colon A_i \to B \mid i \in I\}$$

of morphisms with a common range in a category $\mathcal{A}$ consists of an object A together with morphisms $g_i\colon A \to A_i$ for all $i \in I$, such that (a) $f_i g_i = f_j g_j$ for all $i, j \in I$, and (b) given any object C and any morphisms $h_i\colon C \to A_i$ for all $i \in I$ such that $f_i h_i = f_j h_j$ for all $i, j \in I$, there exists a unique morphism $h\colon C \to A$ such that $h_i = g_i h$ for all $i \in I$. The dual concept is a *pushout* of a collection of morphisms with a common domain.

Let $\{f_i\colon G_i \to H \mid i \in I\}$ be a nonempty collection of morphisms with a common range in the category of partially ordered abelian groups. If

$$G = \{x \in \prod G_i \mid f_i(x_i) = f_j(x_j) \text{ for all } i, j \in I\},$$

then G together with the projection maps $p_i\colon G \to G_i$ is a pullback of the maps f_i in the category of partially ordered abelian groups.

Let $\{f_i\colon (G_i, u_i) \to (H, v) \mid i \in I\}$ be a nonempty collection of morphisms with a common range in the category of partially ordered abelian groups with order-unit. If $(K, u) = \prod (G_i, u_i)$ and

$$G = \{x \in K \mid f_i(x_i) = f_j(x_j) \text{ for all } i, j \in I\},$$

then (G, u) together with the projection maps $p_i\colon G \to G_i$ is a pullback of the maps f_i in the category of partially ordered abelian groups with order-unit.

Note that G can be identified with the convex subgroup generated by u inside a pullback of the positive homomorphisms $f_i\colon G_i \to H$ in the category of partially ordered abelian groups.

We now turn to the construction of pushouts in the category of partially ordered abelian groups. For use in this construction, we adopt the convention that to write $\sum_{i \in I} x_i = x$ in an abelian group G means that there is a finite subset $J \subseteq I$ such that $x_i = 0$ for all $i \in I - J$ and $\sum_{i \in J} x_i = x$.

PROPOSITION 1.17. *Let $\{f_i\colon H \to G_i \mid i \in I\}$ be a nonempty collection of morphisms with a common domain in the category of partially ordered abelian groups. Set $K = \bigoplus G_i$, and for each $i \in I$ let $q_i\colon G_i \to K$ be the injection map. Set*

$$L = \left\{ \sum_{i \in I} q_i f_i(x_i) \mid \text{ each } x_i \in H \text{ and } \sum_{i \in I} x_i = 0 \right\},$$

and let M be the convex subgroup of K generated by L. If $f_i^{-1}(G_i^+) = H^+$ for all $i \in I$, then $M = L$.

Set $G = K/M$, and let $p\colon K \to G$ be the quotient map. Then G together with the maps $pq_i\colon G_i \to G$ is a pushout for the maps f_i in the category of partially ordered abelian groups.

PROOF. It is clear that L is a subgroup of K. Assume for the moment that $f_i^{-1}(G_i^+) = H^+$ for all $i \in I$, and consider any $y \in K$ and $z \in L$ such that $0 \le y \le z$. Then there exist elements $w_i \in H$ for all $i \in I$ such that $\sum w_i = 0$ and $z = \sum q_i f_i(w_i)$, that is, $z_i = f_i(w_i)$ for each $i \in I$. Hence, $0 \le y_i \le f_i(w_i)$ for all $i \in I$, so that $w_i \in f_i^{-1}(G_i^+)$. By assumption, each $w_i \in H^+$. Since $\sum w_i = 0$, it follows that each $w_i = 0$. As a result, each $y_i = 0$, whence $y = 0 \in L$. Thus in this case L is a convex subgroup of K, and hence $M = L$.

We now return to the general case. Since M is a convex subgroup of K, the group G, with the quotient ordering, is a partially ordered abelian group. Given any $i, j \in I$ and any $x \in H$, we have

$$q_i f_i(x) + q_j f_j(-x) \in L,$$

whence $pq_i f_i(x) + pq_j f_j(-x) = 0$. Hence, $pq_i f_i = pq_j f_j$ for all $i, j \in I$.

Consider a partially ordered abelian group G' and positive homomorphisms $g_i\colon G_i \to G'$ such that $g_i f_i = g_j f_j$ for all $i, j \in I$. There exists a unique positive homomorphism $k\colon K \to G'$ such that each $g_i = kq_i$.

Any element $y \in L$ has the form $y = \sum q_i f_i(x_i)$ where the $x_i \in H$ for each $i \in I$ and $\sum x_i = 0$. Choose a particular index $j \in I$. Then

$$k(y) = \sum_{i \in I} kq_i f_i(x_i) = \sum_{i \in I} g_i f_i(x_i) = \sum_{i \in I} g_j f_j(x_i) = g_j f_j \left(\sum_{i \in I} x_i \right) = 0.$$

Thus $L \subseteq \ker(k)$. As $\ker(k)$ is a convex subgroup of K, it follows that $M \subseteq \ker(k)$. Hence, there is a unique positive homomorphism $g\colon G \to G'$ such that $gp = k$, and $gpq_i = kq_i = g_i$ for all $i \in I$.

Finally, consider any positive homomorphism $h\colon G \to G'$ such that $hpq_i = g_i$ for all $i \in I$. Then $hp = k$ (by the uniqueness of k), whence $h = g$. Thus g is unique.

Therefore G is a pushout of the maps f_i. □

PROPOSITION 1.18. *Let $\{f_i\colon (H, v) \to (G_i, u_i) \mid i \in I\}$ be a nonempty collection of morphisms with a common domain in the category of partially ordered* *abelian groups with order-unit. Let G together with maps $g_i\colon G_i \to G$ for all $i \in I$ be a pushout of the positive homomorphisms $f_i\colon H \to G_i$ in the category of partially ordered abelian groups. There exists a unique order-unit $u \in G$ such that $g_i(u_i) = u$ for all $i \in I$, and (G, u) together with the maps g_i is a pushout of the maps f_i in the category of partially ordered abelian groups with order-unit.*

PROOF. Since $g_i(u_i) = g_i f_i(v) = g_j f_j(v) = g_j(u_j)$ for all $i, j \in I$, the existence and uniqueness of a positive element $u \in G$ such that $g_i(u_i) = u$ for all $i \in I$ is clear. To show that u is an order-unit in G, we adopt the notation of Proposition 1.17, so that each $g_i = pq_i$.

Any element $x \in G$ has the form $p(y)$ for some $y \in K$. For each $i \in I$, there exists a nonnegative integer n_i such that $y_i \leq n_i u_i$. Since all but finitely many of the y_i are zero, we may assume that all but finitely many of the n_i are zero. Now

$$y = \sum_{i \in I} q_i(y_i) \leq \sum_{i \in I} n_i q_i(u_i),$$

and consequently

$$x = p(y) \leq \sum_{i \in I} n_i p q_i(u_i) = \sum_{i \in I} n_i g_i(u_i) = \left(\sum_{i \in I} n_i \right) u.$$

Thus u is an order-unit in G.

By definition of u, each of the maps g_i is a normalized positive homomorphism from (G_i, u_i) to (G, u).

Consider an object (G', u') and morphisms $h_i\colon (G_i, u_i) \to (G', u')$ in the category of partially ordered abelian groups with order-unit, such that $h_i f_i = h_j f_j$ for all $i, j \in I$. Since G is a pushout of the f_i in the category of partially ordered abelian groups, there exists a unique positive homomorphism $h\colon G \to G'$ such that $h_i = hg_i$ for all $i \in I$. For any $i \in I$, we have $h(u) = hg_i(u_i) = h_i(u_i) = u'$, so that h is a normalized positive homomorphism from (G, u) to (G', u').

Therefore (G, u) is a pushout of the maps f_i. $\square$

The existence of coproducts in the category of partially ordered abelian groups with order-unit is an easy consequence of the existence of pushouts, as follows.

PROPOSITION 1.19. *Let $\{(G_i, u_i) \mid i \in I\}$ be a nonempty collection of partially ordered abelian groups with order-unit. For each $i \in I$, let f_i be the unique normalized positive homomorphism from $(\mathbb{Z}, 1)$ to (G_i, u_i). In the category of partially ordered abelian groups with order-unit, any pushout of the morphisms f_i is also a coproduct of the objects (G_i, u_i).*

PROOF. Let (G, u) together with morphisms $g_i\colon (G_i, u_i) \to (G, u)$ for all $i \in I$ be a pushout of the f_i in the category of partially ordered abelian groups with order-unit. Consider an object (H, v) and any morphisms

$$h_i\colon (G_i, u_i) \to (H, v)$$

for all $i \in I$. For all $i, j \in I$, we have $h_i f_i(1) = v = h_j f_j(1)$, and hence $h_i f_i = h_j f_j$. Consequently, there exists a unique morphism $h\colon (G, u) \to (H, v)$ such that $h_i = hg_i$ for all $i \in I$.

Thus (G, u) is a coproduct of the (G_i, u_i). □

• **Additional Concepts.**

DEFINITION. Suppose that $\{G_i \mid i \in I\}$ is a nonempty family of partially ordered abelian groups, indexed by a totally ordered set I. Let G be the abelian group $\bigoplus G_i$. Given $x, y \in G$, we have $x_i = 0 = y_i$ for all but finitely many $i \in I$, and so if $x \neq y$ the set $\{i \in I \mid x_i \neq y_i\}$ has a unique minimal element j. Define $x < y$ if and only if $x_j < y_j$. The corresponding relation $\leq$ is a translation-invariant partial order on G known as the *lexicographic ordering.* Equipped with this ordering, G becomes a partially ordered abelian group called the *lexicographic direct sum* of the family $\{G_i\}$.

In specifying the lexicographic direct sum of a finite family of partially ordered abelian groups, the index set is assumed to be an initial segment of $\mathbb{N}$ if not otherwise specified. For example, the lexicographic direct sum of H, K, L is assumed to be the lexicographic direct sum of groups G_1, G_2, G_3 indexed by the totally ordered set $\{1, 2, 3\}$, where $G_1 = H$ and $G_2 = K$ while $G_3 = L$. Note that the ordering of the index set corresponds to the order in which the groups H, K, L are listed.

DEFINITION. Again let $\{G_i \mid i \in I\}$ be a nonempty family of partially ordered abelian groups indexed by a totally ordered set I, and let G be the abelian group $\bigoplus G_i$. This time define $x \leq y$ for elements $x, y \in G$ if and only if either $x = y$ or else $x_j < y_j$ where j is the *largest* index for which $x_j \neq y_j$. This defines a translation-invariant partial order on G known as the *reverse lexicographic ordering*, and when G is equipped with this ordering it is called the *reverse lexicographic direct sum* of the family $\{G_i\}$.

Of course, the reverse lexicographic direct sum of the family $\{G_i \mid i \in I\}$ is isomorphic to the lexicographic direct sum of the family $\{G_i \mid i \in I^*\}$ where I^* is the dual of I. In particular, the reverse lexicographic direct sum of groups H and K is isomorphic to the lexicographic direct sum of K and H.

In a similar vein, a lexicographic ordering may be defined on a direct product of partially ordered abelian groups, provided the index set is well-ordered, as follows.

DEFINITION. Let $\{G_i \mid i \in I\}$ be a family of partially ordered abelian groups, indexed by a well-ordered set I. Let G be the abelian group $\prod G_i$. For $x, y \in G$, define $x \leq y$ if and only if either $x = y$ or else $x_j < y_j$ where j is the least index for which $x_j \neq y_j$. This defines a translation-invariant partial order on G known as the *lexicographic ordering*, and when G is equipped with this ordering it is called the *lexicographic direct product* of the family $\{G_i\}$.

DEFINITION. Let G be a partially ordered abelian group, and let n be a positive integer. We say that G is *n-perforated* if there exists an element $x \in G$ such that $nx \geq 0$ but $x \not\geq 0$; otherwise, G is *n-unperforated.* If G is n-perforated for some positive integer n, then G is *perforated*, while if G is n-unperforated for all positive integers n, then G is *unperforated.*

For example, if the abelian group $\mathbb{Z}$ is made into a partially ordered abelian group with positive cone $\{0, 2, 4, 6, \ldots\}$, then $\mathbb{Z}$ is 2-perforated. (The elements 1,3,5,... that were "omitted" from the positive cone may be viewed as a row of "perforations" in the positive cone.) On the other hand, any partially ordered rational vector space is unperforated. In any partially ordered abelian group, a strictly negative element $x < 0$ satisfies $nx \leq x < 0$ for all positive integers n, so that $nx \not\geq 0$. Consequently, all totally ordered abelian groups are unperforated. More generally, all lattice-ordered abelian groups are unperforated, as we prove in Proposition 1.22.

Note that any unperforated partially ordered abelian group must be torsion-free as an abelian group.

PROPOSITION 1.20. *Let G be a partially ordered abelian group, let H be an ideal of G, and let n be a positive integer. If G is n-unperforated, then so is G/H.*

PROOF. Consider any element $x \in G$ such that $n(x+H) \geq 0$. Since $0+H \leq nx + H$, we must have $0 \leq nx + a$ for some $a \in H^+$. Then $a \leq na$, and so $0 \leq n(x+a)$. As G is n-unperforated, $0 \leq x+a$, whence $x+H \geq 0$. Thus G/H is n-unperforated. □

LEMMA 1.21. *Let G be a lattice-ordered abelian group, and let $a_1, \ldots, a_n$ be elements of G. Then*

$$\sum_{i=1}^{n}(a_i \wedge 0) = \left[\bigwedge_{A\in\mathcal{A}}\left(\sum_{i\in A} a_i\right)\right] \wedge 0 \quad \text{and} \quad \sum_{i=1}^{n}(a_i \vee 0) = \left[\bigvee_{A\in\mathcal{A}}\left(\sum_{i\in A} a_i\right)\right] \vee 0,$$

where $\mathcal{A}$ is the collection of all nonempty subsets of $\{1, 2, \ldots, n\}$.

PROOF. We prove only the first identity, as the second follows from the first by applying the order anti-automorphism $z \mapsto -z$. In the case $n = 1$, there is nothing to prove.

Now let $n > 1$, and assume that

$$\sum_{i=1}^{n-1}(a_i \wedge 0) = \left[\bigwedge_{B\in\mathcal{B}}\left(\sum_{i\in B} a_i\right)\right] \wedge 0,$$

where $\mathcal{B}$ is the collection of all nonempty subsets of $\{1, 2, \ldots, n-1\}$. With the aid of Proposition 1.4, we compute that

$$\begin{aligned}\sum_{i=1}^{n}(a_i \wedge 0) &= \left(\left[\bigwedge_{B\in\mathcal{B}}\left(\sum_{i\in B} a_i\right)\right] \wedge 0\right) + (a_n \wedge 0)\\ &= \left(\bigwedge_{B\in\mathcal{B}}\left[\left(\sum_{i\in B} a_i\right) + (a_n \wedge 0)\right]\right) \wedge (0 + [a_n \wedge 0])\\ &= \left(\bigwedge_{B\in\mathcal{B}}\left[\left(a_n + \sum_{i\in B} a_i\right) \wedge \left(\sum_{i\in B} a_i\right)\right]\right) \wedge a_n \wedge 0\\ &= \left[\bigwedge_{A\in\mathcal{A}}\left(\sum_{i\in A} a_i\right)\right] \wedge 0,\end{aligned}$$

completing the induction step. □

In particular, given a lattice-ordered abelian group G, an element $x \in G$, and a positive integer n, Lemma 1.21 shows that

$$n(x\wedge 0) = nx \wedge (n-1)x \wedge \cdots \wedge x \wedge 0 \quad \text{and} \quad n(x \vee 0) = nx \vee (n-1)x \vee \cdots \vee x \vee 0.$$

PROPOSITION 1.22. *If G is a lattice-ordered abelian group, then G is unperforated.*

PROOF. If $x \in G$ and $n \in \mathbb{N}$ such that $nx \geq 0$, then $nx \wedge 0 = 0$. As a result, it follows from Lemma 1.21 that

$$\begin{aligned}n(x \wedge 0) &= nx \wedge (n-1)x \wedge \cdots \wedge x \wedge 0\\ &= (n-1)x \wedge (n-2)x \wedge \cdots \wedge x \wedge 0 = (n-1)(x \wedge 0),\end{aligned}$$

whence $x \wedge 0 = 0$, and thus $x \geq 0$. Therefore G is unperforated. □

DEFINITION. A partially ordered abelian group G is said to be *archimedean* provided that whenever $x, y \in G$ such that $nx \leq y$ for all positive integers n, then $x \leq 0$.

For example, any subgroup of any $\mathbb{R}^X$ is archimedean. However, if $\mathbb{R}^X$ is given the strict ordering, then as long as X has more than one element, $\mathbb{R}^X$ is not archimedean. For instance, in $\mathbb{R}^2$ we have $n(-1,0) \ll (1,1)$ for all $n \in \mathbb{N}$, yet $(-1,0) \not\ll 0$. The archimedean condition is closely related to the existence of a large supply of real-valued positive homomorphisms. As we shall prove later (Theorem 4.14), if G is an archimedean partially ordered abelian group which contains an order-unit, then G is isomorphic to a subgroup of $\mathbb{R}^X$ (with the pointwise ordering) for some set X.

Several other conditions may be found in the literature under the name "archimedean". Aside from those relevant only to totally ordered groups, these conditions are mainly intended for use in lattice-ordered abelian groups, for which they are equivalent to the condition defined above. For one example, see the proposition below. However, the reader should be aware that most of these alternate conditions are, for general partially ordered abelian groups, weaker than the condition introduced above. For instance, any $\mathbb{R}^X$ with the strict ordering satisfies the condition $(*)$ in the following proposition.

PROPOSITION 1.23. *A lattice-ordered abelian group G is archimedean if and only if the following condition holds:*

$$(*) \qquad \text{If } x, y \in G^+ \text{ and } nx \leq y \text{ for all } n \in \mathbb{N}, \text{ then } x = 0.$$

PROOF. Any archimedean partially ordered abelian group satisfies $(*)$. Conversely, assume that $(*)$ holds in G, and consider any elements $a, b \in G$ such that $na \leq b$ for all $n \in \mathbb{N}$. Set $x = a \vee 0$ and $y = b \vee 0$. In view of Lemma 1.21,

$$nx = na \vee (n-1)a \vee \cdots \vee a \vee 0 \leq b \vee 0 = y$$

for all $n \in \mathbb{N}$. Applying $(*)$, we obtain $x = 0$, and hence $a \leq 0$. Thus G is archimedean. □

PROPOSITION 1.24. *If G is an archimedean directed abelian group, then G is unperforated.*

PROOF. Let $x \in G$ and $k \in \mathbb{N}$ such that $kx \geq 0$. As G is directed, there exists an element $y \in G$ such that $nx \geq y$ for each $n = 0, 1, \ldots, k-1$. Since $qkx \geq 0$ for all positive integers q, it follows that $nx \geq y$ for all positive integers n. Then $x \geq 0$, because G is archimedean. Therefore G is unperforated. □

• **Notes.** The study of partially ordered and lattice-ordered real vector spaces was launched in the late 1930's by Freudenthal [**41**], Kantorovitch [**78**, **80**, **81**], and Riesz [**109**]. Initial studies of partially ordered abelian groups were undertaken in the early 1940's by Clifford [**21**] and Everett and Ulam [**40**]. In this chapter of our book, all the results are fairly basic and may be considered folklore. We shall not attribute them to any particular source.

CHAPTER 2

Interpolation

In full generality, the axioms defining a partially ordered abelian group are too weak to lead to much of any detailed structure theory. To try to build a structure theory, we should expect to require some additional axiom(s) of a form that provide solutions to some variety of basic computations. One time-honored choice is the assumption of the existence of finite infima and suprema, the usefulness of which is reflected by the sheer size of the literature on lattice-ordered abelian groups. In place of lattice-theoretic assumptions, we shall impose a weaker axiom, the Riesz interpolation property, which will become our basic computational tool.

This chapter contains an introduction to the Riesz interpolation property and some of its more general consequences, beginning with some equivalent forms and alternate computational consequences of it. In order to ensure that this interpolation property is available in suitably general circumstances, we prove that it is preserved in ideals, quotient groups modulo ideals, direct products, and direct sums, as well as in certain lexicographic direct products, lexicographic direct sums, pullbacks, and pushouts. The final section of the chapter is devoted to investigating the group of all homomorphisms from a directed abelian group with interpolation into a Dedekind complete lattice-ordered abelian group, culminating in a characterization of the subgroup generated by the positive homomorphisms, and the result that this subgroup is itself a Dedekind complete lattice-ordered abelian group.

• Riesz Interpolation and Decomposition Properties.

PROPOSITION 2.1. *For a partially ordered abelian group G, the following conditions are equivalent:*

(a) *Given x_1, x_2, y_1, y_2 in G such that $x_i \leq y_j$ for all i, j, there exists z in G such that $x_i \leq z \leq y_j$ for all i, j.*

(b) *Given x, y_1, y_2 in G^+ such that $x \leq y_1 + y_2$, there exist x_1, x_2 in G^+ such that $x = x_1 + x_2$ and $x_j \leq y_j$ for each j.*

(c) *Given* x_1, x_2, y_1, y_2 *in* G^+ *such that* $x_1+x_2 = y_1+y_2$, *there exist* $z_{11}, z_{12}, z_{21}, z_{22}$ *in* G^+ *such that* $x_i = z_{i1} + z_{i2}$ *for each* i *and* $y_j = z_{1j} + z_{2j}$ *for each* j.

PROOF. (a)⇒(b): By assumption, $0 \le x$ and $0 \le y_1$, while $x - y_2 \le y_1$. Since $y_2 \ge 0$, we also have $x - y_2 \le x$. By (a), there must be some $x_1 \in G$ such that $0 \le x_1 \le x$ and $x - y_2 \le x_1 \le y_1$. Set $x_2 = x - x_1$. Then x_1 and x_2 are elements of G^+ such that $x = x_1 + x_2$ and $x_1 \le y_1$. Since $x - y_2 \le x_1$, we also have $x_2 = x - x_1 \le y_2$.

(b)⇒(a): Note that $y_j - x_i \in G^+$ for all i, j and that

$$y_2 - x_1 \le (y_2 - x_1) + (y_1 - x_2) = (y_1 - x_1) + (y_2 - x_2).$$

By (b), there exist $z_1, z_2 \in G^+$ such that $y_2 - x_1 = z_1 + z_2$ and each $z_j \le y_j - x_j$. Set $z = x_1 + z_1$, and note that $x_1 \le z$. Since $z_1 \le y_1 - x_1$, we have $z \le y_1$. Since $y_2 - x_1 = z_1 + z_2$, we have $z = y_2 - z_2 \le y_2$. Finally, since $z_2 \le y_2 - x_2$, we have $x_2 \le y_2 - z_2 = z$. Therefore $x_i \le z \le y_j$ for all i, j.

(b)⇒(c): As $x_2 \ge 0$, we have $x_1 \le y_1 + y_2$. By (b), there exist z_{11}, z_{12} in G^+ such that $x_1 = z_{11} + z_{12}$ and each $z_{1j} \le y_j$. Set $z_{2j} = y_j - z_{1j}$ for each j, so that $z_{2j} \in G^+$ and $y_j = z_{1j} + z_{2j}$. Since

$$x_1 + x_2 = y_1 + y_2 = z_{11} + z_{21} + z_{12} + z_{22} = x_1 + z_{21} + z_{22},$$

we also have $x_2 = z_{21} + z_{22}$.

(c)⇒(b): Set $w_1 = x$ and $w_2 = y_1 + y_2 - x$, so that w_1, w_2 are elements of G^+ satisfying $w_1 + w_2 = y_1 + y_2$. By (c), there exist $z_{11}, z_{12}, z_{21}, z_{22}$ in G^+ such that each $w_i = z_{i1} + z_{i2}$ and each $y_j = z_{1j} + z_{2j}$. Then z_{11} and z_{12} are elements of G^+ such that $x = w_1 = z_{11} + z_{12}$. Since each $z_{2j} \ge 0$, we also have $z_{1j} \le z_{1j} + z_{2j} = y_j$. □

DEFINITION. A partially ordered set X is said to satisfy the *Riesz interpolation property* (or just *interpolation*, for short) provided X satisfies the condition given in Proposition 2.1(a): given any x_1, x_2, y_1, y_2 in X such that $x_i \le y_j$ for all i, j, there exists $z \in X$ such that $x_i \le z \le y_j$ for all i, j.

For example, any lattice satisfies this property, since $x_i \le x_1 \vee x_2 \le y_j$ for all i, j, and also $x_i \le y_1 \wedge y_2 \le y_j$ for all i, j. For another example, let G be the vector space $\mathbb{R}^2$, made into a partially ordered real vector space with positive cone

$$G^+ = \{(0,0)\} \cup \{(x,y) \in G \mid 2x > y > 0\}.$$

The interpolation property holds in G, as illustrated in Figure 1, where the elements $z \in G$ to be interpolated between x_1, x_2 and y_1, y_2 are to be found in the interior of the shaded parallelogram.

DEFINITION. A partially ordered abelian group G is said to satisfy the *Riesz decomposition properties* provided conditions (b) and (c) of Proposition 2.1 hold in G. We define an *interpolation group* to be any partially ordered abelian group that satisfies the Riesz interpolation property and, consequently, the Riesz decomposition properties as well.

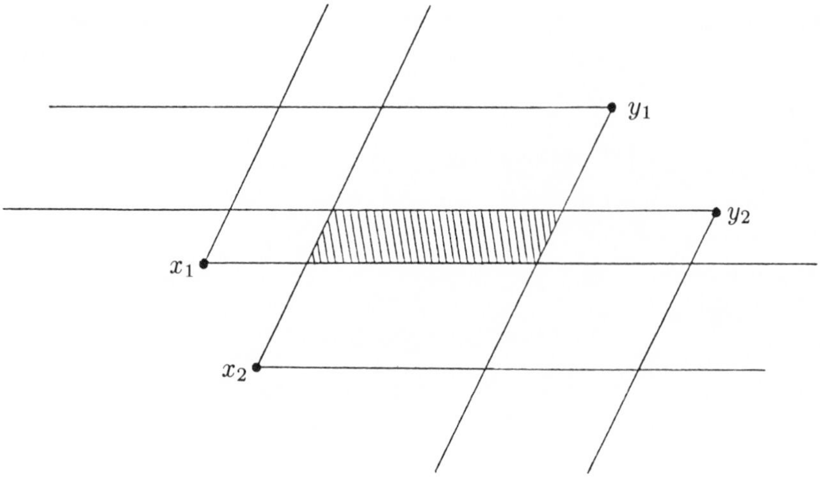

FIGURE 1

For example, any lattice-ordered abelian group is an interpolation group. For an example in which interpolation does not hold, make the abelian group $G = \mathbb{Z}^2$ into a partially ordered abelian group with positive cone

$$G^+ = \{(a,b) \in G \mid 2a \geq b \geq 0\}.$$

Set $x_1 = (0,0)$ and $x_2 = (0,1)$, while $y_1 = (1,1)$ and $y_2 = (1,2)$. Then $x_i \leq y_j$ for all i, j, but there is no element $z \in G$ such that $x_i \leq z \leq y_j$ for all i, j.

Observe that the class of interpolation groups is closed under the formation of direct products, direct sums, and direct limits. Note also that any convex subgroup of an interpolation group is itself an interpolation group.

In many instances, elements to which interpolation is to be applied may not be conveniently indexed. To indicate the assertion that each element of a set X is less than or equal to each element of a set Y, we list the elements of X and Y in separate columns, with the symbol $\leq$ placed between the columns. In this notation, the interpolation property may be expressed as follows: given any elements x, y, z, w in G such that

$$\begin{matrix} x & & z \\ & \leq & \\ y & & w \end{matrix}$$

there exists an element v in G such that

$$\begin{matrix} x & & & & z \\ & \leq & v & \leq & \\ y & & & & w. \end{matrix}$$

This notation also avoids ambiguities that appear when writing multiple inequalities on a single line. For example, the expression

$$a \leq b, c \leq d$$

might mean either that [$a \leq b$ and $c \leq d$] or that

$$a \quad \leq \quad \begin{matrix} b \\ c \end{matrix} \quad \leq \quad d.$$

PROPOSITION 2.2. *Let G be an interpolation group.*

(a) *Given $x_1, \dots, x_n$ and $y_1, \dots, y_k$ in G such that $x_i \leq y_j$ for all i, j, there exists z in G such that $x_i \leq z \leq y_j$ for all i, j.*

(b) *Given $x, y_1, \dots, y_k$ in G^+ such that $x \leq y_1 + \cdots + y_k$, there exist $x_1, \dots, x_k$ in G^+ such that $x = x_1 + \cdots + x_k$ and $x_j \leq y_j$ for all j.*

(c) *Given $x_1, \dots, x_n$ and $y_1, \dots, y_k$ in G^+ such that*

$$x_1 + \cdots + x_n = y_1 + \cdots + y_k,$$

there exist z_{ij} in G^+ (for $i = 1, \dots, n$ and $j = 1, \dots, k$) such that

$$x_i = z_{i1} + \cdots + z_{ik} \quad \text{and} \quad y_j = z_{1j} + \cdots + z_{nj}$$

for all i, j.

PROOF. (a) If $n = 1$, just take $z = x_1$, while if $k = 1$, take $z = y_1$. Thus we may assume that $n \geq 2$ and $k \geq 2$, the case $n = k = 2$ holding by hypothesis. Now let $n + k > 4$, and assume that the result holds whenever the total number of x_i and y_j is less than $n + k$. Either $n > 2$ or $k > 2$, say $n > 2$, the proof for the case $k > 2$ being dual. By the induction hypothesis, there exists $w \in G$ such that $x_i \leq w \leq y_j$ for all $i = 1, \dots, n-1$ and all $j = 1, \dots, k$. Since $2 + k < n + k$, another use of the induction hypothesis provides us with an element $z \in G$ such that

$$\begin{matrix} w \\ x_n \end{matrix} \quad \leq \quad z \quad \leq \quad y_j$$

for all $j = 1, \dots, k$. Then $x_i \leq z \leq y_j$ for all i, j. The proof in case $k > 2$ is similar, completing the induction step.

(b) If $k = 1$ there is nothing to prove, while the case $k = 2$ holds by Proposition 2.1. Now let $k > 2$, and assume that the result holds whenever the number of y_j is less than k. Since

$$x \leq (y_1 + y_2) + y_3 + y_4 + \cdots + y_k,$$

there must exist elements $x_{12}, x_3, x_4, \dots, x_k$ in G^+ such that

$$x = x_{12} + x_3 + x_4 + \cdots + x_k,$$

while $x_{12} \leq y_1 + y_2$ and $x_j \leq y_j$ for each $j = 3, \dots, k$. Then there exist elements $x_1, x_2 \in G^+$ such that $x_{12} = x_1 + x_2$ and $x_j \leq y_j$ for each $j = 1, 2$. As $x = x_1 + \cdots + x_k$, the induction step is complete.

(c) If $n = 1$, there is nothing to prove. Now let $n > 1$, and assume that the result holds whenever the number of x_i is less than n. Since

$$x_1 \leq x_1 + \cdots + x_n = y_1 + \cdots + y_k,$$

it follows from (b) that $x_1 = z_{11} + \cdots + z_{1k}$ for some elements $z_{1j} \in G^+$ such that each $z_{1j} \leq y_j$. Then each $y_j - z_{1j}$ is in G^+, and

$$x_2 + \cdots + x_n = y_1 + \cdots + y_k - x_1 = (y_1 - z_{11}) + \cdots + (y_k - z_{1k}).$$

By the induction hypothesis, there exist elements $z_{ij} \in G^+$ for $i = 2, \ldots, n$ and $j = 1, \ldots, k$ such that $x_i = z_{i1} + \cdots + z_{ik}$ for all $i = 2, \ldots, n$ and

$$y_j - z_{1j} = z_{2j} + \cdots + z_{nj}$$

for all $j = 1, \ldots, k$. This completes the induction step. □

• Ideals and Quotient Groups.

PROPOSITION 2.3. *Let G be a partially ordered abelian group, and let H be an ideal of G. If G has interpolation, then so do H and G/H.*

PROOF. As H is a convex subgroup of G, it clearly has interpolation. Now consider any elements x_1, x_2, y_1, y_2 in G such that $x_i + H \leq y_j + H$ for all i, j. For each i, j, there is some element $a_{ij} \in H$ such that $x_i \leq y_j + a_{ij}$. Since H is directed, there is an element $a \in H$ such that all $a_{ij} \leq a$. Now $x_i \leq y_j + a$ for all i, j, and hence there exists $z \in G$ such that $x_i \leq z \leq y_j + a$ for all i, j. Consequently,

$$x_i + H \leq z + H \leq y_j + H$$

for all i, j, proving that G/H has interpolation. □

The converse of Proposition 2.3 fails in general, as shown in Example 2.7.

PROPOSITION 2.4. *If G is an interpolation group, then any sum of ideals of G is an ideal, and any finite intersection of ideals of G is an ideal.*

PROOF. Let $\{H_i \mid i \in I\}$ be a collection of ideals of G, and set $H = \sum H_i$. Since each H_i is generated (as a subgroup) by a subset of G^+, so is H. Thus H is a directed subgroup of G.

Given $x \in G$ and $y \in H$ with $0 \leq x \leq y$, we may write $y = \sum y_i$ where each $y_i \in H_i$ and all but finitely many of the y_i are zero. Since each H_i is directed, there is some $z_i \in H_i^+$ for which $y_i \leq z_i$. Thus there is no loss of generality in assuming that each $y_i \in H_i^+$. By Riesz decomposition, there exist elements $x_i \in G^+$ for each i such that $x = \sum x_i$ and each $x_i \leq y_i$. Since H_i is convex, it follows that $x_i \in H_i$, and hence $x \in H$. Thus H is convex and so is an ideal of G.

Now let $\{K_1, \ldots, K_n\}$ be a finite collection of ideals of G, and set $K = K_1 \cap \cdots \cap K_n$. As each K_i is a convex subgroup of G, so is K. Given $x_1, x_2 \in K$, choose $y_j \in K_j$ for each $j = 1, \ldots, n$ such that $x_i \leq y_j$ for all i, j. Then there exists $z \in G$ such that $x_i \leq z \leq y_j$ for all i, j. As each K_j is convex, it follows from the relations $x_1 \leq z \leq y_j$ that $z \in K_j$. Thus $z \in K$, proving that K is directed. Therefore K is an ideal of G. □

It follows from Proposition 2.4 that in the lattice of ideals of an interpolation group G, finite infima are given by intersections, while arbitrary suprema are given by sums. In view of the following result, we conclude that the lattice of ideals of G is distributive.

PROPOSITION 2.5. *If* H, K_1, K_2 *are ideals in an interpolation group* G, *then*
(a) $H \cap (K_1 + K_2) = (H \cap K_1) + (H \cap K_2)$.
(b) $H + (K_1 \cap K_2) = (H + K_1) \cap (H + K_2)$.

PROOF. (a) Obviously $(H \cap K_1) + (H \cap K_2) \subseteq H \cap (K_1 + K_2)$. Now consider any element u from $H \cap (K_1 + K_2)$, and write

$$u = x_3 - y_3 = (y_1 - x_1) + (y_2 - x_2),$$

where $x_3, y_3 \in H^+$ and $x_i, y_i \in K_i^+$ for $i = 1, 2$. Then

$$x_1 + x_2 + x_3 = y_1 + y_2 + y_3.$$

By Riesz decomposition, there exist elements $z_{ij} \in G^+$ (for $i, j = 1, 2, 3$) such that

$$x_i = z_{i1} + z_{i2} + z_{i3} \quad \text{and} \quad y_j = z_{1j} + z_{2j} + z_{3j}$$

for all i, j. For $i = 1, 2$ and all j, we have $0 \le z_{ij} \le x_i$, whence $z_{ij} \in K_i$. For all j, we have $0 \le z_{3j} \le x_3$, and so $z_{3j} \in H$. Similarly, $z_{ij} \in K_j$ for $j = 1, 2$ and all i, while $z_{i3} \in H$ for all i.

Hence $z_{i3}, z_{3i} \in H \cap K_i$ for $i = 1, 2$. Set $v_1 = z_{31} - z_{13}$ and $v_2 = z_{32} - z_{23}$. Then each $v_i \in H \cap K_i$, and

$$u = x_3 - y_3 = (z_{31} + z_{32} + z_{33}) - (z_{13} + z_{23} + z_{33}) = v_1 + v_2,$$

proving that u lies in $(H \cap K_1) + (H \cap K_2)$.

(b) Obviously $H + (K_1 \cap K_2) \subseteq (H + K_1) \cap (H + K_2)$. Now consider any element u from $(H + K_1) \cap (H + K_2)$, and write

$$u = (x_3 - y_3) + (x_1 - y_1) = (y_4 - x_4) + (y_2 - x_2),$$

where $x_3, x_4, y_3, y_4 \in H^+$ and $x_i, y_i \in K_i^+$ for $i = 1, 2$. Then

$$x_1 + x_2 + x_3 + x_4 = y_1 + y_2 + y_3 + y_4,$$

and hence there exist elements $z_{ij} \in G^+$ (for $i, j = 1, 2, 3, 4$) such that

$$x_i = z_{i1} + z_{i2} + z_{i3} + z_{i4} \quad \text{and} \quad y_j = z_{1j} + z_{2j} + z_{3j} + z_{4j}$$

for all i, j. Note that $0 \le z_{ij} \le x_i$ and $0 \le z_{ij} \le y_j$ for all i, j. For $i = 1, 2$ and all j, it follows that $z_{ij} \in K_i$, while for $i = 3, 4$ and all j, it follows that $z_{ij} \in H$. Similarly, $z_{ij} \in K_j$ for $j = 1, 2$ and all i, while $z_{ij} \in H$ for $j = 3, 4$ and all i. Set

$$v = x_3 - y_3 + z_{13} + z_{14} - z_{31} - z_{41}$$

and $w = z_{12} - z_{21}$. Then $v \in H$ and $w \in K_1 \cap K_2$, and

$$\begin{aligned} u &= x_3 - y_3 + x_1 - y_1 \\ &= x_3 - y_3 + (z_{11} + z_{12} + z_{13} + z_{14}) - (z_{11} + z_{21} + z_{31} + z_{41}) = v + w, \end{aligned}$$

proving that u lies in $H + (K_1 \cap K_2)$. $\square$

PROPOSITION 2.6. *Let $\{H_i \mid i \in I\}$ be a nonempty family of ideals in an interpolation group G. If $G = \bigoplus H_i$ as a group, then also $G = \bigoplus H_i$ as an ordered group.*

PROOF. We must show that $G^+ = \sum H_i^+$. Obviously $\sum H_i^+ \subseteq G^+$.

Now consider any $x \in G^+$. Since $G = \sum H_i$, there exist distinct indices $i(1), \dots, i(n)$ in I and elements $y_j \in H_{i(j)}$ for each $j = 1, \dots, n$ such that $x = y_1 + \cdots + y_n$. Each $y_j \le z_j$ for some $z_j \in H_{i(j)}^+$. Then $x \le z_1 + \cdots + z_n$, and so $x = x_1 + \cdots + x_n$ for some $x_j \in G^+$ with each $x_j \le z_j$. As each $H_{i(j)}$ is convex, we conclude that each $x_j \in H_{i(j)}^+$, whence $x \in \sum H_i^+$.

Therefore $G^+ = \sum H_i^+$, as required. □

• **Extensions.** Given a partially ordered abelian group G and an ideal H in G, we may refer to G as *an extension of H by G/H* when we view G as having been built from H and G/H. (See Chapter 17 for an introduction to the general theory of extensions.) For G to be an interpolation group, it is necessary that H and G/H be interpolation groups (Proposition 2.3). However, these conditions are not sufficient, as the following example shows; in other words, extensions of interpolation groups by interpolation groups need not be interpolation groups. In Proposition 2.8, we derive a criterion for an extension to be an interpolation group.

EXAMPLE 2.7. *There exists a directed, unperforated, partially ordered abelian group G with an ideal H such that H and G/H each have interpolation but G does not.*

PROOF. We make the abelian group $G = \mathbb{Z}^2$ into a partially ordered abelian group with positive cone

$$G^+ = \{(x,0) \in G \mid x \ge 0\} \cup \{(x,y) \in G \mid x > 0 \text{ and } y > 0\}.$$

It is clear that G is unperforated. Given any element (x,y) in G, we may choose a positive integer n such that $n > x$ and $n > y$, so that $(x,y) \le n(1,1)$. Thus $(1,1)$ is an order-unit in G, and hence G is directed.

If $H = \mathbb{Z} \times \{0\}$, then clearly H is a directed subgroup of G. Given $(x,y) \in G$ and $(z,0) \in H$ such that $(0,0) \le (x,y) \le (z,0)$, we must have $0 \le y \le 0$, so that $y = 0$ and $(x,y) \in H$. Thus H is convex and so is an ideal of G. Observing that $H \cong \mathbb{Z} \cong G/H$ as ordered groups, we see that H and G/H each have interpolation.

Finally, observe that

$$\begin{matrix} (0,0) \\ (0,1) \end{matrix} \le \begin{matrix} (1,1) \\ (1,2) \end{matrix}$$

in G, and suppose that there exists $(x,y) \in G$ satisfying

$$\begin{matrix} (0,0) \\ (0,1) \end{matrix} \le (x,y) \le \begin{matrix} (1,1) \\ (1,2). \end{matrix}$$

As $(0,1) \le (x,y) \le (1,1)$, we must have $y = 1$ and either $x = 0$ or $x = 1$. However, $(0,0) \not\le (0,1)$ and $(1,1) \not\le (1,2)$, and hence no such (x,y) exists in G. Therefore G is not an interpolation group. □

Given a partially ordered abelian group G with an ideal H such that H and G/H each have interpolation, we can at least reduce the amount of work required to check whether or not G has interpolation, by means of the following proposition.

PROPOSITION 2.8. *Let G be a partially ordered abelian group, let H be an ideal of G, and assume that H and G/H each have interpolation. Then G has interpolation if and only if the following condition holds:*

(*) *Given any $x_1, y_1 \in H$ and $x_2, y_2 \in G$ with $x_i \le y_j$ for all i, j, there exists $z \in H$ such that $z \le y_2$ and each $x_i \le z$.*

PROOF. First assume that G has interpolation. Given $x_1, y_1 \in H$ and $x_2, y_2 \in G$ with $x_i \le y_j$ for all i, j, there exists $z \in G$ such that $x_i \le z \le y_j$ for all i, j. As $x_1 \le z \le y_1$ and H is convex, we must also have $z \in H$, establishing (*).

Conversely, assume that (*) holds. We first check the following property:

(I) Given any $x_1, y_1 \in H$ and $x_2, x_3, y_2 \in G$ with $x_i \le y_j$ for all i, j, there exists $z \in H$ such that $z \le y_2$ and each $x_i \le z$.

First, (*) provides us with an element $u_1 \in H$ such that $u_1 \le y_2$ and $x_i \le u_1$ for $i = 1, 2$. Another application of (*) provides us with an element $u_2 \in H$ such that $u_2 \le y_2$ and $x_i \le u_2$ for $i = 1, 3$. Since H is directed, there is some $v \in H$ for which each $u_1 \le v$. Now

$$\begin{matrix} u_1 & & v \\ & \le & \\ u_2 & & y_2 \end{matrix}$$

with $u_1, v \in H$, and so a third application of (*) yields an element $z \in H$ such that $z \le y_2$ and each $u_i \le z$. As $x_i \le u_1 \le z$ for $i = 1, 2$ and $x_3 \le u_2 \le z$, we have $x_i \le z$ for $i = 1, 2, 3$, and (I) is proved.

Next, we improve (I) as follows:

(II) Given any $x_1, y_1 \in H$ and $x_2, x_3, y_2, y_3 \in G$ with $x_i \le y_j$ for all i, j, there exists $z \in H$ such that $x_i \le z \le y_j$ for all i, j.

Using (I) twice, we obtain elements $u_2, u_3 \in H$ such that $x_i \le u_j \le y_j$ for $i = 1, 2, 3$ and $j = 2, 3$. Then

$$\begin{matrix} -y_1 & & -x_1 \\ -u_2 & \le & -x_2 \\ -u_3 & & -x_3. \end{matrix}$$

Applying (I) twice more, we obtain elements $v_2, v_3 \in H$ such that

$$\begin{matrix} -y_1 & & \\ -u_2 & \le & \begin{matrix} v_2 \\ v_3 \end{matrix} \\ -u_3 & & \end{matrix}$$

and $v_i \leq -x_i$ for $i = 2, 3$. Now

$$\begin{matrix} x_1 & & y_1 \\ -v_2 & \leq & u_2 \\ -v_3 & & u_3. \end{matrix}$$

Since these elements all lie in H, which has interpolation, there exists $z \in H$ such that

$$\begin{matrix} x_1 & & & & y_1 \\ -v_2 & \leq & z & \leq & u_2 \\ -v_3 & & & & u_3. \end{matrix}$$

Then $x_i \leq -v_i \leq z \leq u_j \leq y_j$ for $i, j = 2, 3$, so that $x_i \leq z \leq y_j$ for all i, j, and (II) is proved.

Finally, consider arbitrary elements a_1, a_2, b_1, b_2 in G such that $a_i \leq b_j$ for all i, j. In G/H, we then have $a_i + H \leq b_j + H$ for all i, j. Since G/H has interpolation, there exists $c \in G$ such that

$$a_i + H \leq c + H \leq b_j + H$$

for all i, j. Hence, there is an element $d \in H^+$ such that $a_i \leq c + d$ for each i and $c \leq b_j + d$ for each j. Then

$$\begin{matrix} -d & & d \\ a_1 - c & \leq & b_1 - c \\ a_2 - c & & b_2 - c. \end{matrix}$$

Applying (II), we obtain $z \in H$ such that $a_i - c \leq z \leq b_j - c$ for all i, j, and thus $a_i \leq c + z \leq b_j$ for all i, j.

Therefore G is an interpolation group. $\square$

COROLLARY 2.9. *Let G be an interpolation group, let H be an ideal of G, and let K be a convex subgroup of H. If H/K has interpolation, then so does G/K.*

PROOF. Note that K is also a convex subgroup of G. We first show that H/K is an ideal of G/K. It is clear that H/K is a directed subgroup of G/K. Given $x \in G/K$ and $y \in H/K$ satisfying $0 \leq x \leq y$, choose $a \in G^+$ and $b \in H$ such that $x = a + K$ and $y = b + K$. Since $x \leq y$, we obtain $a \leq b + c$ for some $c \in K$. Then $0 \leq a \leq b + c$ with $b + c \in H$, whence $a \in H$, and so $x \in H/K$. Thus H/K is a convex subgroup of G/K, and so is an ideal.

Observe that $(G/K)/(H/K) \cong G/H$ (as ordered groups), so we have that $(G/K)/(H/K)$ has interpolation. Hence, by Proposition 2.8 we need only check that given any $x_1, y_1 \in H/K$ and $x_2, y_2 \in G/K$ with $x_i \leq y_j$ for all i, j, there exists $z \in H/K$ such that $z \leq y_2$ and each $x_i \leq z$.

Choose elements $a_1, b_1 \in H$ and $a_2, b_2 \in G$ such that each $x_i = a_i + K$ and each $y_j = b_j + K$. Since $x_1 \leq y_2$, we have $a_1 \leq b_2 + k$ for some $k \in K$. As long as a_1 is not changed, there is no loss of generality in replacing b_2 by $b_2 + k$. Hence, we may assume that $a_1 \leq b_2$. Similarly, after modifying a_2 and then modifying b_1, we may assume that $a_2 \leq b_2$ and $a_2 \leq b_1$. (However, we may not modify a_1 at this point.)

Since a_1 and b_1 lie in H, which is directed, there is some $c \in H$ such that $a_1 \leq c$ and $b_1 \leq c$. As $a_2 \leq b_1 \leq c$, we thus have

$$\begin{matrix} a_1 & & c \\ & \leq & \\ a_2 & & b_2. \end{matrix}$$

Using interpolation in G, we obtain an element $d \in G$ such that

$$\begin{matrix} a_1 & & & & c \\ & \leq & d & \leq & \\ a_2 & & & & b_2. \end{matrix}$$

Since $a_1 \leq d \leq c$ and $a_1, c \in H$, it follows that $d \in H$. Therefore the coset $z = d+K$ is an element of H/K such that $z \leq y_2$ and each $x_i \leq z$, as desired. □

• **Products, Pullbacks, and Pushouts.** We have already noted that any direct product or direct sum of interpolation groups is an interpolation group. For lexicographic direct products and lexicographic direct sums, this sometimes fails, as we shall see shortly (Example 2.13). To obtain positive results, we assume that the interpolation groups in question either are directed or satisfy the following strong form of the interpolation property.

DEFINITION. A partially ordered set X is said to satisfy *strict interpolation* provided that given any x_1, x_2, y_1, y_2 in X satisfying $x_i < y_j$ for all i, j, there exists $z \in X$ such that $x_i < z < y_j$ for all i, j.

For example, $\mathbb{Q}$ and $\mathbb{R}$ satisfy strict interpolation, but $\mathbb{Z}$ does not.

PROPOSITION 2.10. *Let $\{G_i \mid i \in I\}$ be a family of interpolation groups, indexed by a well-ordered set I, and let G be the lexicographic direct product of this family. For each $i \in I$, assume either that G_i satisfies strict interpolation or that G_j is directed for all $j > i$. Then G is an interpolation group.*

PROOF. Consider any elements x_1, x_2, y_1, y_2 in G such that $x_i \leq y_j$ for all i, j. If $x_1 = y_1$, then $x_i \leq y_1 \leq y_j$ for all i, j, and so we may assume that $x_1 < y_1$. Similarly, there is no loss of generality in assuming that $x_i < y_j$ for all i, j.

Let k be the least index such that $x_{ik} \neq y_{jk}$ for some i, j. After renumbering, we may assume that $x_{1k} \neq y_{1k}$. Then $x_{1k} < y_{1k}$, and $x_{in} = y_{jn}$ for all i, j and all $n < k$, that is, $x_{1n} = x_{2n} = y_{1n} = y_{2n}$ for all $n < k$. Consequently, $x_{ik} \leq y_{jk}$ for all i, j.

If $x_{1k} < x_{2k}$, then $x_1 < x_2$, in which case $x_i \leq x_2 \leq y_j$ for all i, j. Thus we may assume that $x_{1k} \not< x_{2k}$, and similarly that $x_{2k} \not< x_{1k}$ while also $y_{1k} \not< y_{2k}$ and $y_{2k} \not< y_{1k}$.

Since $x_{1k} < y_{1k}$ but $y_{2k} \not< y_{1k}$, we have $x_{1k} \neq y_{2k}$, and hence $x_{1k} < y_{2k}$. Then since $x_{1k} \not< x_{2k}$, we have $x_{2k} \neq y_{jk}$ for each j, so that $x_{2k} < y_{jk}$ for each j. Thus $x_{ik} < y_{jk}$ for all i, j.

Suppose that there exists $z_k \in G_k$ such that $x_{ik} < z_k < y_{jk}$ for all i, j. In this case, setting $z_n = x_{1n}$ for all $n < k$ and $z_n = 0$ for all $n > k$, we obtain components for an element $z \in G$ such that $x_i < z < y_j$ for all i, j.

Now suppose that there is no element $u \in G_k$ such that $x_{ik} < u < y_{jk}$ for all i, j. In particular, G_k does not satisfy strict interpolation. By hypothesis, G_n is directed for all $n > k$.

Since G_k is an interpolation group, there at least exists an element $z_k \in G_k$ such that $x_{ik} \leq z_k \leq y_{jk}$ for all i, j; however, z_k must lie in the set $\{x_{1k}, x_{2k}, y_{1k}, y_{2k}\}$. Set $z_n = x_{1n}$ for all $n < k$, so that $x_{in} = z_n = y_{jn}$ for all $n < k$ and all i, j.

If either $x_{1k} = z_k$ or $x_{2k} = z_k$, then $z_k < y_{jk}$ for each j. For all $n > k$, use the directedness of G_n to choose an element $z_n \in G_n$ such that each $x_{in} \leq z_n$. Then we have components for an element $z \in G$ such that $x_i \leq z < y_j$ for all i, j.

If either $z_k = y_{1k}$ or $z_k = y_{2k}$, then each $x_{ik} < z_k$. In this case, for all $n > k$ choose an element $z_n \in G_n$ such that $z_n \leq y_{jn}$ for each j. This provides us with components for an element $z \in G$ such that $x_i < z \leq y_j$ for all i, j.

Thus in every case there exists $z \in G$ satisfying $x_i \leq z \leq y_j$ for all i, j. □

COROLLARY 2.11. *Let $\{G_i \mid i \in I\}$ be a nonempty family of interpolation groups, indexed by a totally ordered set I, and let G be the lexicographic direct sum of this family. For each $i \in I$, assume either that G_i satisfies strict interpolation or that G_j is directed for all $j > i$. Then G is an interpolation group.*

PROOF. Either repeat the proof of Proposition 2.10, mutatis mutandis, or else observe that G is isomorphic to the direct limit of the lexicographic direct products of finite subfamilies of the G_i, and use Proposition 2.10 to see that these finite lexicographic direct products all have interpolation. □

COROLLARY 2.12. *Let G be the lexicographic direct product of partially ordered abelian groups G_1 and G_2. Then G is an interpolation group if and only if:*

(a) *Both G_1 and G_2 are interpolation groups.*

(b) *Either G_1 satisfies strict interpolation or G_2 is directed.*

PROOF. If (a) and (b) hold, then G has interpolation by Proposition 2.10. Conversely, assume that G is an interpolation group. Observe that $\{0\} \times G_2$ is a convex subgroup of G, and so it must satisfy interpolation. Since $\{0\} \times G_2 \cong G_2$ (as ordered groups), G_2 satisfies interpolation.

Given x_1, x_2, y_1, y_2 in G_1 satisfying $x_i \leq y_j$ for all i, j, we have $(x_i, 0) \leq (y_j, 0)$ in G for all i, j. Then there exists $(z, w) \in G$ such that

$$(x_i, 0) \leq (z, w) \leq (y_j, 0)$$

for all i, j. Hence, $x_i \leq z \leq y_j$ for all i, j. Thus G_1 satisfies interpolation. (Although G_1 is isomorphic to $G/(\{0\} \times G_2)$, Proposition 2.3 does not apply unless G_2 is directed.)

If G_2 is not directed, choose elements u_1 and u_2 in G_2 which have no common upper bound in G_2. Given any x_1, x_2, y_1, y_2 in G_1 with $x_i < y_j$ for all i, j, we have $(x_i, u_k) < (y_j, -u_k)$ in G for all i, j, k. Hence, there is some (z, w) in G such that

$$(x_i, u_k) \le (z, w) \le (y_j, -u_k)$$

for all i, j, k. If $x_1 = z$, then $(x_1, u_k) \le (x_1, w)$ for each k and so each $u_k \le w$, which is impossible. Consequently, $x_1 < z$, and similarly $x_2 < z$. If $z = y_1$, then $(y_1, w) \le (y_1, -u_k)$ for each k, whence $w \le -u_k$ and so each $u_k \le -w$, again impossible. Hence, $z < y_1$, and similarly $z < y_2$. Thus $x_i < z < y_j$ for all i, j, proving that G_1 satisfies strict interpolation. □

EXAMPLE 2.13. *There exists an interpolation group G such that the lexicographic direct product of G with itself does not satisfy interpolation.*

PROOF. Let G denote the abelian group $\mathbb{Z}^2$, made into a partially ordered abelian group with positive cone

$$G^+ = \{(x, 0) \in G \mid x \ge 0\}.$$

Then G is an interpolation group which is not directed and which does not satisfy strict interpolation. Apply Corollary 2.12. □

Turning now to pullbacks, we first use the following example to see that pullbacks of interpolation groups do not always have interpolation. Observe that in the example, $G_1 \cap G_2$ is a pullback of the inclusion maps $G_i \to H$.

EXAMPLE 2.14. *There exists an interpolation group H with subgroups G_1 and G_2 such that G_1 and G_2 have interpolation but $G_1 \cap G_2$ does not.*

PROOF. Let H be the lexicographic direct product of $\mathbb{Q}$ with $\mathbb{Z}^2$. Since $\mathbb{Q}$ satisfies strict interpolation, H has interpolation by Proposition 2.10.

Let G_1 be the lexicographic direct product of $\mathbb{Z}$ with $\mathbb{Z}^2$, and let G_2 be the lexicographic direct product of $\mathbb{Q}$ with $\mathbb{Z}(1, -1)$. Since $\mathbb{Z}^2$ is directed and $\mathbb{Q}$ has strict interpolation, Proposition 2.10 shows that G_1 and G_2 have interpolation.

Now $G_1 \cap G_2$ is the lexicographic direct product of $\mathbb{Z}$ with $\mathbb{Z}(1, -1)$. However, $\mathbb{Z}$ does not satisfy strict interpolation, and $\mathbb{Z}(1, -1)$ is not directed. Thus by Corollary 2.12, $G_1 \cap G_2$ does not have interpolation. □

In the positive direction, pullbacks of interpolation groups along quotient maps are again interpolation groups, as follows.

PROPOSITION 2.15. *Let $\{f_i \colon G_i \to H \mid i \in I\}$ be a nonempty family of morphisms in the category of partially ordered abelian groups, and let G be a pullback of this family. For each $i \in I$, assume that $\ker(f_i)$ is directed and that $f_i(G_i^+) = H^+$. Then if each G_i is an interpolation group, so is G.*

PROOF. Note that each f_i induces an isomorphism of $G_i / \ker(f_i)$ onto $f_i(G_i)$ (as ordered groups). Hence, for any $x, y \in G_i$, we have $f_i(x) \le f_i(y)$ if and only if $x \le y + a$ for some $a \in \ker(f_i)$.

Recall that G may be identified with the subgroup of $\prod G_i$ consisting of those elements x satisfying $f_i(x_i) = f_j(x_j)$ for all $i, j \in I$.

Now consider any $x_1, x_2, y_1, y_2 \in G$ such that $x_i \leq y_j$ for all i, j. Choose a particular index $k \in I$, and use interpolation in G_k to obtain an element $u_k \in G_k$ such that $x_{ik} \leq u_k \leq y_{jk}$ for all i, j. Given any $n \in I$ with $n \neq k$, note that

$$f_k(u_k - x_{1k}) \in H^+ = f_n(G_n^+)$$

while also $f_k(x_{1k}) = f_n(x_{1n})$, and hence $f_k(u_k) = f_n(u_n)$ for some $u_n \in G_n$. Thus we obtain components for an element $u \in G$. Now $x_i - u \leq y_j - u$ for all i, j, and it suffices to find an element $v \in G$ satisfying $x_i - u \leq v \leq y_j - u$ for all i, j. Hence, there is no loss of generality in assuming that $x_{ik} \leq 0 \leq y_{jk}$ for all i, j.

Given any $n \in I$, we claim that some z_n in $\ker(f_n)$ satisfies $x_{in} \leq z_n \leq y_{jn}$ for all i, j. First observe that

$$f_n(x_{in}) = f_k(x_{ik}) \leq 0 \leq f_k(y_{jk}) = f_n(y_{jn})$$

for all i, j. Consequently, there are elements a_1, a_2, b_1, b_2 in $\ker(f_n)$ such that $x_{in} \leq a_i$ for each i and $0 \leq y_{jn} + b_j$ for each j. Since $\ker(f_n)$ is directed, there exists $a \in \ker(f_n)$ such that each $a_i \leq a$, and there exists $b \in \ker(f_n)$ such that each $b_j \leq b$ and also $-a \leq b$. Thus

$$\begin{matrix} x_{1n} & & y_{1n} \\ x_{2n} & \leq & y_{2n} \\ -b & & a. \end{matrix}$$

As G_n has interpolation, there is some $z_n \in G_n$ satisfying

$$\begin{matrix} x_{1n} & & & & y_{1n} \\ x_{2n} & \leq & z_n & \leq & y_{2n} \\ -b & & & & a. \end{matrix}$$

Since $-b$ and a each lie in $\ker(f_n)$, so does z_n, which establishes the claim.

Since $f_m(z_m) = 0 = f_n(z_n)$ for all $m, n \in I$, these z_n are the components for an element $z \in G$. As $x_{in} \leq z_n \leq y_{jn}$ for all i, j, n, we also have $x_i \leq z \leq y_j$ for all i, j. Thus G is an interpolation group. □

COROLLARY 2.16. *Let $\{f_i\colon (G_i, u_i) \to (H, v) \mid i \in I\}$ be a nonempty family of morphisms in the category of partially ordered abelian groups with order-unit, and let (G, u) be a pullback of this family. For each $i \in I$, assume that* $\ker(f_i)$ *is directed, and that $f_i(G_i^+) = H^+$. Then if each G_i is an interpolation group, so is G.*

PROOF. Let G' be a pullback of the maps $f_i\colon G_i \to H$ in the category of partially ordered abelian groups. Then G may be identified with the convex subgroup of G' generated by u. Since G' is an interpolation group by Proposition 2.15, so is G. □

In the case of pushouts of interpolation groups, conditions similar to those used in Proposition 2.15 can be used to ensure that the resulting pushouts are interpolation groups. We do not need a separate result for pushouts in the category of partially ordered abelian groups with order-unit, because of Proposition 1.18.

PROPOSITION 2.17. *Let $\{G_i \mid i \in I\}$ be a nonempty family of interpolation groups, let H be an interpolation group, and for each $i \in I$ let $f_i \colon H \to G_i$ be an isomorphism of H onto an ideal of G_i (as ordered groups). If G is a pushout of the family $\{f_i \mid i \in I\}$, then G is an interpolation group.*

PROOF. We may assume that G was constructed using the notation of Proposition 1.17. Note that $f_i^{-1}(G_i^+) = H^+$ for all $i \in I$, so that $M = L$.

Set $N = \bigoplus f_i(H)$, and observe that N is an ideal of K containing L. Note also that as each G_i has interpolation, so does K.

Pick an index $j \in I$, and observe that $N = q_j f_j(H) \oplus L$ as abelian groups. Hence, $q_j f_j$ induces a group isomorphism of H onto N/L. In addition, we observe that $N^+ = q_j f_j(H^+) + L$. Thus $H \cong N/L$ as ordered groups, whence N/L has interpolation.

By Corollary 2.9, the group $G = K/L$ must have interpolation. □

• **2-Unperforated Interpolation Groups.** At this point we pause to derive a few computational results which will be useful later. Although these results are not needed at this stage, their proofs are fairly elementary, and we derive them here in order to provide the reader with a bit of practice in the use of the interpolation and decomposition properties.

LEMMA 2.18. *Let G be an interpolation group, let $p \in G$, and let $z \in G^+$. If $p \le z$ and $2p \le z$, then there exists $q \in G^+$ such that $p \le q$ and $2q \le z$.*

PROOF. We have

$$\begin{matrix} 0 \\ p \end{matrix} \le \begin{matrix} z \\ z-p \end{matrix}$$

and hence there exists $r \in G$ such that

$$\begin{matrix} 0 \\ p \end{matrix} \le r \le \begin{matrix} z \\ z-p. \end{matrix}$$

It follows that

$$\begin{matrix} 0 \\ p \end{matrix} \le \begin{matrix} r \\ z-r, \end{matrix}$$

and so there exists $q \in G$ such that

$$\begin{matrix} 0 \\ p \end{matrix} \le q \le \begin{matrix} r \\ z-r. \end{matrix}$$

Then $2q \le r + (z - r) = z$, and the proof is complete. □

PROPOSITION 2.19. *Let G be a 2-unperforated interpolation group, let $x, y, z \in G^+$, and let n be a positive integer. If $2^n x \leq 2^n y + z$, then $x = v + w$ for some $v, w \in G^+$ such that $v \leq y$ and $2^n w \leq z$.*

PROOF. First assume that $n = 1$, so that $2x \leq 2y + z$. Set $p = x - y$, whence $2p \leq z$. As $2p \leq z \leq 2z$ and G is 2-unperforated, we have $p \leq z$ as well. By Lemma 2.18, there exists $q \in G^+$ such that $p \leq q$ and $2q \leq z$. Then $x \leq y + q$, and hence $x = v + w$ for some $v, w \in G^+$ such that $v \leq y$ and $w \leq q$. In addition, $2w \leq 2q \leq z$.

Now let $n > 1$, and assume that the result holds for lower powers of 2. Since

$$2^{n-1}(2x) \leq 2^{n-1}(2y) + z,$$

the induction hypothesis implies that $2x = p + q$ for some $p, q \in G^+$ such that $p \leq 2y$ and $2^{n-1}q \leq z$. Then $2x \leq 2y + q$. By the first case proved, $x = v + w$ for some $v, w \in G^+$ such that $v \leq y$ and $2w \leq q$. As $2^n w \leq 2^{n-1}q \leq z$, this completes the induction step. □

COROLLARY 2.20. *Let G be a directed 2-unperforated interpolation group. Let $x, y \in G$, let $z \in G^+$, and let n be a positive integer. If $2^n x \leq 2^n y + z$, then $x = v + w$ for some $v \in G$ and $w \in G^+$ such that $v \leq y$ and $2^n w \leq z$.*

PROOF. Since G is directed, there exists $u \in G$ such that $u \leq x$ and $u \leq y$. Set $x' = x - u$ and $y' = y - u$. Then x' and y' are elements of G^+ such that $2^n x' \leq 2^n y' + z$. By Proposition 2.19, $x' = v' + w$ for some $v', w \in G^+$ such that $v' \leq y'$ and $2^n w \leq z$. Set $v = v' + u$. Then $x = v + w$ and $v \leq y$. □

PROPOSITION 2.21. *Let G be a 2-unperforated interpolation group, let $z \in G^+$, and let n be a positive integer. Set*

$$X = \{x \in G^+ \mid 2^n x \leq z\} \quad \text{and} \quad Y = \{y \in G^+ \mid 2^n y \geq z\}.$$

Then X is upward directed and Y is downward directed.

PROOF. Given $x_1, x_2 \in X$, set $z' = z - 2^n x_2$. Then $z' \in G^+$ and $2^n x_1 \leq 2^n x_2 + z'$. By Proposition 2.19, $x_1 = v + w$ for some $v, w \in G^+$ such that $v \leq x_2$ and $2^n w \leq z'$. Set $x = x_2 + w$, observing that each $x_i \leq x$. Since $2^n w \leq z - 2^n x_2$, we also have $2^n x \leq z$, so that $x \in X$. Thus X is upward directed.

Given $y_1, y_2 \in Y$, set $u = y_1 + y_2 + z$, and note that each $y_i \leq u$, while also $z \leq u \leq 2^n u$. Then $v = 2^n u - z$ is an element of G^+, and the set

$$W = \{w \in G^+ \mid 2^n w \leq v\}$$

is upward directed, by the previous paragraph. Each of the elements $u - y_i$ lies in G^+, and since $2^n y_i \geq z$ we see that $2^n(u - y_i) \leq v$. Hence, there exists $w \in W$ such that each $u - y_i \leq w$. Set $y = u - w$, so that each $y_i \geq y$. As $2^n w \leq v = 2^n u - z$, we have $2^n y \geq z$. In particular, $2^n y \geq 0$, so that $y \geq 0$ (because G is 2-unperforated). Thus $y \in Y$, proving that Y is downward directed. □

COROLLARY 2.22. *Let G be a directed 2-unperforated interpolation group. Let $c \in G$, let n be a positive integer, and set*

$$A = \{a \in G \mid 2^n a \leq c\} \quad \text{and} \quad B = \{b \in G \mid 2^n b \geq c\}.$$

Then A is upward directed and B is downward directed.

PROOF. Given $a_1, a_2 \in A$, choose $x \in G^+$ such that $c + x \geq 0$ and each $a_i + x \geq 0$, and note that $c + 2^n x \geq 0$ as well. Since

$$2^n(a_i + x) \leq c + 2^n x$$

for each i, Proposition 2.21 says that there exists $y \in G^+$ such that $2^n y \leq c + 2^n x$ and each $a_i + x \leq y$. Then $a = y - x$ is an element of A such that each $a_i \leq a$, proving that A is upward directed.

Similarly, the set $\{d \in G \mid 2^n d \leq -c\}$ is upward directed, from which we conclude that B is downward directed. □

Not every interpolation group is 2-unperforated, as the following example shows, and we shall see that neither Proposition 2.19 nor Proposition 2.21 holds for this example. On the other hand, in directed unperforated interpolation groups, Propositions 2.19 and 2.21 hold with 2^n replaced by an arbitrary positive integer, as shown in Propositions 3.22 and 3.23.

EXAMPLE 2.23. *There exists a directed interpolation group G such that G is torsion-free as an abelian group, but G is 2-perforated.*

PROOF. Set $G_1 = \mathbb{Q}$, and make the abelian group $G_2 = \mathbb{Z}$ into a partially ordered abelian group with positive cone $G_2^+ = \{0, 2, 4, 6, \ldots\}$. Note that G_1 and G_2 are interpolation groups. (Although G_2 is 2-perforated, it is not directed.)

Now let G be the lexicographic direct product of G_1 with G_2. As G_1 satisfies strict interpolation, G satisfies interpolation, by Proposition 2.10. It is clear that G is directed and that G is a torsion-free abelian group. Finally, $2(0,1) \in G^+$ but $(0,1) \notin G^+$. Thus G is 2-perforated. □

In the interpolation group G of Example 2.23, set $x = (1,1)$ and $y = (1,0)$, while $z = (0,2)$. Then $x, y, z \in G^+$ and $2x = 2y + z$. However, there do not exist elements $v, w \in G^+$ such that $x = v + w$ while $v \leq y$ and $2w \leq z$. Thus Proposition 2.19 does not hold in G.

Now let $a_1 = (1,0)$ and $a_2 = (1,1)$, while $b = (2,2)$. Then a_1, a_2, b are elements of G^+ such that $2a_i \leq b$ for each i. However, there does not exist any element $a \in G^+$ such that $2a \leq b$ and each $a_i \leq a$. Thus Proposition 2.21 does not hold in G.

• **Relatively Bounded Homomorphisms.** A natural dual object for a partially ordered abelian group G is the group $\operatorname{Hom}_{\mathbb{Z}}(G, \mathbb{R})$ of all group homomorphisms from G to $\mathbb{R}$. Provided G is directed, $\operatorname{Hom}_{\mathbb{Z}}(G, \mathbb{R})$ may be made into a partially ordered abelian group in which the positive cone is the set of all positive homomorphisms. The main objective of this section is to prove that if G is additionally an interpolation group, then the subgroup of $\mathrm{Hom}_{\mathbf{Z}}(G, \mathbf{R})$ generated by the positive homomorphisms is a Dedekind complete lattice-ordered abelian group, which will allow us to apply lattice-theoretic methods when investigating positive real-valued homomorphisms on G. As the development is the same for $\mathrm{Hom}_{\mathbf{Z}}(G, H)$ with H a Dedekind complete lattice-ordered abelian group, we proceed at that level of generality.

DEFINITION. Let G and H be partially ordered abelian groups, and let A denote the abelian group $\mathrm{Hom}_{\mathbf{Z}}(G, H)$ of all group homomorphisms from G to H. Given any $f, g \in A$, we define $f \leq^+ g$ if and only if $g - f$ is a positive homomorphism. The relation $\leq^+$ is a translation-invariant pre-order on A, using which A becomes a pre-ordered abelian group whose positive cone consists of all positive homomorphisms from G to H.

We shall always assume, unless otherwise specified, that $\mathrm{Hom}_{\mathbf{Z}}(G, H)$ has been equipped with $\leq^+$. We observe that if G is directed, then $\mathrm{Hom}_{\mathbf{Z}}(G, H)$ is a *partially* ordered abelian group. Namely, if f and g are homomorphisms from G to H such that $f \leq^+ g$ and $g \leq^+ f$, then $(g - f)(x) \geq 0$ and $(f - g)(x) \geq 0$ for all $x \in G^+$, whence f and g agree on G^+. If G is directed, then it is generated (as a group) by G^+, and so $f = g$, proving that in this case $\leq^+$ is a partial order.

In general, $\mathrm{Hom}_{\mathbf{Z}}(G, H)$ is not directed, and so we work instead with the subgroup generated by the positive homomorphisms. In order to derive a criterion that indicates when a homomorphism lies in this subgroup, we develop a lemma which will aid in the construction of positive homomorphisms.

DEFINITION. Let G and H be partially ordered abelian groups. A map $d \colon G^+ \to H$ is said to be *subadditive* provided that $d(0) = 0$ and $d(x + y) \leq d(x) + d(y)$ for all $x, y \in G^+$.

LEMMA 2.24. *Let G be a directed interpolation group, let H be a Dedekind complete lattice-ordered abelian group, and let $d \colon G^+ \to H$ be a subadditive map. For all $x \in G^+$, assume that the set*

$$D(x) = \{d(x_1) + \cdots + d(x_n) \mid x = x_1 + \cdots + x_n \text{ with all } x_i \in G^+\}$$

is bounded above in H. Then there exists a group homomorphism $f \colon G \to H$ such that $f(x) = \bigvee D(x)$ for all $x \in G^+$.

PROOF. As H is Dedekind complete, we may define $f(x) = \bigvee D(x)$ for all $x \in G^+$. We need only show that $f(0) = 0$ and that f is additive on G^+, since then f extends uniquely to a group homomorphism from G to H. As $d(0) = 0$, it is clear that $f(0) = 0$.

Consider any $x, y \in G^+$. Given any decompositions

$$x = x_1 + \cdots + x_n \quad \text{and} \quad y = y_1 + \cdots + y_k$$

with the $x_i, y_j \in G^+$, we have $x + y = x_1 + \cdots + x_n + y_1 + \cdots + y_k$, and hence

$$\sum d(x_i) + \sum d(y_j) \leq f(x + y).$$

Thus $a+b \leq f(x+y)$ for all $a \in D(x)$ and $b \in D(y)$. In view of Proposition 1.4, it follows that

$$\begin{aligned} f(x)+f(y) &= \Big[\bigvee D(x)\Big] + f(y) = \bigvee_{a\in D(x)} [a+f(y)] \\ &= \bigvee_{a\in D(x)} \Big(a + \Big[\bigvee D(y)\Big]\Big) = \bigvee_{a\in D(x)} \bigvee_{b\in D(y)} (a+b) \\ &\leq f(x+y). \end{aligned}$$

Conversely, given $x+y = z_1+\cdots+z_n$ with the $z_i \in G^+$, we must have

$$x = x_1+\cdots+x_n \quad \text{and} \quad y = y_1+\cdots+y_n$$

for some elements $x_i, y_i \in G^+$ such that $x_i+y_i = z_i$ for all i. Then

$$\sum d(z_i) \leq \sum [d(x_i)+d(y_i)] = \Big[\sum d(x_i)\Big] + \Big[\sum d(y_i)\Big] \leq f(x)+f(y).$$

Thus $f(x+y) \leq f(x)+f(y)$.

Therefore $f(x+y) = f(x)+f(y)$ for all $x, y \in G^+$, as required. $\square$

DEFINITION. Let X and Y be partially ordered sets. A map $f\colon X \to Y$ is said to be *relatively bounded* provided that given any subset W of X which is bounded (above and below) in X, the set $f(W)$ is bounded in Y.

PROPOSITION 2.25. *Let G be a directed interpolation group, let H be a Dedekind complete lattice-ordered abelian group, and let $f\colon G \to H$ be a group homomorphism. Then f is relatively bounded if and only if $f = g-h$ for some positive homomorphisms $g, h\colon G \to H$.*

PROOF. First assume that there exist positive homomorphisms $g, h\colon G \to H$ such that $f = g-h$. Given any bounded set $W \subseteq G$, there exist elements $a, b \in G$ such that $a \leq b$ and $W \subseteq [a,b]$. Since g and h are order-preserving, it follows that $g(a) \leq g(b)$ and $h(a) \leq h(b)$, while also

$$g(W) \subseteq [g(a), g(b)] \quad \text{and} \quad h(W) \subseteq [h(a), h(b)].$$

Then $g(a)-h(b) \leq g(b)-h(a)$ and

$$f(W) \subseteq [g(a)-h(b), g(b)-h(a)],$$

proving that f is relatively bounded.

Conversely, assume that f is relatively bounded. Set $d(x) = f(x) \vee 0$ for all $x \in G^+$, observing that $d(0) = 0$. For all $x, y \in G^+$, we have

$$d(x+y) = [f(x)+f(y)] \vee 0 \leq [f(x) \vee 0] + [f(y) \vee 0] = d(x)+d(y),$$

so that d is a subadditive map from G^+ to H. Now define

$$D(x) = \{d(x_1)+\cdots+d(x_n) \mid x = x_1+\cdots+x_n \text{ with all } x_i \in G^+\}$$

for each $x \in G^+$. We claim that $D(x)$ is bounded above in H.

There exist elements $a \leq b$ in H such that $f([0,x]) \subseteq [a,b]$. Given any decomposition $x = x_1 + \cdots + x_n$ with all the $x_i \in G^+$, we infer from Lemma 1.21 that

$$d(x_1) + \cdots + d(x_n) = \sum_{i=1}^{n} [f(x_i) \vee 0] = \left[\bigvee_{A \in \mathcal{A}} \left(\sum_{i \in A} f(x_i) \right) \right] \vee 0,$$

where $\mathcal{A}$ is the collection of all nonempty subsets of $\{1, 2, \ldots, n\}$. For all $A \in \mathcal{A}$, we have

$$0 \leq \sum_{i \in A} x_i \leq x \quad \text{and} \quad \sum_{i \in A} f(x_i) = f\left(\sum_{i \in A} x_i \right) \leq b.$$

Hence, $d(x_1) + \cdots + d(x_n) \leq b \vee 0$. Thus $b \vee 0$ is an upper bound for $D(x)$, and the claim is proved.

By Lemma 2.24, there exists a group homomorphism $g \colon G \to H$ such that $g(x) = \bigvee D(x)$ for all $x \in G^+$. Since $g(x) \geq d(x) \geq 0$ for all $x \in G^+$, we see that g is a positive homomorphism. In addition, $g(x) \geq d(x) \geq f(x)$ for all $x \in G^+$, so that $g \geq^+ f$. Therefore $h = g - f$ is a positive homomorphism, and $f = g - h$. □

COROLLARY 2.26. *Let G be a directed interpolation group, let H be a Dedekind complete lattice-ordered abelian group, and let B be the set of all relatively bounded group homomorphisms from G to H. Then B is an ideal of* $\operatorname{Hom}_{\mathbf{Z}}(G, H)$.

PROOF. In view of Proposition 2.25, B equals the subgroup of $\operatorname{Hom}_{\mathbf{Z}}(G, H)$ generated by the positive homomorphisms. Thus B is a directed subgroup of $\operatorname{Hom}_{\mathbf{Z}}(G, H)$.

Given $f \in \operatorname{Hom}_{\mathbf{Z}}(G, H)$ and $g \in B$ satisfying $0 \leq^+ f \leq^+ g$, write $g = g_1 - g_2$ for some positive homomorphisms $g_i \colon G \to H$. Since $f \leq^+ g \leq^+ g_1$, we have $f = g_1 - (g_1 - f)$ where g_1 and $g_1 - f$ are positive homomorphisms, and hence $f \in B$. Therefore B is a convex subgroup of $\operatorname{Hom}_{\mathbf{Z}}(G, H)$ and so is an ideal. □

THEOREM 2.27. *Let G be a directed interpolation group, and let H be a Dedekind complete lattice-ordered abelian group.*

(a) *The group B of all relatively bounded group homomorphisms from G to H is a Dedekind complete lattice-ordered abelian group.*

(b) *If $\{f_i \mid i \in I\}$ is a nonempty subset of B which is bounded above, and if $d(x) = \bigvee f_i(x)$ for all $x \in G^+$, then*

$$\left(\bigvee f_i \right)(x) = \bigvee \{ d(x_1) + \cdots + d(x_n) \mid x = x_1 + \cdots + x_n \textit{ with all } x_i \in G^+ \}$$

for all $x \in G^+$.

(c) *If $\{f_i \mid i \in I\}$ is a nonempty subset of B which is bounded below, and if $e(x) = \bigwedge f_i(x)$ for all $x \in G^+$, then for all $x \in G^+$*

$$\left(\bigwedge f_i \right)(x) = \bigwedge \{ e(x_1) + \cdots + e(x_n) \mid x = x_1 + \cdots + x_n \textit{ with all } x_i \in G^+ \}.$$

PROOF. Consider a nonempty subset $\{f_i \mid i \in I\}$ in B which has an upper bound g in B. Given any $x \in G^+$, we have $f_i(x) \leq g(x)$ for all $i \in I$, and hence we may define $d(x) = \bigvee f_i(x)$. This defines a map $d\colon G^+ \to H$, and we note that d is subadditive. For any $x \in G^+$ and any decomposition $x = x_1 + \cdots + x_n$ with all $x_i \in G^+$, we have

$$d(x_1) + \cdots + d(x_n) \leq g(x_1) + \cdots + g(x_n) = g(x).$$

Hence, $g(x)$ is an upper bound for the set

$$D(x) = \{d(x_1) + \cdots + d(x_n) \mid x = x_1 + \cdots + x_n \text{ with all } x_i \in G^+\}.$$

Now by Lemma 2.24 there exists a group homomorphism $f\colon G \to H$ such that $f(x) = \bigvee D(x)$ for all $x \in G^+$. In particular, for $x \in G^+$ and $i \in I$ we have $f_i(x) \leq d(x) \leq f(x)$, and hence all $f_i \leq^+ f$. The maps $f - f_i$ are positive homomorphisms and so lie in B, whence $f \in B$. If $h \in B$ such that all $f_i \leq^+ h$, then $d(x) \leq h(x)$ for any $x \in G^+$. As above, it follows that $h(x)$ is an upper bound for $D(x)$, whence $f(x) \leq h(x)$. Thus $f \leq^+ h$, which proves that f is the supremum of the f_i.

Therefore suprema exist in B and have the form described in (b). On applying the order anti-automorphism $z \mapsto -z$, we see that infima exist in B and have the form described in (c). Therefore B is Dedekind complete. Inasmuch as B is directed (Proposition 2.25), we conclude that B is lattice-ordered. □

COROLLARY 2.28. *If G is a directed interpolation group, then the group B of all relatively bounded group homomorphisms from G to $\mathbb{R}$ is a Dedekind complete lattice-ordered real vector space. Given any $f_1, \ldots, f_n$ in B,*

$$\left(\bigvee f_i\right)(x) = \sup\{f_1(x_1) + \cdots + f_n(x_n) \mid x = x_1 + \cdots + x_n \text{ with all } x_i \in G^+\}$$

$$\left(\bigwedge f_i\right)(x) = \inf\{f_1(x_1) + \cdots + f_n(x_n) \mid x = x_1 + \cdots + x_n \text{ with all } x_i \in G^+\}$$

for all $x \in G^+$.

PROOF. Since $\mathbb{R}$ is a Dedekind complete lattice-ordered abelian group, so is B, by Theorem 2.27. It is clear that B is also a partially ordered real vector space.

Now consider any $f_1, \ldots, f_n$ in B, and set $f = f_1 \vee \cdots \vee f_n$. If $x \in G^+$ and $x = x_1 + \cdots + x_n$ with all $x_i \in G^+$, then

$$f_1(x_1) + \cdots + f_n(x_n) \leq f(x_1) + \cdots + f(x_n) = f(x).$$

Given any real number $\varepsilon > 0$, Theorem 2.27 shows that there exists a decomposition $x = y_1 + \cdots + y_k$ with all $y_i \in G^+$ such that

$$\sum_{j=1}^{k} \max\{f_1(y_j), \ldots, f_n(y_j)\} > f(x) - \varepsilon.$$

After adding zero terms if necessary, there is no loss of generality in assuming that at least n of the y_j are zero. Consequently, we may partition $\{1, 2, \ldots, k\}$ into n pairwise disjoint nonempty sets $J(1), \ldots, J(n)$ such that

$$\max\{f_1(y_j), \ldots, f_n(y_j)\} = f_i(y_j)$$

for all $j \in J(i)$. Defining

$$x_i = \sum_{j \in J(i)} y_j \in G^+$$

for each $i = 1, \ldots, n$, we obtain $x = x_1 + \cdots + x_n$ and

$$\sum_{i=1}^{n} f_i(x_i) = \sum_{i=1}^{n} \sum_{j \in J(i)} f_i(y_j) = \sum_{j=1}^{k} \max\{f_1(y_j), \ldots, f_n(y_j)\} > f(x) - \varepsilon.$$

Therefore $f(x)$ equals the given supremum.

The formula for $(f_1 \wedge \cdots \wedge f_n)(x)$ follows via an application of the order anti-automorphism $z \mapsto -z$. □

In the situation of Theorem 2.27 and Corollary 2.28, $\mathrm{Hom}_{\mathbb{Z}}(G, H)$ is Dedekind complete but not necessarily lattice-ordered. For example, let G be the group of all eventually constant sequences of integers, with the pointwise ordering, and let G_0 be the subgroup of all eventually zero sequences in G. Define a group homomorphism $f_0 \colon G_0 \to \mathbb{R}$ by the rule

$$f_0(x_1, x_2, \ldots, x_n, 0, 0, 0, \ldots) = x_1 + 2x_2 + \cdots + nx_n.$$

Since G_0 is a direct summand of G (as abelian groups), f_0 extends to a group homomorphism $f \colon G \to \mathbb{R}$. As each of the elements

$$(0, 0, \ldots, 0, 1, 0, 0, \ldots)$$

lies below the element $(1, 1, 1, \ldots)$, there does not exist a positive homomorphism $g \colon G \to \mathbb{R}$ such that $f \leq^+ g$. Thus f and 0 have no common upper bound in $\mathrm{Hom}_{\mathbb{Z}}(G, \mathbb{R})$.

• **Notes.** The 2×2 Riesz decomposition property (Proposition 2.1(c)) was introduced in 1940 by Riesz [**109**, p. 175], who also noted that this property implies the $n \times k$ decomposition property (Proposition 2.2(c)) [**109**, pp. 175, 176]. (Riesz studied these properties in an abelian semigroup with cancellation, which is equivalent to working in the positive cone of a partially ordered abelian group.) The remainder of Propositions 2.1 and 2.2 was proved by Birkhoff [**11**, Theorem 49]. The second part of Proposition 2.3 is due to Fuchs [**43**, Proposition 5.3], along with Propositions 2.4 and 2.5 [**43**, Propositions 5.4, 5.5, Theorem 5.6], as well as the case of Proposition 2.6 involving a finite family of ideals [**43**, Proposition 5.8]. Proposition 2.8 was proved by the author and Handelman [**58**, Lemma 6.1].

The main case of Corollary 2.11, namely, that every lexicographic direct sum of directed interpolation groups is an interpolation group, was proved by Elliott [**37**, Theorem 3.10; **38**, Theorem 3.10]. Propositions 2.19 and 2.21 are due to the author, Handelman, and Lawrence [**60**, Lemmas I.5.7, I.6.4]. Theorem 2.27 was first proved for B the group of differences of positive homomorphisms from G to H, in two special cases, one by Kantorovitch [**81**, Theorem 6], who proved it for the case in which G and H are lattice-ordered real vector spaces, and one by Riesz [**109**, Théorème 1], whose result is equivalent to the case in which $H = \mathbb{R}$.

CHAPTER 3

Dimension Groups

A class of interpolation groups whose structure is quite transparent consists of the finite direct products of copies of $\mathbb{Z}$ (with the product ordering). These partially ordered abelian groups are known as simplicial groups. One natural means of constructing more complicated interpolation groups is to form direct limits of simplicial groups. Such direct limits provide examples of dimension groups, which are defined as directed unperforated interpolation groups. A major structural theorem for this class of interpolation groups is the converse of the previous observation: namely, all dimension groups can be obtained as direct limits of simplicial groups. As a corollary, we find that the dimension groups may also be characterized as those partially ordered abelian groups that can be obtained as direct limits of lattice-ordered abelian groups.

The chapter begins with some generalities about the class of dimension groups. In particular, we show that this class is closed under ideals, quotients by ideals, lexicographic direct products and sums, and certain pullbacks and pushouts. Next, some basic properties of simplicial groups are developed, leading to a characterization of simplicial groups as those interpolation groups with order-units in which the positive cones satisfy the descending chain condition. The final section of the chapter culminates with the representation of dimension groups as direct limits of simplicial groups.

• Dimension Groups.

DEFINITION. A *dimension group* is any directed, unperforated, interpolation group.

For example, any lattice-ordered abelian group is a dimension group (directedness and interpolation are clear; the lack of perforation is proved in Proposition 1.22). On the other hand, $\mathbb{Q}^2$ when equipped with the strict ordering is a dimension group which is not lattice-ordered. Note that the class of dimension groups is closed under direct products, direct sums, and direct limits. In particular, any direct limit of lattice-ordered abelian groups is a dimension group. We shall show in Theorem 3.21 that, conversely, any dimension group is isomorphic to a direct limit of lattice-ordered abelian groups.

To illustrate the independence of the three conditions used in defining dimension groups, we recall that a directed interpolation group can even be torsion-free as an abelian group without being unperforated (Example 2.23) and that directed unperforated partially ordered abelian groups need not satisfy interpolation (Example 2.7). Finally, any nonzero torsion-free abelian group, when equipped with the discrete ordering, is an unperforated interpolation group that is not directed.

PROPOSITION 3.1. *If H is an ideal in a dimension group G, then H and G/H are dimension groups.*

PROOF. Since H is an ideal of G, it is directed, and since G is unperforated, so is H. The interpolation property is inherited by H from G because H is a convex subgroup of G. Thus H is a dimension group.

As G is directed, so is G/H. By Propositions 1.20 and 2.3, since G is an unperforated interpolation group, G/H has the same properties. Thus G/H is a dimension group. □

That the converse of Proposition 3.1 fails in general may be seen in Example 2.7. We present an alternate example here, one that will also be used later to exhibit other properties.

EXAMPLE 3.2. *There exists a directed real vector space G with an ideal H such that H and G/H are dimension groups, but G is not an interpolation group.*

PROOF. Let G be the real vector space $\mathbb{R}^3$, made into a partially ordered vector space with positive cone

$$\begin{aligned} G^+ = \{(0,0,0)\} &\cup \{(x,y,0) \in G \mid x > 0 \text{ and } y > 0\} \\ &\cup \{(x,y,z) \in G \mid x+y > 0 \text{ and } z > 0\}. \end{aligned}$$

Note that $(1,1,1)$ is an order-unit in G, so that G is directed.

Now set $H = \{(x,y,z) \in G \mid z = 0\}$. As an ordered vector space, H is isomorphic to $\mathbb{R}^2$ with the strict ordering. Thus H is a dimension group. It is clear that H is an ideal of G. Next, observe that $G/H \cong \mathbb{R}$ as ordered vector spaces. Thus G/H is a dimension group.

Finally, we note that

$$\begin{matrix} (0,0,0) \\ (1,-1,0) \end{matrix} < \begin{matrix} (2,1,0) \\ (1,0,1) \end{matrix}$$

in G, but there does not exist any element $(x,y,z) \in G$ satisfying the following inequality:

$$\begin{matrix} (0,0,0) \\ (1,-1,0) \end{matrix} \leq (x,y,z) \leq \begin{matrix} (2,1,0) \\ (1,0,1). \end{matrix}$$

Therefore G is not an interpolation group. □

• Products, Pullbacks, and Pushouts.

PROPOSITION 3.3. *Any lexicographic direct product or lexicographic direct sum of dimension groups is a dimension group.*

PROOF. Let $\{G_i \mid i \in I\}$ be a family of dimension groups indexed by a well-ordered set I, and let G be the lexicographic direct product of this family. As each G_i is a directed interpolation group, Proposition 2.10 shows that G is an interpolation group. Given an element $x \in G$ such that $x \not\geq 0$, there is an index $j \in I$ such that $x_j \not\geq 0$ and $x_i = 0$ for all $i < j$. Since G_j is unperforated, $nx_j \not\geq 0$ for all $n \in \mathbb{N}$, and hence $nx \not\geq 0$ for all $n \in \mathbb{N}$. Thus G is unperforated.

To check that G is directed, it suffices to show that any element $x \in G$ lies below a positive element. This is clear for $x = 0$, so assume that $x \neq 0$, and let j be the least index such that $x_j \neq 0$. Now G_j is nonzero and directed, whence $G_j^+ \neq \{0\}$. Choose $a \in G_j$ such that $a > 0$, and then choose $b \in G_j$ with $a \leq b$ and $x_j + a \leq b$, so that $0 < b$ and $x_j < b$. Defining $y \in G$ so that $y_j = b$ and $y_i = 0$ for all $i \neq j$, we obtain an element such that $0 < y$ and $x < y$. Therefore G is directed, and hence is a dimension group.

The proof that lexicographic direct sums of dimension groups are dimension groups is essentially the same, using Corollary 2.11 in place of Proposition 2.10. □

PROPOSITION 3.4. *Let $\{f_i \colon G_i \to H \mid i \in I\}$ be a nonempty family of morphisms in the category of partially ordered abelian groups, and let G be a pullback of this family. For each $i \in I$, assume that* $\ker(f_i)$ *is directed and that $f_i(G_i^+) = H^+$. Then if each G_i is a dimension group, so is G.*

PROOF. By Proposition 2.15, G is an interpolation group. As G may be identified with a subgroup of $\prod G_i$, it is unperforated.

Given an element $x \in G$, choose an index $j \in I$, and choose an element $y_j \in G_j^+$ such that $x_j \leq y_j$. For each $i \in I - \{j\}$, choose $y_i \in G_i^+$ such that $f_i(y_i) = f_j(y_j)$, and note that $f_i(x_i) = f_j(x_j) \leq f_j(y_j) = f_i(y_i)$. Since $f_i(G_i^+) = H^+$ and $\ker(f_i)$ is directed, there exists an element z_i in $\ker(f_i)^+$ such that $x_i \leq y_i + z_i$. As $f_i(y_i + z_i) = f_i(y_i) = f_j(y_j)$, there is no loss of generality in replacing y_i by $y_i + z_i$, so that now $x_i \leq y_i$. Thus we obtain components for an element $y \in G^+$ such that $x \leq y$, proving that G is directed. □

PROPOSITION 3.5. *Let $\{f_i \colon (G_i, u_i) \to (H, v) \mid i \in I\}$ be a nonempty family of morphisms in the category of partially ordered abelian groups with order-unit, and let (G, u) be a pullback of this family. For each $i \in I$, assume that* $\ker(f_i)$ *is directed and that $f_i(G_i^+) = H^+$. Then if each G_i is a dimension group, so is G.*

PROOF. Since G has an order-unit, G is directed, and since G is isomorphic to a subgroup of $\prod G_i$, it is unperforated. By Corollary 2.16, G has interpolation. □

LEMMA 3.6. *Let H be an ideal in a directed interpolation group G, and let n be a positive integer. If H and G/H are both n-unperforated, then so is G.*

PROOF. As G is directed, it suffices to show that whenever $x, y \in G^+$ with $nx \leq ny$, then $x \leq y$.

Now $n(x+H) \leq n(y+H)$ in G/H and so $x+H \leq y+H$, because G/H is n-unperforated, whence $x \leq y+a$ for some $a \in H^+$. Then $x = z+b$ for some $z, b \in G^+$ such that $z \leq y$ and $b \leq a$. As $0 \leq b \leq a$, we must have $b \in H^+$. We also have $y-z \in G^+$ and $nb \leq n(y-z)$, and it is enough to show that $b \leq y-z$. Hence, there is no loss of generality in replacing x and y by b and $y-z$.

Thus we may assume that $x \in H^+$. Since $nx \leq ny$, there exists a decomposition $nx = x_1 + \cdots + x_n$ where each $x_i \in G^+$ and each $x_i \leq y$. Also, each $x_i \leq nx$. Then there is some $w \in G$ such that

$$x_i \quad \leq \quad w \quad \leq \quad \begin{matrix} y \\ nx \end{matrix}$$

for all i. Note that as $0 \leq w \leq nx$, we have $w \in H$. In addition,

$$nx = x_1 + \cdots + x_n \leq nw,$$

and therefore $x \leq w \leq y$, because H is n-unperforated. □

PROPOSITION 3.7. *Let $\{G_i \mid i \in I\}$ be a nonempty family of dimension groups, let H be a dimension group, and for each $i \in I$ let $f_i \colon H \to G_i$ be an isomorphism of H onto an ideal of G_i (as ordered groups). If G is a pushout of the family $\{f_i \mid i \in I\}$, then G is a dimension group.*

PROOF. By Proposition 2.17, G is an interpolation group. As in Proposition 1.17, we identify G with K/L, where $K = \bigoplus G_i$ and L is a convex subgroup of K. As each G_i is directed, so is K, and hence G is directed. If $N = \bigoplus f_i(H)$, then (as observed in the proof of Proposition 2.17) N is an ideal of K containing L such that $N/L \cong H$ as ordered groups. Then N/L is unperforated, while $G/(N/L)$, being isomorphic to $\bigoplus[G_i/f_i(H)]$, is unperforated by Proposition 1.20. In view of Lemma 3.6, we conclude that G must be unperforated. □

• Simplicial Groups.

DEFINITION. A *simplicial group* is any partially ordered abelian group that is isomorphic (as an ordered group) to $\mathbb{Z}^n$ for some nonnegative integer n. A *simplicial basis* for a simplicial group G is any basis $\{x_1, \ldots, x_n\}$ for G as a free abelian group such that also $G^+ = \sum \mathbb{Z}^+ x_i$. By convention, the empty set is considered to be a simplicial basis for the zero simplicial group.

PROPOSITION 3.8. *If G is a simplicial group with simplicial basis X, then the ideals of G are exactly the subgroups of G generated by subsets of X.*

PROOF. There is no loss of generality in assuming that $G = \mathbb{Z}^n$ for some positive integer n and that X is the standard basis $\{x_1, \ldots, x_n\}$ for G, so that $x_{ii} = 1$ for all i, while $x_{ij} = 0$ for all $i \neq j$.

Any subset Y of X has the form $Y = \{x_k \mid k \in K\}$ where K is a subset of $\{1, \ldots, n\}$, and the subgroup H of G generated by Y may be described as

$$H = \{z \in G \mid z_i = 0 \text{ for all } i \notin K\}.$$

It is clear from this description that H is an ideal of G.

Conversely, consider an ideal H of G, set

$$K = \{k \in \{1, \ldots, n\} \mid x_k \in H\},$$

and set $Y = \{x_k \mid k \in K\}$. If H is not generated (as a subgroup of G) by Y, then there exists an element $z \in H$ such that $z_j \neq 0$ for some $j \notin K$. Since H is directed, $z = v - w$ for some $v, w \in H^+$, and at least one of v_j, w_j is nonzero, say $v_j \neq 0$. Then $v_j \geq 1$, from which it follows that $v \geq x_j \geq 0$. But $x_j \notin H$, contradicting the fact that H is a convex subgroup of G. Therefore H must be generated by Y, as desired. □

COROLLARY 3.9. *If H is an ideal in a simplicial group G, then H and G/H are simplicial groups, and there exists an ideal K in G such that $G = H \oplus K$ (as ordered groups).*

PROOF. Choose a simplicial basis X for G. By Proposition 3.8, there is a subset $Y \subseteq X$ which generates H as a subgroup of G. Then H is a free abelian group with basis Y, and

$$H^+ = \sum_{y \in Y} \mathbb{Z}^+ y,$$

whence H is a simplicial group.

If K is the subgroup of G generated by $X - Y$, then K is an ideal of G by Proposition 3.8, and K is a simplicial group. Since $G = H \oplus K$ as a group and $G^+ = H^+ + K^+$, we also see that $G = H \oplus K$ as an ordered group. Consequently, $G/H \cong K$ (as ordered groups), and thus G/H is a simplicial group. □

Note that if G is the lexicographic direct product of $\mathbb{Z}$ with itself, then G has an ideal H such that H and G/H are simplicial groups, yet G itself is not simplicial.

DEFINITION. Let X be a partially ordered set with a least element 0. An *atom* of X is any element $x \in X$ such that $x > 0$ but there exist no elements $y \in X$ satisfying $x > y > 0$.

Note that in a simplicial group G, the elements of a simplicial basis for G are atoms of G^+.

LEMMA 3.10. *If G is an interpolation group, then the set of atoms of G^+ is $\mathbb{Z}$-linearly independent.*

PROOF. Suppose that $m_1x_1 + \cdots + m_kx_k = 0$ for some nonzero integers m_i and some distinct atoms x_i of G^+. If all the $m_i > 0$, then

$$x_1 \leq m_1x_1 \leq m_1x_1 + \cdots + m_kx_k = 0,$$

which is impossible. Thus at least one of the m_i is negative. Similarly, at least one of the m_i is positive.

Now after renumbering the m_i and the x_i, we may assume that there is a positive integer n with $1 \leq n < k$ such that $m_i > 0$ for all $i \leq n$ while $m_i < 0$ for all $i > n$. As a result,

$$x_1 \leq m_1x_1 \leq m_1x_1 + \cdots + m_nx_n = (-m_{n+1})x_{n+1} + \cdots + (-m_k)x_k.$$

By Riesz decomposition, $x_1 = y_1 + \cdots + y_t$ for some elements $y_j \in G^+$ such that each $y_j \leq x_{i(j)}$ for some $i(j)$ in $\{n+1, \ldots, k\}$. Choose an index j such that $y_j \neq 0$, so that $x_1 \geq y_j > 0$ and $x_{i(j)} \geq y_j > 0$. Since x_1 and $x_{i(j)}$ are atoms of G^+, it follows that $x_1 = y_j = x_{i(j)}$, which is a contradiction (because $i(j) > 1$). Therefore the atoms of G^+ must be $\mathbb{Z}$-linearly independent. $\square$

LEMMA 3.11. *Let G be an interpolation group, let X be a set of atoms of G^+, and let H be the subgroup of G generated by X. Then H is an ideal of G, and*

$$H^+ = \sum_{x \in X} \mathbb{Z}^+ x.$$

PROOF. Obviously H is a directed subgroup of G. Given $y \in G$ and $z \in H$ with $0 \leq y \leq z$, write $z = m_1x_1 + \cdots + m_kx_k$ for some integers m_i and some elements $x_i \in X$. Set $w = \sum |m_i|x_i$, noting that $y \leq z \leq w$, and write $w = w_1 + \cdots + w_n$ for some elements $w_i \in X$. Then $y = y_1 + \cdots + y_n$ for some elements $y_i \in G^+$ such that each $y_i \leq w_i$. As each w_i is an atom of G^+, either $y_i = 0$ or $y_i = w_i$, and so $y_i \in H$. Hence, $y \in H$, which proves that H is a convex subgroup of G. Therefore H is an ideal of G.

Any element $u \in H^+$ may be written as $u = n_1u_1 + \cdots + n_ku_k$ for some integers n_i and some distinct elements $u_i \in X$. Write each $n_i = s_i - t_i$ for some positive integers s_i, t_i, and note that

$$(s_1u_1 + \cdots + s_ku_k) - (t_1u_1 + \cdots + t_ku_k) = u \geq 0.$$

If $t_1 > s_1$, then

$$u_1 \leq (t_1 - s_1)u_1 \leq (t_1 - s_1)u_1 + t_2u_2 + \cdots + t_ku_k \leq s_2u_2 + \cdots + s_ku_k.$$

In this case, it follows from Riesz decomposition that $u_1 = u_j$ for some $j \geq 2$, which is false. Thus $t_1 \leq s_1$, so that $n_1 \geq 0$. Similarly, the remaining $n_i \geq 0$, and hence

$$u \in \sum_{x \in X} \mathbb{Z}^+ x.$$

Therefore H^+ has the required form. $\square$

PROPOSITION 3.12. *Let G be an interpolation group, and let H be the subgroup of G generated by finitely many distinct atoms $x_1, \ldots, x_n$ of G^+. Then H is an ideal of G, and H is a simplicial group with simplicial basis $\{x_1, \ldots, x_n\}$.*

PROOF. Lemmas 3.10 and 3.11. $\square$

DEFINITION. A partially ordered set X satisfies the *descending chain condition*, abbreviated DCC, if and only if there does not exist an infinite strictly descending chain $x_1 > x_2 > \cdots$ in X.

THEOREM 3.13. *A partially ordered abelian group G is isomorphic (as an ordered group) to a direct sum of copies of $\mathbb{Z}$ if and only if G is a directed interpolation group and G^+ satisfies the DCC.*

PROOF. If G is isomorphic (as an ordered group) to a direct sum of copies of $\mathbb{Z}$, then the required properties are clear. Conversely, assume that G is a directed interpolation group and that G^+ satisfies the DCC. Note from the DCC that every nonzero element of G^+ lies above an atom of G^+.

We claim that any element $x \in G^+$ is a sum (possibly empty) of atoms. If not, $x > 0$, and so there is an atom $x_1 \in G^+$ such that $x > x_1$. Then $x - x_1 > 0$, whence there is an atom $x_2 \in G^+$ such that $x - x_1 > x_2$, that is, $x > x_1 + x_2$. Continuing in this manner, we obtain an infinite sequence $x_1, x_2, \ldots$ of atoms of G^+ such that $x > x_1 + \cdots + x_n$ for all n. But then there is an infinite descending chain

$$x > x - x_1 > \cdots > x - (x_1 + \cdots + x_n) > \cdots$$

in G^+, contradicting the DCC.

Thus every element of G^+ is a sum of atoms, as claimed. Hence

$$G^+ = \sum_{x \in X} \mathbb{Z}^+ x,$$

where X is the set of atoms of G^+. Since G is directed, it follows that G is generated, as a group, by X. As X is $\mathbb{Z}$-linearly independent, by Lemma 3.10, we conclude that

$$G = \bigoplus_{x \in X} \mathbb{Z}x \cong \bigoplus_{x \in X} \mathbb{Z}$$

as ordered groups. □

COROLLARY 3.14. *A partially ordered abelian group G is simplicial if and only if G is an interpolation group with an order-unit and G^+ satisfies the DCC.*

PROOF. This is immediate using Theorem 3.13 and the observation that no infinite direct sum of copies of $\mathbb{Z}$ has an order-unit. □

• **Direct Limits of Simplicial Groups.** In order to represent a given dimension group G as a direct limit of simplicial groups, we require a means of constructing a direct system of simplicial groups and positive homomorphisms suitably closely related to G. It is easy enough to construct closely related simplicial groups, since given any finite subset $\{x_1, \ldots, x_n\}$ in G^+ there is a natural positive homomorphism $f \colon \mathbb{Z}^n \to G$ sending the members of a simplicial basis for $\mathbb{Z}^n$ to the elements x_i. Any element of the kernel of f constitutes a relation among the basis elements of $\mathbb{Z}^n$ which must be annihilated in some simplicial group appearing later in the direct system. This can be accomplished by factoring f through a positive homomorphism $\mathbb{Z}^n \to \mathbb{Z}^m$ whose kernel equals $\ker(f)$. In order to construct such factorizations, we require the following strong decomposition property.

PROPOSITION 3.15. *Let G be a dimension group, and let*

$$p_1x_1 + \cdots + p_nx_n = 0$$

for some integers p_i and some elements $x_i \in G^+$. Then there exist elements $y_1, \dots, y_t$ in G^+ and nonnegative integers q_{ij} (for $i = 1, \dots, n$ and $j = 1, \dots, t$) such that

$$x_i = q_{i1}y_1 + \cdots + q_{it}y_t \quad \textit{and} \quad p_1q_{1j} + \cdots + p_nq_{nj} = 0$$

for all $i = 1, \dots, n$ and $j = 1, \dots, t$.

PROOF. We first dispose of the case in which all the $p_i \geq 0$. If any $p_i > 0$, then

$$0 \leq x_i \leq p_ix_i \leq p_1x_1 + \cdots + p_nx_n = 0,$$

and hence $x_i = 0$. In particular, if all the $p_i > 0$, then all the $x_i = 0$. In this case, just take $t = 1$ and $y_1 = 0$, and set $q_{i1} = 0$ for all i. If not all the p_i are strictly positive, we may renumber the p_i and the x_i so that $p_1 = \cdots = p_t = 0$ and $x_{t+1} = \cdots = x_n = 0$, for some t in $\{1, \dots, n\}$. In this case, take $y_j = x_j$ for $j = 1, \dots, t$, setting $q_{jj} = 1$ for $j = 1, \dots, t$ while all other $q_{ij} = 0$.

In case all the $p_i \leq 0$, we simply apply the results of the previous paragraph to the relation

$$(-p_1)x_1 + \cdots + (-p_n)x_n = 0.$$

For the general case, we assign a "degree" to the coefficient list $p_1, \dots, p_n$ and proceed by an induction on degree. Let us say that the *degree* of a list $p_1, \dots, p_n$ of integers is the ordered pair (p, l) where p is the maximum of the values $|p_i|$ and l is the number of times p appears in the list $|p_1|, \dots, |p_n|$. These degrees lie in the set $\mathbb{Z}^+ \times \mathbb{N}$, which we order lexicographically. As the lexicographic product of $\mathbb{Z}^+$ with $\mathbb{N}$ is well-ordered, an induction indexed by this set is legitimate.

If the coefficient list of the given relation has degree $(0, l)$, then all the $p_i = 0$, a case which we have already covered.

Now suppose that the coefficient list has degree (p, l) with $p > 0$ and that the proposition holds in all cases of lower degree. Because the cases in which either all the $p_i \geq 0$ or all the $p_i \leq 0$ have already been covered, we may assume that at least one $p_i > 0$ and at least one $p_i < 0$.

After renumbering the p_i and the x_i, and multiplying through by -1 if necessary, we may assume that $p_1 = p$ and that there is an integer m with $1 \leq m < n$ such that $p_i \geq 0$ for all $i \leq m$ while $p_i < 0$ for all $i > m$. Note that $p \geq -p_i > 0$ for all $i = m+1, \dots, n$. Now

$$\begin{aligned} px_1 &= p_1x_1 \leq p_1x_1 + \cdots + p_mx_m \\ &= -p_{m+1}x_{m+1} - \cdots - p_nx_n \leq p(x_{m+1} + \cdots + x_n), \end{aligned}$$

and hence $x_1 \leq x_{m+1} + \cdots + x_n$, because G is unperforated. By Riesz decomposition, $x_1 = z_{m+1} + \cdots + z_n$ for some elements $z_i \in G^+$ such that $z_i \leq x_i$ for each $i = m+1, \ldots, n$.

Observe that

$$p\left(\sum_{i=m+1}^{n} z_i\right) + \sum_{i=2}^{n} p_i x_i = \sum_{i=1}^{n} p_i x_i = 0,$$

and consequently

$$(*) \qquad \sum_{i=m+1}^{n} (p+p_i) z_i + \sum_{i=2}^{m} p_i x_i + \sum_{i=m+1}^{n} p_i(x_i - z_i) = 0.$$

Since $p > p + p_i \geq 0$ for all $i = m+1, \ldots, n$, the integer p appears in the list

$$|p + p_{m+1}|, \ldots, |p + p_n|, |p_2|, \ldots, |p_n|$$

exactly $l - 1$ times. Thus $(*)$ is a relation whose coefficient list has degree less than (p, l). By the induction hypothesis, there exist elements $y_1, \ldots, y_t$ in G^+ and nonnegative integers r_{ij} (for $i = m+1, \ldots, n$ and $j = 1, \ldots, t$) and s_{ij} (for $i = 2, \ldots, n$ and $j = 1, \ldots, t$) such that

$$\begin{aligned} z_i &= r_{i1} y_1 + \cdots + r_{it} y_t \quad (\text{for } i = m+1, \ldots, n), \\ x_i &= s_{i1} y_1 + \cdots + s_{it} y_t \quad (\text{for } i = 2, \ldots, m), \\ x_i - z_i &= s_{i1} y_1 + \cdots + s_{it} y_t \quad (\text{for } i = m+1, \ldots, n), \end{aligned}$$

while also

$$(p + p_{m+1}) r_{m+1,j} + \cdots + (p + p_n) r_{nj} + p_2 s_{2j} + \cdots + p_n s_{nj} = 0$$

for all $j = 1, \ldots, t$.

We define nonnegative integers q_{ij} for $i = 1, \ldots, n$ and $j = 1, \ldots, t$ such that

$$\begin{aligned} q_{1j} &= r_{m+1,j} + \cdots + r_{nj}, \\ q_{ij} &= s_{ij} \quad (\text{for } i = 2, \ldots, m), \\ q_{ij} &= r_{ij} + s_{ij} \quad (\text{for } i = m+1, \ldots, n). \end{aligned}$$

It follows that

$$\begin{aligned} x_1 &= \sum_{i=m+1}^{n} z_i = \sum_{i=m+1}^{n} \sum_{j=1}^{t} r_{ij} y_j = \sum_{j=1}^{t} q_{1j} y_j, \\ x_i &= \sum_{j=1}^{t} s_{ij} y_j = \sum_{j=1}^{t} q_{ij} y_j \quad (\text{for } i = 2, \ldots, m), \\ x_i = z_i + (x_i - z_i) &= \sum_{j=1}^{t} r_{ij} y_j + \sum_{j=1}^{t} s_{ij} y_j = \sum_{j=1}^{t} q_{ij} y_j \quad (\text{for } i = m+1, \ldots, n). \end{aligned}$$

In addition, for all $j = 1, \ldots, t$ we compute that

$$\sum_{i=1}^{n} p_i q_{ij} = p\left(\sum_{i=m+1}^{n} r_{ij}\right) + \sum_{i=2}^{m} p_i s_{ij} + \sum_{i=m+1}^{n} p_i(r_{ij} + s_{ij})$$
$$= \sum_{i=m+1}^{n} (p + p_i) r_{ij} + \sum_{i=2}^{m} p_i s_{ij} = 0.$$

This completes the induction step, and the proposition is proved. □

There is a near converse to Proposition 3.15; namely, any directed partially ordered abelian group G satisfying the decomposition property described in the proposition must be a dimension group. For instance, consider any $x \in G$ and $n \in \mathbb{N}$ such that $nx \geq 0$. Since G is directed, $x = x_1 - x_2$ for some $x_1, x_2 \in G^+$. Set $x_3 = nx$, so that $x_3 \in G^+$ and $nx_1 - nx_2 - x_3 = 0$. If the decomposition property of Proposition 3.15 holds in G, there exist elements $y_1, \ldots, y_t$ in G^+ and nonnegative integers q_{ij} (for $i = 1, 2, 3$ and $j = 1, \ldots, t$) such that

$$x_i = q_{i1} y_1 + \cdots + q_{it} y_t$$

for $i = 1, 2, 3$ and $nq_{1j} - nq_{2j} - q_{3j} = 0$ for $j = 1, \ldots, t$. Then $nq_{1j} \geq nq_{2j}$ for each j, so that $q_{1j} \geq q_{2j}$. Consequently, $x_1 \geq x_2$, and hence $x \geq 0$, proving that G is unperforated. In a similar fashion, the Riesz decomposition properties follow from the decomposition property of Proposition 3.15.

PROPOSITION 3.16. *Let G_1 be a simplicial group, let G be a dimension group, and let $g_1 : G_1 \to G$ be a positive homomorphism. Then there exist a simplicial group G_2 and positive homomorphisms*

$$G_1 \xrightarrow{h} G_2 \xrightarrow{g_2} G$$

such that $g_1 = g_2 h$ and $\ker(g_1) = \ker(h)$.

PROOF. We claim initially that given $a_1, \ldots, a_k$ in $\ker(g_1)$, there exist a simplicial group G_2 and positive homomorphisms $h \colon G_1 \to G_2$ and $g_2 \colon G_2 \to G$ such that $g_1 = g_2 h$ and each a_i lies in $\ker(h)$.

First suppose that $k = 1$. If $G_1 = \{0\}$, then we need only take $G_2 = \{0\}$ and let h, g_2 be the appropriate zero maps. Now assume that G_1 is nonzero, and let $\{e_1, \ldots, e_n\}$ be a simplicial basis for G_1. Set $x_i = g_1(e_i)$ for each $i = 1, \ldots, n$, and note that $x_i \in G^+$. Write $a_1 = p_1 e_1 + \cdots + p_n e_n$ for some integers p_i, and observe that

$$p_1 x_1 + \cdots + p_n x_n = g_1(a_1) = 0.$$

According to Proposition 3.15, there exist elements $y_1, \ldots, y_t$ in G^+ and nonnegative integers q_{ij} (for $i = 1, \ldots, n$ and $j = 1, \ldots, t$) such that

$$x_i = q_{i1} y_1 + \cdots + q_{it} y_t \quad \text{and} \quad p_1 q_{1j} + \cdots + p_n q_{nj} = 0$$

for all i, j.

Set $G_2 = \mathbb{Z}^t$, and let $\{f_1, \ldots, f_t\}$ be a simplicial basis for G_2. Define group homomorphisms $h\colon G_1 \to G_2$ and $g_2\colon G_2 \to G$ so that

$$h(e_i) = q_{i1}f_1 + \cdots + q_{it}f_t$$

for $i = 1, \ldots, n$ and $g_2(f_j) = y_j$ for $j = 1, \ldots, t$. As each $q_{ij} \in \mathbb{Z}^+$ and each $y_j \in G^+$, we see that h and g_2 are positive homomorphisms. Since

$$g_2h(e_i) = g_2(q_{i1}f_1 + \cdots + q_{it}f_t) = q_{i1}y_1 + \cdots + q_{it}y_t = x_i = g_1(e_i)$$

for all $i = 1, \ldots, n$, we obtain $g_2h = g_1$. Also,

$$h(a_1) = h\left(\sum_{i=1}^{n} p_ie_i\right) = \sum_{i=1}^{n}\sum_{j=1}^{t} p_iq_{ij}f_j = \sum_{j=1}^{t}\left(\sum_{i=1}^{n} p_iq_{ij}\right)f_j = 0,$$

so that $a_1 \in \ker(h)$.

Now let $k > 1$, and assume that there exist a simplicial group G_3 and positive homomorphisms $h'\colon G_1 \to G_3$ and $g_3\colon G_3 \to G$ such that $g_1 = g_3h'$ and $a_1, \ldots, a_{k-1}$ lie in $\ker(h')$. As $g_3h'(a_k) = g_1(a_k) = 0$, the element $h'(a_k)$ lies in $\ker(g_3)$. Hence, by the first case of the claim, there exist a simplicial group G_2 and positive homomorphisms $h''\colon G_3 \to G_2$ and $g_2\colon G_2 \to G$ such that $g_3 = g_2h''$ and $h'(a_k) \in \ker(h'')$. Set $h = h''h'$, which is a positive homomorphism from G_1 to G_2 such that $g_2h = g_2h''h' = g_3h' = g_1$. Since $a_1, \ldots, a_{k-1}$ lie in $\ker(h')$, they also lie in $\ker(h)$. Because $h'(a_k)$ lies in $\ker(h'')$, the element a_k lies in $\ker(h)$. This completes the induction step, proving the claim.

Since G_1 is finitely generated as an abelian group, all its subgroups are finitely generated. Hence, we may choose elements $a_1, \ldots, a_k$ that generate $\ker(g_1)$ as a group. By the claim above, there exist a simplicial group G_2 and positive homomorphisms $h\colon G_1 \to G_2$ and $g_2\colon G_2 \to G$ such that $g_1 = g_2h$ and $a_1, \ldots, a_k$ all lie in $\ker(h)$. Thus $\ker(g_1) \subseteq \ker(h)$. The reverse inclusion follows from the factorization $g_1 = g_2h$. Therefore $\ker(g_1) = \ker(h)$. □

With the aid of Proposition 3.16, we can now present any dimension group as a direct limit of simplicial groups. We first deal with the countable case, which is simpler than the general case.

THEOREM 3.17. *Any countable dimension group is isomorphic to a direct limit of a countable sequence of simplicial groups (in the category of partially ordered abelian groups).*

PROOF. Given a countable dimension group G, write $G^+ = \{x_1, x_2, \ldots\}$ (with repetitions allowed). We construct simplicial groups $G_1, G_2, \ldots$ and positive homomorphisms $g_n\colon G_n \to G$ and $h_n\colon G_n \to G_{n+1}$ for all n such that

(a) $x_n \in g_n(G_n^+)$ for all n;

(b) $g_nh_{n-1} = g_{n-1}$ for all $n > 1$;

(c) $\ker(g_{n-1}) = \ker(h_{n-1})$ for all $n > 1$.

First set $G_1 = \mathbb{Z}$, and define $g_1\colon G_1 \to G$ so that $g_1(1) = x_1$.

Now assume that groups $G_1, \ldots, G_n$ and maps $g_1, \ldots, g_n$ and $h_1, \ldots, h_{n-1}$ have been constructed, for some n, satisfying conditions (a), (b), (c). The direct product $H = G_n \times \mathbb{Z}$ is a simplicial group, and we may define a positive homomorphism $g\colon H \to G$ by the rule

$$g(a, m) = g_n(a) + mx_{n+1}.$$

By Proposition 3.16, there exist a simplicial group G_{n+1} and positive homomorphisms $h\colon H \to G_{n+1}$ and $g_{n+1}\colon G_{n+1} \to G$ such that $g = g_{n+1}h$ and $\ker(g) = \ker(h)$. Note that

$$g_{n+1}h(0,1) = g(0,1) = x_{n+1}$$

with $h(0,1) \in G^+_{n+1}$, so that $x_{n+1} \in g_{n+1}(G^+_{n+1})$. If $q\colon G_n \to H$ is the natural injection map, then the map $h_n = hq$ is a positive homomorphism from G_n to G_{n+1} such that

$$g_{n+1}h_n = g_{n+1}hq = gq = g_n.$$

Consequently, $\ker(h_n) \subseteq \ker(g_n)$. On the other hand, since $gq = g_n$ we have

$$q(\ker(g_n)) \subseteq \ker(g) = \ker(h),$$

whence $\ker(g_n) \subseteq \ker(hq) = \ker(h_n)$. Thus $\ker(g_n) = \ker(h_n)$, so that the groups $G_1, \ldots, G_{n+1}$ and the maps $g_1, \ldots, g_{n+1}$ and $h_1, \ldots, h_n$ satisfy conditions (a), (b), (c). This completes the induction step.

Now let K be a direct limit (in the category of partially ordered abelian groups) of the sequence

$$G_1 \xrightarrow{h_1} G_2 \xrightarrow{h_2} G_3 \xrightarrow{h_3} \cdots$$

and for all $n \in \mathbb{N}$ let $q_n\colon G_n \to K$ be the natural map. In view of condition (b), there exists a unique positive homomorphism $g\colon K \to G$ such that $gq_n = g_n$ for all n. Using condition (a), we obtain

$$x_n \in g_n(G^+_n) = gq_n(G^+_n) \subseteq g(K^+)$$

for all n, and hence $g(K^+) = G^+$. In particular, it follows that g is surjective (because G is directed).

Given $x \in \ker(g)$, write $x = q_n(y)$ for some $n \in \mathbb{N}$ and some $y \in G_n$. Then

$$g_n(y) = gq_n(y) = g(x) = 0,$$

and so $h_n(y) = 0$, by condition (c). Hence, $x = q_{n+1}h_n(y) = 0$. Thus g is injective.

Therefore g is a group isomorphism satisfying $g(K^+) = G^+$, and hence g is an isomorphism in the category of partially ordered abelian groups. □

COROLLARY 3.18. *Any countable dimension group with order-unit is isomorphic to a direct limit of a countable sequence of simplicial groups with order-unit (in the category of partially ordered abelian groups with order-unit).*

PROOF. If (G, u) is a countable dimension group with order-unit, then by Theorem 3.17 G is isomorphic to a direct limit of a sequence

$$G_1 \xrightarrow{h_1} G_2 \xrightarrow{h_2} G_3 \xrightarrow{h_3} \cdots$$

of simplicial groups and positive homomorphisms (in the category of partially ordered abelian groups). We may identify G with the direct limit of this sequence. For all $n \in \mathbb{N}$, let $q_n : G_n \to G$ be the natural map.

Now $u \in q_m(G_m^+)$ for some m. As G is a direct limit of the sequence

$$G_m \xrightarrow{h_m} G_{m+1} \xrightarrow{h_{m+1}} G_{m+2} \xrightarrow{h_{m+2}} \cdots,$$

there is no loss of generality in assuming that $m = 1$. Choose $u_1 \in G_1^+$ such that $q_1(u_1) = u$, and define $u_n \in G_n^+$ for all $n > 1$ so that $u_{n+1} = h_n(u_n)$ for all n. Then $q_n(u_n) = u$ for all n.

For all $n \in \mathbb{N}$, let H_n be the convex subgroup of G_n generated by u_n. Then H_n is an ideal of G_n and u_n is an order-unit in H_n. By Corollary 3.9, H_n is a simplicial group. Given $x \in H_n$, we have $-tu_n \leq x \leq tu_n$ for some positive integer t, whence

$$-tu_{n+1} \leq h_n(x) \leq tu_{n+1},$$

and so $h_n(x) \in H_{n+1}$. Thus $h_n(H_n) \subseteq H_{n+1}$.

Now each h_n restricts to a positive homomorphism $h'_n : H_n \to H_{n+1}$, and we have a sequence

$$(*) \qquad (H_1, u_1) \xrightarrow{h'_1} (H_2, u_2) \xrightarrow{h'_2} (H_3, u_3) \xrightarrow{h'_3} \cdots$$

of objects and morphisms in the category of partially ordered abelian groups with order-unit. We claim that (G, u) is a direct limit of this sequence. It suffices to show that

$$G = \bigcup_{n=1}^{\infty} q_n(H_n) \quad \text{and} \quad G^+ = \bigcup_{n=1}^{\infty} q_n(H_n^+).$$

Given any $x \in G^+$, we have $x = q_n(y)$ for some $n \in \mathbb{N}$ and some $y \in G_n^+$. In addition, $x \leq tu$ for some positive integer t, so that $q_n(y) \leq q_n(tu_n)$. After increasing n if necessary, we may assume that $y \leq tu_n$, whence $y \in H_n^+$. Thus $G^+ = \bigcup q_n(H_n^+)$. As G is directed, it follows that $G = \bigcup q_n(H_n)$.

Therefore (G, u) is a direct limit of the sequence $(*)$, as claimed. □

THEOREM 3.19. *Any dimension group is isomorphic to a direct limit of a direct system of simplicial groups (in the category of partially ordered abelian groups).*

PROOF. Let G be a dimension group. As the zero partially ordered abelian group is simplicial, we may assume that G is nonzero. Since G is directed, there exists a nonzero element $x \in G^+$, and the elements $x, 2x, 3x, \ldots$ are all distinct, because G is a torsion-free abelian group. Thus G^+ is infinite.

Let $\mathcal{A}$ be the family of all nonempty finite subsets of G^+, viewed as a directed set under inclusion. We construct simplicial groups G_A for all $A \in \mathcal{A}$, positive homomorphisms $g_A : G_A \to G$ for all $A \in \mathcal{A}$, and positive homomorphisms $h_{AB} : G_B \to G_A$ for all proper containments $A \supset B$ in $\mathcal{A}$, such that

(a) $A \subseteq g_A(G_A^+)$ for all $A \in \mathcal{A}$;

(b) $g_A h_{AB} = g_B$ whenever $A \supset B$ in $\mathcal{A}$;

(c) $h_{AB}h_{BC} = h_{AC}$ whenever $A \supset B \supset C$ in $\mathcal{A}$;

(d) $\ker(g_B) = \ker(h_{AB})$ whenever $A \supset B$ in $\mathcal{A}$.

We proceed by induction on cardinality.

For any singleton $A = \{a\}$ in $\mathcal{A}$, set $G_A = \mathbb{Z}$ and define the positive homomorphism $g_A\colon G_A \to G$ so that $g_A(1) = a$.

Now assume, for some integer $n > 1$, that the appropriate simplicial groups and positive homomorphisms have been constructed for all sets in $\mathcal{A}$ with cardinality less than n. Let $A \in \mathcal{A}$ with $\operatorname{card}(A) = n$.

Let $\mathcal{B}$ be the collection of all nonempty proper subsets of A, and set

$$H = \bigoplus_{B \in \mathcal{B}} G_B,$$

which is a simplicial group. For all $B \in \mathcal{B}$, let $q_B\colon G_B \to H$ be the injection map. There exists a unique positive homomorphism $g\colon H \to G$ such that $gq_B = g_B$ for all $B \in \mathcal{B}$. According to Proposition 3.16, we may choose a simplicial group G_A and positive homomorphisms $h\colon H \to G_A$ and $g_A\colon G_A \to G$ such that $g = g_A h$ and $\ker(g) = \ker(h)$.

For each $x \in A$, we have $\{x\} \in \mathcal{B}$, and hence

$$\{x\} \subseteq g_{\{x\}}(G^+_{\{x\}}) = gq_{\{x\}}(G^+_{\{x\}}) = g_A h q_{\{x\}}(G^+_{\{x\}}) \subseteq g_A(G^+_A).$$

Thus $A \subseteq g_A(G^+_A)$.

For each $B \in \mathcal{B}$, define $h_{AB} = hq_B$, which is a positive homomorphism from G_B to G_A such that

$$g_A h_{AB} = g_A h q_B = g q_B = g_B.$$

If $B \supset C$ in $\mathcal{A}$, then $gq_B h_{BC} = g_B h_{BC} = g_C = gq_C$. Hence,

$$(q_B h_{BC} - q_C)(G_C) \subseteq \ker(g) = \ker(h),$$

and consequently $h_{AB} h_{BC} = hq_B h_{BC} = hq_C = h_{AC}$.

For each $B \in \mathcal{B}$, we have $\ker(h_{AB}) \subseteq \ker(g_B)$ because $g_A h_{AB} = g_B$. On the other hand, as $g_B = gq_B$ we have

$$q_B(\ker(g_B)) \subseteq \ker(g) = \ker(h),$$

and hence $\ker(g_B) \subseteq \ker(hq_B) = \ker(h_{AB})$. Thus $\ker(g_B) = \ker(h_{AB})$.

This completes the induction step of the construction.

Because of condition (c), the collection $\{G_A, h_{AB}\}$ is a direct system of simplicial groups and positive homomorphisms, indexed by the directed set $\mathcal{A}$. Let K be a direct limit of this system (in the category of partially ordered abelian groups), and for all $A \in \mathcal{A}$ let $q_A\colon G_A \to K$ be the natural map. In view of condition (b), there exists a unique positive homomorphism $g\colon K \to G$ such that $gq_A = g_A$ for all $A \in \mathcal{A}$. Using condition (a), we obtain

$$A \subseteq g_A(G^+_A) = gq_A(G^+_A) \subseteq g(K^+)$$

for all $A \in \mathcal{A}$, and hence $g(K^+) = G^+$. In particular, it follows that g is surjective.

Given $x \in \ker(g)$, write $x = q_B(y)$ for some $B \in \mathcal{A}$ and some $y \in G_B$. Then

$$g_B(y) = gq_B(y) = g(x) = 0.$$

As G^+ is infinite, there exists $A \in \mathcal{A}$ such that $A \supset B$. Then

$$y \in \ker(g_B) = \ker(h_{AB})$$

by condition (d), and hence $x = q_A h_{AB}(y) = 0$. Thus g is injective.

Therefore g is a group isomorphism satisfying $g(K^+) = G^+$, and hence g is an isomorphism in the category of partially ordered abelian groups. □

COROLLARY 3.20. *Any dimension group with order-unit is isomorphic to a direct limit of a direct system of simplicial groups with order-unit* (*in the category of partially ordered abelian groups with order-unit*).

PROOF. Analogous to Corollary 3.18. □

In particular, we may use Theorem 3.19 to see that the dimension groups are exactly the direct limits of lattice-ordered abelian groups, as follows.

THEOREM 3.21. *A partially ordered abelian group G is a dimension group if and only if G is isomorphic to a direct limit of a direct system of lattice-ordered abelian groups* (*in the category of partially ordered abelian groups*).

PROOF. If G is a dimension group, apply Theorem 3.19 together with the observation that all simplicial groups are lattice-ordered. For the converse, since the class of dimension groups is closed under direct limits, we need only observe that any lattice-ordered abelian group H is a dimension group. But H is obviously a directed interpolation group, and H is unperforated by Proposition 1.22. Thus H is a dimension group. □

Another consequence of Theorem 3.19 is that finitary computational implications which are valid in all simplicial groups must also be valid in all dimension groups. We illustrate this principle with the following versions of Propositions 2.19 and 2.21.

PROPOSITION 3.22. *Let G be a dimension group, let $x, y, z \in G$, and let k be a positive integer.*

(a) *If $x, y, z \in G^+$ and $kx \le ky + z$, then $x = v + w$ for some $v, w \in G^+$ such that $v \le y$ and $kw \le z$.*

(b) *If $z \in G^+$ and $kx \le ky + z$, then $x = v + w$ for some $v \in G$ and $w \in G^+$ such that $v \le y$ and $kw \le z$.*

PROOF. As these properties are preserved in direct limits of partially ordered abelian groups, we need only verify them for all simplicial groups, because of Theorem 3.19. Since these properties are also preserved in direct products of partially ordered abelian groups, we may further reduce to the case that $G = \mathbb{Z}$.

Properties (a) and (b) now have identical proofs. Namely, if $x \le y$, take $v = x$ and $w = 0$, while if $x > y$, take $v = y$ and $w = x - y$. □

Proposition 3.22 may also be obtained from the strong decomposition property of Proposition 3.15, with more computational effort.

PROPOSITION 3.23. *Let G be a dimension group, let $z \in G$, and let k be a positive integer. Set*

$$X = \{x \in G \mid kx \leq z\} \quad \text{and} \quad Y = \{y \in G \mid ky \geq z\}.$$

Then X is upward directed and Y is downward directed.

PROOF. It suffices to prove that whenever $x_1, x_2 \in G$ with $kx_i \leq z$ for each i, there exists $x \in G$ such that $kx \leq z$ and each $x_i \leq x$. As in Proposition 3.22, we may reduce to the case that $G = \mathbb{Z}$. In that case, take $x = \max\{x_1, x_2\}$. □

• **Notes.** The term "dimension group" was first introduced by Elliott [**35**, Theorem 5.6] to denote any partially ordered abelian group isomorphic to a direct limit of a sequence of simplicial groups. In later usage, arbitrary direct limits of simplicial groups were allowed. After the characterization of direct limits of simplicial groups as directed unperforated interpolation groups, by Effros, Handelman, and Shen [**30**, Theorem 2.2], the latter characterization became the more convenient condition with which to define dimension groups.

Part of Proposition 3.1, namely that any quotient of an unperforated interpolation group by an ideal is again an unperforated interpolation group, was proved by Teller [**119**, Lemma 1.2]. Example 3.2 was constructed by Handelman [**67**, Section VII]. The dimension group cases of Lemma 3.10 and Proposition 3.12 are due to Shen [**115**, Proposition 2.1 and Theorem 2.3]. Corollary 3.14 was proved by the author and Handelman [**57**, Proposition 4.3].

Proposition 3.15 was derived by Effros, Handelman, and Shen in the course of proving [**30**, Theorem 2.2]. A version of Proposition 3.16 for divisible dimension groups was obtained by Shen in the course of proving [**115**, Theorem 3.5]. Theorem 3.19 was first proved for divisible dimension groups, by Shen [**115**, Theorem 3.5], who also proved that in general the theorem would follow from an appropriate version of Proposition 3.16 [**115**, Theorem 3.1]. The full theorem was established by Effros, Handelman, and Shen [**30**, Theorem 2.2].

Propositions 3.22 and 3.23 are due to Handelman (unpublished).

CHAPTER 4

States

While the representation of dimension groups as direct limits of simplicial groups, developed in the previous chapter, does in one sense provide a complete description of dimension groups, it does not provide a description that makes all calculations within dimension groups easy, and of course it reveals nothing about perforated interpolation groups. A general method for investigating interpolation groups is to study the positive homomorphisms into $\mathbb{R}$. This can be done most effectively for a partially ordered abelian group (G, u) with order-unit, since in that case every nonzero positive homomorphism from G to $\mathbb{R}$ is a positive multiple of a normalized positive homomorphism from (G, u) to $(\mathbb{R}, 1)$. Such normalized positive homomorphisms are called states on (G, u), and sufficiently many states exist to provide a substantial amount of information about G. The set S of all states on (G, u) may be viewed as a compact convex subset of the product space $\mathbb{R}^G$, and the "double dual" object in this context, namely, the collection of all affine continuous real-valued functions on S, is a partially ordered real Banach space. The relationship between G and this "double dual" provides a very useful technique for investigating the structure of G.

In this chapter we introduce states and derive some of their basic properties, particularly existence, extension, and uniqueness theorems, descriptions of the ranges of values of states on given group elements, and, as a consequence, conditions under which information from the states yields inequalities in the group. In particular, the states completely determine the ordering in the group if and only if the group is archimedean. We leave the development of the "double dual" until after introducing the general theory of compact convex sets.

• Existence.

DEFINITION. Let (G, u) be a partially ordered abelian group with order-unit. A *state on* (G, u) is any normalized positive homomorphism from (G, u) to $(\mathbb{R}, 1)$, that is, any additive map $s \colon G \to \mathbb{R}$ such that $s(G^+) \subseteq \mathbb{R}^+$ and $s(u) = 1$.

For instance, let X be a compact topological space, let $G = C(X, \mathbb{R})$, and let u be the constant function 1, which is an order-unit in G. Then evaluation at any point of X defines a state on (G, u). Also, if μ is a nonnegative Borel measure on X such that $\mu(X) = 1$, then integration with respect to μ defines a state on (G, u).

The main technique for proving the existence of states on (G, u) involves extending states to G from subgroups of G that contain u. We first consider the corresponding problem of extending positive homomorphisms.

LEMMA 4.1. *Let G be a partially ordered abelian group, let H be a subgroup of G, and let $x \in G$. Let $f\colon H \to \mathbb{R}$ be a positive homomorphism, and set*

$$p = \sup\{f(y)/m \mid y \in H;\ m \in \mathbb{N};\ y \le mx\},$$
$$r = \inf\{f(z)/n \mid z \in H;\ n \in \mathbb{N};\ nx \le z\}.$$

(a) $-\infty \le p \le r \le \infty$.

(b) *If there exists a positive homomorphism $g\colon H + \mathbb{Z}x \to \mathbb{R}$ extending f, then $p \le g(x) \le r$.*

(c) *If there exists a real number q such that $p \le q \le r$, then f extends to a positive homomorphism $g\colon H + \mathbb{Z}x \to \mathbb{R}$ such that $g(x) = q$.*

PROOF. (a) If no element of $\mathbb{N}x$ lies above any element of H, then $p = -\infty$, while if no element of $\mathbb{N}x$ lies below any element of H, then $r = \infty$. In either of these cases, $p \le r$. If there exist $y, z \in H$ and $m, n \in \mathbb{N}$ such that $y \le mx$ and $nx \le z$, then $ny \le mnx \le mz$ and so $nf(y) \le mf(z)$, whence $f(y)/m \le f(z)/n$. Thus $p \le r$ in all cases.

(b) If $y \in H$ and $m \in \mathbb{N}$ with $y \le mx$, then $f(y) = g(y) \le mg(x)$, whence $f(y)/m \le g(x)$. Thus $p \le g(x)$. If $z \in H$ and $n \in \mathbb{N}$ with $nx \le z$, then $ng(x) \le g(z) = f(z)$, whence $g(x) \le f(z)/n$. Thus $g(x) \le r$.

(c) We claim that if $w \in H$ and $k \in \mathbb{Z}$ such that $w+kx \ge 0$, then $f(w)+kq \ge 0$. If $k = 0$, then $w \ge 0$, and so $f(w) + kq = f(w) \ge 0$, because f is positive. If $k > 0$, then as $-w \le kx$ we have $f(-w)/k \le p \le q$, whence $f(w) + kq \ge 0$. If $k < 0$, then as $(-k)x \le w$ we have $q \le r \le f(w)/(-k)$, and again $f(w)+kq \ge 0$. Thus the claim is proved.

In particular, if $w \in H$ and $k \in \mathbb{Z}$ with $w + kx = 0$, then since $w + kx \ge 0$ and $-w-kx \ge 0$, we obtain $f(w)+kq \ge 0$ and $f(-w)-kq \ge 0$, so that $f(w)+kq = 0$. Therefore f extends to a well-defined group homomorphism $g\colon H+\mathbb{Z}x \to \mathbb{R}$ such that $g(x) = q$. The claim above shows that g is positive. □

PROPOSITION 4.2. *Let G be a partially ordered abelian group, let H be a subgroup of G, and assume that each element of G is bounded above by an element of H. Then any positive homomorphism $f\colon H \to \mathbb{R}$ extends to a positive homomorphism $G \to \mathbb{R}$.*

PROOF. Let $\mathcal{K}$ be the collection of all pairs (K, g) such that K is a subgroup of G containing H and $g\colon K \to \mathbb{R}$ is a positive homomorphism extending f. For (K, g) and (K', g') in $\mathcal{K}$, define $(K, g) \le (K', g')$ if and only if $K \subseteq K'$ and g' extends g. By Zorn's Lemma, there exists a maximal element (K, g) in $\mathcal{K}$, and we need only show that $K = G$.

If there exists an element $x \in G - K$, set

$$p = \sup\{g(y)/m \mid y \in K;\ m \in \mathbb{N};\ y \leq mx\},$$
$$r = \inf\{g(z)/n \mid z \in K;\ n \in \mathbb{N};\ nx \leq z\}.$$

By hypothesis, there exist elements $u, v \in H$ such that $x \leq u$ and $-x \leq v$. Since $u \in K$ and $x \leq u$, we obtain $r \leq g(u) < \infty$. Since $-v \in K$ and $-v \leq x$, we obtain $p \geq g(-v) > -\infty$. According to Lemma 4.1, $p \leq r$, and so $p, r \in \mathbb{R}$. Hence, Lemma 4.1 shows that g extends to a positive homomorphism $h\colon K + \mathbb{Z}x \to \mathbb{R}$. But then $(K + \mathbb{Z}x, h)$ is an element of $\mathcal{K}$ strictly above (K, g), contradicting the maximality of (K, g).

Therefore $K = G$, as desired. □

Without the boundedness hypothesis, Proposition 4.2 can easily fail. For example, let G be the lexicographic direct product of $\mathbb{Z}$ with itself, let H be the subgroup $\{0\} \times \mathbb{Z}$ of G, and define a positive homomorphism $f\colon H \to \mathbb{R}$ so that $f(0,1) = 1$. Since $(0,n) \leq (1,0)$ for all $n \in \mathbb{Z}$, there is no way to extend f to a positive homomorphism $G \to \mathbb{R}$.

COROLLARY 4.3. *Let (G, u) be a partially ordered abelian group with order-unit, and let H be a subgroup of G such that $u \in H$. Then any state on (H, u) extends to a state on (G, u).*

PROOF. Given any $x \in G$, we have $x \leq nu$ for some $n \in \mathbb{N}$, and $nu \in H$. Thus Proposition 4.2 applies. □

COROLLARY 4.4. *Let (G, u) be a partially ordered abelian group with order-unit. There exists a state on (G, u) if and only if G is nonzero.*

PROOF. If s is a state on (G, u), then $s(u) = 1$ and so $u \neq 0$. Conversely, assume that G is nonzero. Since u is an order-unit, $u \neq 0$, and then $nu \geq u > 0$ for all positive integers n, while $mu \leq -u < 0$ for all negative integers m. Hence,

$$\mathbb{Z}u \cap G^+ = \mathbb{Z}^+ u,$$

and $ku \neq 0$ for all nonzero integers k. Thus there exists a group isomorphism $t\colon \mathbb{Z}u \to \mathbb{Z}$ such that $t(u) = 1$, and t is a state on $(\mathbb{Z}u, u)$. By Corollary 4.3, t extends to a state on (G, u). □

On a partially ordered abelian group with order-unit, the nonzero real-valued positive homomorphisms are just the positive multiples of states, as follows.

LEMMA 4.5. *Let (G, u) be a partially ordered abelian group with order-unit, and let $f\colon G \to \mathbb{R}$ be a positive homomorphism.*

(a) *If f is nonzero, then $f(u) > 0$, and $f = f(u)s$ for some state s on (G, u).*

(b) *If G is nonzero, then $f = \alpha s$ for some nonnegative real number α and some state s on (G, u).*

PROOF. (a) By assumption, $f(x) \neq 0$ for some $x \in G$. Choose $n \in \mathbb{N}$ such that $-nu \leq x \leq nu$, and observe that $-nf(u) \leq f(x) \leq nf(u)$, whence $f(u) > 0$. Now the map $s = f(u)^{-1}f$ is a state on (G, u), and $f = f(u)s$.

(b) If $f \neq 0$, apply (a). If $f = 0$, then $f = 0s$ for any state s on (G, u), and by Corollary 4.4 there is at least one state on (G, u). □

In the case of a nonzero interpolation group (G, u) with order-unit, we shall now determine that the relatively bounded real-valued homomorphisms on G are exactly the linear combinations of states on (G, u). As the relatively bounded real-valued homomorphisms on G form a Dedekind complete lattice-ordered real vector space (Corollary 2.28), we thus have the option of applying lattice-theoretic methods to the study of states on interpolation groups.

PROPOSITION 4.6. *Let (G, u) be a nonzero interpolation group with order-unit, and let f be a group homomorphism from G to $\mathbb{R}$. Then f is relatively bounded if and only if $f = \alpha s - \beta t$ for some $\alpha, \beta \in \mathbb{R}^+$ and some states s, t on (G, u).*

PROOF. As G has an order-unit, it is directed. Apply Proposition 2.25 and Lemma 4.5. □

• **Values of States.** Given a partially ordered abelian group (G, u) with order-unit, we use Lemma 4.1 to obtain a description of the set of values assumed by the states of (G, u) at any particular element of G. We then investigate the extent to which the values of states of (G, u) determine the ordering in G.

PROPOSITION 4.7. *Let (G, u) be a nonzero partially ordered abelian group with order-unit, let $x \in G$, and set*

$$p = \sup\{k/m \mid k \in \mathbb{Z};\ m \in \mathbb{N};\ ku \leq mx\},$$
$$r = \inf\{l/n \mid l \in \mathbb{Z};\ n \in \mathbb{N};\ nx \leq lu\}.$$

(a) $-\infty < p \leq r < \infty$.

(b) *If s is any state on (G, u), then $p \leq s(x) \leq r$.*

(c) *If q is any real number such that $p \leq q \leq r$, then there exists a state s on (G, u) such that $s(x) = q$.*

PROOF. Note that since G is nonzero, $ju < 0$ for all negative integers j.

(a) There exist positive integers k, l such that $-x \leq ku$ and $x \leq lu$, whence $p \geq -k > -\infty$ and $r \leq l < \infty$. Given $k, l \in \mathbb{Z}$ and $m, n \in \mathbb{N}$ such that $ku \leq mx$ and $nx \leq lu$, we have $knu \leq mnx \leq lmu$, and so $(lm - kn)u \geq 0$. Consequently, $lm - kn \geq 0$, and hence $k/m \leq l/n$. Thus $p \leq r$.

(b) If $k \in \mathbb{Z}$ and $m \in \mathbb{N}$ with $ku \leq mx$, then $k \leq ms(x)$ (because $s(u) = 1$), whence $k/m \leq s(x)$. Thus $p \leq s(x)$. If $l \in \mathbb{Z}$ and $n \in \mathbb{N}$ such that $nx \leq lu$, then $ns(x) \leq l$, whence $s(x) \leq l/n$. Thus $s(x) \leq r$.

(c) Set $H = \mathbb{Z}u$. Since $u > 0$, there exists a state f on (H, u), and we observe that

$$p = \sup\{f(y)/m \mid y \in H;\ m \in \mathbb{N};\ y \leq mx\},$$
$$r = \inf\{f(z)/n \mid z \in H;\ n \in \mathbb{N};\ nx \leq z\}.$$

By Lemma 4.1, f extends to a positive homomorphism $g\colon H+\mathbb{Z}x \to \mathbb{R}$ such that $g(x)=q$. Then g is a state on $(H+\mathbb{Z}x, u)$, and by Corollary 4.3, g extends to a state s on (G,u). $\square$

COROLLARY 4.8. *Let (G,u) be a nonzero partially ordered abelian group with order-unit, let $x \in G^+$, and set*

$$p' = \sup\{k/m \mid k \in \mathbb{Z}^+;\ m \in \mathbb{N};\ ku \le mx\},$$
$$r' = \inf\{l/n \mid l,n \in \mathbb{N} \text{ and } nx \le lu\}.$$

(a) $0 \le p' \le r' < \infty$.

(b) *If s is any state on (G,u), then $p' \le s(x) \le r'$.*

(c) *If q is any real number such that $p' \le q \le r'$, then there exists a state s on (G,u) such that $s(x)=q$.*

PROOF. Since $0u \le x$, we have $p' \ge 0$. For the rest, we need only show that p' and r' coincide with the values p and r defined in Proposition 4.7.

Obviously $p' \le p$. Now consider any $k \in \mathbb{Z}$ and $m \in \mathbb{N}$ such that $ku \le mx$. If $k \ge 0$, then $k/m \le p'$ by definition of p', while if $k<0$, then $k/m < 0 \le p'$. Thus $p \le p'$.

Obviously $r \le r'$. Now consider any $l \in \mathbb{Z}$ and $n \in \mathbb{N}$ such that $nx \le lu$. If $l>0$, then $r' \le l/n$ by definition of r'. If $l<0$, then $x \le nx \le lu < 0$, which is impossible. If $l=0$, then $x \le nx \le lu = 0$, whence $x=0$. In this case, $n'x = 0 < u$ for all $n' \in \mathbb{N}$, so that $r' \le 1/n'$ for all $n' \in \mathbb{N}$, and hence $r' \le 0 = l/n$. Therefore $r' \le l/n$ in all cases, proving that $r' \le r$. $\square$

PROPOSITION 4.9. *Let (G,u) be a nonzero partially ordered abelian group with order-unit, and let $x \in G$. Then $s(x) \ge 0$ for all states s on (G,u) if and only if there exist positive integers $m_1, m_2, \dots$ such that $m_n(nx+u) > 0$ for all $n \in \mathbb{N}$.*

PROOF. First assume that there exist positive integers $m_1, m_2, \dots$ such that $m_n(nx+u)>0$ for all n. Given any state s on (G,u), we then have

$$m_n(ns(x)+1) \ge 0$$

for all n, whence $s(x) \ge -1/n$ for all n, and thus $s(x) \ge 0$.

Conversely, assume that $s(x) \ge 0$ for all states s on (G,u). Set

$$p = \sup\{k/m \mid k \in \mathbb{Z};\ m \in \mathbb{N};\ ku \le mx\}.$$

By Proposition 4.7, there exists a state s on (G,u) such that $s(x)=p$, and hence $p \ge 0$. Now given any $n \in \mathbb{N}$, there must exist $k \in \mathbb{Z}$ and $m \in \mathbb{N}$ such that $ku \le mx$ and $k/m > -1/n$. Then $kn > -m$, whence $mnx \ge knu > -mu$, and therefore $m(nx+u) > 0$. $\square$

COROLLARY 4.10. *Let (G,u) be a nonzero unperforated partially ordered abelian group with order-unit, and let $x \in G$. Then $s(x) \ge 0$ for all states s on (G,u) if and only if $nx+u>0$ for all $n \in \mathbb{N}$.* $\square$

COROLLARY 4.11. *Let (G, u) be a nonzero unperforated partially ordered abelian group with order-unit, and let $y \in G^+$. Then $s(y) = 0$ for all states s on (G, u) if and only if $ny < u$ for all $n \in \mathbb{N}$.*

PROOF. Since $y \in G^+$, we have $s(y) = 0$ for all states s on (G, u) if and only if $s(-y) \geq 0$ for all states s on (G, u). By Corollary 4.10, this happens if and only if $-ny + u > 0$ for all $n \in \mathbb{N}$. □

A problem parallel to Proposition 4.9 is to characterize those elements of G at which all states have strictly positive values. This problem has a neater solution, and one which has more widespread usefulness.

THEOREM 4.12. *Let (G, u) be a partially ordered abelian group with order-unit, and let $x \in G$. Then $s(x) > 0$ for all states s on (G, u) if and only if there exists a positive integer m such that mx is an order-unit in G.*

PROOF. We may assume that G is nonzero. If mx is an order-unit for some $m \in \mathbb{N}$, then $kmx \geq u$ for some $k \in \mathbb{N}$. In this case, for any state s on (G, u) we have $kms(x) \geq 1$, and thus $s(x) > 0$.

Conversely, assume that $s(x) > 0$ for all states s on (G, u). In view of Proposition 4.7, there exist $k \in \mathbb{Z}$ and $m \in \mathbb{N}$ such that $ku \leq mx$ and $k/m > 0$. Then $k > 0$ and $mx \geq ku \geq u > 0$. For any element $y \in G$, there exists a positive integer n such that $y \leq nu$, whence $y \leq n(mx)$. Therefore mx is an order-unit in G. □

COROLLARY 4.13. *Let (G, u) be an unperforated partially ordered abelian group with order-unit, and let $x \in G$. Then $s(x) > 0$ for all states s on (G, u) if and only if x is an order-unit in G.*

PROOF. If mx is an order-unit in G for some $m \in \mathbb{N}$, then it follows from the inequality $mx \geq 0$ that $x \geq 0$. Hence, we conclude that x is an order-unit. □

In particular, Theorem 4.12 and Corollary 4.13 show that if x is an element in a nonzero partially ordered abelian group G with an order-unit u, satisfying $s(x) > 0$ for all states s on (G, u), then $mx > 0$ for some $m \in \mathbb{N}$, and further $x > 0$ in the case that G is unperforated. Thus the values of the states on (G, u) exert considerable influence on the ordering in G. The family of states on (G, u) precisely determines the ordering in G if and only if G is archimedean, as follows.

THEOREM 4.14. *For a partially ordered abelian group (G, u) with order-unit, the following conditions are equivalent:*

(a) *$G^+ = \{x \in G \mid s(x) \geq 0$ for all states s on $(G, u)\}$.*

(b) *G is isomorphic (as an ordered group) to a subgroup of $\mathbb{R}^X$, for some nonempty set X.*

(c) *G is archimedean.*

PROOF. We may assume that G is nonzero.

(a)⇒(b): Let X be the set of states on (G, u) (then X is nonempty by Corollary 4.4), and let $\phi\colon G \to \mathbb{R}^X$ be the evaluation map, so that $\phi(x)(s) = s(x)$

for all $x \in G$ and $s \in X$. By (a), for any $x, y \in G$ we have $x \leq y$ if and only if $\phi(x) \leq \phi(y)$. In particular, it follows that ϕ is injective. Thus ϕ provides an isomorphism of G onto the subgroup $\phi(G) \subseteq \mathbb{R}^X$ (as ordered groups).

(b)⇒(c): This is clear.

(c)⇒(a): By definition, $s(x) \geq 0$ for all $x \in G^+$ and all states s on (G, u). Conversely, consider an element $x \in G$ such that $s(x) \geq 0$ for all states s on (G, u). Since G is unperforated (by Proposition 1.24), Corollary 4.10 shows that $nx + u > 0$ for all $n \in \mathbb{N}$. Then $n(-x) < u$ for all $n \in \mathbb{N}$, whence $-x \leq 0$ by the archimedean assumption, and therefore $x \in G^+$. □

Thus any archimedean partially ordered abelian group with order-unit possesses enough states to completely determine the ordering. One might also expect any archimedean partially ordered abelian group without an order-unit to possess enough positive homomorphisms into $\mathbb{R}$ to determine the ordering. However, there exist nonzero archimedean partially ordered abelian groups that have no nonzero positive homomorphisms into $\mathbb{R}$ (Example 9.6).

• **Uniqueness.** With further applications of Proposition 4.7 and Corollary 4.8, we can derive criteria for a partially ordered abelian group with order-unit to have exactly one state, as follows.

LEMMA 4.15. *Let (G, u) be a partially ordered abelian group with order-unit, and let s, t be states on (G, u). If either $s \leq^+ t$ or $t \leq^+ s$, then $s = t$.*

PROOF. By symmetry, we may assume that $s \leq^+ t$. Set $f = t - s$, and observe that f is a positive homomorphism from G to $\mathbb{R}$ such that $f(u) = 0$. Given any $x \in G$, we have $-nu \leq x \leq nu$ for some $n \in \mathbb{N}$, whence

$$0 = f(-nu) \leq f(x) \leq f(nu) = 0$$

and so $f(x) = 0$. Therefore $f = 0$ and $s = t$. □

PROPOSITION 4.16. *Let (G, u) be a nonzero partially ordered abelian group with order-unit. For all $x \in G^+$, set*

$$f^*(x) = \inf\{l/n \mid l, n \in \mathbb{N} \text{ and } nx \leq lu\},$$
$$f_*(x) = \sup\{k/m \mid k \in \mathbb{Z}^+;\ m \in \mathbb{N};\ ku \leq mx\}.$$

Then the following conditions are equivalent:

(a) *There is a unique state on (G, u).*

(b) $f^* = f_*$. *(In this case, f^* extends to a state on (G, u).)*

(c) f^* *is an additive map.*

(d) f_* *is an additive map.*

PROOF. There is at least one state on (G, u), by Corollary 4.4.

(a)⇒(b): Given any $x \in G^+$, Corollary 4.8 shows that there exist states s, t on (G, u) such that $s(x) = f^*(x)$ and $t(x) = f_*(x)$. As $s = t$, we obtain $f^*(x) = f_*(x)$.

(b)⇒(a): Consider any states s, t on (G, u). For any $x \in G^+$, Corollary 4.8 shows that

$$f_*(x) \quad \leq \quad \begin{matrix} s(x) \\ t(x) \end{matrix} \quad \leq \quad f^*(x),$$

and thus $s(x) = t(x) = f^*(x)$. Thus s, t, f^* all agree on G^+. In particular, the states s and t are extensions of f^*. As G is directed, we also have $s = t$.

(b)⇒(c): Since f^* extends to a group homomorphism, it must be additive.

(c)⇒(a): Observe that $f^*(0) = 0$. Then, as f^* is additive, it extends to a group homomorphism $f\colon G \to \mathbb{R}$. By Corollary 4.8, $f^*(x) \geq 0$ for all $x \in G^+$, so that f is a positive homomorphism. Since $nu \not\leq lu$ for any positive integers $n > l$, we see that $f^*(u) = 1$. Thus f is a state on (G, u).

Now consider an arbitrary state s on (G, u), and note from Corollary 4.8 that $s(x) \leq f^*(x)$ for all $x \in G^+$, that is, $s \leq^+ f$. By Lemma 4.15, $s = f$. Thus f is the unique state on (G, u).

(b)⇒(d)⇒(a): These implications are proved in the same manner as the implications (b)⇒(c)⇒(a). □

COROLLARY 4.17. *Let (G, u) be a nonzero totally ordered abelian group with order-unit. Then there exists a unique state s on (G, u), and*

$$\begin{aligned} s(x) &= \inf\{l/n \mid l, n \in \mathbb{N} \text{ and } nx \leq lu\} \\ &= \sup\{k/m \mid k \in \mathbb{Z}^+;\ m \in \mathbb{N};\ ku \leq mx\} \end{aligned}$$

for all $x \in G^+$.

PROOF. Define f^* and f_* as in Proposition 4.16. If there is more than one state on (G, u), then by the proposition, $f^*(x) \neq f_*(x)$ for some $x \in G^+$. In view of Corollary 4.8, $0 \leq f_*(x) < f^*(x)$, and hence we may choose $k, n \in \mathbb{N}$ such that

$$f_*(x) < k/n < f^*(x).$$

Since $k/n < f^*(x)$, we have $nx \not\leq ku$, and consequently $ku \leq nx$. But then $k/n \leq f_*(x)$, a contradiction.

Therefore there is a unique state s on (G, u). By Proposition 4.16, the restriction of s to G^+ coincides with f^*, which equals f_*. This yields the desired formulas for s. □

• Additional Uniqueness Criteria.

THEOREM 4.18. *For a nonzero partially ordered abelian group (G, u) with order-unit, the following conditions are equivalent:*

(a) *There exists a unique state on (G, u).*

(b) *Given any positive integers $a > b$ and any elements $x, y \in G^+$ for which $x + y$ is an order-unit, there exists a positive integer n such that either $nbx < nay$ or $nby < nax$.*

(c) *There exist positive integers $a \geq b$ such that given any element $x \in G^+$ and any positive integer l, there is some $n \in \mathbb{N}$ for which either $nbx \leq na(lu)$ or $nb(lu) \leq nax$.*

PROOF. (a)⇒(b): Let s be the unique state on (G, u). As $x + y$ is an order-unit, $s(x + y) > 0$. Thus either $s(x) > 0$ or $s(y) > 0$; say $s(x) > 0$.

If $bs(x) < as(y)$, then $s(ay - bx) > 0$. In this case, Theorem 4.12 shows that $n(ay - bx) > 0$ for some $n \in \mathbb{N}$. If $bs(x) \geq as(y)$, then since $s(x) > 0$ and $s(y) \geq 0$, we have

$$as(x) > bs(x) \geq as(y) \geq bs(y),$$

and hence $s(ax - by) > 0$. In this case, $n(ax - by) > 0$ for some $n \in \mathbb{N}$, by a second application of Theorem 4.12.

(b)⇒(c): For instance, choose $a = 2$ and $b = 1$. Given $x \in G^+$ and $l \in \mathbb{N}$, the sum $x + lu$ is an order-unit, and so condition (b) applies.

(c)⇒(a): If there exist distinct states s, t on (G, u), choose an element $z \in G$ such that $s(z) \neq t(z)$. If necessary, interchange s and t, so that $s(z) > t(z)$. Since

$$s(z + iu) = s(z) + i > t(z) + i = t(z + iu)$$

for any $i \in \mathbb{N}$, we may replace z by any $z + iu$. Hence, there is no loss of generality in assuming that $z \in G^+$. Choose $j \in \mathbb{N}$ such that

$$j(s(z) - t(z)) > 2a^2/b^2,$$

and then replace z by jz. Thus we may assume, without loss of generality, that

$$s(z) - t(z) > 2a^2/b^2.$$

Now let k be the largest nonnegative integer such that $mku \leq mz$ for some $m \in \mathbb{N}$. (Since $z \leq k_0 u$ for some $k_0 \in \mathbb{N}$, we must have $k \leq k_0$.) Set

$$x = mz - mku,$$

observing that $x \in G^+$ and

$$s(x) \geq s(x) - t(x) = ms(z) - mt(z) > 2a^2 m/b^2.$$

Define f^* and f_* as in Proposition 4.16. If $f_*(x) > m$, then there exist $k' \in \mathbb{Z}^+$ and $m' \in \mathbb{N}$ such that $k'u \leq m'x$ and $k'/m' > m$. Then $mm' < k'$, whence

$$mm'u < k'u \leq m'x = mm'z - mm'ku$$

and so $mm'(k+1)u < mm'z$, contradicting the maximality of k. Thus $f_*(x) \leq m$.

As $0 < b/a \leq 1$, there exists a positive integer l such that

$$m < bl/a \leq m + 1 \leq 2m,$$

and we note that $l \leq 2am/b$. According to (c), there is some $n \in \mathbb{N}$ such that either $nbx \leq nalu$ or $nblu \leq nax$. If $nblu \leq nax$, then

$$f_*(x) \geq bl/a > m,$$

which is false. If $nbx \le nalu$, then

$$f^*(x) \le al/b \le 2a^2m/b^2.$$

Since $s(x) \le f^*(x)$ (by Corollary 4.8) and $s(x) > 2a^2m/b^2$, this too is false.

Therefore there is a unique state on (G, u). □

An alternate version of Theorem 4.18, in which the given conditions need only be checked within the interval $[0, u]$, may be proved for interpolation groups, as follows.

THEOREM 4.19. *Let (G, u) be a nonzero interpolation group with order-unit. Then there exists a unique state on (G, u) if and only if there exist positive integers $a \ge b$ such that:*

(∗) *Given any elements $u_1, u_2 \in G^+$ with $u_1 + u_2 = u$, there is some positive integer n for which either $nbu_1 \le nau_2$ or $nbu_2 \le nau_1$.*

PROOF. If there is a unique state on (G, u), then (∗) holds for any positive integers $a > b$, by Theorem 4.18. Conversely, assume that we are given positive integers $a \ge b$ for which (∗) holds, and suppose that there exist distinct states s_1, s_2 on (G, u).

Let B denote the group of all relatively bounded homomorphisms from G to $\mathbb{R}$. As G is a directed interpolation group, Corollary 2.28 shows that B is a lattice-ordered real vector space. Since $s_1 \ne s_2$, we have $s_1 \not\le^+ s_2$ and $s_2 \not\le^+ s_1$ (by Lemma 4.15), whence each $s_i >^+ s_1 \wedge s_2$. Set

$$t_i = s_i - (s_1 \wedge s_2)$$

for $i = 1, 2$. Then t_1, t_2 are nonzero positive homomorphisms, and $t_1 \wedge t_2 = 0$ because of Proposition 1.4. As t_i is a positive homomorphism, $\ker(t_i)$ is a convex subgroup of G, and hence $u \notin \ker(t_i)$ (because $\ker(t_i) \ne G$). Thus each $t_i(u) > 0$. Set

$$\varepsilon = b(\min\{t_1(u), t_2(u)\})/(a+b) > 0.$$

Since $(t_1 \wedge t_2)(u) = 0$, Corollary 2.28 says that $u = u_1 + u_2$ for some u_1, u_2 in G^+ such that $t_1(u_1) + t_2(u_2) < \varepsilon$. Hence, each $t_i(u_i) < \varepsilon$. According to (∗), there is some $n \in \mathbb{N}$ for which either $nbu_1 \le nau_2$ or $nbu_2 \le nau_1$. If $nbu_1 \le nau_2$, then

$$nbu = nbu_1 + nbu_2 \le n(a+b)u_2,$$

and so $t_2(u_2) \ge bt_2(u)/(a+b) \ge \varepsilon$, which is impossible. Similarly, the inequality $nbu_2 \le nau_1$ also leads to a contradiction.

Therefore there is a unique state on (G, u). □

The criteria developed in Theorems 4.18 and 4.19 for there to be only one state on (G, u) may be viewed as weak forms of the condition that G be totally ordered. Certainly if G is totally ordered, then there is a unique state on (G, u) (Corollary 4.17), but it is not necessary for G to be totally ordered in order to have a unique state, as the following example shows.

EXAMPLE 4.20. *There exists a dimension group (G, u) with order-unit such that (G, u) has only one state but G is not totally ordered.*

PROOF. Make the abelian group $H = \mathbb{Z}$ into a partially ordered abelian group using the discrete ordering, and let G be the lexicographic direct product of $\mathbb{Q}$ with H. Since $\mathbb{Q}$ has strict interpolation and H has interpolation, G has interpolation by Proposition 2.10. Since $\mathbb{Q}$ is directed, so is G. Finally, it is clear that G is unperforated and thus that G is a dimension group. Obviously G is not totally ordered.

Set $u = (1, 0)$, and note that u is an order-unit in G. Consider any state s on (G, u). For $k \in \mathbb{Z}$ and $n \in \mathbb{N}$, we have $n(k/n, 0) = ku$ and so $ns(k/n, 0) = k$, whence $s(k/n, 0) = k/n$. In other words, $s(a, 0) = a$ for all $a \in \mathbb{Q}$. For any $b \in H$, we have $-u \leq n(0, b) \leq u$ for all $n \in \mathbb{N}$, whence $-1 \leq ns(0, b) \leq 1$ for all $n \in \mathbb{N}$, and so $s(0, b) = 0$. Thus $s(a, b) = a$ for all $(a, b) \in G$, and therefore s is unique. □

• **Discrete States.** One factor influencing the behavior of a state s on a partially ordered abelian group (G, u) with order-unit is its image $s(G)$, which is an additive subgroup of $\mathbb{R}$. As the following lemma shows, there are two very different types of additive subgroups of $\mathbb{R}$, and we shall find it important to know which type of subgroup the image of a state is.

LEMMA 4.21. *Any additive subgroup of $\mathbb{R}$ is either a cyclic group or a dense subset of $\mathbb{R}$.*

PROOF. Suppose that G is an additive subgroup of $\mathbb{R}$ which is not dense in $\mathbb{R}$. Then G cannot contain arbitrarily small positive real numbers, and so there exists a positive real number ε such that the open interval $(0, \varepsilon)$ is disjoint from G. We claim that there is a least positive real number in G. If not, there exist positive real numbers $a_1 > a_2 > \cdots$ that all lie in G. However, each of the positive real numbers $a_i - a_{i+1}$ lies in G, and all but finitely many of them must be less than ε, which is impossible.

Thus there does exist a least positive real number $a \in G$. Any real number b may be expressed in the form $na + c$ where $n \in \mathbb{Z}$ and $0 \leq c < a$. If $b \in G$, then also $c \in G$, whence $c = 0$ by the minimality of a, and hence $b = na$. Therefore $G = \mathbb{Z}a$. □

Any cyclic additive subgroup of $\mathbb{R}$ is a discrete space in the relative topology. Hence, such subgroups are sometimes called "discrete subgroups" of $\mathbb{R}$, and we carry over this terminology for use with states.

DEFINITION. Let (G, u) be a partially ordered abelian group with order-unit. A state s on (G, u) is called a *discrete state* if and only if $s(G)$ is a cyclic subgroup of $\mathbb{R}$.

Note that if s is a discrete state on (G, u), then since the number $1 = s(u)$ lies in $s(G)$, a generator for $s(G)$ must be a reciprocal of an integer. Hence, $s(G) = (1/n)\mathbb{Z}$ for some positive integer n. (We prefer to write $(1/n)\mathbb{Z}$ rather than $\mathbb{Z}(1/n)$ for the group generated by $1/n$ to avoid confusion with the ring $\mathbb{Z}[1/n]$ consisting of rational numbers whose denominators are powers of n.)

For example, if $G = \mathbb{Z}^2$ and $u = (m,n)$ for some positive integers m and n, then we may define states s and t on (G,u) by the rules

$$s(x,y) = x/m \quad \text{and} \quad t(x,y) = y/n.$$

These states are both discrete, since $s(G) = (1/m)\mathbb{Z}$ and $t(G) = (1/n)\mathbb{Z}$. However, not all states on (G,u) are discrete. For instance, we may define a state r on (G,u) by the rule

$$r(x,y) = (x + \alpha y)/(m + n\alpha),$$

where α is a positive irrational number, and r is an indiscrete state, because its image

$$r(G) = (m + n\alpha)^{-1}\mathbb{Z} + \alpha(m + n\alpha)^{-1}\mathbb{Z}$$

is a dense subgroup of $\mathbb{R}$.

Arguments involving discrete states on an interpolation group can often be reduced to the case of a simplicial group, because of the following result.

PROPOSITION 4.22. *Let (G,u) be an interpolation group with order-unit, let s be a state on (G,u), and set*

$$H = \{x - y \mid x, y \in \ker(s)^+\}.$$

Then H is an ideal of G. If s is a discrete state, then G/H is a simplicial group.

PROOF. Since $\ker(s)$ is a convex subgroup of G, Proposition 1.9 shows that H is an ideal of G. Thus G/H is an interpolation group. The coset $v = u + H$ is an order-unit in G/H, and s induces a state t on G/H, where $t(x + H) = s(x)$ for all $x \in G$. Note that

$$\ker(t)^+ = \{0\},$$

by the definition of H. In particular, it follows that whenever $a, b \in G/H$ with $a > b$, then $t(a - b) > 0$, and hence $t(a) > t(b)$.

Now assume that s is discrete. Then

$$t(G/H) = s(G) = (1/m)\mathbb{Z}$$

for some $m \in \mathbb{N}$. If there is an infinite descending chain $a_1 > a_2 > \cdots$ in $(G/H)^+$, then we obtain an infinite descending chain

$$t(a_1) > t(a_2) > \cdots$$

in $(1/m)\mathbb{Z}^+$, which is impossible. Thus $(G/H)^+$ satisfies the DCC. As G/H is also an interpolation group with an order-unit, we conclude from Corollary 3.14 that G/H is simplicial. □

• **Notes.** The use of the term "state" can be traced back to quantum mechanics. States on partially ordered abelian groups with order-unit were introduced as natural generalizations of states on partially ordered real vector spaces with order-unit, which in turn generalized states on C^*-algebras. In a unital C^*-algebra A, the set A_{sa} of self-adjoint elements can be made into a partially ordered real vector space with positive cone $\{a^*a \mid a \in A\}$ and order-unit 1. By the Gelfand-Naimark representation theory, the states on $(A_{\text{sa}}, 1)$ are exactly the functionals of the form $a \mapsto (\phi(a)x, x)$, where ϕ is a *-algebra representation of A on a Hilbert space H and x is a unit-vector in H. In classical quantum mechanics, the unit-vectors $x \in H$ correspond to physical states of a physical system, the self-adjoint operators b on H correspond to observable quantities, and the inner product (bx, x) measures the expected value for the observable b with the system in the state x.

Proposition 4.2 was first proved for partially ordered normed linear spaces $H \subseteq G$ such that H contains an interior point of G^+, by Krein [**85**, Theorem I]. A special case was proved by Kantorovitch [**79**, Lemma, p. 535], namely, the case in which G is the space of all bounded real-valued functions on the open interval $(0, 1)$ and H is any linear subspace of G that contains 1; however, Kantorovitch's proof is valid for any partially ordered real vector spaces $H \subseteq G$ such that H contains an order-unit of G. A more general version of the proposition, for partially ordered abelian semigroups with zero, was proved by Aumann [**7**, Satz 3]. Corollary 4.3 was established, independently of the partially ordered vector space antecedents, by the author and Handelman [**56**, Theorem 3.2], who used it to obtain Corollary 4.4 [**56**, Corollary 3.3].

Proposition 4.7 is a generalization of Corollary 4.8, which was proved by the author and Handelman [**56**, Lemma 4.1], along with Proposition 4.9 [**56**, Lemma 6.1]. Theorem 4.12 is a slight generalization of Corollary 4.13, which is due to Effros, Handelman, and Shen [**30**, Theorem 1.4]. The implication (c)$\Rightarrow$(a) in Theorem 4.14 was proved by the author, Handelman, and Lawrence [**60**, Proposition I.5.3]. Proposition 4.16, and a slightly weaker version of Theorem 4.18, was proved by the author and Handelman [**56**, Proposition 4.2 and Theorem 4.3]. Theorem 4.19 is the interpolation group version of a result first proved for pseudo-rank functions on unit-regular rings, by the author [**49**, Theorem 18.16]. Proposition 4.22 was established by the author and Handelman in the course of proving [**57**, Lemma 4.4].

CHAPTER 5

Compact Convex Sets

To make the most effective use of states on a partially ordered abelian group (G, u) with order-unit, the collection of all states on (G, u) should be endowed with some kind of structure. This can be done by viewing the states as members of the space $\mathbb{R}^G$ of all real-valued functions on G. When $\mathbb{R}^G$ is equipped with the product topology, it becomes a linear topological space, and the collection of all states on (G, u) becomes a compact convex subset of $\mathbb{R}^G$. This allows us to apply the general theory of compact convex sets to the study of states on (G, u).

With this purpose in mind, we set aside partially ordered abelian groups for a period, in order to present some of the basic theory of compact convex sets. We first outline a natural category of convex sets and a natural category of compact convex sets, discussing isomorphisms, inverse limits, products, and coproducts in these categories. Then we turn to the notions of faces and extreme points of convex sets. (Roughly speaking, a face of a convex set is a convex subset that is closed under convex decompositions, and an extreme point of a convex set is a point that is indecomposable with respect to convex combinations.) Some work on the problem of separating disjoint compact convex sets by closed hyperplanes is required in order to prove the existence of extreme points. This is the famous Krein-Mil′man Theorem, which may be viewed as showing that most compact convex sets are "generated" (as compact convex sets) by their extreme points. At the end of the chapter, we investigate the general properties of a specific but ubiquitous compact convex set, namely, the set of all probability measures on a compact Hausdorff space.

In several later chapters, we shall return to the general theory of compact convex sets and develop further portions of that theory, as needed for application to state spaces.

• Basic Definitions.

DEFINITION. Let E be a real vector space. A *convex combination* of points $x_1, \dots, x_n$ in E is any linear combination of the form $\alpha_1 x_1 + \cdots + \alpha_n x_n$ such that each $\alpha_i \geq 0$ and $\alpha_1 + \cdots + \alpha_n = 1$. If all the $\alpha_i > 0$, this is called a *positive convex combination.* A *convex set* in E is any subset of E which is closed under convex combinations. The *convex hull* of a subset $X \subseteq E$ is the smallest convex subset of E that contains X.

Several observations concerning these definitions are easily proved. First, a set $K \subseteq E$ is convex if and only if $\alpha x + (1-\alpha)y \in K$ for all $x, y \in K$ and all real numbers $\alpha \in [0,1]$, that is, if and only if K contains all line segments whose endpoints are in K. Second, the convex hull of a set $X \subseteq E$ is just the collection of all convex combinations of points of X. Finally, if $K_1, \ldots, K_n$ are convex subsets of E, then the convex hull of $K_1 \cup \cdots \cup K_n$ is just the collection of all convex combinations of the form $\alpha_1 x_1 + \cdots + \alpha_n x_n$ where each $x_i \in K_i$. We leave the proofs of these statements as exercises for the reader.

DEFINITION. Let E be a real vector space. An *affine combination* of points $x_1, \ldots, x_n$ in E is any linear combination of the form $\alpha_1 x_1 + \cdots + \alpha_n x_n$ such that $\alpha_1 + \cdots + \alpha_n = 1$. An *affine subspace* of E is any subset of E that is closed under affine combinations. The *affine span* of a subset $X \subseteq E$ is the smallest affine subspace of E that contains X.

Note that the affine subspaces of E are just the translates of linear subspaces of E, that is, subsets of the form $L + x$ where $x \in E$ and L is a linear subspace of E.

DEFINITION. Let E be a real vector space. A subset $X \subseteq E$ is *affinely dependent* if some point in X can be expressed as an affine combination of other points of X. Otherwise, X is *affinely independent.* If A is an affine subspace of E, the *dimension* of A is the vector space dimension of any linear subspace of E of which A is a translate. The *dimension* of any convex set $K \subseteq E$ is the dimension of the affine span of K.

Given $y \in X \subseteq E$, we observe that X is affinely independent if and only if the set $\{x - y \mid x \in X\}$ is linearly independent. Note that points $x_1, \ldots, x_n$ in E are affinely dependent if and only if, after possible renumbering, some convex combination of $x_1, \ldots, x_k$ equals some convex combination of $x_{k+1}, \ldots, x_n$. If A is a finite-dimensional affine subspace of E, then the dimension of A is exactly one less than the cardinality of any maximal affinely independent subset of A.

DEFINITION. Let E_1 and E_2 be real vector spaces, and for $i = 1, 2$ let K_i be a convex subset of E_i. A map $f\colon K_1 \to K_2$ is said to be *affine* if and only if f preserves convex combinations. If f is also a bijection (in which case f^{-1} is an affine map), f is called an *affine isomorphism.* The convex sets K_1 and K_2 are *affinely isomorphic* if and only if there exists an affine isomorphism of K_1 onto K_2.

Given an affine map $f\colon E_1 \to E_2$ between real vector spaces, it is an easy exercise to check that $f - f(0)$ (i.e., the map sending any $x \in E_1$ to $f(x) - f(0)$) is a linear map. Hence, f is linear if and only if $f(0) = 0$. Moreover, it follows from the linearity of $f - f(0)$ that f preserves affine combinations.

DEFINITION. By *the category of convex sets* we mean the category whose objects are all convex subsets of real vector spaces and whose morphisms are all affine maps between these objects.

Note that isomorphisms in this category are exactly the affine isomorphisms defined above.

Most of the convex sets we shall work with will lie in real linear topological spaces. As we shall have no need of complex linear topological spaces, we omit the adjective "real".

DEFINITION. By *the category of compact convex sets* we shall mean the category whose objects are all compact convex subsets of *Hausdorff* linear topological spaces and whose morphisms are all affine continuous maps between these objects. An isomorphism in this category is any affine isomorphism which is also a homeomorphism; such maps are called *affine homeomorphisms*. Hence, to abbreviate the statement that objects K and L are isomorphic in this category, we say that K and L are *affinely homeomorphic*.

PROPOSITION 5.1. *In a linear topological space, the closure of any convex set is convex.*

PROOF. Let K be a convex subset of a linear topological space, let x and y be points of the closure $\overline{K}$, and let α and β be nonnegative real numbers whose sum is 1. The points x and y may be expressed as limits of nets $\{x_i\}$ and $\{y_j\}$ of points of K, and there is no loss of generality in assuming that these nets have the same index set. Then $\{\alpha x_i + \beta y_i\}$ is a net of points of K converging to $\alpha x + \beta y$, and hence $\alpha x + \beta y$ lies in $\overline{K}$. □

PROPOSITION 5.2. *If $K_1, \dots, K_n$ are compact convex subsets of a linear topological space, then the convex hull of $K_1 \cup \dots \cup K_n$ is compact.*

PROOF. Let K denote the convex hull of $K_1 \cup \dots \cup K_n$, and set

$$S = \{(\alpha_1, \dots, \alpha_n) \in \mathbb{R}^n \mid \text{ each } \ \alpha_i \geq 0 \text{ and } \alpha_1 + \dots + \alpha_n = 1\}.$$

Define a map $f\colon S \times K_1 \times \dots \times K_n \to K$ according to the rule

$$f((\alpha_1, \dots, \alpha_n), x_1, \dots, x_n) = \alpha_1 x_1 + \dots + \alpha_n x_n,$$

and observe that f is a continuous surjection. As $S \times K_1 \times \dots \times K_n$ is compact, we conclude that K must be compact. □

• **Categorical Concepts.** Products in the category of convex sets may be constructed as cartesian products. Namely, given a convex set K_i in a real vector space E_i for each index i in some index set I, the cartesian product $\prod K_i$ is a convex subset of the product vector space $\prod E_i$, and $\prod K_i$ together with the natural projection maps is a product for the family $\{K_i \mid i \in I\}$ in the category of convex sets. Inverse limits in this category may be constructed in the same manner, as convex subsets of cartesian products. Coproducts will be constructed using the following notion.

DEFINITION. Let K be a convex set, and let $\{K_i \mid i \in I\}$ be a nonempty family of convex subsets of K. We say that K is the *direct convex sum* of the

K_i provided

(a) K equals the convex hull of $\bigcup_{i\in I} K_i$;

(b) Whenever $i(1),\ldots,i(n)$ are distinct indices in I and

$$\alpha_1 x_1 + \cdots + \alpha_n x_n = \beta_1 y_1 + \cdots + \beta_n y_n$$

are convex combinations in K with $x_j, y_j \in K_{i(j)}$ for $j=1,\ldots,n$, then $\alpha_j = \beta_j$ for all j and $x_j = y_j$ for those j such that $\alpha_j > 0$.

Roughly speaking, K is the direct convex sum of the K_i if and only if every element of K may be expressed, as uniquely as possible, as a convex combination of points from the K_i.

PROPOSITION 5.3. *Let $\{K_i \mid i \in I\}$ be a nonempty family of convex sets. Then there exists a convex set K together with injective affine maps $q_i\colon K_i \to K$ for all $i \in I$ such that K is the direct convex sum of the $q_i(K_i)$.*

PROOF. Each K_i is a convex subset of a real vector space E_i. Set $E'_i = E_i \times \mathbb{R}$ for each $i \in I$, and set $E = \prod_{i\in I} E'_i$. For each $j \in I$, define a map $q_j\colon K_j \to E$ such that $(q_j(x))_j = (x,1)$ for all $x \in K_j$ while $(q_j(x))_i = 0$ for all $x \in K_j$ and all $i \neq j$, and observe that q_j is affine. If K is the convex hull of $\bigcup_{i\in I} q_i(K_i)$, the required properties are clear. $\square$

PROPOSITION 5.4. *Let $\{K_i \mid i\in I\}$ be a nonempty family of convex sets, let K be a convex set, and for each $i \in I$ let $q_i\colon K_i \to K$ be an affine map. Then K together with the q_i is a coproduct for the K_i in the category of convex sets if and only if each q_i is injective and K is the direct convex sum of the $q_i(K_i)$.*

PROOF. First assume that each q_i is injective and that K is the direct convex sum of the $q_i(K_i)$. Consider a convex set L and affine maps $r_i\colon K_i \to L$ for each $i \in I$. Given a point $x \in K$, choose a convex combination

$$x = \alpha_1 q_{i(1)}(x_1) + \cdots + \alpha_n q_{i(n)}(x_n)$$

where the $i(j)$ are distinct elements of I and each $x_j \in K_{i(j)}$, and set

$$r(x) = \alpha_1 r_{i(1)}(x_1) + \cdots + \alpha_n r_{i(n)}(x_n) \in L.$$

That $r(x)$ is well defined follows from our assumptions on K and the q_i. Given a positive convex combination $\alpha x + \alpha' x'$ in K, we may choose convex combinations

$$x = \alpha_1 q_{i(1)}(x_1) + \cdots + \alpha_n q_{i(n)}(x_n), \qquad x' = \alpha'_1 q_{i(1)}(x'_1) + \cdots + \alpha'_n q_{i(n)}(x'_n)$$

where the $i(j)$ are distinct elements of I while also $x_j, x'_j \in K_{i(j)}$ and $\alpha_j + \alpha'_j > 0$ for all j. Set $\beta_j = \alpha\alpha_j + \alpha'\alpha'_j$ for each j, and note that $\beta_j > 0$. If $\gamma_j = \alpha\alpha_j\beta_j^{-1}$ and $\gamma'_j = \alpha'\alpha'_j\beta_j^{-1}$ for each j, then

$$\begin{aligned}
\alpha x + \alpha' x' &= \sum_{j=1}^{n} (\alpha\alpha_j q_{i(j)}(x_j) + \alpha'\alpha'_j q_{i(j)}(x'_j)) \\
&= \sum_{j=1}^{n} \beta_j(\gamma_j q_{i(j)}(x_j) + \gamma'_j q_{i(j)}(x'_j)) = \sum_{j=1}^{n} \beta_j q_{i(j)}(\gamma_j x_j + \gamma'_j x'_j)
\end{aligned}$$

and each $\gamma_j x_j + \gamma_j' x_j' \in K_{i(j)}$. Also, $\beta_1 + \cdots + \beta_n = 1$, and so the last summation is a convex combination. Hence,

$$\begin{aligned} r(\alpha x + \alpha' x') &= \sum_{j=1}^{n} \beta_j r_{i(j)}(\gamma_j x_j + \gamma_j' x_j') = \sum_{j=1}^{n} (\beta_j \gamma_j r_{i(j)}(x_j) + \beta_j \gamma_j' r_{i(j)}(x_j')) \\ &= \sum_{j=1}^{n} \alpha \alpha_j r_{i(j)}(x_j) + \sum_{j=1}^{n} \alpha' \alpha_j' r_{i(j)}(x_j') = \alpha r(x) + \alpha' r(x'), \end{aligned}$$

proving that r is an affine map. It is clear that $rq_i = r_i$ for all $i \in I$, and that r is the only affine map from K to L with this property. Thus $\{K, q_i\}$ is a coproduct for the K_i.

Conversely, assume that $\{K, q_i\}$ is a coproduct for the K_i. By Proposition 5.3, there exists a convex set K' together with injective affine maps $q_i' \colon K_i \to K'$ for all $i \in I$ such that K' is the direct convex sum of the $q_i'(K_i)$. Using the portion of the proposition already proved, $\{K', q_i'\}$ is a coproduct for the K_i. Consequently, by the uniqueness properties of coproducts, there exists an affine isomorphism $r \colon K' \to K$ such that $rq_i' = q_i$ for all i, from which we conclude that each q_i is injective and that K is the direct convex sum of the $q_i(K_i)$. □

The existence of arbitrary coproducts in the category of convex sets is given by Propositions 5.3 and 5.4.

In the category of compact convex sets, products may be constructed as cartesian products equipped with the product topology. Similarly, inverse limits may be constructed as compact convex subsets of cartesian products.

PROPOSITION 5.5. *Let $K_1, \dots, K_n$ be compact convex subsets of Hausdorff linear topological spaces $E_1, \dots, E_n$. Then there exists a compact convex set K in a Hausdorff linear topological space E together with injective affine continuous maps $q_i \colon K_i \to K$ for $i = 1, \dots, n$ such that K is the direct convex sum of $q_1(K_1), \dots, q_n(K_n)$.*

PROOF. Construct a convex set K and injective affine maps q_i as in the proof of Proposition 5.3. Then K is the direct convex sum of the $q_i(K_i)$. If $E_1', \dots, E_n', E$ are given the product topologies, then E is a Hausdorff linear topological space and each q_i is continuous. As the K_i are compact, so are the $q_i(K_i)$. By Proposition 5.2, K is compact. □

PROPOSITION 5.6. *Let $K_1, \dots, K_n, K$ be compact convex subsets of Hausdorff linear topological spaces, and for each $i \in I$ let $q_i \colon K_i \to K$ be an affine continuous map. Then K together with the q_i is a coproduct for the K_i in the category of compact convex sets if and only if $q_1, \dots, q_n$ are all injective and K is the direct convex sum of $q_1(K_1), \dots, q_n(K_n)$.*

PROOF. If the q_i are injective and K is the direct convex sum of the $q_i(K_i)$, then by Proposition 5.4 $\{K, q_i\}$ is at least a coproduct for the K_i in the category of convex sets. Hence, given a compact convex set L (in a Hausdorff linear topological space) and affine continuous maps $r_i \colon K_i \to L$ for each i, there exists a unique affine map $r \colon K \to L$ such that $rq_i = r_i$ for all i.

We claim that r is continuous. Set

$$S = \{(\alpha_1, \dots, \alpha_n) \in \mathbb{R}^n \mid \text{ each } \alpha_i \geq 0 \text{ and } \alpha_1 + \cdots + \alpha_n = 1\},$$

let $p \colon S \times K_1 \times \cdots \times K_n \to K$ be the map given by the rule

$$p((\alpha_1, \dots, \alpha_n), x_1, \dots, x_n) = \alpha_1 q_1(x_1) + \cdots + \alpha_n q_n(x_n),$$

and observe that p is a continuous surjection. The map rp may be expressed as the composition of the maps

$$S \times K_1 \times \cdots \times K_n \xrightarrow{f} S \times L^n \xrightarrow{g} L$$

given by the rules

$$f((\alpha_1, \dots, \alpha_n), x_1, \dots, x_n) = ((\alpha_1, \dots, \alpha_n), r_1(x_1), \dots, r_n(x_n)),$$
$$g((\alpha_1, \dots, \alpha_n), y_1, \dots, y_n) = \alpha_1 y_1 + \cdots + \alpha_n y_n.$$

As f and g are continuous, rp is continuous. Consequently, if X is any closed subset of L, then $p^{-1}r^{-1}(X)$ is a closed subset of $S \times K_1 \times \cdots \times K_n$ and hence is compact. Since p is a continuous surjection, $r^{-1}(X)$ equals the compact set $p(p^{-1}r^{-1}(X))$, whence $r^{-1}(X)$ is closed in K (because K is Hausdorff). Thus r is continuous, as claimed.

Therefore $\{K, q_i\}$ is a coproduct for the K_i in the category of compact convex sets. The converse follows as in Proposition 5.4, using Proposition 5.5 in place of Proposition 5.3. □

From Propositions 5.5 and 5.6 we obtain the existence of finite coproducts in the category of compact convex sets.

To illustrate the difference between products and coproducts in the category of compact convex sets, consider the unit interval $[0, 1]$ in $\mathbb{R}$. The product of $[0, 1]$ with itself is affinely homeomorphic to a square, while the coproduct of $[0, 1]$ with itself is affinely homeomorphic to a tetrahedron.

• Extreme Points and Faces.

DEFINITION. An *extreme point* of a convex set K is any point $x \in K$ which does not lie in the interior of any line segment with endpoints in K. In other words, given any convex combination $x = \alpha_1 x_1 + \alpha_2 x_2$ with $x_1, x_2 \in K$, we must have $\alpha_1 = 1$ or $\alpha_2 = 1$ or $x_1 = x_2 = x$. The collection of all extreme points of K is called the *extreme boundary* of K, denoted $\partial_e K$.

For example, if K is a plane convex polygon, then the extreme points of K are just its vertices. In this case $\partial_e K$ is finite, and K equals the convex hull of $\partial_e K$. If K is a closed disc in the plane, then $\partial_e K$ is the topological boundary of K, which is a circle. In this case $\partial_e K$ is infinite, but again K equals the convex hull of $\partial_e K$. If K is an open disc in the plane, then K has no extreme points, and so K does not equal the convex hull of $\partial_e K$.

For each of these examples, $\partial_e K$ is compact. For an example in which $\partial_e K$ is not compact, let K be the convex hull in $\mathbb{R}^3$ of the set

$$X = \{(1,0,1),(1,0,-1)\} \cup \{(x,y,0) \in \mathbb{R}^3 \mid x^2+y^2=1\}.$$

In this case $\partial_e K = X - \{(1,0,0)\}$, which is not compact.

DEFINITION. A *face* of a convex set K is a convex subset F of K (possibly empty) such that any line segment with endpoints in K whose interior meets F must be contained in F. In other words, given any positive convex combination $x = \alpha_1 x_1 + \alpha_2 x_2$ with $x \in F$ and $x_1, x_2 \in K$, we must have $x_1, x_2 \in F$. Note that the intersection of any family of faces of K is a face of K. Hence, given any subset $X \subseteq K$ there is a smallest face of K that contains X, called the *face generated by X in K*.

Note that if $x \in K$, then $\{x\}$ is a face of K if and only if x is an extreme point of K. Note also that any face of a face of K is a face of K. In particular, if F is any face of K, then $\partial_e F = F \cap \partial_e K$.

If $f \colon K \to L$ is an affine map between convex sets, then for any face F of L, the inverse image $f^{-1}(F)$ is a face of K. In particular, $f^{-1}(\{x\})$ is a face of K for any $x \in \partial_e K$. Hence, if L is a closed interval $[a,b]$ in $\mathbb{R}$, then $f^{-1}(\{a\})$ and $f^{-1}(\{b\})$ are faces of K.

If K is a plane convex polygon, then K has four types of faces. One type consists of the empty set, another type consists of K itself, and a third type consists of the singletons $\{x\}$ where x is a vertex of K. The remaining faces of K are the line segments connecting pairs of adjacent vertices of K. For comparison, note that a closed disc in the plane has only three of these types of faces, since it has no one-dimensional faces.

PROPOSITION 5.7. *Let X be a subset of a convex set K. Let F be the set of those points $x \in K$ for which there exists a positive convex combination $\alpha x + \beta y = z$ with $y \in K$ and z in the convex hull of X. Then F equals the face generated by X in K.*

PROOF. Let Y denote the convex hull of X, and note that any face G of K that contains X must contain Y. Given $x \in F$, there exists a positive convex combination $\alpha x + \beta y = z$ with $y \in K$ and $z \in Y$. Then $z \in G$, and hence $x \in G$ (because G is a face), proving that $F \subseteq G$. Thus F is contained in all faces of K that contain X. It remains to show that F is a face of K containing X. Given any $x \in X$, the positive convex combination $\frac{1}{2}x + \frac{1}{2}x$ lies in Y, whence $x \in F$. Thus $X \subseteq F$.

Now consider a convex combination $x = \alpha_1 x_1 + \alpha_2 x_2$ with each $x_i \in F$. If either $\alpha_1 = 0$ or $\alpha_2 = 0$, then obviously $x \in F$; hence, we may assume that each $\alpha_i > 0$. For each i, there exists a positive convex combination $\beta_i x_i + \gamma_i y_i = z_i$ with $y_i \in K$ and $z_i \in Y$. Since $0 < \beta_i < 1$, each $\beta_i^{-1} > 1$, and consequently $\alpha_1\beta_1^{-1} + \alpha_2\beta_2^{-1} > 1$. Set

$$\alpha = (\alpha_1\beta_1^{-1} + \alpha_2\beta_2^{-1})^{-1},$$

and note that $0 < \alpha < 1$. Since $\alpha\alpha_1\beta_1^{-1} + \alpha\alpha_2\beta_2^{-1} = 1$, the point

$$z = \alpha\alpha_1\beta_1^{-1}z_1 + \alpha\alpha_2\beta_2^{-1}z_2$$

lies in Y. Set $\beta = 1 - \alpha$. We have

$$\begin{aligned}\beta^{-1}\alpha(\alpha_1\beta_1^{-1}\gamma_1 + \alpha_2\beta_2^{-1}\gamma_2) &= \beta^{-1}\alpha(\alpha_1\beta_1^{-1}(1-\beta_1) + \alpha_2\beta_2^{-1}(1-\beta_2)) \\ &= \beta^{-1}\alpha(\alpha_1\beta_1^{-1} + \alpha_2\beta_2^{-1} - \alpha_1 - \alpha_2) \\ &= \beta^{-1}\alpha(\alpha^{-1} - 1) = \beta^{-1}(1-\alpha) = 1,\end{aligned}$$

and hence the point

$$y = \beta^{-1}\alpha(\alpha_1\beta_1^{-1}\gamma_1 y_1 + \alpha_2\beta_2^{-1}\gamma_2 y_2)$$

lies in K. Inasmuch as

$$\begin{aligned}\alpha x + \beta y &= \alpha(\alpha_1 x_1 + \alpha_2 x_2) + \alpha(\alpha_1\beta_1^{-1}\gamma_1 y_1 + \alpha_2\beta_2^{-1}\gamma_2 y_2) \\ &= \alpha(\alpha_1\beta_1^{-1}(\beta_1 x_1 + \gamma_1 y_1) + \alpha_2\beta_2^{-1}(\beta_2 x_2 + \gamma_2 y_2)) \\ &= \alpha(\alpha_1\beta_1^{-1}z_1 + \alpha_2\beta_2^{-1}z_2) = z,\end{aligned}$$

it follows that $x \in F$. Thus F is a convex subset of K.

Finally, consider a positive convex combination $\alpha_1 x_1 + \alpha_2 x_2 = x$ such that $x_1, x_2 \in K$ and $x \in F$. Then there exists a positive convex combination $\alpha x + \beta y = z$ with $y \in K$ and $z \in Y$. Set $\gamma = 1 - \alpha\alpha_1$, and note that $0 < \gamma < 1$. Since

$$\gamma^{-1}\alpha\alpha_2 + \gamma^{-1}\beta = \gamma^{-1}\alpha(1-\alpha_1) + \gamma^{-1}(1-\alpha) = \gamma^{-1}(1-\alpha\alpha_1) = 1,$$

the point $w = \gamma^{-1}\alpha\alpha_2 x_2 + \gamma^{-1}\beta y$ lies in K. In addition,

$$\alpha\alpha_1 x_1 + \gamma w = \alpha\alpha_1 x_1 + \alpha\alpha_2 x_2 + \beta y = \alpha x + \beta y = z,$$

whence $x_1 \in F$. Similarly, $x_2 \in F$.

Therefore F is a face of K. □

COROLLARY 5.8. *Let K be a convex set, and let $z \in K$. Then the face generated by z in K consists precisely of those points $x \in K$ for which there exists a positive convex combination $\alpha x + \beta y = z$ with $y \in K$.* □

• **Separation by Hyperplanes.** In order to deal effectively with compact convex sets, we need to guarantee a large supply of extreme points, as well as a large supply of affine continuous real-valued functions. This can be achieved by separation results of the following kind: given disjoint compact convex sets K and L in a linear topological space E, find a continuous linear functional f on E and a real number α such that $f \ll \alpha$ on K and $f \gg \alpha$ on L. If such f and α exist, then K is said to lie on the "negative side" of the hyperplane $f^{-1}(\{\alpha\})$ while L is said to lie on the "positive side", and K and L have been separated by this hyperplane. In order to prove the existence of sufficiently many continuous linear functionals to separate disjoint compact convex sets in this manner, we shall have to assume that the topology on E has a basis consisting of open convex sets. However, this assumption is not needed for the preliminary steps.

We begin by setting up a situation in which continuous linear functionals may be obtained from the Hahn-Banach Theorem. Recall that a *sublinear functional* on a real vector space E is a map $p\colon E \to \mathbb{R}$ such that $p(x+y) \le p(x) + p(y)$ for all $x, y \in E$ and $p(\alpha x) = \alpha p(x)$ for all $x \in E$ and $\alpha \in \mathbb{R}^+$. (In particular, $p(0) = 0$.) For instance, the norm on a real normed linear space E is a sublinear functional on E.

PROPOSITION 5.9. *Let A be an open convex subset of a linear topological space E, such that $0 \in A$. For all $x \in E$, set*

$$p(x) = \inf\{\alpha \in \mathbb{R} \mid \alpha > 0 \text{ and } \alpha^{-1}x \in A\}.$$

Then p is a nonnegative, continuous, sublinear functional on E such that

$$A = \{x \in E \mid p(x) < 1\} \quad \text{and} \quad \overline{A} = \{x \in E \mid p(x) \le 1\}.$$

PROOF. Given $x \in E$, the sequence $x, \frac{1}{2}x, \frac{1}{3}x, \ldots$ converges to 0, whence $\frac{1}{n}x$ lies in A for some $n \in \mathbb{N}$, and so $p(x) \le n$. Thus $0 \le p(x) < \infty$. If λ is a positive real number, then the set of positive real numbers α such that $\alpha^{-1}(\lambda x) \in A$ coincides with the set of real numbers of the form $\lambda\beta$, where β is a positive real number such that $\beta^{-1}x \in A$, and hence $p(\lambda x) = \lambda p(x)$. This relation also holds for $\lambda = 0$, since $p(0) = 0$.

Given $x, y \in E$ and a positive real number ε, there exist positive real numbers α, β such that $\alpha^{-1}x$ and $\beta^{-1}y$ are in A while also

$$\alpha < p(x) + \varepsilon/2 \quad \text{and} \quad \beta < p(y) + \varepsilon/2.$$

Set $\gamma = \alpha + \beta > 0$, and observe that

$$\gamma^{-1}(x+y) = (\alpha\gamma^{-1})(\alpha^{-1}x) + (\beta\gamma^{-1})(\beta^{-1}y),$$

which lies in A because A is convex. As a result,

$$p(x+y) \le \gamma < p(x) + p(y) + \varepsilon.$$

Thus $p(x+y) \le p(x) + p(y)$, proving that p is a nonnegative sublinear functional on E.

Consider a point $x \in A$. Since the sequence $2x, \frac{3}{2}x, \frac{4}{3}x, \ldots$ converges to x, we must have $((n+1)/n)x$ in A for some $n \in \mathbb{N}$. Hence, $p(x) \le n/(n+1) < 1$. Conversely, consider a point $y \in E$ such that $p(y) < 1$. Then there exists a positive real number $\alpha < 1$ such that $\alpha^{-1}y \in A$. Since y may now be written as a convex combination

$$y = \alpha(\alpha^{-1}y) + (1-\alpha)0$$

with $\alpha^{-1}y$ and 0 in A, it follows that $y \in A$. Therefore

$$A = \{x \in E \mid p(x) < 1\}.$$

Given $x \in E$ with $p(x) \le 1$, there exists a sequence $\alpha_1, \alpha_2, \ldots$ of positive real numbers such that each $\alpha_n < (n+1)/n$ and each $\alpha_n^{-1}x \in A$. For each n, set $\beta_n = \max\{1, \alpha_n\}$. Then $\beta_n^{-1}x$ is a convex combination of $\alpha_n^{-1}x$ and 0, whence

$\beta_n^{-1}x \in A$. The β_n converge to 1 and the points $\beta_n^{-1}x$ converge to x, so that $x \in \overline{A}$. Conversely, consider a point $y \in \overline{A}$. Given any positive real number ε, the set

$$U = \{u \in E \mid -\varepsilon^{-1}u \in A\} = -\varepsilon A$$

is an open neighborhood of 0 in E, and so $y + U$ is an open neighborhood of y. Hence, there exists $z \in U$ such that $y + z \in A$. Then we may write the point $(1+\varepsilon)^{-1}y$ as a convex combination

$$(1+\varepsilon)^{-1}y = (1+\varepsilon)^{-1}(y+z) + \varepsilon(1+\varepsilon)^{-1}(-\varepsilon^{-1}z)$$

with $y+z$ and $-\varepsilon^{-1}z$ in A, whence $(1+\varepsilon)^{-1}y \in A$, and so $p(y) \leq 1+\varepsilon$. Therefore

$$\overline{A} = \{x \in E \mid p(x) \leq 1\}.$$

It remains to prove that p is continuous. Thus consider the sets

$$V = \{v \in E \mid p(v) < \lambda\} \quad \text{and} \quad W = \{w \in E \mid p(w) > \lambda\},$$

for any $\lambda \in \mathbb{R}$. If $\lambda \leq 0$, then V is empty, while if $\lambda > 0$, then $V = \lambda A$. In either case, V is open. If $\lambda < 0$, then $W = E$, while if $\lambda > 0$, then $W = \lambda(E - \overline{A})$. If $\lambda = 0$, then

$$W = \bigcup_{n=1}^{\infty} \left\{ w \in E \mid p(w) > \frac{1}{n} \right\} = \bigcup_{n=1}^{\infty} \frac{1}{n}(E - \overline{A}).$$

Thus W is open in all cases. Therefore p is continuous. □

The sublinear functional p constructed in Proposition 5.9 is called *the Minkowski functional associated with A*. For present purposes, the continuity of p and the description of $\overline{A}$ in terms of p are not needed, but this information will prove useful later.

PROPOSITION 5.10. *Let A be an open convex subset of a linear topological space E, and let B be an affine subspace of E. If A and B are disjoint, then there exists an affine continuous map $f\colon E \to \mathbb{R}$ such that $f \ll 1$ on A while $f = 1$ on B.*

PROOF. We may obviously assume that there exists a point $a \in A$ and that B is nonempty. The translate $A - a$ is an open convex subset of E containing the origin, and the translate $B - a$ is an affine subspace of E that is disjoint from $A - a$. If there exists an affine continuous map $h\colon E \to \mathbb{R}$ such that $h \ll 1$ on $A - a$ and $h = 1$ on $B - a$, then the rule $f(x) = h(x-a)$ defines an affine continuous map $f\colon E \to \mathbb{R}$ such that $f \ll 1$ on A and $f = 1$ on B.

Thus there is no loss of generality in assuming that $0 \in A$. Define the nonnegative sublinear functional p on E as in Proposition 5.9.

Now $B = L + b$ for some linear subspace L in E and some point $b \in B$. Since A and B are disjoint, $0 \notin B$, and so $b \notin L$. Hence, there exists a linear functional g on the subspace $M = L + \mathbb{R}b$ such that $g = 0$ on L and $g(b) = 1$. We claim that $g \leq p$ on M.

Consider $x \in M$, and write $x = y + \alpha b$ for some $y \in L$ and $\alpha \in \mathbb{R}$. If $\alpha \leq 0$, then

$$g(x) = \alpha \leq 0 \leq p(x).$$

Now assume that $\alpha > 0$. Since the point $\alpha^{-1}x = \alpha^{-1}y + b$ lies in B, it is not in A, and hence $p(\alpha^{-1}x) \geq 1$. Thus

$$p(x) = \alpha p(\alpha^{-1}x) \geq \alpha = g(x)$$

in this case. Therefore $g \leq p$ on M, as claimed.

By the Hahn-Banach Theorem, g extends to a linear functional f on E such that $f \leq p$. Then $f \leq p \ll 1$ on A, while $f = 1$ on B because $f = g = 0$ on L and $f(b) = g(b) = 1$. Since $f(x) < 1 = f(b)$ for all $x \in A$, the open set $A - b$ is disjoint from the kernel of f. Consequently, $\ker(f)$ is not dense in E and so must be closed in E. Therefore f is continuous. □

LEMMA 5.11. *Any nonzero continuous linear functional f on a linear topological space E is an open map.*

PROOF. Choose a point $x \in E$ such that $f(x) = 1$, and define a linear map $g\colon \mathbb{R} \to E$ such that $g(1) = x$. Then g is a continuous linear map, and fg equals the identity map on $\mathbb{R}$. Given any open set U in E, it follows from the choice of g that

$$f(U) = g^{-1}(U + \ker(f)).$$

Since $U + \ker(f)$ is a union of translates of U, it is open in E, and therefore $g^{-1}(U + \ker(f))$ is open in $\mathbb{R}$. Thus $f(U)$ is open. □

THEOREM 5.12. *Let A and B be disjoint nonempty open convex subsets of a linear topological space E. Then there exist $\alpha \in \mathbb{R}$ and a continuous linear functional f on E such that $f \ll \alpha$ on A and $f \gg \alpha$ on B.*

PROOF. Set $C = \{a - b \mid a \in A \text{ and } b \in B\}$. Since C is a union of translates of A, it is open. It is clear that C is convex, and $0 \notin C$ because A and B are disjoint.

Applying Proposition 5.10 to the open convex set C and the affine subspace $\{0\}$, we obtain an affine continuous function $g\colon E \to \mathbb{R}$ such that $g \ll 1$ on C and $g(0) = 1$. Then the rule $f(x) = g(x) - 1$ defines a continuous linear functional f on E such that $f \ll 0$ on C. Since C is nonempty, f is nonzero.

Now $f(a) < f(b)$ for all $a \in A$ and $b \in B$, so that every number in $f(A)$ is less than every number in $f(B)$. Since $f(A)$ and $f(B)$ are open subsets of $\mathbb{R}$ (by Lemma 5.11), we conclude that there exists $\alpha \in \mathbb{R}$ such that $f(a) < \alpha < f(b)$ for all $a \in A$ and $b \in B$. □

To obtain from Theorem 5.12 the corresponding result for disjoint nonempty compact convex sets A and B, we need to fit A and B inside disjoint open convex sets. This requires an assumption on the existence of open convex sets in E, as follows.

DEFINITION. A linear topological space E is said to be *locally convex* provided the collection of all open convex subsets of E forms a basis for the topology on E. We use the phrase *locally convex space* as an abbreviation for "locally convex linear topological space". Similarly, we use the phrase *locally convex Hausdorff space* as an abbreviation for "locally convex Hausdorff linear topological space".

For example, any normed linear space E is a locally convex Hausdorff space, since every open ball in E is an open convex subset. Also, for any set X the product space $\mathbb{R}^X$ (with the product topology) is a locally convex Hausdorff space.

LEMMA 5.13. *Let A and B be disjoint convex subsets of a locally convex Hausdorff space E. If A is closed and B is compact, there exist disjoint open convex sets A' and B' in E such that $A \subseteq A'$ and $B \subseteq B'$.*

PROOF. Let $\mathcal{N}$ denote the collection of all open neighborhoods of 0 in E. For each $b \in B$, choose $U_b \in \mathcal{N}$ such that $b + U_b$ is disjoint from A. By continuity of addition, we may then choose $V_b \in \mathcal{N}$ such that $V_b + V_b \subseteq U_b$. The open sets $b + V_b$ cover B, and hence, by compactness, there exist finitely many points $b(1), \ldots, b(n)$ in B such that

$$B \subseteq [b(1) + V_{b(1)}] \cup \cdots \cup [b(n) + V_{b(n)}].$$

Set $V = V_{b(1)} \cap \cdots \cap V_{b(n)}$. We claim that $B + V$ is disjoint from A. Given points $b \in B$ and $v \in V$, we have $b = b(i) + y$ for some i and some point $y \in V_{b(i)}$. Since

$$y + v \in V_{b(i)} + V \subseteq V_{b(i)} + V_{b(i)} \subseteq U_{b(i)},$$

the point $b + v$ lies in $b(i) + U_{b(i)}$, whence $b + v \notin A$. Therefore $B + V$ is disjoint from A, as claimed.

By continuity of subtraction, there is some $W \in \mathcal{N}$ such that $x - y \in V$ for all $x, y \in W$. Since $B + V$ is disjoint from A, it follows that $B + W$ and $A + W$ are disjoint. As E is locally convex, W contains a convex open neighborhood C of 0. Then $A + C$ and $B + C$ are disjoint convex subsets of E such that $A \subseteq A + C$ and $B \subseteq B + C$. Each of the sets $A + C$ and $B + C$ is a union of translates of the open set C, and therefore $A + C$ and $B + C$ are open. □

THEOREM 5.14. *Let A and B be disjoint nonempty convex subsets of a locally convex Hausdorff space E. If A is closed and B is compact, there exists a continuous linear functional f on E such that*

$$\sup\{f(x) \mid x \in A\} < \inf\{f(y) \mid y \in B\}.$$

PROOF. In view of Lemma 5.13 and Theorem 5.12, there exist $\alpha \in \mathbb{R}$ and a continuous linear functional f on E such that $f \ll \alpha$ on A and $f \gg \alpha$ on B. Then

$$\sup\{f(x) \mid x \in A\} \leq \alpha < \inf\{f(y) \mid y \in B\},$$

because B is compact. □

COROLLARY 5.15. *If E is a locally convex Hausdorff space, there exist enough continuous linear functionals on E to separate the points of E.* □

• Existence of Extreme Points.

LEMMA 5.16. *Let K be a compact convex subset of a linear topological space, let $f\colon K \to \mathbb{R}$ be an affine continuous function, and set*

$$m = \sup\{f(x) \mid x \in K\}.$$

Then $f^{-1}(\{m\})$ is a nonempty closed face of K.

PROOF. Since K is compact, m is finite and $f^{-1}(\{m\})$ is a nonempty closed subset of K. We may view f as an affine map of K into $(-\infty, m]$. As m is an extreme point of $(-\infty, m]$, we conclude that $f^{-1}(\{m\})$ must be a face of K. □

THEOREM 5.17 [KREIN-MIL'MAN THEOREM]. *If K is a compact convex subset of a locally convex Hausdorff space E, then K equals the closure of the convex hull of $\partial_e K$.*

PROOF. Let $\mathcal{F}$ denote the collection of all nonempty closed faces of K. If $\{F_i \mid i \in I\}$ is any nonempty chain in $\mathcal{F}$, then $\bigcap F_i$ is certainly a closed face of K. Any intersection of finitely many of the F_i is nonempty, and so $\bigcap F_i$ is nonempty because K is compact. Hence, every nonempty chain in $\mathcal{F}$ has a lower bound in $\mathcal{F}$. By Zorn's Lemma (going downward!), it follows that any member of $\mathcal{F}$ contains a minimal member of $\mathcal{F}$.

We claim that any minimal F in $\mathcal{F}$ must have the form $\{w\}$ for some w in $\partial_e K$. If there exist distinct points $x, y \in F$, then by Corollary 5.15 there exists a continuous linear functional f on E such that $f(x) < f(y)$. Set

$$m = \sup\{f(z) \mid z \in F\}.$$

Since the restriction of f to F is an affine continuous function, Lemma 5.16 shows that $F \cap f^{-1}(\{m\})$ is a nonempty closed face of F, and thus a nonempty closed face of K. However, since $F \cap f^{-1}(\{m\})$ does not contain the point x, it is properly contained in F, contradicting the minimality of F. Therefore F must be a singleton, say $F = \{w\}$, and then $w \in \partial_e K$ because F is a face of K. This establishes the claim.

Set J equal to the closure of the convex hull of $\partial_e K$, so that J is a compact convex subset of K. If there exists a point v in $K - J$, then by Theorem 5.14 there exists a continuous linear functional g on E such that

$$\sup\{g(x) \mid x \in J\} < g(v).$$

Set $n = \sup\{g(y) \mid y \in K\}$. By Lemma 5.16, $K \cap g^{-1}(\{n\})$ is a nonempty closed face of K. As observed above, $K \cap g^{-1}(\{n\})$ must contain a minimal member of $\mathcal{F}$, and this face must have the form $\{w\}$ for some $w \in \partial_e K$. Then $g(w) = n$. On the other hand, since $w \in \partial_e K \subseteq J$ we also have $g(w) < g(v) \leq n$, which is a contradiction. Therefore $J = K$. □

We may view the Krein-Mil′man Theorem as saying that any compact convex subset K of a locally convex Hausdorff space E is "generated" (as a compact convex set) by its extreme boundary, since K is the smallest compact convex subset of E that contains $\partial_e K$. As a result, many questions concerning the behavior of K may be settled by examining the behavior of its extreme points. The following corollaries provide several examples of such a procedure.

COROLLARY 5.18. *Let K be a compact convex subset of a locally convex Hausdorff space, and let F be a closed face of K. Then F equals the closure of the convex hull of $F \cap \partial_e K$.*

PROOF. Since F is a compact convex set, it equals the closure of the convex hull of $\partial_e F$. In addition, $\partial_e F = F \cap \partial_e K$ because F is a face of K. □

Conversely, if K is a compact convex set, it might be asked whether the closure of the convex hull of any subset of $\partial_e K$ must be a face of K. In general, the answer is negative, as may be seen in the case that K is a square in the plane.

COROLLARY 5.19. *Let K be a compact convex subset of a locally convex Hausdorff space, and let $f\colon K \to \mathbb{R}$ be an affine continuous function. Then there exist extreme points x and y in K such that*

$$f(x) = \sup\{f(z) \mid z \in K\} \quad \text{and} \quad f(y) = \inf\{f(z) \mid z \in K\}.$$

PROOF. If $m = \sup\{f(z) \mid z \in K\}$, then by Lemma 5.16 $f^{-1}(\{m\})$ is a nonempty closed face of K. In view of Corollary 5.18, there must exist a point x in $f^{-1}(\{m\}) \cap \partial_e K$. The extreme point y may be obtained by applying this argument to the affine continuous function $-f$. □

COROLLARY 5.20. *Let K be a compact convex subset of a locally convex Hausdorff space, and let $f, g\colon K \to \mathbb{R}$ be affine continuous functions.*

(a) *If $f \le g$ on $\partial_e K$, then $f \le g$.*

(b) *If $f \ll g$ on $\partial_e K$, then $f \ll g$.*

(c) *If $f = g$ on $\partial_e K$, then $f = g$.*

PROOF. The affine continuous function $f - g$ attains its maximum value at some point x in $\partial_e K$, by Corollary 5.19. Hence, if $f \le g$ on $\partial_e K$, then $f(x) - g(x) \le 0$, and so $f - g \le 0$. Likewise, if $f \ll g$ on $\partial_e K$, then $f(x) - g(x) < 0$, and so $f - g \ll 0$. This proves (a) and (b). Condition (c) follows automatically from (a). □

The most effective uses of the Krein-Mil′man Theorem of course occur in applications to compact convex sets whose extreme boundaries are known. The following theorem, which is a partial converse to the Krein-Mil′man Theorem, is an aid in identifying extreme boundaries.

THEOREM 5.21. *Let K be a compact convex subset of a locally convex Hausdorff space E, and let $X \subseteq K$. If K equals the closure of the convex hull of X, then $\partial_e K$ is contained in the closure of X.*

PROOF. Without loss of generality, we may assume that X is closed in K. Assume that there exists an extreme point y of K that does not lie in X.

Given any $x \in X$, Corollary 5.15 provides us with a continuous linear functional f on E such that $f(x) < f(y)$. Hence, there exist real numbers α such that x lies in the open set $f^{-1}(-\infty, \alpha)$ while $f(y) > \alpha$. As X is compact, it can be covered by finitely many open sets of this form. Thus there exist continuous linear functionals $f_1, \ldots, f_n$ on E and real numbers $\alpha_1, \ldots, \alpha_n$ such that

$$X \subseteq \bigcup_{i=1}^{n} f_i^{-1}(-\infty, \alpha_i)$$

and each $\alpha_i < f_i(y)$.

Set $H_i = K \cap f_i^{-1}(-\infty, \alpha_i]$ for each $i = 1, \ldots, n$, and note that H_i is a compact convex subset of K. If H is the convex hull of $H_1 \cup \cdots \cup H_n$, then Proposition 5.2 shows that H is compact. Hence, as H contains X, it must contain the closure of the convex hull of X, so that $H = K$. Thus $y \in H$. Consequently, y is a convex combination of points from $H_1 \cup \cdots \cup H_n$. Since y is an extreme point of K, it follows that $y \in H_i$ for some i. But then $f_i(y) \leq \alpha_i$, which is false.

Therefore $\partial_e K \subseteq X$. □

• **Probability Measures.** In order to have some convenient examples of infinite-dimensional compact convex sets to keep in mind, we discuss the compact convex set of all probability measures on a compact Hausdorff space.

DEFINITION. Recall that a *probability measure* on a compact Hausdorff space X is any nonnegative regular Borel measure μ on X such that $\mu(X) = 1$. We use $M_1^+(X)$ to denote the set of all probability measures on X.

The Riesz Representation Theorem allows us to identify the set $M(X)$ of all finite signed regular Borel measures on X with the dual of the real Banach space $C(X, \mathbb{R})$, by setting

$$\mu(f) = \int_X f \, d\mu$$

for all $\mu \in M(X)$ and all $f \in C(X, \mathbb{R})$. (We adopt the same notation for any real-valued function f on X that is integrable with respect to μ.) For our purposes, the most convenient topology on the dual space $C(X, \mathbb{R})^*$ is the weak* topology. Recall that in this topology, a net $\{p_\alpha\}$ in $C(X, \mathbb{R})^*$ converges to an element p if and only if $p_\alpha(f) \to p(f)$ for all $f \in C(X, \mathbb{R})$.

PROPOSITION 5.22. *Let X be a compact Hausdorff space. With the weak* topology, $C(X, \mathbb{R})^*$ is a locally convex Hausdorff space, and $M_1^+(X)$ is a compact convex subset of $C(X, \mathbb{R})^*$.*

PROOF. Set $E = C(X, \mathbb{R})$, and note that E^* is clearly Hausdorff in the weak* topology. The natural basic open sets for this topology, namely,

$$\{q \in E^* : |q(f_i) - p(f_i)| < \varepsilon \text{ for each } i = 1, \ldots, n\}$$

(where $p \in E^*$ and $f_1, \ldots, f_n \in E$, while ε is a positive real number), are convex subsets of E^*. Thus the weak* topology on E^* is locally convex.

It is clear that $M_1^+(X)$ is a convex subset of E^*. Observe that $M_1^+(X)$ equals the intersection of the sets

$$\{p \in E^* \mid p(1) = 1\} \quad \text{and} \quad \{p \in E^* \mid p(f) \geq 0\}$$

(for all those $f \in E$ satisfying $f \geq 0$), each of which is weak*-closed in E^*. Hence, $M_1^+(X)$ is weak*-closed in E^*. Any function f in the unit ball of E satisfies $-1 \leq f \leq 1$, whence $-1 \leq p(f) \leq 1$ for each $p \in M_1^+(X)$. As a result, $M_1^+(X)$ is contained in the unit ball of E^*, which is weak*-compact by the Banach-Alaoglu Theorem. Therefore $M_1^+(X)$ is weak*-compact. □

Given any compact Hausdorff space X, we shall always assume that $M_1^+(X)$ has been equipped with the weak* topology. In case X is a compact metric space, it is useful to know that $M_1^+(X)$ is then metrizable, as the following proposition shows. In particular, it follows that $M_1^+(X)$ is separable, since any compact metric space is separable.

PROPOSITION 5.23. *If X is a compact metric space, then $C(X, \mathbb{R})$ is separable and $M_1^+(X)$ is metrizable.*

PROOF. Set $E = C(X, \mathbb{R})$. Let d be the metric on X, and let $\{x_1, x_2, \ldots\}$ be a countable dense subset of X. Define functions $f_1, f_2, \ldots$ in E by $f_n(x) = d(x, x_n)$ for all $x \in X$, let A be the rational subalgebra of E generated by $\{1, f_1, f_2, \ldots\}$, and note that A is countable. Given distinct points $x, y \in X$, choose $n \geq \mathbb{N}$ such that $d(x, x_n) < d(x, y)/2$. Then

$$d(y, x_n) > d(x, y)/2 > d(x, x_n),$$

so that $f_n(x) \neq f_n(y)$. Thus A separates the points of X. Now let B denote the real linear subspace of E spanned by A, and note that A is dense in B. In addition, B is a real subalgebra of E that contains the constant functions and separates the points of X. By the Stone-Weierstrass Theorem, B is dense in E, and hence A is dense in E. Thus E is separable.

Write A in the form $\{g_1, g_2, \ldots\}$, and then define

$$e(p, q) = \sum_{n=1}^{\infty} \frac{2^{-n}|p(g_n) - q(g_n)|}{1 + |p(g_n) - q(g_n)|}$$

for all $p, q \in E^*$. Then e is a symmetric nonnegative real-valued function on $E^* \times E^*$. If $p, q \in E^*$ such that $e(p, q) = 0$, then $p(g_n) = q(g_n)$ for all n, and hence $p = q$, because A is dense in E. We leave to the reader the exercise of checking that e satisfies the triangle inequality, so that e is a metric on E^*.

We claim that the e-topology on E^* is weaker than the weak* topology. Thus consider a net $\{p_\alpha\}$ in E^* which converges to some $p \in E^*$ in the weak* topology. Given any positive real number ε, choose $k \in \mathbb{N}$ such that

$$\sum_{n=k+1}^{\infty} 2^{-n} < \varepsilon/2.$$

For each $n = 1, \ldots, k$, there exists an index β_n such that

$$|p_\alpha(g_n) - p(g_n)| < \varepsilon/2$$

for all $\alpha \geq \beta_n$. Choose an index β such that $\beta \geq \beta_n$ for all $n = 1, \ldots, k$. For all $\alpha \geq \beta$, we have

$$e(p_\alpha, p) \leq \sum_{n=1}^{k} 2^{-n}|p_\alpha(g_n) - p(g_n)| + \sum_{n=k+1}^{\infty} 2^{-n} < \varepsilon,$$

proving that $p_\alpha \to p$ in the e-topology. Thus the e-topology is weaker than the weak* topology, as claimed.

Consequently, the set-theoretic identity map

$$(M_1^+(X), \text{weak}^* \text{ topology}) \to (M_1^+(X), e\text{-topology})$$

is a continuous bijection. As $M_1^+(X)$ is compact in the weak* topology and Hausdorff in the e-topology, this map must be a homeomorphism. Therefore the weak* topology on $M_1^+(X)$ coincides with the e-topology. □

We next turn to the task of identifying the extreme boundary of $M_1^+(X)$.

DEFINITION. Let X be a compact Hausdorff space, and let $x \in X$. The *point mass at x* is the probability measure ε_x on X defined so that

$$\varepsilon_x(A) = 1 \text{ if } x \in A \quad \text{while} \quad \varepsilon_x(A) = 0 \text{ if } x \notin A$$

for all Borel subsets $A \subseteq X$.

Note that $\varepsilon_x(f) = f(x)$ for all $f \in C(X, \mathbb{R})$.

PROPOSITION 5.24. *If X is a compact Hausdorff space, the rule $\delta(x) = \varepsilon_x$ defines a homeomorphism δ of X onto $\partial_e M_1^+(X)$.*

PROOF. Given $x \in X$, let us consider a positive convex combination $\varepsilon_x = \alpha_1\mu_1 + \alpha_2\mu_2$ where μ_1 and μ_2 are in $M_1^+(X)$. For each $i = 1, 2$, we have

$$\mu_i(X - \{x\}) \leq \alpha_i^{-1}\varepsilon_x(X - \{x\}) = 0,$$

whence $\mu_i(\{x\}) = 1$, and so $\mu_i = \varepsilon_x$. Thus ε_x is an extreme point of $M_1^+(X)$. Therefore δ is a map of X into $\partial_e M_1^+(X)$. It is clear that δ is injective. Consider a net $\{x_i\}$ in X which converges to a point $x \in X$. For any function f in $C(X, \mathbb{R})$, we have $f(x_i) \to f(x)$, which may be rewritten as $\delta(x_i)(f) \to \delta(x)(f)$. Since $M_1^+(X)$ has the weak* topology, $\delta(x_i) \to \delta(x)$ in $M_1^+(X)$. Thus δ is continuous.

Now δ is a continuous bijection of X onto $\delta(X)$. By compactness, δ is actually a homeomorphism of X onto $\delta(X)$. Hence, it only remains to show that $\partial_e M_1^+(X)$ is contained in $\delta(X)$.

Given any $\lambda \in \partial_e M_1^+(X)$, we first observe that λ only assumes the values 0 and 1. For otherwise, X may be expressed as the union of two disjoint Borel sets A_1 and A_2 such that each $\lambda(A_i) > 0$. Then, setting

$$\lambda_i(A) = \lambda(A \cap A_i)/\lambda(A_i)$$

for each $i = 1, 2$ and all Borel sets $A \subseteq X$, we obtain distinct probability measures λ_1, λ_2 in $M_1^+(X)$ such that $\lambda = \lambda(A_1)\lambda_1 + \lambda(A_2)\lambda_2$, contradicting the assumption that λ is an extreme point of $M_1^+(X)$.

Next, observe that if A_1 and A_2 are any Borel subsets of X such that each $\lambda(A_i) = 1$, then $\lambda(A_1 \cap A_2) = 1$. For if not, then $\lambda(A_1 \cap A_2) = 0$ and so

$$\begin{aligned}\lambda((A_1 \cup A_2) - (A_1 \cap A_2)) &= \lambda(A_1 - (A_1 \cap A_2)) + \lambda(A_2 - (A_1 \cap A_2)) \\ &= \lambda(A_1) + \lambda(A_2) = 2,\end{aligned}$$

which is impossible.

Thus if $\mathcal{A}$ is the collection of those closed subsets A of X for which $\lambda(A) = 1$, then $\mathcal{A}$ is closed under finite intersections. In particular, $\mathcal{A}$ satisfies the Finite Intersection Property. As X is compact, it follows that the set $Y = \bigcap \mathcal{A}$ is a nonempty closed subset of X. No compact subset of $X - Y$ can belong to $\mathcal{A}$, whence $\lambda(B) = 0$ for all compact subsets $B \subseteq X - Y$, and so $\lambda(X - Y) = 0$ (because λ is regular). Thus $\lambda(Y) = 1$.

If there exist distinct points $y_1, y_2 \in Y$, choose a continuous function f from Y to $[0, 1]$ such that $f(y_1) = 0$ and $f(y_2) = 1$. Then the sets

$$Y_1 = f^{-1}([0, \tfrac{1}{2}]) \quad \text{and} \quad Y_2 = f^{-1}([\tfrac{1}{2}, 1])$$

are proper closed subsets of Y such that $Y_1 \cup Y_2 = Y$. Neither Y_i can belong to $\mathcal{A}$, whence each $\lambda(Y_i) = 0$. But then $\lambda(Y) = 0$, which is false.

Therefore $Y = \{y\}$ for some point $y \in X$, and consequently $\lambda = \varepsilon_y$. Thus $\partial_e M_1^+(X)$ is indeed contained in $\delta(X)$, as desired. □

DEFINITION. We refer to the map δ in Proposition 5.24 as *the natural homeomorphism of X onto $\partial_e M_1^+(X)$*.

In particular, Proposition 5.24 shows that $\partial_e M_1^+(X)$ is always compact, a property not shared by all compact convex sets, as noted above.

• Faces of Probability Measures.

PROPOSITION 5.25. *Let X be a compact Hausdorff space. Then the closed faces of $M_1^+(X)$ are exactly the subsets of $M_1^+(X)$ of the form*

$$\sigma(A) = \{\mu \in M_1^+(X) \mid \mu(A) = 1\},$$

where A is a closed subset of X.

PROOF. Given a closed set $A \subseteq X$, it is clear that $\sigma(A)$ is a convex subset of $M_1^+(X)$. Now consider a positive convex combination $\alpha_1\mu_1 + \alpha_2\mu_2 = \mu$ where $\mu_1, \mu_2 \in M_1^+(X)$ and $\mu \in \sigma(A)$. Then

$$\alpha_1\mu_1(A) + \alpha_2\mu_2(A) = \mu(A) = 1.$$

Since each $\mu_i(A) \leq 1$, it follows that each $\mu_i(A) = 1$, so that $\mu_i \in \sigma(A)$. Thus $\sigma(A)$ is a face of $M_1^+(X)$.

Consider any λ in $M_1^+(X) - \sigma(A)$, so that $\lambda(X - A) > 0$. As $X - A$ is open, there must be a compact set $B \subseteq X - A$ such that $\lambda(B) > 0$. Choose a continuous function $f\colon X \to [0,1]$ such that $f = 0$ on A and $f = 1$ on B. Then $\lambda(f) \geq \lambda(B) > 0$, whence λ is contained in the open set

$$U = \{\mu \in M_1^+(X) \mid \mu(f) > 0\}.$$

Observing that $\mu(f) = 0$ for all $\mu \in \sigma(A)$, we see that U is disjoint from $\sigma(A)$. This proves that $\sigma(A)$ is closed in $M_1^+(X)$.

Therefore $\sigma(A)$ is a closed face of $M_1^+(X)$.

Now consider any closed face F in $M_1^+(X)$. If δ is the natural homeomorphism of X onto $\partial_e M_1^+(X)$, then the set $A = \delta^{-1}(F)$ is a closed subset of X. Observe that

$$\begin{aligned}\sigma(A) \cap \partial_e M_1^+(X) &= \{\varepsilon_x \mid x \in X \text{ and } \varepsilon_x(A) = 1\}\\ &= \{\varepsilon_x \mid x \in A\} = F \cap \partial_e M_1^+(X).\end{aligned}$$

By Corollary 5.18, the closure of the convex hull of $\sigma(A) \cap \partial_e M_1^+(X)$ equals $\sigma(A)$, while the closure of the convex hull of $F \cap \partial_e M_1^+(X)$ equals F. Therefore $\sigma(A) = F$. □

COROLLARY 5.26. *Let X be a compact Hausdorff space. Then the closed faces of $M_1^+(X)$ are exactly those subsets of $M_1^+(X)$ which can be expressed as the closures of the convex hulls of subsets of $\partial_e M_1^+(X)$.*

PROOF. Any closed face F of $M_1^+(X)$ equals the closure of the convex hull of $F \cap \partial_e M_1^+(X)$, by Corollary 5.18. Conversely, let Y be a subset of $\partial_e M_1^+(X)$, and let F be the closure of the convex hull of Y. As F also equals the closure of the convex hull of the closure of Y, there is no loss of generality in assuming that Y is closed.

If δ is the natural homeomorphism of X onto $\partial_e M_1^+(X)$, then the set $A = \delta^{-1}(Y)$ is a closed subset of X. Define $\sigma(A)$ as in Proposition 5.25. Then

$$\sigma(A) \cap \partial_e M_1^+(X) = \{\varepsilon_x \mid x \in A\} = Y.$$

Applying Corollary 5.18 once again, we conclude that $\sigma(A)$ equals the closure of the convex hull of Y, that is, $\sigma(A) = F$. Therefore F is a closed face of $M_1^+(X)$. □

In general, the closure of a face of a compact convex set K need not be a face of K, as we shall see in Example 6.10. However, this does not happen in any $M_1^+(X)$, as we now show with the aid of our descriptions of the closed faces of $M_1^+(X)$.

COROLLARY 5.27. *Let X be a compact Hausdorff space. If F is any face of $M_1^+(X)$, then the closure of F is also a face of $M_1^+(X)$.*

PROOF. For each $\mu \in M_1^+(X)$, define $s(\mu)$ to be the set of those points $x \in X$ such that $\mu(U) > 0$ for all open neighborhoods U of x. Then set

$$A_0 = \bigcup_{\mu \in F} s(\mu),$$

and let A be the closure of A_0. We shall show that the closure of F coincides with $\sigma(A)$ (in the notation of Proposition 5.25).

For any $\mu \in F$, we have $s(\mu) \subseteq A$. Hence, any point in $X - A$ has an open neighborhood U such that $\mu(U) = 0$. As a result, $\mu(C) = 0$ for all compact subsets $C \subseteq X - A$, and so $\mu(X - A) = 0$. Then $\mu(A) = 1$, so that $\mu \in \sigma(A)$. Thus $F \subseteq \sigma(A)$, and consequently $\overline{F} \subseteq \sigma(A)$.

The set $\overline{F}$ is a closed convex subset of $M_1^+(X)$. To show that $\sigma(A) \subseteq \overline{F}$, it suffices to show that all extreme points of $\sigma(A)$ lie in $\overline{F}$, because $\sigma(A)$ equals the closure of the convex hull of $\partial_e\sigma(A)$. Since $\sigma(A)$ is a face of $M_1^+(X)$, we have

$$\partial_e\sigma(A) = \sigma(A) \cap \partial_e M_1^+(X) = \{\varepsilon_x \mid x \in A\},$$

which is the closure of the set $\{\varepsilon_x \mid x \in A_0\}$ (by virtue of Proposition 5.24). Thus we need only show that $\varepsilon_x \in \overline{F}$ for any $x \in A_0$.

We must show that any open neighborhood V of ε_x meets F. As V contains an open convex neighborhood of ε_x, there is no loss of generality in assuming that V itself is convex. By compactness, there exists a neighborhood of ε_x whose closure is contained in V, and this neighborhood in turn contains an open convex neighborhood W of ε_x. Note that $\overline{W}$ is a compact convex neighborhood of ε_x. As $\overline{W} \subseteq V$, we need only show that $\overline{W}$ meets F.

Because of Proposition 5.24, the set

$$U = \{y \in X \mid \varepsilon_y \in W\}$$

is an open neighborhood of x. Now

$$\partial_e\sigma(\overline{U}) = \sigma(\overline{U}) \cap \partial_e M_1^+(X) = \{\varepsilon_y \mid y \in \overline{U}\} \subseteq \overline{W}.$$

Since $\overline{W}$ is closed and convex, we obtain $\sigma(\overline{U}) \subseteq \overline{W}$ from the Krein-Mil′man Theorem. As $x \in A_0$, there is some $\mu \in F$ such that $x \in s(\mu)$, and so $\mu(\overline{U}) \geq \mu(U) > 0$. If $\mu(\overline{U}) = 1$, then $\mu \in \sigma(\overline{U})$ and hence $\mu \in \overline{W}$. Thus $\overline{W} \cap F$ is nonempty in this case.

If $\mu(\overline{U}) = \alpha < 1$, then we may define measures μ_1, μ_2 in $M_1^+(X)$ such that

$$\mu_1(B) = \alpha^{-1}\mu(B \cap \overline{U}) \quad \text{and} \quad \mu_2(B) = (1 - \alpha)^{-1}\mu(B \cap (X - \overline{U}))$$

for all Borel sets $B \subseteq X$. Then $\alpha\mu_1 + (1 - \alpha)\mu_2 = \mu$, and hence $\mu_1 \in F$, because F is a face of $M_1^+(X)$. As $\mu_1(\overline{U}) = 1$, we obtain $\mu_1 \in \sigma(\overline{U})$ and $\mu_1 \in \overline{W}$. Thus $\overline{W} \cap F$ is nonempty in this case also.

Therefore $\varepsilon_x \in \overline{F}$ for any $x \in A_0$, and consequently $\sigma(A) \subseteq \overline{F}$. Hence, $\overline{F} = \sigma(A)$, and thus by Proposition 5.25 $\overline{F}$ is a closed face of $M_1^+(X)$. □

• **Notes.** Coproducts in the category of compact convex sets (Propositions 5.5 and 5.6) were constructed by Semadeni [**113**, Theorem 7]. Proposition 5.7 was observed by Alfsen [**1**, p. 99].

Proposition 5.9 was first proved for $E = \mathbb{R}^3$ by Minkowski [**95**, §§2,3], and later for E any normed linear space by Ascoli [**5**, pp. 48,49]. A version of Proposition 5.10, in which E is a normed linear space, B is a linear subspace of E, and A

is the interior of a closed convex subset of E, was proved by Mazur [**93**, Satz 1]. Theorem 5.12, in the case that E is a normed linear space and A and B are interiors of closed convex subsets of E, was proved by Eidelheit [**34**, Satz]. Theorem 5.14 is a well-known standard result for which no historical references seem to be available.

Theorem 5.17 was first proved by Krein and Mil′man in the case that K is a bounded regularly convex (equivalently, bounded weak* compact convex) subset of the dual of a real Banach space [**87**, Theorem]. The case in which E is any linear topological space in which points can be separated from compact convex sets by closed hyperplanes was proved by Kelley [**82**]. Theorem 5.21 is a descendent of the following theorem of Mil′man: if X is a bounded subset of the dual of a complex Banach space E, and if K is the smallest regularly convex set containing X, then $\partial_e K$ is contained in the weak* closure of X [**94**, Theorem 1]. (A set $M \subseteq E^*$ is regularly convex provided that for each $f \in E^* - M$, there exists $x \in E$ such that $\mathrm{Re}(f(x))$ is strictly less than $\inf\{\mathrm{Re}(g(x)) \mid g \in M\}$.)

Propositions 5.22, 5.23, 5.24 are all folklore. Corollaries 5.26 and 5.27 are due to Alfsen [**1**, Proposition 4 and Theorem 1].

CHAPTER 6

State Spaces

Just as the linear transformations from a vector space into its scalar field form a useful dual object for the vector space, the positive homomorphisms from a partially ordered abelian group G into $\mathbb{R}$ form a useful dual object for G. (The choice of $\mathbb{R}$ rather than $\mathbb{Z}$ or $\mathbb{Q}$ as an appropriate target for positive homomorphisms defined on G is dictated by the observation that far more positive homomorphisms can usually be defined from G to $\mathbb{R}$ than from G to $\mathbb{Z}$ or $\mathbb{Q}$, as evidenced by extension results such as Proposition 4.2 and Corollary 4.3.) When G has an order-unit u, all the nonzero positive homomorphisms from G to $\mathbb{R}$ may be normalized into states, and hence we concentrate on the collection of states on (G, u). This set, known as the state space of (G, u), turns out to be a compact convex subset of the linear topological space $\mathbb{R}^G$.

The construction of state spaces defines a contravariant functor from the category of partially ordered abelian groups with order-unit into the category of compact convex sets, providing a rough "duality" between the two categories, with which we may attempt to transform questions concerning partially ordered abelian groups with order-unit into "dual" questions concerning compact convex sets. In order to transform questions (or answers) in the reverse direction, a second "duality" functor is required, from the category of compact convex sets into the category of partially ordered abelian groups with order-unit. This functor, and the resulting "double dual" mechanism, will be introduced in the following chapter.

This chapter is concerned with the construction of the state space functor and with a preliminary view of how the structure of the state space, as a compact convex set, reflects the structure of the partially ordered abelian group with order-unit from which it arose. Categorically, we show that the state space functor converts quotient objects into subobjects, finite products into finite coproducts, and direct limits into inverse limits. Techniques are introduced for describing the face of a state space generated by a given family of states and for deciding when states lie in disjoint faces of a state space. As any state space is "generated" by the extremal states (because of the Krein-Mil′man Theorem), the structure of a state space is heavily influenced by the behavior of the extremal states. Although the development of further machinery is required in order to work effectively with general extremal states, it is relatively easy to examine discrete extremal states on interpolation groups, and the final section of the chapter is devoted to this. In particular, we prove that the quotient of an interpolation group modulo the kernel of a discrete extremal state is naturally isomorphic (as an ordered group) to the image of the state and that any discrete state on an interpolation group is a rational convex combination of discrete extremal states.

• Basic Structure.

DEFINITION. Let (G, u) be a partially ordered abelian group with order-unit. The *state space of* (G, u), denoted $S(G, u)$, is the set of all states on (G, u).

Note from Corollary 4.4 that $S(G, u)$ is nonempty as long as G is nonzero.

We view $S(G, u)$ as a subset of the real vector space $\mathbb{R}^G$ of all real-valued functions on G. In addition, we equip $\mathbb{R}^G$ with the product topology, so that it becomes a linear topological space. We always assume that $S(G, u)$ has been endowed with the relative topology from $\mathbb{R}^G$. The topology on $S(G, u)$ may be described in terms of convergence by observing that a net $\{s_\alpha\}$ in $S(G, u)$ converges to a state s in $S(G, u)$ if and only if $s_\alpha(x) \to s(x)$ for all $x \in G$. Of course, since G is directed, it suffices to know that $s_\alpha(x) \to s(x)$ for all $x \in G^+$. A description of the topology on $S(G, u)$ in terms of open sets is given in the following result.

PROPOSITION 6.1. *Let (G, u) be a partially ordered abelian group with order-unit. The topology on $S(G, u)$ has a subbasis of open sets of the form*

$$\{s \in S(G, u \mid s(x) < k\}$$

for $x \in G^+$ and $k \in \mathbb{N}$.

PROOF. On intersecting $S(G, u)$ with the standard subbasic open sets in $\mathbb{R}^G$, we find that $S(G, u)$ has a subbasis consisting of open sets of the form

$$V(y, \alpha) = \{s \in S(G, u) \mid s(y) < \alpha\}, \qquad W(y, \alpha) = \{s \in S(G, u) \mid s(y) > \alpha\}$$

for $y \in G$ and $\alpha \in \mathbb{R}$. Note that each $W(y, \alpha) = V(-y, -\alpha)$. Thus the $V(y, \alpha)$ form a subbasis for $S(G, u)$.

Now consider any $y \in G$ and $\alpha \in \mathbb{R}$ and any $s \in V(y, \alpha)$. Choose a positive integer m such that $y + mu \in G^+$, and note that $s(y + mu) \geq 0$. There exist $k, n \in \mathbb{N}$ such that

$$s(y + mu) < k/n < \alpha + m,$$

and we observe that

$$s \in V(ny + mnu, k) \subseteq V(y, \alpha).$$

Therefore the $V(x, k)$, for $x \in G^+$ and $k \in \mathbb{N}$, form a subbasis for $S(G, u)$. $\square$

PROPOSITION 6.2. *Let (G,u) be a partially ordered abelian group with order-unit. Then $\mathbb{R}^G$ is a locally convex Hausdorff space, and $S(G,u)$ is a compact convex subset of $\mathbb{R}^G$.*

PROOF. It is clear that $\mathbb{R}^G$ is a locally convex Hausdorff space and that $S(G,u)$ is a convex subset of $\mathbb{R}^G$.

For each $x \in G$, choose $n_x \in \mathbb{N}$ such that $-n_x u \le x \le n_x u$. For all states s on (G,u), we have $-n_x \le s(x) \le n_x$, and hence

$$S(G,u) \subseteq \prod_{x\in G} [-n_x, n_x].$$

This product of closed intervals is compact by Tychonoff's Theorem. In addition, $S(G,u)$ is an intersection of closed subsets of $\mathbb{R}^G$, namely the sets

$$\{f \in \mathbb{R}^G \mid f(x+y) = f(x) + f(y)\} \quad (\text{for } x, y \in G),$$
$$\{f \in \mathbb{R}^G \mid f(x) \ge 0\} \quad (\text{for } x \in G^+), \quad \text{and} \quad \{f \in \mathbb{R}^G \mid f(u) = 1\}.$$

Therefore $S(G,u)$ is a closed subset of the given product of intervals, and consequently it is compact. □

In particular, Proposition 6.2 allows us to apply the Krein-Mil′man Theorem and its corollaries to any state space. The following improvements on Theorem 4.12 and Corollary 4.13 represent the most direct application of this method.

THEOREM 6.3. *Let (G,u) be a partially ordered abelian group with order-unit, and let $x \in G$. Then $s(x) > 0$ for all extremal states s on (G,u) if and only if there exists a positive integer m such that mx is an order-unit in G.*

PROOF. We may assume that G is nonzero. Define a map $p\colon S(G,u) \to \mathbb{R}$ by the rule $p(s) = s(x)$, and observe that p is affine and continuous. If $s(x) > 0$ for all extremal states s on (G,u), then $p \gg 0$ on $\partial_e S(G,u)$. According to Corollary 5.20, $p \gg 0$, that is, $s(x) > 0$ for all states s on (G,u). Then Theorem 4.12 shows that there exists $m \in \mathbb{N}$ such that mx is an order-unit in G. □

COROLLARY 6.4. *Let (G,u) be an unperforated partially ordered abelian group with order-unit, and let $x \in G$. Then $s(x) > 0$ for all extremal states s on (G,u) if and only if x is an order-unit in G.* □

A similar improvement may be made to Theorem 4.14. Specifically, if (G,u) is an archimedean partially ordered abelian group with order-unit, then

$$G^+ = \{x \in G \mid s(x) \ge 0 \text{ for all } s \in \partial_e S(G,u)\}.$$

PROPOSITION 6.5. *Let (G,u) be a partially ordered abelian group with order-unit, let $s_1, s_2, \dots$ be states on (G,u), and let $\alpha_1, \alpha_2, \dots$ be nonnegative real numbers such that $\sum \alpha_k = 1$. Then there is a state s on (G,u) such that $s(x) = \sum \alpha_k s_k(x)$ for all $x \in G$, and $\sum \alpha_k s_k \to s$ in $\mathbb{R}^G$. If $\alpha_1 > 0$, then*

$$\sum_{k=1}^{n} \alpha_k(\alpha_1 + \cdots + \alpha_n)^{-1} s_k \to s \quad (\text{as } n \to \infty) \text{ in } S(G,u).$$

PROOF. Any x in G satisfies $-mu \leq x \leq mu$ for some $m \in \mathbb{N}$, whence $|s_k(x)| \leq m$ for all $k = 1, 2, \ldots$. Hence, the series $\sum \alpha_k s_k(x)$ converges absolutely, to some real number $s(x)$. This defines a map $s\colon G \to \mathbb{R}$, and it is clear that s is a state on (G, u). Since

$$\sum \alpha_k s_k(x) \to s(x)$$

for all $x \in G$, the series $\sum \alpha_k s_k$ converges to s in $\mathbb{R}^G$.

If $\alpha_1 > 0$, then the functions

$$t_n = \sum_{k=1}^{n} \alpha_k (\alpha_1 + \cdots + \alpha_n)^{-1} s_k$$

are states on (G, u). Since $\sum \alpha_k s_k \to s$ in $\mathbb{R}^G$ and $(\sum \alpha_k)^{-1} \to 1$ in $\mathbb{R}$, we conclude that $t_n \to s$ in $S(G, u)$. □

DEFINITION. In the situation of Proposition 6.5, we write $s = \sum \alpha_k s_k$, and we say that s is a *σ-convex combination* of the s_k. In case all the $\alpha_k > 0$, we say that s is a *positive σ-convex combination* of the s_k.

DEFINITION. Let (G, u) be a partially ordered abelian group with order-unit, and let X be a subset of $S(G, u)$. By the *kernel of* X, denoted $\ker(X)$, we mean the intersection of the kernels of all the states in X, that is,

$$\ker(X) = \{x \in G \mid s(x) = 0 \text{ for all } s \in X\}.$$

Note that $\ker(X)$ is a convex subgroup of G. Hence, by Proposition 1.9, the subgroup of G generated by $\ker(X)^+$ is an ideal of G. In general, $\ker(X)$ itself is not an ideal of G. For example, if $G = \mathbb{Z}^2$ and $u = (1, 1)$, then the rule $s(x, y) = (x + y)/2$ defines a state s on (G, u), and $\ker(s) = \mathbb{Z}(1, -1)$, which is not directed.

• **Some Examples.** We investigate a few cases of partially ordered abelian groups with order-unit for which the state space can be conveniently described. In general, a state space need not enjoy any special properties, since all compact convex subsets of locally convex Hausdorff spaces appear as state spaces (Theorem 7.1).

PROPOSITION 6.6. *Let (G, u) be a simplicial group with order-unit. Choose a simplicial basis $\{x_1, \ldots, x_n\}$ for G, and write $u = a_1 x_1 + \cdots + a_n x_n$ for some integers a_i. Then each $a_i > 0$. There are states $s_1, \ldots, s_n$ on (G, u) defined by the rule*

$$s_j(c_1 x_1 + \cdots + c_n x_n) = c_j / a_j$$

(for all $c_1, \ldots, c_n \in \mathbb{Z}$). These s_j are affinely independent,

$$\partial_e S(G, u) = \{s_1, \ldots, s_n\},$$

and $S(G, u)$ equals the convex hull of $\{s_1, \ldots, s_n\}$.

PROOF. Since u is an order-unit, $x_1 + \cdots + x_n \leq ku$ for some $k \in \mathbb{N}$, whence

$$(ka_1 - 1)x_1 + \cdots + (ka_n - 1)x_n = ku - (x_1 + \cdots + x_n) \geq 0,$$

and so $ka_i - 1 \geq 0$ for each i. Thus each $a_i > 0$.

It is clear that $s_1, \ldots, s_n \in S(G, u)$. If $\alpha_1 s_1 + \cdots + \alpha_n s_n = 0$ for some real numbers $\alpha_1, \ldots, \alpha_n$, then on evaluating at $x_1, \ldots, x_n$ we obtain $\alpha_j / a_j = 0$ for each j, so that each $\alpha_j = 0$. Thus the s_j are linearly independent and hence affinely independent.

Given $s \in S(G, u)$, set $\alpha_j = a_j s(x_j)$ for each $j = 1, \ldots, n$. These α_j are nonnegative real numbers such that

$$\alpha_1 + \cdots + \alpha_n = s(a_1 x_1) + \cdots + s(a_n x_n) = s(u) = 1.$$

For each $i = 1, \ldots, n$, we have

$$(\alpha_1 s_1 + \cdots + \alpha_n s_n)(x_i) = \alpha_i / a_i = s(x_i).$$

Hence, $\alpha_1 s_1 + \cdots + \alpha_n s_n = s$, because the x_i generate G (as a group). Therefore $S(G, u)$ equals the convex hull of $\{s_1, \ldots, s_n\}$.

In particular, every extreme point of $S(G, u)$ is a convex combination of the s_j, from which it follows that any extreme point of $S(G, u)$ equals one of the s_j. Conversely, consider a positive convex combination $s_k = \alpha_1 t_1 + \alpha_2 t_2$ for some k and some t_1, t_2 in $S(G, u)$. Write each t_i as a convex combination

$$t_i = \beta_{i1} s_1 + \cdots + \beta_{in} s_n,$$

so that

$$s_k = (\alpha_1 \beta_{11} + \alpha_2 \beta_{21}) s_1 + \cdots + (\alpha_1 \beta_{1n} + \alpha_2 \beta_{2n}) s_n.$$

Since the s_j are linearly independent, $\alpha_1 \beta_{1j} + \alpha_2 \beta_{2j} = 0$ for all $j \neq k$, and consequently $\beta_{1j} = \beta_{2j} = 0$ for all $j \neq k$. Thus $t_1 = t_2 = s_k$, proving that s_k is an extreme point of $S(G, u)$.

Therefore $\partial_e S(G, u) = \{s_1, \ldots, s_n\}$. □

In particular, Proposition 6.6 shows that the state space of any simplicial group with order-unit is a classical simplex, an observation to which we shall return in Chapter 10.

Most state spaces, however, do not have affinely independent extreme boundaries. For example, let

$$G = \{(x_1, x_2, x_3, x_4) \in \mathbb{Z}^4 \mid x_1 + x_2 = x_3 + x_4\},$$

and set $u = (1, 1, 1, 1)$, which is an order-unit in G. Each of the coordinate projections $s_i \colon G \to \mathbb{Z}$ is a state on (G, u), and since all states on (G, u) extend to states on $(\mathbb{Z}^4, u)$ (by Corollary 4.3), we find that $S(G, u)$ equals the convex hull of $\{s_1, s_2, s_3, s_4\}$. It follows easily that

$$\partial_e S(G, u) = \{s_1, s_2, s_3, s_4\}.$$

However, by definition of G we have

$$\tfrac{1}{2}s_1 + \tfrac{1}{2}s_2 = \tfrac{1}{2}s_3 + \tfrac{1}{2}s_4$$

and so these extremal states are affinely dependent.

LEMMA 6.7. *If G is a directed real vector space, then any positive group homomorphism $f\colon G \to \mathbb{R}$ is a linear map.*

PROOF. Given any $x \in G$ and any integers m, n with $n \neq 0$, we have

$$nf((m/n)x) = f(mx) = mf(x),$$

and hence $f((m/n)x) = (m/n)f(x)$. Thus f is at least $\mathbb{Q}$-linear.

Now consider any $y \in G^+$ and any $\alpha \in \mathbb{R}$. If p and q are any rational numbers such that $p \leq \alpha \leq q$, then $py \leq \alpha y \leq qy$ (because $y \geq 0$), whence

$$pf(y) = f(py) \leq f(\alpha y) \leq f(qy) = qf(y).$$

Since $f(y) \geq 0$, it follows that $f(\alpha y) = \alpha f(y)$. As every element of G is a difference of positive elements, we conclude that f is $\mathbb{R}$-linear. □

In particular, if (G, u) is a partially ordered real vector space with order-unit, then Lemma 6.7 shows that all states on (G, u) are linear.

PROPOSITION 6.8. *If X is any compact Hausdorff space, then the constant function 1 is an order-unit in $C(X, \mathbb{R})$, and*

$$S(C(X, \mathbb{R}), 1) = M_1^+(X)$$

(*as compact convex sets*).

PROOF. That 1 is an order-unit in $C(X, \mathbb{R})$ follows from the fact that all functions in $C(X, \mathbb{R})$ are bounded. Observe that

$$M_1^+(X) = \{p \in C(X, \mathbb{R})^* \mid p \geq^+ 0 \text{ and } p(1) = 1\},$$

and that the weak* topology on $M_1^+(X)$ coincides with the relative product topology inherited via the inclusions

$$M_1^+(X) \subseteq C(X, \mathbb{R})^* \subseteq \mathbb{R}^{C(X,\mathbb{R})}.$$

Thus $M_1^+(X)$ is a convex subset of $S(C(X, \mathbb{R}), 1)$, and the topology on $M_1^+(X)$ coincides with the relative topology from $S(C(X, \mathbb{R}), 1)$.

It only remains to show that any state s in $S(C(X, \mathbb{R}), 1)$ lies in $M_1^+(X)$. By Lemma 6.7, s is a linear map. For any g in the unit ball of $C(X, \mathbb{R})$, we have $-1 \leq g \leq 1$ and so $-1 \leq s(g) \leq 1$, whence $|s(g)| \leq 1$. Thus s is bounded, and hence $s \in C(X, \mathbb{R})^*$. Since $s \geq^+ 0$ and $s(1) = 1$, we conclude that $s \in M_1^+(X)$, as desired. □

PROPOSITION 6.9. *Let X be a nonempty set, and let G be an additive subgroup of $\mathbb{R}^X$ such that $1 \in G$ and all functions in G are bounded. Let G_1 denote*

G equipped with the pointwise ordering, and let G_2 denote G equipped with the strict ordering. Then 1 is an order-unit in each G_i, and $S(G_1, 1) = S(G_2, 1)$.

PROOF. The boundedness hypothesis ensures that 1 is an order-unit in each G_i. Every group homomorphism from G to $\mathbb{R}$ that is positive with respect to $\leq$ is also positive with respect to $\lll$. Hence, $S(G_1, 1)$ is contained in $S(G_2, 1)$. Conversely, consider any s in $S(G_2, 1)$. Given $x \in G_1^+$, we have $x \gg -1/n$ for all $n \in \mathbb{N}$, so that $nx \gg -1$ for all $n \in \mathbb{N}$. Then $ns(x) \geq -1$ for all $n \in \mathbb{N}$, whence $s(x) \geq 0$. Therefore $s \in S(G_1, 1)$. $\square$

For instance, if X is a compact Hausdorff space and G_2 denotes $C(X, \mathbb{R})$ equipped with the strict ordering, then the combination of Propositions 6.8 and 6.9 shows that $S(G_2, 1) = M_1^+(X)$.

EXAMPLE 6.10. *There exists a dimension group (G, u) with order-unit such that $\partial_e S(G, u)$ is not compact, and also $S(G, u)$ has a face whose closure is not a face of $S(G, u)$.*

PROOF. Set $G = \{x \in \mathbb{R}^{\mathbb{N}} \mid x_n \to (x_1 + x_2)/2\}$, and set $u = (1, 1, 1, \ldots)$. As G is a partially ordered real vector space, it is unperforated. Clearly u is an order-unit in G, and hence, in particular, G is directed.

Given $x_1, x_2, y_1, y_2 \in G$ satisfying $x_i \leq y_j$ for all i, j, set

$$a_n = \max\{x_{1n}, x_{2n}\} \quad \text{and} \quad b_n = \min\{y_{1n}, y_{2n}\}$$

for all $n \in \mathbb{N}$. Then $a_n \leq b_n$ for all n, and

$$a_n \to \max\{(x_{11} + x_{12})/2, (x_{21} + x_{22})/2\} \leq (a_1 + a_2)/2$$
$$b_n \to \min\{(y_{11} + y_{12})/2, (y_{21} + y_{22})/2\} \geq (b_1 + b_2)/2.$$

Hence, there exist real numbers $z_3, z_4, \ldots$ such that $a_n \leq z_n \leq b_n$ for all $n = 3, 4, \ldots$ and $z_n \to (a_1 + a_2)/2$. Setting $z_1 = a_1$ and $z_2 = a_2$, we obtain an element $z \in G$ such that $x_i \leq z \leq y_j$ for all i, j. Thus G has interpolation and so is a dimension group.

Each of the coordinate projections $s_n \colon G \to \mathbb{R}$ is a state on (G, u). Since

$$s_n(x) = x_n \to (x_1 + x_2)/2 = (s_1(x) + s_2(x))/2$$

for all $x \in G$, we see that $s_n \to \frac{1}{2}s_1 + \frac{1}{2}s_2$ in $S(G, u)$. We claim that each s_n is in $\partial_e S(G, u)$, whence $\partial_e S(G, u)$ is not closed in $S(G, u)$ and hence is not compact.

Thus consider any positive convex combination $s_k = \alpha_1 t_1 + \alpha_2 t_2$ where $k \in \mathbb{N}$ and t_1, t_2 lie in $S(G, u)$. Note from Lemma 6.7 that t_1 and t_2 must be linear. If $k > 2$, we may define an element $v \in G^+$ so that $v_k = 1$ while $v_n = 0$ for all $n \neq k$. Then $u - v$ is an element of G^+ such that

$$\alpha_1 t_1(u - v) + \alpha_2 t_2(u - v) = s_k(u - v) = 0.$$

Since each $\alpha_i > 0$ and each $t_i(u - v) \geq 0$, it follows that each $t_i(u - v) = 0$. Hence, each $t_i(v) = 1$. Given any $x \in G$, choose $m \in \mathbb{N}$ such that $-mu \leq x \leq mu$. Then

$$-m(u - v) \leq x - x_k v \leq m(u - v),$$

whence each $t_i(x - x_k v) = 0$, and so

$$t_i(x) = x_k t_i(v) = x_k = s_k(x).$$

Thus $t_1 = t_2 = s_k$; so s_k is indeed an extreme point of $S(G,u)$. We leave to the reader the similar task of showing that s_1 and s_2 are extreme points of $S(G,u)$.

Now each $s_n \in \partial_e S(G,u)$ while $\frac{1}{2}s_1 + \frac{1}{2}s_2 \notin \partial_e S(G,u)$; therefore $\partial_e S(G,u)$ is not compact.

Finally, let F be the convex hull of $\{s_3, s_4, s_5, \dots\}$ in $S(G,u)$. Define the element $w = (1, -1, 0, 0, 0, \dots)$ in G, and note that $s_3, s_4, s_5, \dots$ are all contained in the closed convex set

$$K = \{s \in S(G,u) \mid s(w) = 0\}.$$

Consequently, $\overline{F} \subseteq K$, from which we see that $s_1, s_2 \notin \overline{F}$. On the other hand, $\frac{1}{2}s_1 + \frac{1}{2}s_2$ is in $\overline{F}$, because the s_n converge to it. Thus $\overline{F}$ is not a face of $S(G,u)$.

It remains to show that F is a face of $S(G,u)$. Consider a positive convex combination $\alpha_1 t_1 + \alpha_2 t_2 = s$ where $t_1, t_2 \in S(G,u)$ and $s \in F$. Choose an integer $k > 2$ such that s is a convex combination of $s_3, \dots, s_k$, and define $p \in G^+$ so that $p_n = 1$ for each $n = 3, \dots, k$ while $p_n = 0$ for $n = 1, 2$ and all $n > k$. As $s_3, \dots, s_k$ all vanish on $u - p$, we infer that t_1 and t_2 both vanish on $u - p$. From this, it follows (as in Proposition 6.6), that t_1 and t_2 are convex combinations of $s_3, \dots, s_k$, whence $t_1, t_2 \in F$. Therefore F is a face of $S(G,u)$. □

• **Functoriality.**

DEFINITION. Let $f\colon (G,u) \to (H,v)$ be a morphism in the category of partially ordered abelian groups with order-unit. If s is any state on (H,v), then sf is a state on (G,u). Hence, the rule $S(f)(s) = sf$ defines a map

$$S(f)\colon S(H,v) \to S(G,u),$$

and we observe that $S(f)$ is affine and continuous. It is clear that the assignment $f \mapsto S(f)$ preserves identity maps and reverses compositions of maps. Therefore the assignments $(G,u) \mapsto S(G,u)$ and $f \mapsto S(f)$ define a contravariant functor S from the category of partially ordered abelian groups with order-unit to the category of compact convex sets, known as *the state space functor.*

We shall derive three main properties of the state space functor. Roughly speaking, these properties are conversion properties: namely, the state space functor converts quotient objects into subobjects, finite products into finite coproducts, and direct limits into inverse limits.

LEMMA 6.11. *Let (G,u) be a partially ordered abelian group with order-unit, and let X be a subset of G^+. Then the set*

$$F = \{s \in S(G,u) \mid X \subseteq \ker(s)\}$$

is a closed face of $S(G,u)$.

PROOF. It is clear that F is a closed convex subset of $S(G,u)$. Consider a positive convex combination $\alpha_1 s_1 + \alpha_2 s_2 = s$ with $s_1, s_2 \in S(G,u)$ and $s \in F$.

Given any $x \in X$, we have

$$\alpha_1 s_1(x) + \alpha_2 s_2(x) = s(x) = 0$$

with each $\alpha_i > 0$ and each $s_i(x) \geq 0$, whence each $s_i(x) = 0$. Thus s_1 and s_2 vanish on X, so that $s_1, s_2 \in F$. Therefore F is a face of $S(G,u)$. $\square$

PROPOSITION 6.12. *Let (G,u) be a partially ordered abelian group with order-unit, let H be an ideal of G, and let*

$$p\colon (G,u) \to (G/H, u+H)$$

be the quotient map. Then the map $S(p)$ provides an affine homeomorphism of $S(G/H, u+H)$ onto the set

$$F = \{s \in S(G,u) \mid H \subseteq \ker(s)\},$$

which is a closed face of $S(G,u)$. Moreover,

$$S(p)(\partial_e S(G/H, u+H)) = \partial_e F = F \cap \partial_e S(G,u).$$

PROOF. Obviously $S(p)$ maps $S(G/H, u+H)$ into F, and since p is surjective, $S(p)$ is injective. Any s in F induces a group homomorphism $t\colon G/H \to \mathbb{R}$ such that $tp = s$. As

$$t((G/H)^+) = tp(G^+) = s(G^+) \subseteq \mathbb{R}^+$$

and $t(u+H) = tp(u) = s(u) = 1$, the function t is a state on $(G/H, u+H)$ such that $S(p)(t) = s$.

Thus $S(p)$ gives a continuous affine bijection of $S(G/H, u+H)$ onto F. By compactness, this map must also be a homeomorphism. Since H is directed, it is generated (as a subgroup of G) by H^+, and hence

$$F = \{s \in S(G,u) \mid H^+ \subseteq \ker(s)\}.$$

Now Lemma 6.11 shows that F is a closed face of $S(G,u)$.

The final statement of the proposition follows from the affine isomorphism of $S(G/H, u+H)$ onto F provided by $S(p)$, and from the fact that F is a face of $S(G,u)$. $\square$

• Products and Limits.

PROPOSITION 6.13. *Let $(G,u) = (G_1,u_1) \times \cdots \times (G_n,u_n)$ be a finite product in the category of partially ordered abelian groups with order-unit, and for $i = 1, \ldots, n$ let $p_i\colon (G,u) \to (G_i,u_i)$ be the projection map. Then $S(G,u)$, together with the maps $S(p_i)$, is a coproduct for $S(G_1,u_1), \ldots, S(G_n,u_n)$ in the category of compact convex sets.*

PROOF. In view of Proposition 5.6, it suffices to prove that the maps $S(p_i)$ are all injective and that $S(G,u)$ is the direct convex sum of the convex subsets

$$F_i = S(p_i)(S(G_i,u_i)).$$

The injectivity of the $S(p_i)$ follows from the surjectivity of the p_i.

For $i = 1,\dots,n$ let $r_i \colon G_i \to G$ be the injection map into the ith component of G, and set $v_i = r_i(u_i) \in G^+$. Note that $v_1 + \cdots + v_n = u$. Now consider any $s \in S(G,u)$, and set $\alpha_i = s(v_i)$ for each $i = 1,\dots,n$. If

$$I = \{i \in \{1,2,\dots,n\} \mid \alpha_i \neq 0\},$$

then $\{\alpha_i \mid i \in I\}$ is a set of positive real numbers such that

$$\sum_{i\in I} \alpha_i = \alpha_1 + \cdots + \alpha_n = s(v_1 + \cdots + v_n) = s(u) = 1.$$

For $i \in I$, observe that the map $s_i = \alpha_i^{-1} s r_i$ is a state on (G_i,u_i). For $i \notin I$, we have $sr_i(u_i) = s(v_i) = \alpha_i = 0$, and hence $sr_i = 0$ (because $sr_i \geq^+ 0$ and u_i is an order-unit in G_i). Consequently,

$$s = s(r_1p_1 + \cdots + r_np_n) = \sum_{i\in I} sr_ip_i = \sum_{i\in I} \alpha_i s_i p_i = \sum_{i\in I} \alpha_i S(p_i)(s_i).$$

Thus $S(G,u)$ equals the convex hull of $F_1 \cup \cdots \cup F_n$.

Finally, consider convex combinations

$$\sum_{j\in J} \alpha_j S(p_j)(s_j) = \sum_{j\in J} \beta_j S(p_j)(t_j)$$

where $J \subseteq \{1,\dots,n\}$ and $s_j, t_j \in S(G_j,u_j)$ for each $j \in J$. Then

$$\sum_{j\in J} \alpha_j s_j p_j = \sum_{j\in J} \beta_j t_j p_j.$$

For each $k \in J$, it follows that

$$\alpha_k s_k = \left(\sum_{j\in J} \alpha_j s_j p_j\right) r_k = \left(\sum_{j\in J} \beta_j t_j p_j\right) r_k = \beta_k t_k.$$

Hence, $\alpha_k = \alpha_k s_k(u_k) = \beta_k t_k(u_k) = \beta_k$, and $s_k = t_k$ if $\alpha_k > 0$.

Therefore $S(G,u)$ is the direct convex sum of the F_i. □

Proposition 6.6 may be partly regarded as a special case of Proposition 6.13, since the coproduct of a finite set of points in the category of compact convex sets is a classical simplex. We may also use Proposition 6.13 in the construction of examples. For instance, if (G,u) is the product of (G_1,u_1) with $(\mathbb{Z},1)$, where (G_1,u_1) is chosen so that $S(G_1,u_1)$ is affinely homeomorphic to a square, then $S(G,u)$ is affinely homeomorphic to a square-base pyramid.

PROPOSITION 6.14. *Let $\{(G_i, u_i), f_{ji}\}$ be a direct system in the category of partially ordered abelian groups with order-unit, indexed by a directed set I. Let (G, u) be a direct limit of this system, and for all $i \in I$ let $q_i\colon (G_i, u_i) \to (G, u)$ be the natural map. Then $S(G, u)$, together with the maps $S(q_i)$, is an inverse limit of the inverse system $\{S(G_i, u_i), S(f_{ji})\}$ in the category of compact convex sets.*

PROOF. The inverse system $\{S(G_i, u_i), S(f_{ji})\}$ has an inverse limit K in the category of compact convex sets. For all $i \in I$, let $p_i\colon K \to S(G_i, u_i)$ be the natural map. Since $S(q_i) = S(q_j f_{ji}) = S(f_{ji})S(q_j)$ whenever $i \leq j$ in I, there exists a unique affine continuous map $g\colon S(G, u) \to K$ such that $p_i g = S(q_i)$ for all $i \in I$, and it suffices to prove that g is an affine homeomorphism of $S(G, u)$ onto K.

Given $s, t \in S(G, u)$ such that $g(s) = g(t)$, we compute that

$$sq_i = S(q_i)(s) = p_i g(s) = p_i g(t) = S(q_i)(t) = tq_i$$

for all $i \in I$, so that s and t agree on each of the subgroups $q_i(G_i)$. As $G = \bigcup q_i(G_i)$, we obtain $s = t$. Thus g is injective.

Now consider any $x \in K$, and set $s_i = p_i(x)$ in $S(G_i, u_i)$ for all $i \in I$. Each s_i is a normalized positive homomorphism from (G_i, u_i) to $(\mathbb{R}, 1)$, and $s_i = S(f_{ji})(s_j) = s_j f_{ji}$ whenever $i \leq j$ in I. Hence, there exists a unique normalized positive homomorphism s from (G, u) to $(\mathbb{R}, 1)$ such that $sq_i = s_i$ for all $i \in I$. In other words, $s \in S(G, u)$ and

$$p_i g(s) = S(q_i)(s) = sq_i = s_i = p_i(x)$$

for all $i \in I$, and so $g(s) = x$, proving that g is surjective.

Therefore g is an affine continuous bijection of $S(G, u)$ onto K. By compactness, g is a homeomorphism. □

• **Faces.** The faces of a convex set may be viewed as analogs of the ideals in algebraic systems such as rings or partially ordered abelian groups. To facilitate working with faces of state spaces, we derive criteria for deciding when a state lies in the face generated by a given family of states and for deciding when finitely many given states lie in pairwise disjoint faces of the state space.

PROPOSITION 6.15. *Let (G, u) be a partially ordered abelian group with order-unit, let $s \in S(G, u)$, and let $X \subseteq S(G, u)$. Then s lies in the face generated by X in $S(G, u)$ if and only if $s \leq^+ \alpha t$ for some positive real number α and some state t in the convex hull of X.*

PROOF. If s is in the face generated by X, then by Proposition 5.7 there exists a positive convex combination $\beta s + \gamma s' = t$, where s' is in $S(G, u)$ and t is in the convex hull of X. Then $\beta s \leq^+ t$, and hence $s \leq^+ \beta^{-1} t$.

Conversely, assume that $s \leq^+ \alpha t$ for some positive real number α and some t in the convex hull of X. Then $\alpha t - s$ is a positive homomorphism from G to $\mathbb{R}$.

Hence, $\alpha t - s = \beta s'$ for some $\beta \in \mathbb{R}^+$ and some $s' \in S(G,u)$. Now $s + \beta s' = \alpha t$ and

$$1 + \beta = s(u) + \beta s'(u) = \alpha t(u) = \alpha,$$

whence $\alpha^{-1} + \alpha^{-1}\beta = 1$. Consequently, $0 < \alpha^{-1} \leq 1$ and

$$\alpha^{-1}s + (1 - \alpha^{-1})s' = t.$$

Since t lies in the face generated by X, so does s. □

PROPOSITION 6.16. *Let (G,u) be an interpolation group with order-unit, let B be the group of all relatively bounded homomorphisms from G to $\mathbb{R}$, and let $s_1, \ldots, s_n$ be states on (G,u). Then the following conditions are equivalent:*

(a) *The s_i lie in pairwise disjoint faces of $S(G,u)$.*

(b) *If i,j are any distinct indices in $\{1,\ldots,n\}$, then $s_i \wedge s_j = 0$ in B.*

(c) *Given any $x \in G^+$ and any positive real number ε, there exist $x_1,\ldots,x_n$ in G^+ such that $x = x_1 + \cdots + x_n$ and $s_i(x - x_i) < \varepsilon$ for all i.*

PROOF. Recall from Corollary 2.28 that B is lattice-ordered and from Proposition 1.7 that B is distributive as a lattice.

(a)⇒(b): As $s_i \wedge s_j$ is a positive homomorphism, $s_i \wedge s_j = \alpha s$ for some $\alpha \in \mathbb{R}^+$ and some $s \in S(G,u)$. If $\alpha > 0$, then $s \leq^+ \alpha^{-1}s_i$ and $s \leq^+ \alpha^{-1}s_j$, and so s lies in the face generated by s_i as well as in the face generated by s_j, because of Proposition 6.15. Since this is impossible, $\alpha = 0$, and thus $s_i \wedge s_j = 0$.

(b)⇒(a): For $i = 1,\ldots,n$, let F_i denote the face generated by s_i in $S(G,u)$. It suffices to show that $F_1,\ldots,F_n$ are pairwise disjoint. If not, there exists $s \in F_i \cap F_j$ for some $i \neq j$. By Proposition 6.15, there is some positive real number α for which $s \leq^+ \alpha s_i$ and $s \leq^+ \alpha s_j$. But then $\alpha^{-1}s \leq^+ s_i \wedge s_j$, contradicting the assumption that $s_i \wedge s_j = 0$. Thus the F_i must be pairwise disjoint.

(b)⇒(c): We proceed by induction on n, the case $n = 1$ being trivial. Next, suppose that $n = 2$. According to Corollary 2.28, $x = x_2 + x_1$ for some $x_1, x_2 \in G^+$ such that $s_1(x_2) + s_2(x_1) < \varepsilon$. Then $s_i(x - x_i) < \varepsilon$ for $i = 1,2$, proving that (c) holds in this case.

Now let $n > 2$, and assume that the implication (b)⇒(c) holds for $n-1$ states. Set $s = (s_2 + \cdots + s_n)/(n-1)$ in $S(G,u)$, and set $t = s_2 \vee \cdots \vee s_n$ in B. Note that

$$(n-1)s = s_2 + \cdots + s_n \leq^+ (n-1)t,$$

whence $s \leq^+ t$. Using lattice distributivity in B, we obtain

$$s_1 \wedge s \leq^+ s_1 \wedge t = (s_1 \wedge s_2) \vee (s_1 \wedge s_3) \vee \cdots \vee (s_1 \wedge s_n) = 0,$$

and so $s_1 \wedge s = 0$. Consequently, $x = x_1 + y$ for some $x_1, y \in G^+$ such that

$$s_1(x - x_1) < \varepsilon/(2n-2) \quad \text{and} \quad s(x - y) < \varepsilon/(2n-2).$$

Then $s_1(x - x_1) < \varepsilon$, and for $i = 2,\ldots,n$ we have

$$s_i(x_1) = s_i(x - y) \leq (n-1)s(x - y) < \varepsilon/2.$$

By the induction hypothesis, $y = x_2 + \cdots + x_n$ for some $x_i \in G^+$ such that $s_i(y - x_i) < \varepsilon/2$ for each $i = 2, \dots, n$. Then $x = x_1 + \cdots + x_n$ and

$$s_i(x - x_i) = s_i(x_1) + s_i(y - x_i) < \varepsilon$$

for $i = 2, \dots, n$, completing the induction step.

(c)⇒(b): Given $x \in G^+$ and a positive real number ε, there exist elements $x_1, \dots, x_n$ in G^+ such that $x = x_1 + \cdots + x_n$ and $s_k(x - x_k) < \varepsilon/2$ for each $k = 1, \dots, n$. If $y = x_j$ and $z = x - x_j$, then $y, z \in G^+$ and $x = y + z$, while also $s_j(z) < \varepsilon/2$. As $0 \le y \le x - x_i$, we also have

$$s_i(y) \le s_i(x - x_i) < \varepsilon/2,$$

and so $s_i(y) + s_j(z) < \varepsilon$.

Now $(s_i \wedge s_j)(x) = 0$, by Corollary 2.28. Since G is generated (as a group) by G^+, we conclude that $s_i \wedge s_j = 0$, as required. □

• **Change of Order-Unit.** We now consider the effect of a change of order-unit on the state space of a partially ordered abelian group. The two state spaces are at least related by a homeomorphism that corresponds faces to faces, as follows. In particular, the lattices of faces of the two state spaces are isomorphic.

PROPOSITION 6.17. *Let G be a partially ordered abelian group with order-units u and v. There exists a homeomorphism*

$$r\colon S(G, u) \to S(G, v)$$

such that $r(s) = s(v)^{-1}s$ for all $s \in S(G, u)$, and the maps r and r^{-1} both preserve extreme points and faces.

PROOF. Choose $n \in \mathbb{N}$ such that $u \le nv$ and $v \le nu$. For any state s in $S(G, u)$, we have $ns(v) \ge s(u) = 1$, and so $s(v) > 0$. Thus the function $r(s) = s(v)^{-1}s$ is a state in $S(G, v)$. This provides us with a map r from $S(G, u)$ to $S(G, v)$, and we observe that r is continuous. Similarly, the rule $r'(t) = t(u)^{-1}t$ defines a continuous map r' from $S(G, v)$ to $S(G, u)$. Observe that

$$r'r(s) = r'(s(v)^{-1}s) = [s(v)^{-1}s(u)]^{-1}s(v)^{-1}s = s$$

for all $s \in S(G, u)$, whence $r'r$ is the identity map on $S(G, u)$. Similarly, rr' is the identity map on $S(G, v)$, and thus r and r' are inverse homeomorphisms.

We claim that whenever s in $S(G, u)$ is a (positive) convex combination of states t_1, t_2 in $S(G, u)$, then $r(s)$ is a (positive) convex combination of $r(t_1)$ and $r(t_2)$ in $S(G, v)$. Given a convex combination $s = \alpha_1 t_1 + \alpha_2 t_2$ in $S(G, u)$, we compute that

$$\begin{aligned} r(s) = s(v)^{-1}s &= [\alpha_1 t_1(v)/s(v)][t_1(v)^{-1}t_1] + [\alpha_2 t_2(v)/s(v)][t_2(v)^{-1}t_2] \\ &= [\alpha_1 t_1(v)/s(v)]r(t_1) + [\alpha_2 t_2(v)/s(v)]r(t_2). \end{aligned}$$

Since $\alpha_1 t_1(v)/s(v)$ and $\alpha_2 t_2(v)/s(v)$ are nonnegative real numbers whose sum is 1, this shows that $r(s)$ is a convex combination of $r(t_1)$ and $r(t_2)$. Moreover, if each $\alpha_i > 0$, then each $\alpha_i t_i(v)/s(v)$ is positive, and hence $r(s)$ is a positive convex combination of $r(t_1)$ and $r(t_2)$. This establishes the claim.

Now let us consider any face F in $S(G,v)$. Given a convex combination $s = \alpha_1 t_1 + \alpha_2 t_2$ with $s \in S(G,u)$ and $t_1, t_2 \in r^{-1}(F)$, the claim shows that $r(s)$ is a convex combination of the points $r(t_1), r(t_2)$ from F, whence $r(s) \in F$, and so $s \in r^{-1}(F)$. Thus $r^{-1}(F)$ is convex. Given a positive convex combination $\alpha_1 s_1 + \alpha_2 s_2 = t$ with $s_1, s_2 \in S(G,u)$ and $t \in r^{-1}(F)$, the claim shows that some positive convex combination of $r(s_1), r(s_2)$ equals the point $r(t)$ in the face F, whence each $r(s_i) \in F$, and so each $s_i \in r^{-1}(F)$. Thus $r^{-1}(F)$ is a face of $S(G,u)$.

Therefore r^{-1} preserves faces. By symmetry, r preserves faces as well. As extreme points correspond to singleton faces, r and r^{-1} also preserve extreme points. $\square$

In particular, the map r in Proposition 6.17 restricts to a homeomorphism of $\partial_e S(G,u)$ onto $\partial_e S(G,v)$.

In the situation of Proposition 6.17, if $nu = kv$ for some $n,k \in \mathbb{N}$, then $r(s) = (k/n)s$ for all $s \in S(G,u)$. In this case, r is an affine homeomorphism. However, in general r is not an affine map, as may be seen in the case that $G = \mathbb{Z}^2$ while $u = (1,2)$ and $v = (2,1)$. The state spaces $S(G,u)$ and $S(G,v)$ are, however, still affinely homeomorphic in this example. That fails in general, as the following example shows.

EXAMPLE 6.18. *There exists a dimension group G with order-units u and v such that $S(G,u)$ is not affinely homeomorphic to $S(G,v)$.*

PROOF. Construct G and u as in Example 6.10. Set

$$H = \{x \in G \mid x_1 = x_2 = 0\},$$
$$K = \{x \in H \mid x_n = 0 \text{ for all but finitely many } n \in \mathbb{N}\},$$

and observe that H and K are ideals of G. We claim that any state s in $S(G,u)$ that vanishes on K must also vanish on H. Given any $x \in H$ and any positive real number ε, choose an integer $p > 1/\varepsilon$, and choose a positive integer m such that $|x_n| < 1/p$ for all $n > m$. (Such an m exists because $x_n \to 0$.) Define $y \in K$ so that $y_n = x_n$ for all $n \le m$ while $y_n = 0$ for all $n > m$. Then $|x_n - y_n| < 1/p$ for all n, whence $-u \le p(x-y) \le u$, and consequently

$$-1 \le ps(x-y) = ps(x) \le 1.$$

Thus $|s(x)| \le 1/p < \varepsilon$, proving that $s(x) = 0$. Therefore s vanishes on H, as claimed.

If $\pi\colon G \to \mathbb{R}^2$ is the map given by the rule $\pi(x) = (x_1, x_2)$, then $\pi(G^+) = (\mathbb{R}^2)^+$ and $\ker(\pi) = H$. Hence, π induces an isomorphism of G/H onto $\mathbb{R}^2$ (as ordered groups). There are exactly two extremal states on $(\mathbb{R}^2, \pi(u))$, and their compositions with π are the states s_1 and s_2. Consequently, any state on (G,u) that vanishes on H must be a convex combination of s_1 and s_2.

We claim that $\partial_e S(G,u) = \{s_1, s_2, s_3, \ldots\}$. Thus consider any extremal state s on (G,u). If s vanishes on H, then s must be a convex combination of s_1 and s_2, whence either $s = s_1$ or $s = s_2$. If s does not vanish on H, then by the claim above, s does not vanish on K either. Hence, there is an element $v \in K$ such that $v_k = 1$ for some $k > 2$ while $v_n = 0$ for all $n \neq k$ and $s(v) > 0$. Given any $x \in G^+$, we have $x_k v \leq x$ and so $x_k s(v) \leq s(x)$, whence

$$s_k(x) = x_k \leq s(v)^{-1} s(x).$$

Thus $s_k \leq^+ s(v)^{-1}s$. According to Proposition 6.15, s_k must lie in the face generated by s in $S(G,u)$. This face is just $\{s\}$, so that $s_k = s$. Therefore $s_1, s_2, s_3, \ldots$ are the only extremal states on (G,u), as claimed.

At this point, we claim that there do not exist extremal states $t_1, t_2, t_3, \ldots$ on (G,u) such that $t_1 \neq t_2$ and $t_n \to \frac{1}{3}t_1 + \frac{2}{3}t_2$. If such extremal states do exist, then as $\frac{1}{3}t_1 + \frac{2}{3}t_2$ is not extremal, the set $\{t_1, t_2, t_3, \ldots\}$ must be infinite. As a result, infinitely many of $s_1, s_2, s_3, \ldots$ appear in this sequence, and hence $t_n \to \frac{1}{2}s_1 + \frac{1}{2}s_2$. Now

$$\tfrac{1}{3}t_1 + \tfrac{2}{3}t_2 = \tfrac{1}{2}s_1 + \tfrac{1}{2}s_2.$$

By inspection, there do not exist $t_1, t_2 \in \partial_e S(G,u)$ with this property, and so the claim holds.

Now set $v = (\frac{2}{3}, \frac{4}{3}, 1, 1, 1, \ldots)$, and observe that v is an order-unit in G. The maps $s_3, s_4, \ldots$ are states on (G,v), and by Proposition 6.17 they are all extremal. In addition, the maps

$$t_1 = \tfrac{3}{2}s_1 = s_1(v)^{-1}s_1 \quad \text{and} \quad t_2 = \tfrac{3}{4}s_2 = s_2(v)^{-1}s_2$$

are distinct extremal states on (G,v), by Proposition 6.17. Hence, $t_1, t_2, s_3, s_4, s_5, \ldots$ are extreme points of $S(G,v)$ such that $t_1 \neq t_2$ and

$$s_n \to \tfrac{1}{2}s_1 + \tfrac{1}{2}s_2 = \tfrac{1}{3}t_1 + \tfrac{2}{3}t_2.$$

As no such sequence of extreme points exists in $S(G,u)$, we conclude that $S(G,u)$ and $S(G,v)$ cannot be affinely homeomorphic. □

• **Discrete States.** We have already seen that associated with a discrete state on an interpolation group is a quotient group that is simplicial (Proposition 4.22). This will enable us to express any such discrete state as a convex combination, with rational coefficients, of discrete extremal states. For a discrete extremal state on an interpolation group, the corresponding quotient group is naturally isomorphic to a discrete additive subgroup of $\mathbb{R}$, as follows.

PROPOSITION 6.19. *Let (G,u) be an interpolation group with order-unit, and let s be a discrete extremal state on (G,u). Then* $\ker(s)$ *is an ideal of G, the quotient group $G/\ker(s)$ is totally ordered, and the group homomorphism of $G/\ker(s)$ into $\mathbb{R}$ induced by s provides an isomorphism of $G/\ker(s)$ onto $s(G)$ (as ordered groups).*

PROOF. Let H be the subgroup of G generated by $\ker(s)^+$. According to Proposition 4.22, H is an ideal of G, and the quotient G/H is simplicial. If

$p\colon G \to G/H$ is the quotient map, then s induces a state t on $(G/H, p(u))$ such that $tp = s$, and we note, by definition of H, that $\ker(t)^+ = \{0\}$.

Choose a simplicial basis $\{x_1, \ldots, x_n\}$ for G/H, and write

$$p(u) = a_1x_1 + \cdots + a_nx_n$$

for some integers a_i. By Proposition 6.6, each $a_i > 0$, and $S(G/H, p(u))$ equals the convex hull of the states $s_1, \ldots, s_n$ defined by the rule

$$s_j(c_1x_1 + \cdots + c_nx_n) = c_j/a_j.$$

Then t is a convex combination of $s_1, \ldots, s_n$, and hence s is a convex combination of $s_1p, \ldots, s_np$. As s is extremal, it follows that $s = s_jp$ for some j, whence $t = s_j$. Since $\ker(s_j)^+$ is nonzero if $n > 1$, we must have $n = 1$.

Thus $G/H = \mathbb{Z}x_1 \cong \mathbb{Z}$ (as ordered groups), and hence G/H is totally ordered. Also, G/H must have only two convex subgroups, namely $\{0\}$ and G/H. As a result, $\ker(t) = \{0\}$, so that $\ker(s) = H$. Therefore $\ker(s)$ is an ideal of G, and $G/\ker(s)$ is totally ordered. The group homomorphism of $G/\ker(s)$ into $\mathbb{R}$ induced by s is just the state $t = s_1$, which is an isomorphism of $G/\ker(s)$ onto $s(G)$ (as ordered groups). □

COROLLARY 6.20. *Let (G, u) be an interpolation group with order-unit, let $s_1, \ldots, s_n$ be distinct discrete extremal states on (G, u), and set*

$$H = \ker(s_1) \cap \cdots \cap \ker(s_n).$$

Then H is an ideal of G, the quotient group G/H is simplicial, and the group homomorphism of G/H into $\mathbb{R}^n$ induced by $(s_1, \ldots, s_n)$ provides an isomorphism of G/H onto the subgroup

$$s_1(G) \times \cdots \times s_n(G) \subseteq \mathbb{R}^n$$

(as ordered groups).

PROOF. Set $H_i = \ker(s_i)$ for each $i = 1, \ldots, n$. By Proposition 6.19, each H_i is an ideal of G, and each s_i induces an isomorphism of G/H_i onto $s_i(G)$ (as ordered groups). In particular, G/H_i has no ideals except itself and $\{0\}$, whence H_i is a maximal proper ideal of G. Also, Proposition 2.4 shows that H is an ideal of G.

If $H_i = H_j$ for some i, j, then s_i induces a state on $(G/H_j, u + H_j)$. As G/H_j is totally ordered, this state must coincide with the state induced by s_j (Corollary 4.17), whence $s_i = s_j$ and so $i = j$. Thus $H_i \neq H_j$ whenever $i \neq j$. In this case, since $H_i + H_j$ is an ideal of G (Proposition 2.4), it follows from the maximality of H_i that $H_i + H_j = G$. Hence, because of Proposition 2.5,

$$H_i + \left(\bigcap_{j \neq i} H_j \right) = \bigcap_{j \neq i} (H_i + H_j) = G$$

for all $i = 1, \ldots, n$. As a result, the natural map

$$G/H \to (G/H_1) \times \cdots \times (G/H_n)$$

is an isomorphism of abelian groups, and hence the map

$$q\colon G/H \to s_1(G) \times \cdots \times s_n(G)$$

induced by $(s_1, \ldots, s_n)$ is a group isomorphism.

Obviously q is a positive homomorphism. Conversely, consider any coset $x+H$ in G/H such that $q(x+H) \geq 0$. Then each $s_i(x) \geq 0$, whence $x + H_i \geq 0$, and so $0 \leq x + a_i$ for some $a_i \in H_i^+$. Now $0 \leq a_i$ and $-x \leq a_i$ for all i, and hence there exists $a \in G$ such that

$$\begin{matrix} 0 \\ -x \end{matrix} \leq a \leq a_i \quad (\text{for } i = 1, \ldots, n).$$

Since $0 \leq a \leq a_i$ for all i, we have $a \in H_i$ for all i, so that $a \in H^+$. Then from the inequality $0 \leq x + a$ we conclude that $x + H \geq 0$.

Therefore q is an isomorphism of ordered groups. As each $s_i(G)$ is cyclic, it follows that G/H is simplicial. □

COROLLARY 6.21. *Let (G,u) be an interpolation group with order-unit, and let s be a discrete state on (G,u). Then s is extremal if and only if given any $x, y \in G^+$, there exists $z \in G^+$ such that*

$$z \leq \begin{matrix} x \\ y \end{matrix} \quad \textit{and} \quad s(z) = \min\{s(x), s(y)\}.$$

PROOF. First assume that s is extremal, and consider any $x, y \in G^+$. Either $s(x) \leq s(y)$ or $s(y) \leq s(x)$; say $s(x) \leq s(y)$. In view of Proposition 6.19, it follows that

$$x + \ker(s) \leq y + \ker(s),$$

and then $x \leq y + a$ for some $a \in \ker(s)^+$. Hence, $x = z + b$ for some $z, b \in G^+$ such that $z \leq y$ and $b \leq a$. Note that $z \leq x$ and that $b \in \ker(s)$, whence

$$s(z) = s(x) = \min\{s(x), s(y)\}.$$

If s is not extremal, there exists a positive convex combination $s = \alpha_1 t_1 + \alpha_2 t_2$ where t_1 and t_2 are distinct states on (G,u). Then $t_1(a) \neq t_2(a)$ for some $a \in G$. Replacing a by $-a$, if necessary, we may assume that $t_1(a) < t_2(a)$. Choose $m \in \mathbb{N}$ such that the element $b = a + mu$ lies in G^+. Since $0 \leq t_1(b) < t_2(b)$, there exist positive integers k, n such that

$$t_1(b) < k/n < t_2(b).$$

Set $x = nb$ and $y = ku$. Then x and y are elements of G^+ such that $t_1(x) < t_1(y)$ and $t_2(y) < t_2(x)$. For any $z \in G^+$ satisfying $z \leq x$ and $z \leq y$, we have

$$s(z) = \alpha_1 t_1(z) + \alpha_2 t_2(z) \leq \alpha_1 t_1(x) + \alpha_2 t_2(y) < \alpha_1 t_1(x) + \alpha_2 t_2(x) = s(x),$$

and similarly $s(z) < s(y)$. Thus $s(z) < \min\{s(x), s(y)\}$. □

Indiscrete extremal states on interpolation groups need not satisfy the condition given in Corollary 6.21. For example, let (G,u) be the dimension group with order-unit constructed in Example 4.20, and let s be the unique state on (G, u). As $S(G, u)$ is a single point, s must be extremal. If $v = (1, 1)$ in G, then $s(u) = s(v) = 1$, but there does not exist an element $z \in G^+$ such that $z \leq u$ and $z \leq v$ while also $s(z) = 1$.

In general, the condition of Corollary 6.21 holds "to within epsilon" for any extremal state on an interpolation group, as we shall prove in Theorem 12.14.

DEFINITION. A *rational convex combination* is any convex combination in which all the coefficients are rational numbers.

PROPOSITION 6.22. *Let (G, u) be an interpolation group with order-unit, and let s be a state on (G, u). Then s is discrete if and only if s is a rational convex combination of discrete extremal states.*

PROOF. Given a rational convex combination $s = \alpha_1 s_1 + \cdots + \alpha_n s_n$ where each s_i is a discrete extremal state on (G, u), choose a positive integer k such that each $\alpha_i \in (1/k)\mathbb{Z}$. Since each of the groups $s_i(G)$ is cyclic, there exists a positive integer m such that each $s_i(G) \subseteq (1/m)\mathbb{Z}$. Then $s(G) \subseteq (1/km)\mathbb{Z}$, and hence $s(G)$ is cyclic. Thus s is discrete.

Conversely, assume that s is discrete, and let H be the subgroup of G generated by $\ker(s)^+$. By Proposition 4.22, H is an ideal of G, and G/H is a simplicial group. Let $p\colon G \to G/H$ be the quotient map, and let t be the state on $(G/H, p(u))$ induced by s, so that $tp = s$.

Let $\{x_1, \ldots, x_n\}$ be a simplicial basis for G/H, and write

$$p(u) = a_1 x_1 + \cdots + a_n x_n$$

for some integers a_i. According to Proposition 6.6, each $a_i > 0$, the extremal states on $(G/H, p(u))$ are exactly the states $t_1, \ldots, t_n$ given by the rule

$$t_j(c_1 x_1 + \cdots + c_n x_n) = c_j / a_j,$$

and $S(G/H, p(u))$ equals the convex hull of $\{t_1, \ldots, t_n\}$. By Proposition 6.12, each of the states $s_j = S(p)(t_j) = t_j p$ is extremal. In addition,

$$s_j(G) = t_j(G/H) = (1/a_j)\mathbb{Z}.$$

Thus s_j is a discrete extremal state of (G, u).

Now there exist convex combinations

$$t = \alpha_1 t_1 + \cdots + \alpha_n t_n \quad \text{and} \quad s = \alpha_1 s_1 + \cdots + \alpha_n s_n.$$

For each $j = 1, \ldots, n$, we observe that

$$\alpha_j = t(a_j x_j) \in t(G/H) = s(G).$$

Since $s(G) = (1/m)\mathbb{Z}$ for some $m \in \mathbb{N}$, we conclude that $\alpha_j \in \mathbb{Q}$. Therefore s is a rational convex combination of the s_j. □

• **Notes.** Theorem 6.3 and Corollary 6.4 are straightforward refinements of Corollary 4.13, which is due to Effros, Handelman, and Shen [**30**, Theorem

1.4]. Example 6.10 is a modification of an example discovered independently by Alfsen [**1**, Theorem 1] and Lindenstrauss [**91**, p. 78]. Proposition 6.15 was proved by the author, Handelman, and Lawrence [**60**, Proposition I.2.4]. The equivalence of conditions (a) and (c) in Proposition 6.16 is due to the author and Handelman [**57**, Lemmas 2.7 and 2.8]. Proposition 6.17 and Example 6.18 are the state space analogs of results proved for pseudo-rank functions on regular rings by the author [**49**, Theorem 17.14 and Example 17.15]. Propositions 6.19 and 6.22 were proved by the author and Handelman (unpublished); that any discrete state on an interpolation group with order-unit is a convex combination of discrete extremal states was derived in the course of proving [**58**, Lemma 2.11].

CHAPTER 7

Representation By Affine Continuous Functions

The inverse of the "duality" between partially ordered abelian groups with order-unit and compact convex sets is the contravariant functor that assigns to each compact convex set K the partially ordered real Banach space $\mathrm{Aff}(K)$ of all affine continuous real-valued functions on K, together with the order-unit 1 in $\mathrm{Aff}(K)$. When restricted to the category of compact convex subsets of locally convex Hausdorff spaces, the composite of this functor followed by the state space functor is naturally equivalent to the identity functor. The opposite composite of these two functors provides us with a natural "double dual" machinery. Namely, any partially ordered abelian group (G,u) with order-unit has, via evaluations, a natural normalized positive homomorphism into $\mathrm{Aff}(S(G,u))$, which provides a representation of G in terms of affine continuous real-valued functions on a compact convex set.

In this chapter we introduce affine continuous function spaces and the affine continuous function representation of a partially ordered abelian group with order-unit. We begin investigating the extent to which this representation is faithful and the consequent exploitation of this representation as a means of translating partially ordered abelian group questions into affine continuous function questions.

• Affine Continuous Function Spaces.

DEFINITION. Given a compact convex set K in a linear topological space, we use $\mathrm{Aff}(K)$ to denote the collection of all affine continuous real-valued functions on K.

Of course $\mathrm{Aff}(K)$ is a linear subspace of $C(K,\mathbb{R})$ and so is a partially ordered real vector space. (By our earlier conventions, we always assume, unless otherwise specified, that $\mathrm{Aff}(K)$ has been equipped with the pointwise ordering.) As with $C(K,\mathbb{R})$, the constant function 1 is an order-unit in $\mathrm{Aff}(K)$. We also equip $\mathrm{Aff}(K)$ with the supremum norm. Since $\mathrm{Aff}(K)$ is easily seen to be a norm-closed subspace of $C(K,\mathbb{R})$, it is thus a real Banach space. We may refer to all this structure on $\mathrm{Aff}(K)$ by saying that it is a "partially ordered real Banach space with order-unit 1". Unlike $C(K,\mathbb{R})$, the space $\mathrm{Aff}(K)$ need not be lattice-ordered. For example, if K is a square in $\mathbb{R}^2$, then $\mathrm{Aff}(K)$ is isomorphic to the partially ordered real vector space

$$A = \{(x_1, x_2, x_3, x_4) \in \mathbb{R}^4 \mid x_1 + x_2 = x_3 + x_4\}$$

(to see this, evaluate at the vertices of the square). In A, the elements $(1, -1, 0, 0)$ and $(-1, 1, 0, 0)$ have no supremum.

We shall mainly be concerned with the situation in which the linear topological space containing K is locally convex and Hausdorff. In this case, there are enough functions in $\mathrm{Aff}(K)$ to separate the points of K, by Corollary 5.15. We shall use this observation in the future without explicit reference. Our first task is to show that the state space of $(\mathrm{Aff}(K), 1)$ may be naturally identified with K. In particular, this proves that every compact convex subset of a locally convex Hausdorff space is affinely homeomorphic to a state space of a partially ordered real vector space with order-unit.

This theorem may be viewed functorially as follows. First, set $A(K) = (\mathrm{Aff}(K), 1)$ for any compact convex set K. Second, for any affine continuous map $p\colon K \to L$ between compact convex sets, let $A(p)$ be the normalized positive homomorphism from $A(L)$ to $A(K)$ defined by composition with p, so that $A(p)(q) = qp$ for all $q \in A(L)$. Then $A(—)$ is a contravariant functor from the category of compact convex sets to the category of partially ordered abelian groups with order-unit. If the domain of A is restricted to the category of compact convex subsets of locally convex Hausdorff spaces, then the composite functor SA (where S denotes the state space functor) is naturally equivalent to the identity functor on the restricted compact convex set category. As we shall have little need of the functorial properties of A, we leave their development to the reader's inclinations.

THEOREM 7.1. *Let K be a compact convex subset of a locally convex Hausdorff space, and let $S = S(\mathrm{Aff}(K), 1)$. Then the evaluation map $\psi\colon K \to S$ is an affine homeomorphism of K onto S.*

PROOF. It is clear that evaluation at any point $x \in K$ defines a state $\psi(x)$ in S (that is, $\psi(x)(f) = f(x)$ for all $f \in \mathrm{Aff}(K)$). Thus ψ is indeed a map from K to S. It is clear that ψ is affine and continuous. Since $\mathrm{Aff}(K)$ separates the points of K, we also see that ψ is injective.

Recall from Proposition 6.8 that the state space of $(C(K, \mathbb{R}), 1)$ is $M_1^+(K)$. Let ρ denote the restriction map

$$M_1^+(K) = S(C(K, \mathbb{R}), 1) \to S(\mathrm{Aff}(K), 1) = S,$$

and note that ρ is affine and continuous. By Corollary 4.3, we know that all states on $(\mathrm{Aff}(K), 1)$ extend to states on $(C(K, \mathbb{R}), 1)$, that is, ρ is surjective.

Given any $s \in \partial_e S$, we observe that $\rho^{-1}(\{s\})$ is a nonempty closed face of $M_1^+(K)$. Hence, by Corollary 5.18, there exists a measure

$$\lambda \in \rho^{-1}(\{s\}) \cap \partial_e M_1^+(K).$$

Proposition 5.24 then shows that $\lambda = \varepsilon_x$ for some point $x \in K$. Then $\rho(\varepsilon_x) = s$ and hence

$$s(f) = \varepsilon_x(f) = f(x) = \psi(x)(f)$$

for all $f \in \mathrm{Aff}(K)$, whence $s = \psi(x)$. Thus $\partial_e S \subseteq \psi(K)$.

As $\psi(K)$ is a compact convex subset of S, we conclude from the Krein-Mil'man Theorem that $\psi(K) = S$. Therefore ψ is an affine continuous bijection of K onto S. By compactness, ψ is a homeomorphism. □

DEFINITION. The evaluation map ψ in Theorem 7.1 is called *the natural affine homeomorphism of K onto $S(\mathrm{Aff}(K), 1)$*.

COROLLARY 7.2. *Let K be a compact convex subset of a locally convex Hausdorff space, and let A denote the real vector space $\mathrm{Aff}(K)$ equipped with the strict ordering. Then the evaluation map $K \to S(A, 1)$ is an affine homeomorphism of K onto $S(A, 1)$.*

PROOF. Proposition 6.9 and Theorem 7.1. □

COROLLARY 7.3. *Let K be a nonempty compact convex subset of a locally convex Hausdorff space, and let p be a continuous linear functional on $\mathrm{Aff}(K)$. Then there exist $\alpha_1, \alpha_2 \in \mathbb{R}^+$ and $x_1, x_2 \in K$ such that*

$$p(f) = \alpha_1 f(x_1) - \alpha_2 f(x_2)$$

for all $f \in \mathrm{Aff}(K)$.

PROOF. By the Hahn-Banach Theorem, p extends to a continuous linear functional q on $C(K, \mathbb{R})$. Viewing q as a signed regular Borel measure on K, take its Hahn decomposition into positive and negative parts. This provides us with positive linear functionals q_1 and q_2 on $C(K, \mathbb{R})$ such that $q = q_1 - q_2$. (Alternatively, check that q is relatively bounded, and apply Proposition 2.25 to the interpolation group $C(K, \mathbb{R})$.) Restricting q_1 and q_2 to $\mathrm{Aff}(K)$, we obtain positive linear functionals p_1 and p_2 on $\mathrm{Aff}(K)$ such that $p = p_1 - p_2$.

By Lemma 4.5, each $p_i = \alpha_i s_i$ for some $\alpha_i \in \mathbb{R}^+$ and some state s_i on $(\mathrm{Aff}(K), 1)$. Since each s_i coincides with evaluation at a point $x_i \in K$ by Theorem 7.1, the desired formula for p is obtained. □

COROLLARY 7.4. *Let K be a nonempty compact convex subset of a locally convex Hausdorff space, and let A be a linear subspace of $\mathrm{Aff}(K)$. If A contains the constant functions and separates the points of K, then A is dense in $\mathrm{Aff}(K)$.*

PROOF. If the closure of A is a proper subspace of $\mathrm{Aff}(K)$, then there exists a nonzero continuous linear functional p on $\mathrm{Aff}(K)$ that vanishes on A. According to Corollary 7.3, there exist $\alpha_1, \alpha_2 \in \mathbb{R}^+$ and $x_1, x_2 \in K$ such that

$$p(f) = \alpha_1 f(x_1) - \alpha_2 f(x_2)$$

for all $f \in \mathrm{Aff}(K)$. Since $1 \in A$, we have $\alpha_1 - \alpha_2 = p(1) = 0$. Hence, $\alpha_1 > 0$. Now

$$\alpha_1(f(x_1) - f(x_2)) = p(f) = 0$$

for all $f \in A$, so that $f(x_1) = f(x_2)$ for all $f \in A$. As A separates the points of K, we conclude that $x_1 = x_2$. But then $p = 0$, which is a contradiction.

Therefore A is dense in $\mathrm{Aff}(K)$. $\square$

COROLLARY 7.5. *Let X be a compact Hausdorff space, and let δ be the natural homeomorphism of X onto $\partial_e M_1^+(X)$. Let K be a compact convex subset of a locally convex Hausdorff space, and let $f\colon X \to K$ be a continuous map. Then there exists a unique affine continuous map $g\colon M_1^+(X) \to K$ such that $g\delta = f$.*

PROOF. Let $f^*\colon \mathrm{Aff}(K) \to C(X, \mathbb{R})$ be the map induced by f, so that $f^*(p) = pf$ for all $p \in \mathrm{Aff}(K)$. Then f^* is a normalized positive homomorphism from $(\mathrm{Aff}(K), 1)$ to $(C(X, \mathbb{R}), 1)$, and hence f^* induces an affine continuous map

$$S(f^*)\colon S(C(X, \mathbb{R}), 1) \to S(\mathrm{Aff}(K), 1).$$

Of course $S(C(X, \mathbb{R}), 1) = M_1^+(X)$ (Proposition 6.8). Now let ψ be the natural affine homeomorphism of K onto $S(\mathrm{Aff}(K), 1)$. Then the map $g = \psi^{-1}S(f^*)$ is an affine continuous map from $M_1^+(X)$ to K.

Given any $x \in X$, we observe that

$$S(f^*)\delta(x) = S(f^*)(\varepsilon_x) = \varepsilon_x f^* = \psi f(x).$$

Thus $S(f^*)\delta = \psi f$, whence $g\delta = f$. The uniqueness of g is immediate from the Krein-Mil'man Theorem. $\square$

As any compact convex set K in a locally convex Hausdorff space may be identified (via Theorem 7.1) with the state space of $(\mathrm{Aff}(K), 1)$, any general result proved for arbitrary state spaces must hold for K. An example of this is the existence of infinite convex combinations in K, as follows.

PROPOSITION 7.6. *Let K be a compact convex subset of a locally convex Hausdorff space E, let $x_1, x_2, \dots$ be points of K, and let $\alpha_1, \alpha_2, \dots$ be nonnegative real numbers such that $\sum \alpha_k = 1$. Then there exists a point $x \in K$ such that $\sum \alpha_k x_k \to x$ in E and $f(x) = \sum \alpha_k f(x_k)$ for all $f \in \mathrm{Aff}(K)$. If $\alpha_1 > 0$, then*

$$\sum_{k=1}^{n} \alpha_k(\alpha_1 + \cdots + \alpha_n)^{-1} x_k \to x$$

(as $n \to \infty$) in K.

PROOF. There is no loss of generality in assuming that $\alpha_1 > 0$. Set $S = S(\mathrm{Aff}(K), 1)$, and let $\psi\colon K \to S$ be the natural affine homeomorphism. By Proposition 6.5, there is a state $s \in S$ such that $s(f) = \sum \alpha_k \psi(x_k)(f)$ for all $f \in \mathrm{Aff}(K)$, and

$$\sum_{k=1}^{n} \alpha_k(\alpha_1 + \cdots + \alpha_n)^{-1} \psi(x_k) \to s$$

in S. Then $s = \psi(x)$ for some $x \in K$ such that $f(x) = \sum \alpha_k f(x_k)$ for all $f \in \text{Aff}(K)$, and

$$\sum_{k=1}^{n} \alpha_k(\alpha_1 + \cdots + \alpha_n)^{-1} x_k \to x$$

in K. As $\alpha_1 + \cdots + \alpha_n \to 1$, we conclude that $\sum \alpha_k x_k \to x$ in E. □

DEFINITION. In the situation of Proposition 7.6, we write $x = \sum \alpha_k x_k$, and we say that x is a *σ-convex combination* of the x_k. In case all the $\alpha_k > 0$, we say that x is a *positive σ-convex combination* of the x_k. A *σ-convex subset* of K is any subset of K that is closed under σ-convex combinations. Note that the intersection of any family of σ-convex subsets of K is σ-convex. Hence, given any subset $X \subseteq K$, there is a smallest σ-convex subset of K that contains X, called the *σ-convex hull of X*.

Note that any ordinary convex combination can be rewritten as a σ-convex combination in which all but finitely many of the coefficients are zero. Hence, every σ-convex subset of K is also a convex subset.

• Affine Representations.

DEFINITION. Let (G, u) be a partially ordered abelian group with order-unit. Evaluation at any point $x \in G$ defines a map

$$\hat{x}\colon S(G,u) \to \mathbb{R}$$

(so that $\hat{x}(s) = s(x)$ for all $s \in S(G,u)$). It is clear from the definitions of the pointwise operations and the product topology on $\mathbb{R}^G$ that $\hat{x}$ is affine and continuous. Hence, the rule $\phi(x) = \hat{x}$ defines a map

$$\phi\colon G \to \text{Aff}(S(G,u)),$$

known as *the natural map from G to* Aff$(S(G,u))$. As each $\hat{x}$ restricts to a continuous real-valued function on the extreme boundary of $S(G,u)$, we likewise obtain a map

$$\psi\colon G \to C(\partial_e S(G,u), \mathbb{R})$$

(where $\psi(x)(s) = s(x)$ for all $x \in G$ and $s \in \partial_e S(G,u)$), known as *the natural map from G to $C(\partial_e S(G,u), \mathbb{R})$*.

We observe that ϕ and ψ are positive homomorphisms and that $\phi(u)$ and $\psi(u)$ are the constant function 1. The map ϕ may be thought of as an "affine representation" of G, in that ϕ does provide a representation (usually not faithful) of G in terms of affine continuous real-valued functions on a compact convex set. This representation is faithful (in the sense of partially ordered abelian groups) if and only if G is archimedean, as follows.

THEOREM 7.7. *For a partially ordered abelian group (G, u) with order-unit, the following conditions are equivalent*:

(a) *The natural map $\phi\colon G \to$* Aff$(S(G,u))$ *provides an isomorphism of G onto a subgroup of* Aff$(S(G,u))$ *(as ordered groups)*.

(b) *The natural map* $\psi\colon G \to C(\partial_e S(G,u), \mathbb{R})$ *provides an isomorphism of* G *onto a subgroup of* $C(\partial_e S(G,u), \mathbb{R})$ *(as ordered groups).*

(c) G *is archimedean.*

PROOF. (a) $\Rightarrow$ (c): Since $\mathrm{Aff}(S(G,u))$ is archimedean, this implication is clear.

(c) $\Rightarrow$ (b): If $x \in G$ such that $\psi(x) \geq 0$, then $\phi(x) \geq 0$ on $\partial_e S(G,u)$. By Corollary 5.20, $\phi(x) \geq 0$, and then Theorem 4.14 shows that $x \geq 0$. Hence,

$$G^+ = \psi^{-1}(C(\partial_e S(G,u), \mathbb{R})^+).$$

In particular, any $x \in \ker(\psi)$ satisfies $\psi(x) \geq 0$ and $\psi(-x) \geq 0$, whence $x \geq 0$ and $-x \geq 0$, and thus $x = 0$. Hence, ψ is injective. Thus ψ provides an isomorphism of G onto $\psi(G)$ (as ordered groups).

(b) $\Rightarrow$ (a): Any $x \in G$ satisfying $\phi(x) \geq 0$ also satisfies $\psi(x) \geq 0$, and so $x \geq 0$, by (b). Hence,

$$G^+ = \phi^{-1}(\mathrm{Aff}(S(G,u))^+).$$

As above, it follows that ϕ provides an isomorphism of G onto $\phi(G)$ (as ordered groups). □

Although the natural map $\phi\colon G \to \mathrm{Aff}(S(G,u))$ does not in general completely determine the ordering on a partially ordered abelian group (G,u) with order-unit, it does have considerable influence on the ordering in G. For instance, we may rewrite Theorems 4.12 and 6.3 and Corollaries 4.13 and 6.4 in terms of ϕ and the natural map from G to $C(\partial_e S(G,u), \mathbb{R})$ as follows.

THEOREM 7.8. *Let* (G,u) *be a partially ordered abelian group with order-unit, let* $x \in G$, *and let*

$$\phi\colon G \to \mathrm{Aff}(S(G,u)) \quad \textit{and} \quad \psi\colon G \to C(\partial_e S(G,u), \mathbb{R})$$

be the natural maps. Then $\phi(x) \gg 0$ *if and only if* $\psi(x) \gg 0$ *if and only if there exists a positive integer* m *such that* mx *is an order-unit in* G. *In the case that* G *is unperforated,* $\phi(x) \gg 0$ *if and only if* $\psi(x) \gg 0$ *if and only if* x *is an order-unit in* G. □

In particular, Theorem 7.8 says that $\phi(x) \gg 0$ implies $mx \geq 0$ for some $m \in \mathbb{N}$. For consequences of the weaker assumption $\phi(x) \geq 0$, see Proposition 4.9 and Corollary 4.10.

At the other end of the representation, the most useful goal would be a complete description of $\phi(G)$. This has been achieved only under fairly restrictive hypotheses on G, such as Dedekind σ-completeness (Corollary 9.15) or norm-completeness (Theorem 15.7). Failing this, it is still quite desirable to have a description of the closure of $\phi(G)$ in $\mathrm{Aff}(S(G,u))$. Such a description is derived for 2-unperforated interpolation groups in Corollary 13.6. In the absence of any divisibility assumptions, $\phi(G)$ should not be expected to be dense in $\mathrm{Aff}(S(G,u))$ (e.g., consider the case that $G = \mathbb{Z}^n$). However, $\phi(G)$ does become dense in $\mathrm{Aff}(S(G,u))$ when suitable denominators are adjointed, as follows.

THEOREM 7.9. *Let (G, u) be a nonzero partially ordered abelian group with order-unit, and let $\phi\colon G \to \mathrm{Aff}(S(G, u))$ be the natural map. Then the subgroup*

$$\{\phi(x)/2^n \mid x \in G \text{ and } n \in \mathbb{N}\}$$

is dense in $\mathrm{Aff}(S(G, u))$. *Moreover, if G is unperforated, then the set*

$$\{\phi(x)/2^n \mid x \in G^+ \text{ and } n \in \mathbb{N}\}$$

is dense in $\mathrm{Aff}(S(G, u))^+$.

PROOF. Set $A = \mathrm{Aff}(S(G, u))$, let B be the linear subspace of A spanned by $\phi(G)$, and let

$$C = \{\phi(x)/2^n \mid x \in G \text{ and } n \in \mathbb{N}\}.$$

Since $\phi(u) = 1$, all the constant functions lie in B. As B separates the points of $S(G, u)$, Corollary 7.4 shows that B is dense in A. Observing that C is dense in B, we conclude that C is dense in A.

Now suppose that G is unperforated. Given any $p \in A^+$ and any positive real number ε, there exist $x \in G$ and $n \in \mathbb{N}$ such that

$$\|\phi(x)/2^n - (p + \varepsilon/2)\| < \varepsilon/2,$$

whence $p \ll \phi(x)/2^n \ll p + \varepsilon$. As $\phi(x) \gg 2^n p \geq 0$, Theorem 7.8 shows that $x \in G^+$. Therefore the functions of the form $\phi(x)/2^n$, where $x \in G^+$ and $n \in \mathbb{N}$, comprise a dense subset of A^+. □

COROLLARY 7.10. *Let (G, u) be a nonzero partially ordered abelian group with order-unit, and let X and Y be disjoint nonempty compact convex subsets of $S(G, u)$. Then there exists $x \in G$ such that*

$$\sup\{s(x) \mid s \in X\} < 0 < \inf\{t(x) \mid t \in Y\}.$$

PROOF. According to Theorem 5.14, there exists a continuous linear functional p on $\mathbb{R}^G$ such that

$$\sup\{p(s) \mid s \in X\} < \inf\{p(t) \mid t \in Y\}.$$

Choose a real number α such that $\sup(p(X)) < \alpha < \inf(p(Y))$. Then $p - \alpha$ restricts to an affine continuous real-valued function q on $S(G, u)$ such that

$$\sup\{q(s) \mid s \in X\} < 0 < \inf\{q(t) \mid t \in Y\}.$$

Choose a positive real number ε such that $q \leq -\varepsilon$ on X while $q \geq \varepsilon$ on Y.

By Theorem 7.9, there exist $x \in G$ and $n \in \mathbb{N}$ such that

$$\|\phi(x)/2^n - q\| < \varepsilon,$$

where $\phi\colon G \to \mathrm{Aff}(S(G, u))$ is the natural map. Then $\phi(x)/2^n \ll q + \varepsilon \leq 0$ on X, so that $\phi(x) \ll 0$ on X, and similarly $\phi(x) \gg 0$ on Y. By compactness,

$$\sup\{s(x) \mid s \in X\} = \sup\{\phi(x)(s) \mid s \in X\} < 0,$$

and similarly $\inf\{t(x) \mid t \in Y\} > 0$. □

COROLLARY 7.11. *Let (G,u) be a nonzero partially ordered abelian group with order-unit, and let X be a nonempty subset of $S(G,u)$. Assume that whenever $x \in G$ satisfies $s(x) > 0$ for all $s \in X$, then $t(x) \geq 0$ for all $t \in S(G,u)$. Then $\partial_e S(G,u)$ is contained in the closure of X.*

PROOF. Let Y be the closure of the convex hull of X. If there exists a state t in $S(G,u)$ that is not in Y, then by Corollary 7.10 there exists $x \in G$ such that $t(x) < 0$ while $s(x) > 0$ for all $s \in Y$. But then $s(x) > 0$ for all $s \in X$, and we have a contradiction. Therefore $Y = S(G,u)$, and the desired conclusion follows from Theorem 5.21. □

• **Order-Unit Norms.**

DEFINITION. Let (G,u) be a partially ordered abelian group with order-unit. For all $x \in G$, define

$$\|x\|_u = \inf\{k/n \mid k,n \in \mathbb{N} \text{ and } -ku \leq nx \leq ku\}.$$

The function $\|\cdot\|_u$ is known as *the order-unit norm relative to* u on G, because of the norm-like properties exhibited in the following proposition. We abbreviate $\|\cdot\|_u$ to $\|\cdot\|$ whenever there is no danger of confusion as to the order-unit.

PROPOSITION 7.12. *Let (G,u) be a partially ordered abelian group with order-unit, and let $\phi\colon G \to \mathrm{Aff}(S(G,u))$ be the natural map. Let $x,y \in G$ and $m \in \mathbb{Z}$.*

(a) $\|mx\| = |m| \cdot \|x\|$.

(b) $\|x+y\| \leq \|x\| + \|y\|$.

(c) *If $-y \leq x \leq y$, then $\|x\| \leq \|y\|$.*

In the case that G is nonzero, then

(d) $\|u\| = 1$.

(e) $\|x\| = \max\{|s(x)| \colon s \in S(G,u)\} = \|\phi(x)\|$.

PROOF. If $G = \{0\}$, then $\|z\| = 0$ for all $z \in G$, in which case (a), (b), (c) are clear. Thus we may assume that G is nonzero. As (a), (b), (c), (d) are immediate consequences of (e) in this case, it suffices to prove (e). (Properties (a), (b), (c), (d) may also be proved directly, by routine calculations.)

Following Proposition 4.7, define

$$p = \sup\{k/m \mid k \in \mathbb{Z}; m \in \mathbb{N}; ku \leq mx\},$$
$$r = \inf\{l/n \mid l \in \mathbb{Z}; n \in \mathbb{N}; nx \leq lu\}.$$

Then $-\infty < p \leq r < \infty$ and $[p,r] = \{s(x) \mid s \in S(G,u)\}$, whence

$$\max\{|p|,|r|\} = \max\{|s(x)| \colon s \in S(G,u)\}.$$

Hence, we need only show that $\|x\| = \max\{|p|,|r|\}$.

If $k,n \in \mathbb{N}$ and $-ku \leq nx \leq ku$, then

$$-k/n \leq p \leq r \leq k/n,$$

and so $\max\{|p|,|r|\} \leq k/n$. Thus $\max\{|p|,|r|\} \leq \|x\|$.

Given any positive real number ε, there exist $k, l \in \mathbb{Z}$ and $m, n \in \mathbb{N}$ such that $ku \leq mx$ and $nx \leq lu$, while also

$$p - \varepsilon < k/m \leq p \quad \text{and} \quad r \leq l/n < r + \varepsilon.$$

Then $|k/m| < |p| + \varepsilon$ and $|l/n| < |r| + \varepsilon$. Setting $h = \max\{|kn|, |lm|\}$, we obtain

$$-hu \leq knu \leq mnx \leq lmu \leq hu,$$

and consequently

$$\|x\| \leq h/mn = \max\{|k/m|, |l/n|\} < \max\{|p|, |r|\} + \varepsilon.$$

Therefore $\|x\| = \max\{|p|, |r|\}$, as desired. $\square$

LEMMA 7.13. *Let (G, u) be a nonzero partially ordered abelian group with order-unit, let $x \in G$, and let $k, n \in \mathbb{N}$.*

(a) *If $\|x\| < k/n$, then $-kmu < mnx < kmu$ for some $m \in \mathbb{N}$.*

(b) *If $\|x\| < k/n$ and G is unperforated, then $-ku < nx < ku$.*

(c) *If $\|x\| \leq k/n$ and G is archimedean, then $-ku \leq nx \leq ku$.*

PROOF. (a) There must exist $l, m \in \mathbb{N}$ such that $-lu \leq mx \leq lu$ and $l/m < k/n$. Then $ln < km$. Since $u > 0$, we find that

$$-kmu < -lnu \leq mnx \leq lnu < kmu.$$

(b) This is immediate from (a).

(c) For all $s \in S(G, u)$, we have $|s(x)| \leq \|x\| \leq k/n$. Hence,

$$s(-ku) = -k \leq s(nx) \leq k = s(ku)$$

for all $s \in S(G, u)$. Then $-ku \leq nx \leq ku$ by Theorem 4.14. $\square$

LEMMA 7.14. *Let (G, u) be a dimension group with order-unit, let $x \in G$, and assume that $\|x\| < k/2^n$ for some positive integers k, n. Then $x = y - z$ for some $y, z \in G^+$ such that $2^n y \leq ku$ and $2^n z \leq ku$.*

PROOF. There is no loss of generality in assuming that G is nonzero. As G is unperforated, $-ku < 2^n x < ku$ by Lemma 7.13. In particular,

$$2^n x \leq 2^n \cdot 0 + ku.$$

According to Corollary 2.20, $x = v + w$ for some $v \in G$ and $w \in G^+$ such that $v \leq 0$ and $2^n w \leq ku$. Now

$$-ku < 2^n x = 2^n v + 2^n w,$$

and hence $2^n(-v) \leq 2^n w + ku$. As the elements $-v, w, ku$ all lie in G^+, Proposition 2.19 says that $-v = a + z$ for some $a, z \in G^+$ such that $a \leq w$ and $2^n z \leq ku$. Set $y = w - a$, so that $y \in G^+$ and

$$x = v + w = w - a - z = y - z.$$

As $a \geq 0$, we also have $2^n y \leq 2^n w \leq ku$. $\square$

PROPOSITION 7.15. *Let (G,u) be a dimension group with order-unit, and let H be an ideal of G. Then*

$$\|x+H\| = \inf\{\|y\| : y \in x+H\}$$

for any $x \in G$.

PROOF. First consider any $y \in x+H$. Whenever $-ku \leq ny \leq ku$ for some positive integers k, n, we have

$$-k(u+H) \leq n(y+H) = n(x+H) \leq k(u+H),$$

and so $\|x+H\| \leq k/n$. Thus $\|x+H\| \leq \|y\|$ for all $y \in x+H$.

Given any positive real number ε, we may choose positive integers k, n such that

$$\|x+H\| < k/2^n < \|x+H\| + \varepsilon.$$

According to Lemma 7.14, $x+H = (x_1+H) - (x_2+H)$ for some $x_1, x_2 \in G^+$ such that

$$2^n(x_i+H) \leq k(u+H)$$

for each i. Hence, there exist elements $a_1, a_2 \in H^+$ such that

$$2^n x_i \leq ku + a_i \leq 2^n a_i + ku$$

for each i. By Proposition 2.19, each $x_i = v_i + w_i$ for some v_i, w_i in G^+ such that $v_i \leq a_i$ and $2^n w_i \leq ku$. As $0 \leq v_i \leq a_i$, each $v_i \in H$. Hence,

$$x+H = (x_1+H) - (x_2+H) = (w_1+H) - (w_2+H),$$

and so the element $y = w_1 - w_2$ lies in $x+H$. Since

$$-ku \leq -2^n w_2 \leq 2^n(w_1 - w_2) = 2^n y \leq 2^n w_1 \leq ku,$$

we conclude that $\|y\| \leq k/2^n < \|x+H\| + \varepsilon$.

Therefore $\|x+H\|$ equals the required infimum. □

DEFINITION. Let (G,u) be a partially ordered abelian group with order-unit. In view of Proposition 7.12, the rule $d(x,y) = \|x-y\|$ defines a pseudo-metric d on G. We refer to the corresponding topology on G as the *norm-topology*. In particular, subsets of G that are closed in this topology are said to be *norm-closed.*

In general, the pseudo-metric d is not a metric (e.g., consider the lexicographic direct product of $\mathbb{Z}$ with itself), and hence the norm-topology on G is not Hausdorff. If v is another order-unit in G, then there exists $n \in \mathbb{N}$ such that $u \leq nv$ and $v \leq nu$. It follows that

$$\|\cdot\|_u \leq n\|\cdot\|_v \quad \text{and} \quad \|\cdot\|_v \leq n\|\cdot\|_u,$$

and hence the $\|\cdot\|_u$-topology on G coincides with the $\|\cdot\|_v$-topology. Thus any properties of the norm-topology on G are independent of the choice of order-unit.

LEMMA 7.16. *Let (G,u) be an archimedean partially ordered abelian group with order-unit. Let $x, x_1, x_2, \ldots$ and $y, y_1, y_2, \ldots$ be elements of G such that $x_n \to x$ and $y_n \to y$ in the norm-topology. If $x_n \leq y_n$ for all n, then $x \leq y$.*

PROOF. We may assume that G is nonzero. If ϕ is the natural map from G to $\mathrm{Aff}(S(G,u))$, then since ϕ preserves norms (Proposition 7.12), we have $\phi(x_n) \to \phi(x)$ and $\phi(y_n) \to \phi(y)$ in $\mathrm{Aff}(S(G,u))$. In addition, $\phi(x_n) \leq \phi(y_n)$ for all n, and hence $\phi(x) \leq \phi(y)$. As G is archimedean, $x \leq y$ by Theorem 7.7. □

PROPOSITION 7.17. *Let (G,u) be an interpolation group with order-unit. Then G is archimedean if and only if G is 2-unperforated and G^+ is norm-closed in G.*

PROOF. If G is archimedean, then G is unperforated by Proposition 1.24, and Lemma 7.16 shows that G^+ is norm-closed in G. Conversely, assume that G is 2-unperforated and that G^+ is norm-closed in G.

Given any $a, b \in G$ such that $na \leq b$ for all $n \in \mathbb{N}$, we must show that $a \leq 0$. Write $a = x - y$ for some $x, y \in G^+$, and choose $p \in \mathbb{N}$ such that $b \leq pu$. Then $nx \leq ny + pu$ for all $n \in \mathbb{N}$, and we must show that $x \leq y$.

For each $k \in \mathbb{N}$, we have $2^k x \leq 2^k y + pu$. According to Proposition 2.19, it follows that $x = v_k + w_k$ for some $v_k, w_k \in G^+$ such that $v_k \leq y$ and $2^k w_k \leq pu$. Since

$$-pu \leq 0 \leq 2^k w_k \leq pu,$$

we find that $\|w_k\| \leq p/2^k$. Hence, $w_k \to 0$, and so $y - v_k \to y - x$. As each $y - v_k$ lies in the norm-closed set G^+, we conclude that $y - x$ is in G^+. Therefore $x \leq y$, as required. □

• Bounded Homomorphisms.

DEFINITION. Let (G,u) be a partially ordered abelian group with order-unit, and let B be a real normed linear space. A group homomorphism $f\colon G \to B$ is *bounded* if and only if there exists $\alpha \in \mathbb{R}^+$ such that $\|f(x)\| \leq \alpha \|x\|_u$ for all $x \in G$. If f is bounded, the least such nonnegative real number α is called the *norm* of f, denoted $\|f\|_u$, and abbreviated to $\|f\|$ when there is no danger of confusion as to the order-unit.

Note that the question of whether or not f is bounded is independent of the choice of order-unit in G. However, the value of the norm of f does depend on the order-unit.

PROPOSITION 7.18. *Let (G,u) be a partially ordered abelian group with order-unit, and let $s\colon G \to \mathbb{R}$ be a group homomorphism. Then s is a state on (G,u) if and only if s is bounded and $\|s\| = s(u) = 1$.*

PROOF. First assume that s is a state on (G,u). Then $|s(x)| \leq \|x\|$ for all $x \in G$, by Proposition 7.12. As a result, s is bounded, and $\|s\| \leq 1$. Since $\|u\| = 1$ and $|s(u)| = 1$, it follows that $\|s\| = 1$.

Conversely, assume that s is bounded and that $\|s\| = s(u) = 1$. Given any $x \in G^+$, choose $m \in \mathbb{N}$ such that $x \leq mu$. Then $0 \leq mu - x \leq mu$, whence $\|mu - x\| \leq m$. Since $\|s\| = 1$, it follows that

$$m - s(x) \leq |s(mu - x)| \leq \|mu - x\| \leq m,$$

and so $s(x) \geq 0$. Thus s is a positive homomorphism, and hence s is a state on (G, u). □

PROPOSITION 7.19. *Let (G, u) be an interpolation group with order-unit, and let $f\colon G \to \mathbb{R}$ be a group homomorphism. Then f is bounded if and only if f is relatively bounded.*

PROOF. First assume that f is bounded. If X is any bounded subset of G, then there is some $k \in \mathbb{N}$ such that $X \subseteq [-ku, ku]$. Consequently, $\|x\| \leq k$ for all $x \in X$, whence $|f(x)| \leq k\|f\|$ for all $x \in X$, and so $f(X)$ is a bounded subset of $\mathbb{R}$. Thus f is relatively bounded.

Conversely, assume that f is relatively bounded. Then there is some $\alpha \in \mathbb{R}^+$ such that $f([0, u]) \subseteq [-\alpha, \alpha]$. Now let $x \in G$, and consider any $k, n \in \mathbb{N}$ such that $-ku \leq nx \leq ku$. As $0 \leq nx + ku \leq 2ku$, we must have

$$nx + ku = y_1 + \cdots + y_{2k}$$

for some elements $y_i \in [0, u]$. Then $|f(y_i)| \leq \alpha$ for each i, while also $|f(u)| \leq \alpha$. Hence,

$$n|f(x)| = |f(y_1 + \cdots + y_{2k} - ku)| \leq |f(y_1)| + \cdots + |f(y_{2k})| + k|f(u)| \leq 3k\alpha,$$

so that $|f(x)| \leq 3\alpha(k/n)$. Therefore $|f(x)| \leq 3\alpha\|x\|$, and so f is bounded. □

An immediate corollary of Propositions 7.19 and 4.6 is that if (G, u) is a nonzero interpolation group with order-unit, then a homomorphism $f\colon G \to \mathbb{R}$ is bounded if and only if $f = \alpha s - \beta t$ for some $\alpha, \beta \in \mathbb{R}^+$ and some $s, t \in S(G, u)$. This result is actually valid for all nonzero partially ordered abelian groups with order-unit, as we shall prove in Corollary 7.21.

PROPOSITION 7.20. *Let (G, u) be a nonzero partially ordered abelian group with order-unit, and let $\phi\colon G \to \mathrm{Aff}(S(G, u))$ be the natural map. Let B be a real Banach space, and let $f\colon G \to B$ be a bounded homomorphism. Then there exists a unique bounded linear transformation*

$$g\colon \mathrm{Aff}(S(G, u)) \to B$$

such that $g\phi = f$, and $\|g\| = \|f\|$.

PROOF. As $\|f(x)\| \leq \|f\| \cdot \|x\| = \|f\| \cdot \|\phi(x)\|$ for all $x \in G$, we see that f vanishes on $\ker(\phi)$. Hence, f induces a homomorphism $f_1\colon \phi(G) \to B$ such that $f_1\phi = f$. It is clear that f_1 is bounded and that $\|f_1\| = \|f\|$.

Let G_2 be the rational linear subspace of $\mathrm{Aff}(S(G, u))$ spanned by $\phi(G)$. Then f_1 extends uniquely to a rational-linear transformation $f_2\colon G_2 \to B$. It is clear that f_2 is bounded and that $\|f_2\| = \|f_1\| = \|f\|$.

In particular, f_2 is uniformly continuous. As G_2 is dense in $\mathrm{Aff}(S(G,u))$ (by Theorem 7.9), f_2 extends uniquely to a continuous map $g\colon \mathrm{Aff}(S(G,u)) \to B$. It follows from the continuity of g that g is linear and bounded, with $\|g\| = \|f_2\| = \|f\|$. Also, $g\phi = f_1\phi = f$. □

COROLLARY 7.21. *Let (G,u) be a nonzero partially ordered abelian group with order-unit, and let $f\colon G \to \mathbb{R}$ be a group homomorphism. Then f is bounded if and only if $f = \alpha s - \beta t$ for some $\alpha, \beta \in \mathbb{R}^+$ and some $s,t \in S(G,u)$.*

PROOF. First assume that $f = \alpha s - \beta t$ for some $\alpha, \beta \in \mathbb{R}^+$ and some s, t in $S(G,u)$. As s and t are bounded with $\|s\| = \|t\| = 1$ (by Proposition 7.18), we see that

$$|f(x)| \leq \alpha|s(x)| + \beta|t(x)| \leq (\alpha + \beta)\|x\|$$

for all $x \in G$, whence f is bounded.

Conversely, assume that f is bounded, and let $\phi\colon G \to \mathrm{Aff}(S(G,u))$ be the natural map. According to Proposition 7.20, there exists a unique bounded linear functional g on $\mathrm{Aff}(S(G,u))$ such that $g\phi = f$. Then Corollary 7.3 shows that there exist $\alpha, \beta \in \mathbb{R}^+$ and $s, t \in S(G,u)$ such that

$$g(q) = \alpha q(s) - \beta q(t)$$

for all $q \in \mathrm{Aff}(K)$. For all $x \in G$, we obtain

$$f(x) = g\phi(x) = \alpha\phi(x)(s) - \beta\phi(x)(t) = \alpha s(x) - \beta t(x),$$

and therefore $f = \alpha s - \beta t$. □

Of course Proposition 4.6 may now be viewed as a corollary of Proposition 7.19 and Corollary 7.21.

• **Notes.** Theorem 7.1 is folklore, sometimes attributed to Kadison because the method used in its proof appeared in [**77**, p. 328]. Corollary 7.4 was proved by Jellett [**75**, Lemma, p. 224].

The implication (c) ⇒ (a) in Theorem 7.7 is a restatement of the corresponding implication in Theorem 4.14, which was proved by the author, Handelman, and Lawrence [**60**, Proposition I.5.3]. Theorem 7.8 is a restatement of Theorems 4.12 and 6.3 and Corollaries 4.13 and 6.4, all stemming from Corollary 4.13, which is due to Effros, Handelman, and Shen [**30**, Theorem 1.4]. Theorem 7.9 was proved by the author, Handelman, and Lawrence [**60**, Proposition I.9.1]. Corollary 7.11 is contained in a result of the author and Warfield [**61**, Corollary 2.5].

Proposition 7.15 was proved by the author, as was Proposition 7.17 [**53**, Proposition 3.3]. Proposition 7.18 was adapted from the corresponding folklore result for states on a partially ordered real vector space with order-unit. Proposition 7.20 was proved by the author and Handelman [**58**, Lemma 2.5]. The decomposition of bounded real-valued homomorphisms given in Corollary 7.21 was first proved for dimension groups by Handelman [**67**, Lemma III.7(b)].

CHAPTER 8

General Comparability

This chapter represents the first stage of the development of the tools needed to obtain a useful description of the natural affine representation

$$\phi\colon (G,u) \to (\mathrm{Aff}(S(G,u),1)$$

for an interpolation group (G,u) with order-unit. The ultimate goal is to prove that when G is unperforated, $\phi(G)$ is dense in the subgroup

$$A = \{p \in \mathrm{Aff}(S(G,u)) \mid p(s) \in s(G) \text{ for all discrete } s \in \partial_e S(G,u)\},$$

and $\phi(G^+)$ is dense in A^+. A number of stages are required to reach these results, the first being the case in which (G,u) satisfies general comparability, meaning, roughly, that G has enough direct product decompositions to be treated in the same fashion as a direct product of totally ordered abelian groups. The second stage consists of transferring the results of the first stage to the case in which G is Dedekind σ-complete, since then (G,u) satisfies general comparability, as we prove in the following chapter. In this case G is actually isomorphic to A, via ϕ. Returning to the general case, we introduce, in a later chapter, a pseudo-norm on G induced by any state s on (G,u), and show that the completion of G with respect to this pseudo-norm is Dedekind complete. The use of this completion represents the third stage of our development, for it allows us to apply the results of the second stage, via the completion, to states in a neighborhood of s. The fourth and final stage consists of course in patching together these local results in a global theorem.

The present chapter begins with a look at the characteristic elements of (G,u), which correspond to the components of u in various direct product decompositions of G, and at the collection of these characteristic elements, which forms a boolean algebra. Then general comparability is introduced, a condition which ensures that for any elements $x, y \in G$, there is a direct product decomposition $G = G_1 \times G_2$ such that "$x \leq y$ on G_1" while "$x \geq y$ on G_2", that is, the G_1-components of x and y satisfy $x_1 \leq y_1$, while the G_2-components of x and y satisfy $x_2 \geq y_2$. In particular, it is shown that any interpolation group with general comparability is lattice-ordered. In the presence of general comparability, the extremal states and the closed faces of the state space are identified.

The extremal states correspond bijectively to the maximal ideals of the boolean algebra of characteristic elements, while the closed faces of the state space all arise naturally from ideals of G. The chapter concludes with the result that when (G, u) has general comparability, $\phi(G)$ is dense in the subgroup A of $\mathrm{Aff}(S(G, u))$ defined above.

• **Characteristic Elements.** Just as a decomposition of a ring with identity into a direct sum of ideals yields a corresponding decomposition of the identity element into an orthogonal sum of central idempotents, a decomposition of a partially ordered abelian group with order-unit into a direct sum of ideals yields a corresponding decomposition of the order-unit into a sum of elements of the following sort.

DEFINITION. Let (G, u) be a partially ordered abelian group with order-unit. A *characteristic element of* (G, u) is any element $e \in G$ such that $0 \leq e \leq u$ and $e \wedge (u - e) = 0$ (that is, $e \wedge (u - e)$ exists and equals 0).

Note that if e is a characteristic element of (G, u), then so is $u - e$. In addition, $e \vee (u - e) = u$, by Proposition 1.6.

For example, if X is a compact Hausdorff space, then the characteristic elements of $(C(X, \mathbb{R}), 1)$ are exactly the characteristic functions of clopen subsets of X. In an arbitrary partially ordered abelian group (G, u) with order-unit, at least 0 and u are characteristic elements.

LEMMA 8.1. *Let (G, u) be a partially ordered abelian group with order-unit, and let $H_1, \ldots, H_n$ be ideals of G such that $G = H_1 \oplus \cdots \oplus H_n$ (as an ordered group). Then there exist unique elements $e_i \in H_i$ for each $i = 1, \ldots, n$ such that $u = e_1 + \cdots + e_n$, and each e_i is a characteristic element of (G, u).*

PROOF. The existence and uniqueness of the e_i follow from the fact that as an abelian group, G is the direct sum of the subgroups H_i. As also $G^+ = H_1^+ + \cdots + H_n^+$, we see that $0 \leq e_i \leq u$ for each i.

Now $0 \leq e_i$ and $0 \leq u - e_i$. Given any $x \in G$ for which $x \leq e_i$ and $x \leq u - e_i$, write $x = x_1 + \cdots + x_n$ where each $x_i \in H_i$, and observe that the elements

$$(e_i - x_i) + \sum_{j \neq i} (-x_j) \quad \text{and} \quad (-x_i) + \sum_{j \neq i} (e_j - x_j)$$

lie in G^+. Consequently, $-x_j \in H_j^+$ for all $j = 1, \ldots, n$, so that each $x_j \leq 0$, and hence $x \leq 0$. Therefore $e_i \wedge (u - e_i) = 0$, proving that e_i is a characteristic element of (G, u). $\square$

In general, characteristic elements in partially ordered abelian groups with order-unit need not arise from decompositions into direct sums of ideals as in Lemma 8.1. However, this does hold for interpolation groups, as follows.

LEMMA 8.2. *Let (G, u) be an interpolation group with order-unit, and let e be a characteristic element of (G, u). Let H be the convex subgroup of G generated* *by e, and let K be the convex subgroup of G generated by $u-e$. Then H and K are ideals of G, and $G = H \oplus K$ (as an ordered group).*

PROOF. As e and $u-e$ are elements of G^+, Proposition 1.9 shows that H and K are ideals of G.

Given any $x \in (H \cap K)^+$, choose $n, k \in \mathbb{N}$ such that $x \leq ne$ and $x \leq k(u-e)$. Then $x = x_1 + \cdots + x_n$ for some elements $x_i \in G^+$ such that each $x_i \leq e$. Since each $x_i \leq x \leq k(u-e)$, each $x_i = x_{i1} + \cdots + x_{ik}$ for some elements $x_{ij} \in G^+$ such that each $x_{ij} \leq u-e$. In addition, each $x_{ij} \leq x_i \leq e$, whence

$$0 \leq x_{ij} \leq e \wedge (u-e) = 0.$$

Then $x_{ij} = 0$ for all i, j, and hence $x = 0$.

Thus $(H \cap K)^+ = \{0\}$. As $H \cap K$ is an ideal of G (by Proposition 2.4), it is directed, and therefore $H \cap K = \{0\}$.

Given any $y \in G^+$, choose $n \in \mathbb{N}$ such that

$$y \leq nu = ne + n(u-e).$$

Then $y = v + w$ for some $v, w \in G^+$ such that $v \leq ne$ and $w \leq n(u-e)$. Hence, $v \in H^+$ and $w \in K^+$, so that $y \in H^+ + K^+$.

Thus $G^+ = H^+ + K^+$. As G is directed, it follows that $G = H + K$. Therefore $G = H \oplus K$ as an ordered group. □

DEFINITION. Let (G, u) be an interpolation group with order-unit, and let e be a characteristic element of (G, u). Let H be the convex subgroup of G generated by e, and let K be the convex subgroup of G generated by $u-e$, so that $G = H \oplus K$ as an ordered group (by Lemma 8.2). We define p_e to be the projection of G onto H with kernel K.

Thus p_e is a positive homomorphism of G into itself such that $p_e(u-e) = 0$ and $p_e(G)$ equals the convex subgroup of G generated by e. Similarly, p_{u-e} is a positive homomorphism of G into itself such that $p_{u-e}(e) = 0$ and $p_{u-e}(G)$ equals the convex subgroup of G generated by $u-e$. Hence,

$$G = p_e(G) \oplus p_{u-e}(G)$$

(as an ordered group). Note that $p_e + p_{u-e}$ equals the identity map on G.

In particular, these projections relative to the characteristic elements 0 and u are easy to describe. Namely, p_0 is the zero map on G, while p_u is the identity map on G.

PROPOSITION 8.3. *Let (G, u) be an interpolation group with order-unit and let e be a characteristic element of (G, u). If $x \in G^+$ and $x \leq nu$ for some $n \in \mathbb{N}$, then $p_e(x) = x \wedge ne$.*

PROOF. As $x \leq nu$, we have $p_e(x) \leq p_e(nu) = ne$. In addition, since $0 \leq x$ we obtain $0 \leq p_{u-e}(x) = x - p_e(x)$, whence $p_e(x) \leq x$. Given $y \in G$ such that $y \leq x$ and $y \leq ne$, we first note that $p_e(y) \leq p_e(x)$. In addition,

$$y - p_e(y) = p_{u-e}(y) \leq p_{u-e}(ne) = 0,$$

and hence $y \leq p_e(y) \leq p_e(x)$. Therefore $p_e(x) = x \wedge ne$. □

COROLLARY 8.4. *Let (G, u) be an interpolation group with order-unit, let H be an ideal of G, and let e be a characteristic element of (G, u). Then $e + H$ is a characteristic element of $(G/H, u + H)$, and*

$$p_{e+H}(x + H) = p_e(x) + H$$

for all $x \in G$.

PROOF. Obviously $0 + H \leq e + H \leq u + H$. According to Proposition 1.13,

$$\begin{aligned} 0 + H &= (e \wedge (u - e)) + H = (e + H) \wedge (u - e + H) \\ &= (e + H) \wedge [(u + H) - (e + H)]. \end{aligned}$$

Thus $e + H$ is a characteristic element of $(G/H, u + H)$.

Given any $x \in G^+$, choose $n \in \mathbb{N}$ such that $x \leq nu$, and note that $x + H \leq n(u + H)$. In view of Propositions 8.3 and 1.13,

$$p_e(x) + H = (x \wedge ne) + H = (x + H) \wedge n(e + H) = p_{e+H}(x + H).$$

Hence, if $\pi \colon G \to G/H$ is the quotient map, $p_{e+H}\pi$ agrees with πp_e on G^+. As G is directed, we conclude that $p_{e+H}\pi = \pi p_e$. □

• **Projection Bases.**

DEFINITION. Let (G, u) be a partially ordered abelian group with order-unit. The *projection base of* (G, u) is the set of all characteristic elements of (G, u), and we denote this set by $B(G, u)$.

We view $B(G, u)$ as a partially ordered set with respect to the ordering inherited from G. Since 0 and u lie in $B(G, u)$, this set is nonempty. Note that 0 is the least element of $B(G, u)$, while u is the greatest element of $B(G, u)$.

LEMMA 8.5. *Let (G, u) be an interpolation group with order-unit, and let e and f be characteristic elements of (G, u).*

(a) *$p_e(f) = p_f(e) = e \wedge f$.*

(b) *$e \wedge f$ is a characteristic element of (G, u).*

(c) *$n(e \wedge f) = ne \wedge nf$ for all $n \in \mathbb{N}$.*

(d) *$p_{e \wedge f} = p_e p_f = p_f p_e$.*

PROOF. (a) As $0 \leq e \leq u$, Proposition 8.3 shows that $p_f(e) = e \wedge f$. Similarly, $p_e(f) = f \wedge e$.

(b) Clearly $0 \leq e \wedge f \leq e \leq u$. By Proposition 1.4,

$$u - (e \wedge f) = (u - e) \vee (u - f) \leq (u - e) + (u - f).$$

Given any $x \in G$ such that $x \leq e \wedge f$ and $x \leq u - (e \wedge f)$, there exists $y \in G$ such that

$$\begin{matrix} x \\ 0 \end{matrix} \leq y \leq \begin{matrix} e \wedge f \\ u - (e \wedge f). \end{matrix}$$

Then $y \leq (u - e) + (u - f)$, whence $y = a + b$ for some $a, b \in G^+$ such that $a \leq u - e$ and $b \leq u - f$. As

$$a \leq y \leq e \wedge f \leq e \quad \text{and} \quad b \leq y \leq e \wedge f \leq f,$$

we must have $a \leq e \wedge (u-e) = 0$ and $b \leq f \wedge (u-f) = 0$, and so $y \leq 0$. Therefore $x \leq 0$, proving that $(e \wedge f) \wedge [u - (e \wedge f)] = 0$.

(c) As $0 \leq nf \leq nu$, Proposition 8.3 shows that $p_e(nf) = ne \wedge nf$. On the other hand, it follows from (a) that

$$p_e(nf) = np_e(f) = n(e \wedge f).$$

(d) Given any $x \in G^+$, choose $n \in \mathbb{N}$ such that $x \leq nu$. Using Proposition 8.3 and property (c), we compute that

$$p_e p_f(x) = p_e(x \wedge nf) = (x \wedge nf) \wedge ne = x \wedge n(e \wedge f) = p_{e \wedge f}(x).$$

Thus $p_e p_f$ and $p_{e\wedge f}$ agree on G^+. As G is directed, we conclude that $p_e p_f = p_{e \wedge f}$. By symmetry, $p_f p_e = p_{f \wedge e}$. □

LEMMA 8.6. *Let (G, u) be an interpolation group with order-unit, and let e and f be characteristic elements of (G, u).*

(a) *$e \vee f$ exists in G and is a characteristic element of (G, u).*

(b) *$p_{e \vee f} = p_e + p_f - p_e p_f$.*

PROOF. (a) As $e \wedge f$ exists by Lemma 8.5, $e \vee f$ exists by Proposition 1.6. In view of Proposition 1.4,

$$u - (e \vee f) = (u - e) \wedge (u - f),$$

which by Lemma 8.5 is a characteristic element of (G, u). Thus $e \vee f$ is a characteristic element of (G, u).

(b) Let i denote the identity map on G. Using Lemma 8.5, we compute that

$$\begin{aligned} i - p_{e \vee f} &= p_{u-(e \vee f)} = p_{(u-e)\wedge(u-f)} = p_{u-e} p_{u-f} \\ &= (i - p_e)(i - p_f) = i - p_e - p_f + p_e p_f. \end{aligned}$$

Therefore $p_{e \vee f} = p_e + p_f - p_e p_f$. □

THEOREM 8.7. *If (G, u) is an interpolation group with order-unit, then $B(G, u)$ is a boolean algebra. If, in addition, G is Dedekind σ-complete (or Dedekind complete), then $B(G, u)$ is σ-complete (or complete).*

PROOF. By construction, $B(G, u)$ is a partially ordered set with a least element 0 and a greatest element u. For any $e, f \in B(G, u)$, Lemmas 8.5 and 8.6 show that $e \wedge f$ and $e \vee f$ exist in G and lie in $B(G, u)$. Hence, $e \wedge f$ and $e \vee f$ are the infimum and supremum of $\{e, f\}$ in $B(G, u)$, proving that $B(G, u)$ is a lattice. In view of Proposition 1.7, this lattice is distributive. For any $e \in B(G, u)$, we have

$$e \wedge (u - e) = 0 \quad \text{and} \quad e \vee (u - e) = u$$

in G and hence also in $B(G, u)$, which shows that the lattice $B(G, u)$ is complemented. Therefore $B(G, u)$ is a boolean algebra.

Now assume that G is Dedekind σ-complete. If $\{e_i \mid i \in I\}$ is a nonempty countable subset of $B(G, u)$, then as $e_i \leq u$ for all $i \in I$, there exists an element

$e = \bigvee e_i$ in G. Note that $0 \le e \le u$. For all $i \in I$, we have $0 \le u - e \le u - e_i$, and consequently $e_i \wedge (u - e) = 0$. Then, by Proposition 1.7,

$$e \wedge (u-e) = \left(\bigvee e_i\right) \wedge (u-e) = \bigvee (e_i \wedge (u-e)) = 0,$$

proving that $e \in B(G,u)$. Thus e is also the supremum of the e_i in $B(G,u)$.

This shows that every nonempty countable subset of $B(G,u)$ has a supremum in $B(G,u)$. The empty set has a supremum in $B(G,u)$ as well, namely 0. As $B(G,u)$ is a boolean algebra, it follows that every countable subset of $B(G,u)$ has an infimum. Therefore $B(G,u)$ is a σ-complete lattice.

If G is Dedekind complete, a similar argument shows that $B(G,u)$ is a complete lattice. □

DEFINITION. Let (G,u) be an interpolation group with order-unit. We use $\operatorname{Max} B(G,u)$ to denote the space of all maximal ideals of the boolean algebra $B(G,u)$, equipped with the usual Stone topology.

Thus $\operatorname{Max} B(G,u)$ is a compact, Hausdorff, totally disconnected space. In case G is Dedekind complete, $\operatorname{Max} B(G,u)$ is extremally disconnected.

• **Comparability.**

DEFINITION. Let (G,u) be an interpolation group with order-unit. We say that (G,u) satisfies *general comparability* if for each $x, y \in G$ there is some $e \in B(G,u)$ such that $p_e(x) \le p_e(y)$ while $p_{u-e}(x) \ge p_{u-e}(y)$.

For example, if (G,u) is any totally ordered abelian group with order-unit, we observe that (G,u) trivially satisfies general comparability. [Namely, any elements $x, y \in G$ satisfy either $x \le y$ or $x \ge y$; hence, either $p_u(x) \le p_u(y)$ and $p_0(x) \ge p_0(y)$, or else $p_0(x) \le p_0(y)$ and $p_u(x) \ge p_u(y)$.] More generally, if (G,u) is a product of totally ordered abelian groups with order-unit (in the category of partially ordered abelian groups with order-unit), then (G,u) satisfies general comparability. In particular, all simplicial groups with order-unit satisfy general comparability. We shall prove in Theorem 9.9 that any Dedekind σ-complete lattice-ordered abelian group with order-unit satisfies general comparability.

On the other hand, if $G = \mathbb{R}^2$ with the strict ordering and $u = (1,1)$, then (G,u) does not satisfy general comparability, because the only characteristic elements of (G,u) are 0 and u.

PROPOSITION 8.8. *Let (G,u) be an interpolation group with order-unit and let H be an ideal of G. If (G,u) satisfies general comparability, then so does $(G/H, u+H)$.*

PROOF. This follows immediately from Corollary 8.4. □

PROPOSITION 8.9. *Let (G,u) be an interpolation group with order-unit, satisfying general comparability. Then G is a lattice. If $x, y \in G$ and $e \in B(G,u)$ such that $p_e(x) \le p_e(y)$ and $p_{u-e}(x) \ge p_{u-e}(y)$, then*

$$x \wedge y = p_e(x) + p_{u-e}(y) \quad \text{and} \quad x \vee y = p_e(y) + p_{u-e}(x).$$

PROOF. Given any $x, y \in G$, there exists $e \in B(G,u)$ such that $p_e(x) \leq p_e(y)$ and $p_{u-e}(x) \geq p_{u-e}(y)$. Set

$$v = p_e(x) + p_{u-e}(y) \quad \text{and} \quad w = p_e(y) + p_{u-e}(x).$$

Then $v \leq p_e(y) + p_{u-e}(y) = y$ and similarly $v \leq x$, while $w \geq y$ and $w \geq x$.

Now consider any $z \in G$ such that $z \leq x$ and $z \leq y$. Then $p_e(z) \leq p_e(x)$ and $p_{u-e}(z) \leq p_{u-e}(y)$, whence

$$z = p_e(z) + p_{u-e}(z) \leq p_e(x) + p_{u-e}(y) = v.$$

Thus $v = x \wedge y$. Similarly, $w = x \vee y$.

Therefore G is a lattice. □

In particular, it follows from Propositions 8.9 and 1.22 that any interpolation group with general comparability is unperforated.

• **Extremal States.** We begin investigating the state space of an interpolation group with general comparability by studying the extremal states. These extremal states are determined by their behavior on characteristic elements, resulting in a homeomorphism between the extreme boundary of the state space and the maximal ideal space of the projection base.

LEMMA 8.10. *Let (G,u) be an interpolation group with order-unit, let e be a characteristic element of (G,u), and let s be a state on (G,u).*

(a) *If $s(e) = 0$, then $sp_e = 0$ and $s = sp_{u-e}$.*

(b) *If $s(e) = 1$, then $sp_{u-e} = 0$ and $s = sp_e$.*

(c) *If $s(e) = \alpha$ where $0 < \alpha < 1$, then the functions*

$$s' = \alpha^{-1} sp_e \quad \text{and} \quad s'' = (1-\alpha)^{-1} sp_{u-e}$$

are distinct states on (G,u) such that $s = \alpha s' + (1-\alpha)s''$.

(d) *If s is extremal, then either $s(e) = 0$ or $s(e) = 1$.*

PROOF. (a) Given any $x \in G$, we have $-me \leq p_e(x) \leq me$ for some $m \in \mathbb{N}$. Since $s(e) = 0$, it follows that $sp_e(x) = 0$. Thus $sp_e = 0$, and consequently $s = sp_{u-e}$, because $p_e + p_{u-e}$ is the identity map on G.

(b) This follows from (a) on replacing e by $u - e$.

(c) It is clear that s' and s'' are positive homomorphisms from G to $\mathbb{R}$ such that $s = \alpha s' + (1-\alpha)s''$. Since

$$s'(u) = \alpha^{-1}s(e) = 1 \quad \text{and} \quad s''(u) = (1-\alpha)^{-1}s(u-e) = 1,$$

we see that s' and s'' are states on (G,u). As $s'(e) = 1$ while $s''(e) = 0$, we also see that $s' \neq s''$.

(d) This follows immediately from (c). □

PROPOSITION 8.11. *Let (G,u) be an interpolation group with order-unit, satisfying general comparability, and let s be an extremal state on (G,u). Then s is a lattice homomorphism,* $\ker(s)$ *is an ideal of G, and $G/\ker(s)$ is totally* *ordered. Moreover, s induces an isomorphism of $G/\ker(s)$ onto $s(G)$ (as ordered groups).*

PROOF. Given any $x, y \in G$, there exists $e \in B(G,u)$ such that $p_e(x) \leq p_e(y)$ while $p_{u-e}(x) \geq p_{u-e}(y)$, and Proposition 8.9 shows that

$$x \wedge y = p_e(x) + p_{u-e}(y) \quad \text{and} \quad x \vee y = p_e(y) + p_{u-e}(x).$$

According to Lemma 8.10, either $s(e) = 0$ or $s(e) = 1$. If $s(e) = 0$, then the lemma shows that $sp_e = 0$ and $s = sp_{u-e}$. As a result,

$$\begin{aligned} s(x) &= sp_{u-e}(x) \geq sp_{u-e}(y) = s(y) \\ s(x \wedge y) &= sp_{u-e}(y) = s(y) = \min\{s(x), s(y)\} \\ s(x \vee y) &= sp_{u-e}(x) = s(x) = \max\{s(x), s(y)\}. \end{aligned}$$

Similarly, if $s(e) = 1$, then $s(x) \leq s(y)$, and $s(x \wedge y) = s(x)$ while $s(x \vee y) = s(y)$. Therefore s is a lattice homomorphism.

Since s is a positive homomorphism, $\ker(s)$ is a convex subgroup of G. Given any $x, y \in \ker(s)$, we have

$$s(x \vee y) = \max\{s(x), s(y)\} = 0,$$

so that $x \vee y \in \ker(s)$. Hence, $\ker(s)$ is directed and thus is an ideal of G.

Given any $x, y \in G$, there exists $e \in B(G,u)$ such that $p_e(x) \leq p_e(y)$ while $p_{u-e}(x) \geq p_{u-e}(y)$, and Lemma 8.10 shows that either $sp_e = 0$ or $sp_{u-e} = 0$. If $sp_e = 0$, then $p_e(x)$ and $p_e(y)$ are in $\ker(s)$, whence

$$x + \ker(s) = p_{u-e}(x) + \ker(s) \geq p_{u-e}(y) + \ker(s) = y + \ker(s).$$

Similarly, if $sp_{u-e} = 0$, then $x + \ker(s) \leq y + \ker(s)$. Therefore $G/\ker(s)$ is totally ordered.

The map $\bar{s}\colon G/\ker(s) \to s(G)$ induced by s is a group isomorphism, as well as a positive homomorphism. Hence, if a and b are elements of $G/\ker(s)$ such that $a \leq b$, then $\bar{s}(a) \leq \bar{s}(b)$. On the other hand, if $a \not\leq b$, then $a > b$, because $G/\ker(s)$ is totally ordered. Then $\bar{s}(a) > \bar{s}(b)$ (since $\bar{s}$ is positive and injective), whence $\bar{s}(a) \not\leq \bar{s}(b)$. Thus $\bar{s}$ is an isomorphism of ordered groups. □

In particular, Proposition 8.11 shows that any interpolation group with general comparability has a good supply of totally ordered quotient groups.

LEMMA 8.12. *Let (G,u) be an interpolation group with order-unit, satisfying general comparability. Let s be a state on (G,u), and set $M = B(G,u) \cap \ker(s)$. Then the following conditions are equivalent*:

(a) *s is extremal.*

(b) *Whenever $t \in S(G,u)$ such that $\ker(t) \supseteq M$, then $t = s$.*

(c) *M is a maximal ideal of $B(G,u)$.*

PROOF. Given $e, f \in M$, observe that $e \vee f \leq e + f$. Then

$$0 \leq s(e \vee f) \leq s(e) + s(f) = 0,$$

whence $s(e \vee f) = 0$ and $e \vee f \in M$. Similarly, if $g \in B(G,u)$ with $g \leq e$, we obtain $s(g) = 0$ and $g \in M$. Thus M is an ideal of $B(G,u)$. As $u \notin M$, this ideal is proper.

(a)⇒(c): For any $e \in B(G,u)$, Lemma 8.10 shows that either $s(e) = 0$ or $s(e) = 1 = s(u)$, whence either $e \in M$ or $u - e \in M$. Therefore M is a maximal ideal of $B(G,u)$.

(c)⇒(b): If $e \in B(G,u) - M$, then $u - e \in M$. Then $u - e \in \ker(t)$ and so $t(e) = t(u) = 1$, whence $e \notin \ker(t)$. Thus $B(G,u) \cap \ker(t) = M$.

Let H be the subgroup of G generated by $\ker(s)^+$. Then H is an ideal of G, by Proposition 1.9, and we claim that G/H is totally ordered. Given any $x, y \in G$, there exists $e \in B(G,u)$ such that $p_e(x) \leq p_e(y)$ while $p_{u-e}(x) \geq p_{u-e}(y)$. As M is a maximal ideal of $B(G,u)$, either $e \in M$ or $u - e \in M$. Then either $e \in H$ or $u - e \in H$, whence either $p_e(G) \subseteq H$ or $p_{u-e}(G) \subseteq H$. If $p_e(G) \subseteq H$, then as $p_{u-e}(x) \geq p_{u-e}(y)$, we obtain $x + H \geq y + H$. Similarly, if $p_{u-e}(G) \subseteq H$, then $x + H \leq y + H$. Thus G/H is totally ordered, as claimed.

Next, we claim that $H \subseteq \ker(t)$. If not, there is an element x in $\ker(s)^+$ that is not in $\ker(t)$. Then $t(x) > 0$, and we may choose $n \in \mathbb{N}$ such that $nt(x) > 1$. Thus $y = nx$ is an element of $\ker(s)^+$ for which $t(y) > 1$. There exists $e \in B(G,u)$ such that

$$p_e(y) \leq p_e(u) = e \quad \text{and} \quad p_{u-e}(y) \geq p_{u-e}(u) = u - e,$$

and either $e \in M$ or $u - e \in M$. If $e \in M$, then $s(e) = 0$ and so $s = sp_{u-e}$, by Lemma 8.10. But then

$$s(y) = sp_{u-e}(y) \geq s(u - e) = 1,$$

contradicting the fact that $y \in \ker(s)$. If $u - e \in M$, then $t(e) = 1$ and $t = tp_e$, by a second application of Lemma 8.10. However, this leads to the relation

$$t(y) = tp_e(y) \leq t(e) = 1,$$

which contradicts the fact that $t(y) > 1$. Thus H must be contained in $\ker(t)$, as claimed.

Now s and t induce states s' and t' on $(G/H, u + H)$ such that $s = s'\pi$ and $t = t'\pi$, where $\pi\colon G \to G/H$ is the quotient map. As G/H is totally ordered, $s' = t'$ by Corollary 4.17. Therefore $s = t$.

(b)⇒(a): Consider a positive convex combination $s = \alpha_1 s_1 + \alpha_2 s_2$ where s_1 and s_2 are states on (G,u). For any $e \in M$, we have

$$\alpha_1 s_1(e) + \alpha_2 s_2(e) = s(e) = 0$$

with each $\alpha_i > 0$ and each $s_i(e) \geq 0$, whence each $s_i(e) = 0$. Then M is contained in each $\ker(s_i)$, and hence $s_1 = s_2 = s$ by (b). Therefore s is extremal. □

PROPOSITION 8.13. *Let (G,u) be an interpolation group with order-unit, satisfying general comparability, and let M be a maximal ideal of $B(G,u)$. Then* *there exists a unique state t on (G,u) such that $B(G,u) \cap \ker(t) = M$, and*

$$t(x) = \inf\{l/n \mid l, n \in \mathbb{N} \text{ and } np_e(x) \leq le \text{ for some } e \in B(G,u) - M\}$$
$$= \sup\{k/m \mid k \in \mathbb{Z}^+, m \in \mathbb{N}, \text{ and } ke \leq mp_e(x) \text{ for some } e \in B(G,u) - M\}$$

for all $x \in G^+$.

PROOF. If K is the convex subgroup of G generated by M, then K is an ideal of G, by Proposition 1.9. As M is upward directed, we see that

$$K = \{x \in G \mid -ne \leq x \leq ne \text{ for some } n \in \mathbb{N} \text{ and } e \in M\} = \bigcup_{e \in M} p_e(G).$$

Given any $f \in K \cap B(G,u)$, it follows that $p_e(f) = f$ for some $e \in M$. As $0 \leq f \leq u$, Proposition 8.3 shows that $p_e(f) = e \wedge f$. Hence, $f \leq e$, and so $f \in M$. Thus $K \cap B(G,u) = M$.

Next, we show that G/K is totally ordered. Given any $x, y \in G$, there exists $e \in B(G,u)$ such that $p_e(x) \leq p_e(y)$ and $p_{u-e}(x) \geq p_{u-e}(y)$. Either $e \in M$ or $u - e \in M$. If $e \in M$, then $p_e(G) \subseteq K$ and so

$$x + K = p_{u-e}(x) + K \geq p_{u-e}(y) + K = y + K.$$

Similarly, if $u - e \in M$, then $x + K \leq y + K$. Thus G/K is totally ordered.

Let $\pi\colon (G,u) \to (G/K, u+K)$ be the quotient map. By Corollary 4.17, there exists a unique state s on $(G/K, u+K)$. Then the function $t = s\pi$ is a state on (G,u) such that $M \subseteq K \subseteq \ker(t)$. For any e in $B(G,u) - M$, we have $u - e \in M$ and so $t(u-e) = 0$, whence $t(e) = 1$ and hence $e \notin \ker(t)$. Thus $B(G,u) \cap \ker(t) = M$.

If $t' \in S(G,u)$ and $B(G,u) \cap \ker(t') = M$, then M is contained in the convex subgroup $\ker(t')$, whence $K \subseteq \ker(t')$. Then t' induces a state s' on $(G/K, u+K)$ such that $s'\pi = t'$. As s is unique, $s' = s$, and so $t' = t$. Therefore t is unique.

The given formulas for t follow directly from the corresponding formulas for s given in Corollary 4.17. □

THEOREM 8.14. *Let (G,u) be an interpolation group with order-unit, satisfying general comparability. Then the rule*

$$\theta(s) = B(G,u) \cap \ker(s)$$

defines a homeomorphism θ of $\partial_e S(G,u)$ onto $\operatorname{Max} B(G,u)$.

PROOF. In view of Lemma 8.12 and Proposition 8.13, we see that θ is a bijection of $\partial_e S(G,u)$ onto $\operatorname{Max} B(G,u)$.

The space $\operatorname{Max} B(G,u)$ has a basis consisting of clopen sets of the form

$$U = \{M \in \operatorname{Max} B(G,u) \mid e \notin M\},$$

for $e \in B(G,u)$. We observe that

$$\theta^{-1}(U) = \{s \in \partial_e S(G,u) \mid s(e) \neq 0\},$$

which is an open subset of $\partial_e S(G,u)$. Thus θ is continuous.

By Proposition 6.1, $\partial_e S(G,u)$ has a subbasis consisting of open sets

$$V = \{s \in \partial_e S(G,u) \mid s(x) < k\},$$

for $x \in G^+$ and $k \in \mathbb{N}$. Set

$$X = \{e \in B(G,u) \mid np_e(x) \leq le \text{ for some } l, n \in \mathbb{N} \text{ with } l/n < k\}.$$

We claim that $\theta(V) = \{M \in \operatorname{Max} B(G,u) \mid X \nsubseteq M\}$.

Given any $M \in \theta(V)$, we have $\theta^{-1}(M)(x) < k$. Hence, by Proposition 8.13, there exist $l, n \in \mathbb{N}$ such that $l/n < k$ and $np_e(x) \leq le$ for some e in $B(G,u) - M$. Then $e \in X$, whence $X \nsubseteq M$. Conversely, given $M \in \operatorname{Max} B(G,u)$ with $X \nsubseteq M$, choose $e \in X - M$. Then there exist $l, n \in \mathbb{N}$ such that $l/n < k$ and $np_e(x) \leq le$. By Proposition 8.13, $\theta^{-1}(M)(x) \leq l/n < k$, whence $\theta^{-1}(M) \in V$. Thus $\theta(V)$ has the form claimed and hence is open in $\operatorname{Max} B(G,u)$.

Therefore θ is an open map, and so is a homeomorphism. □

COROLLARY 8.15. *If (G,u) is an interpolation group with order-unit, satisfying general comparability, then $\partial_e S(G,u)$ is compact and totally disconnected.* □

A consequence of Corollary 8.15, in light of Theorem 10.17 and Corollary 11.20, is that for any interpolation group (G,u) with order-unit, satisfying general comparability, $S(G,u)$ is affinely homeomorphic to $M_1^+(\partial_e S(G,u))$.

• **Closures of Faces.** The identification of the extremal states on an interpolation group with general comparability, achieved in the previous section, allows us to identify the closed faces of the state space, as well as the closures of arbitrary faces, as follows.

THEOREM 8.16. *Let (G,u) be an interpolation group with order-unit, satisfying general comparability. Let $X \subseteq S(G,u)$, and set*

$$V = \{s \in S(G,u) \mid \ker(X)^+ \subseteq \ker(s)\},$$

$$W = \{s \in S(G,u) \mid B(G,u) \cap \ker(X) \subseteq \ker(s)\}.$$

Then $V = W$ and V is a closed face of $S(G,u)$. Moreover, V equals the closure of the face generated by X in $S(G,u)$.

PROOF. By Lemma 6.11, V and W are closed faces of $S(G,u)$. Obviously $V \subseteq W$. Now let Y denote the closure of the face generated by X in $S(G,u)$. As V contains X, it must contain the face generated by X, and hence $Y \subseteq V$.

Now $Y \subseteq V \subseteq W$, and it only remains to show that $W \subseteq Y$. Since Y is a closed convex set and W equals the closure of the convex hull of $\partial_e W$, it suffices to prove that $\partial_e W \subseteq Y$. Thus consider any $t \in \partial_e W$, and note that $t \in \partial_e S(G,u)$, because W is a face of $S(G,u)$. By Lemma 8.12, the set

$$M = B(G,u) \cap \ker(t)$$

is a maximal ideal of $B(G,u)$, and then Proposition 8.13 provides us with formulas for t on G^+.

Set $A = B(G,u) - M$, which is a nonempty downward-directed subset of $B(G,u)$. As $t \in W$, we have

$$M = B(G,u) \cap \ker(t) \supseteq B(G,u) \cap \ker(X),$$

whence A is disjoint from $\ker(X)$. Hence, given any $e \in A$, we may choose a state $s_e \in X$ such that $s_e(e) > 0$. Then the function $t_e = s_e(e)^{-1} s_e p_e$ is a state on (G,u) such that $t_e(e) = 1$. By Lemma 8.10, $t_e = t_e p_e$. Since $t_e \leq^+ s_e(e)^{-1} s_e$, Proposition 6.15 says that t_e lies in the face generated by X in $S(G,u)$, and hence $t_e \in Y$.

Now $\{t_e \mid e \in A\}$ is a net in Y, and we claim that $t_e \to t$ in $S(G,u)$. It is enough to show that $t_e(x) \to t(x)$ for any $x \in G^+$.

Given any positive real number ε, by Proposition 8.13 there exist $k \in \mathbb{Z}^+$ and $l, m, n \in \mathbb{N}$ such that

$$k/m > t(x) - \varepsilon \quad \text{and} \quad l/n < t(x) + \varepsilon,$$

while also $kf \leq mp_f(x)$ and $np_g(x) \leq lg$ for some $f, g \in A$. Note that $f \wedge g$ is in A. For any $e \in A$ with $e \leq f \wedge g$, we have

$$ke = p_e(kf) \leq mp_e p_f(x) = mp_e(x) \quad \text{and} \quad np_e(x) = np_e p_g(x) \leq p_e(lg) = le,$$

whence $k = kt_e(e) \leq mt_e p_e(x) = mt_e(x)$ and similarly $nt_e(x) \leq l$. As a result,

$$t(x) - \varepsilon < k/m \leq t_e(x) \leq l/n < t(x) + \varepsilon.$$

Thus $|t_e(x) - t(x)| < \varepsilon$ for all $e \in A$ such that $e \leq f \wedge g$.

Therefore $t_e(x) \to t(x)$ for any $x \in G^+$, as claimed. Hence, $t_e \to t$ in $S(G,u)$. Since Y is closed in $S(G,u)$, we conclude that $t \in Y$.

Thus $\partial_e W \subseteq Y$, and hence $W \subseteq Y$, by the Krein-Mil'man Theorem. Therefore $Y = V = W$. □

COROLLARY 8.17. *Let (G,u) be an interpolation group with order-unit, satisfying general comparability, and let $X \subseteq S(G,u)$. If $\ker(X)^+ = \{0\}$, then $S(G,u)$ equals the closure of the face generated by X in $S(G,u)$.* □

COROLLARY 8.18. *Let (G,u) be an interpolation group with order-unit, satisfying general comparability. Then the closure of any face of $S(G,u)$ is a face of $S(G,u)$. Moreover, the closed faces of $S(G,u)$ are exactly the sets*

$$F_H = \{s \in S(G,u) \mid H \subseteq \ker(s)\},$$

where H is any ideal of G.

PROOF. That closures of faces are faces in $S(G,u)$ is immediate from Theorem 8.16. We have noted, in Proposition 6.12, that any F_H is a closed face of $S(G,u)$. Conversely, let F be a closed face of $S(G,u)$, and H the subgroup of G generated by $\ker(F)^+$. Then H is an ideal of G, by Proposition 1.9, and

$$F = \{s \in S(G,u) \mid \ker(F)^+ \subseteq \ker(s)\} = F_H,$$

because of Theorem 8.16. □

• **Functional Representations.** Since the extreme boundary of the state space of an interpolation group with general comparability is compact and totally disconnected (Corollary 8.15), it is relatively easy to patch together local results on such an extreme boundary. Using a patching argument, we can approximate any "allowable" continuous real-valued function on the extreme boundary of the state space by evaluation at a group element, where the "allowable" functions are those having suitably restricted values at discrete extremal states.

PROPOSITION 8.19. *Let (G, u) be an interpolation group with order-unit, satisfying general comparability, and let p and q be functions in $C(\partial_e S(G, u), \mathbb{R})$ such that $p \ll q$. Assume that $(p(s), q(s)) \cap s(G)$ is nonempty for all discrete extremal states s on (G, u). Then there exists $x \in G$ such that*

$$p(s) < s(x) < q(s) \quad \text{for all } s \in \partial_e S(G, u).$$

PROOF. Set $X = \partial_e S(G, u)$. For any $y \in G$, the set

$$V_y = \{s \in X \mid p(s) < s(y) < q(s)\}$$

is an open subset of X. We claim that these open sets cover X.

Given $s \in X$, we have $p(s) < q(s)$, so that $(p(s), q(s))$ is a nonempty open interval in $\mathbb{R}$. If s is discrete, then $(p(s), q(s)) \cap s(G)$ is nonempty by hypothesis, while if s is not discrete, then $(p(s), q(s)) \cap s(G)$ is nonempty because $s(G)$ is dense in $\mathbb{R}$. In either case, there is some $y \in G$ such that $s(y) \in (p(s), q(s))$, whence $s \in V_y$.

Thus the open sets V_y cover X, as claimed. As X is compact and totally disconnected (Corollary 8.15), we may write X as a disjoint union of clopen subsets $U_1, \dots, U_n$ such that each $U_i \subseteq V_{x(i)}$ for some $x(i) \in G$.

Let $\theta \colon X \to \operatorname{Max} B(G, u)$ be the homeomorphism given in Theorem 8.14. Then $\theta(U_1), \dots, \theta(U_n)$ are pairwise disjoint clopen subsets of $\operatorname{Max} B(G, u)$. Thus, there exist elements $e(1), \dots, e(n)$ in $B(G, u)$ such that

$$\theta(U_i) = \{M \in \operatorname{Max} B(G, u) \mid e(i) \notin M\}$$

for all i, and $e(i) \wedge e(j) = 0$ for all $i \neq j$. Then each

$$U_i = \{s \in X \mid s(e(i)) \neq 0\},$$

and we note from Lemma 8.10 that $s(e(i)) = 1$ and $s = sp_{e(i)}$ for all $s \in U_i$. For all $i \neq j$, we see by Lemma 8.5 that $p_{e(i)} p_{e(j)} = p_{e(i) \wedge e(j)} = 0$.

Set $x = p_{e(1)}(x(1)) + \cdots + p_{e(n)}(x(n))$. Any $s \in X$ lies in some U_j, and so

$$s(x) = sp_{e(j)}(x) = sp_{e(j)}(x(j)) = s(x(j)).$$

Since $U_j \subseteq V_{x(j)}$, we also have $p(s) < s(x(j)) < q(s)$. Therefore $p(s) < s(x) < q(s)$. $\square$

THEOREM 8.20. *Let (G, u) be an interpolation group with order-unit, satisfying general comparability, and let $\psi \colon G \to C(\partial_e S(G,u), \mathbb{R})$ be the natural map. Set*

$$B = \{p \in C(\partial_e S(G,u), \mathbb{R}) \mid p(s) \in s(G) \text{ for all discrete } s \in \partial_e S(G,u)\}.$$

Then $\psi(G)$ is a dense subgroup of B.

PROOF. Obviously $\psi(G) \subseteq B$. Given $p \in B$ and a positive real number ε, note that the intersection $(p(s) - \varepsilon, p(s) + \varepsilon) \cap s(G)$ is nonempty for all discrete extremal states s on (G, u), by definition of B. Hence, Proposition 8.19 shows that there exists $x \in G$ such that

$$p(s) - \varepsilon < s(x) < p(s) + \varepsilon$$

for all $s \in \partial_e S(G,u)$. Then $p - \varepsilon \ll \psi(x) \ll p + \varepsilon$, whence $||\psi(x) - p|| < \varepsilon$. Therefore $\psi(G)$ is dense in B. □

In the particular case of Theorem 8.20 in which there are no discrete extremal states on (G, u), it follows that $\psi(G)$ is dense in $C(\partial_e S(G,u), \mathbb{R})$.

COROLLARY 8.21. *Let (G, u) be an interpolation group with order-unit, satisfying general comparability, and let $\phi \colon G \to \mathrm{Aff}(S(G,u))$ be the natural map. Set*

$$A = \{p \in \mathrm{Aff}(S(G,u)) \mid p(s) \in s(G) \text{ for all discrete } s \in \partial_e S(G,u)\}.$$

Then $\phi(G)$ is a dense subgroup of A.

PROOF. Obviously $\phi(G) \subseteq A$. Now consider any $p \in A$ and any positive real number ε. If q is the restriction of p to $\partial_e S(G,u)$, then by Theorem 8.20 there exists $x \in G$ such that $||\psi(x) - q|| < \varepsilon$. Hence,

$$|\ \phi(x)(s) - p(s)\ |=|\ \psi(x)(s) - q(s)\ |\leq ||\psi(x) - q|| < \varepsilon$$

for all $s \in \partial_e S(G,u)$, so that $p - \varepsilon \ll \phi(x) \ll p + \varepsilon$ on $\partial_e S(G,u)$. According to Corollary 5.20, $p - \varepsilon \ll \phi(x) \ll p + \varepsilon$, and thus $||\phi(x) - p|| < \varepsilon$. Therefore $\phi(G)$ is dense in A. □

• **Notes.** Characteristic elements were introduced (without being named) by Freudenthal [**41**, (5.2)] relative to a lattice-ordered real vector space G containing an element $u > 0$ such that $u \wedge x > 0$ for all elements $x > 0$ in G. He proved that when G is Dedekind σ-complete, the collection of characteristic elements is closed under arbitrary countable infima and suprema [**41**, (5.4), (7.1)]. Characteristic elements in this context, assuming G to be Dedekind complete, were named "quasi-units" by Vulich [**120**, p. 850], who observed that in this case the collection of characteristic elements is closed under arbitrary infima and suprema [**120**, p. 851].

Most of the results of this chapter were proved by the author, Handelman, and Lawrence [**60**, Chapters I.3 and I.4]. For instance, Lemma 8.2 and Proposition 8.3 are [**60**, Lemma I.3.1 and Proposition I.3.2(e)], while Theorem 8.7 is [**60**, Proposition I.3.5]. Proposition 8.9 is contained in [**60**, Lemma I.4.4], and Proposition 8.11 is [**60**, Propositions I.4.6(a)(b) and I.4.7(b)]. Proposition 8.13 and Theorem 8.14 are [**60**, Theorems I.4.11 and I.4.12], while Theorem 8.16 is [**60**, Theorem I.4.14].

Proposition 8.19, Theorem 8.20, and Corollary 8.21 are new to the extent of their derivations directly from the general comparability property. More general results of this type (see Chapter 13) were known earlier.

CHAPTER 9

Dedekind σ-Completeness

The purpose of this chapter is to develop a complete continuous function representation for any Dedekind σ-complete lattice-ordered abelian group (G, u) with order-unit. First, we prove that G is archimedean, so that the natural map

$$\psi: G \to C(\partial_e S(G, u), \mathbb{R})$$

provides an isomorphism of G onto $\psi(G)$ (as ordered groups). Second, we prove that (G, u) satisfies general comparability, and hence the results of the previous chapter show that $\psi(G)$ is dense in the subgroup

$$B = \{p \in C(\partial_e S(G, u), \mathbb{R}) \mid p(s) \in s(G) \text{ for all discrete } s \in \partial_e S(G, u)\}.$$

Third, we show that $C(\partial_e S(G, u), \mathbb{R})$ is a Dedekind σ-complete lattice-ordered real vector space and that ψ preserves countable infima and suprema. This result, in conjunction with the denseness of $\psi(G)$ in B, allows us to conclude that $\psi(G)$ equals B. Therefore ψ provides an isomorphism of G onto B (as ordered groups). As an easy corollary, we find that the natural map $\phi: G \to \text{Aff}(S(G, u))$ provides an isomorphism of G onto the corresponding subgroup of $\text{Aff}(S(G, u))$, so that G also has a complete affine function representation.

We begin the chapter by considering some examples of Dedekind σ-complete lattice-ordered abelian groups built from continuous real-valued functions. In particular, for a compact Hausdorff space X, the lattice-ordered real vector space $C(X, \mathbb{R})$ is Dedekind σ-complete if and only if X is basically disconnected. Following the examples, we derive the archimedean and general comparability properties for a Dedekind σ-complete lattice-ordered abelian group (G, u) with order-unit. The final section of the chapter is devoted to the continuous function representation of G and the affine function representation of G.

• **Prototypical Examples.** Prototypical examples of Dedekind σ-complete lattice-ordered abelian groups may be constructed from continuous real-valued functions on the following kind of topological spaces.

DEFINITION. A topological space X is said to be *basically disconnected* provided the closure of every open F_σ subset of X is open.

As a reminder of the relevance of this topological property to σ-completeness properties, recall that a boolean algebra is σ-complete if and only if its maximal ideal space is basically disconnected.

LEMMA 9.1. *Let X be a compact, Hausdorff, basically disconnected space, and let f be a bounded real-valued function on X. For each $x \in X$, let $\mathcal{N}(x)$ be the collection of open neighborhoods of x, and define*

$$g(x) = \inf_{U \in \mathcal{N}(x)} \sup_{y \in U} f(y).$$

Assume for all $\alpha \in \mathbb{R}$ that $f^{-1}(\alpha, \infty)$ is an open F_σ subset of X. Then g is a continuous real-valued function on X.

PROOF. As f is bounded, it is clear that g is real-valued. To prove continuity, it suffices to show that $g^{-1}(-\infty, \alpha)$ and $g^{-1}(\alpha, \infty)$ are open in X for any $\alpha \in \mathbb{R}$.

Given any $x \in g^{-1}(-\infty, \alpha)$, there exists $U \in \mathcal{N}(x)$ such that

$$\sup\{f(y) \mid y \in U\} < \alpha.$$

Since U is also an open neighborhood of any $z \in U$, it follows that

$$g(z) \leq \sup\{f(y) \mid y \in U\} < \alpha$$

for all $z \in U$. Hence, $x \in U \subseteq g^{-1}(-\infty, \alpha)$, proving that $g^{-1}(-\infty, \alpha)$ is open.

For all $n \in \mathbb{N}$, set

$$\begin{aligned} A_n &= g^{-1}\left[\alpha + \tfrac{1}{n}, \infty\right), \\ B_n &= f^{-1}\left[\alpha + \tfrac{1}{n}, \infty\right), \\ C_n &= f^{-1}\left(\alpha + \tfrac{1}{n}, \infty\right). \end{aligned}$$

Given $x \in X$, we have $g(x) \geq \alpha + \frac{1}{n}$ if and only if every open neighborhood of x contains a point y such that $f(y) \geq \alpha + \frac{1}{n}$. Hence, $A_n = \overline{B}_n$. In addition, we note that $C_n \subseteq B_n \subseteq C_{n+1}$ for all n, whence $\overline{C}_n \subseteq \overline{B}_n \subseteq \overline{C}_{n+1}$ for all n. Thus

$$g^{-1}(\alpha, \infty) = \bigcup_{n=1}^{\infty} A_n = \bigcup_{n=1}^{\infty} \overline{B}_n = \bigcup_{n=1}^{\infty} \overline{C}_n.$$

By hypothesis, each C_n is an open F_σ subset of X. Then each $\overline{C}_n$ is open, because X is basically disconnected, and thus $g^{-1}(\alpha, \infty)$ is open.

Therefore g is continuous. □

THEOREM 9.2. *Let X be a compact, Hausdorff, basically disconnected space, and let A be either $\mathbb{R}$ or a discrete additive subgroup of $\mathbb{R}$. Then $C(X, A)$ is a Dedekind σ-complete lattice-ordered abelian group.*

PROOF. Observe that infima and suprema of bounded subsets of A lie in A.

Set $G = C(X, A)$, and consider any sequence $\{f_1, f_2, \ldots\}$ of functions in G which has an upper bound f' in G. Define

$$f(x) = \sup\{f_n(x) \mid n \in \mathbb{N}\}$$

for all $x \in X$. Then $f_1 \leq f \leq f'$, and so f is a bounded function from X to A.

Now define the function g as in Lemma 9.1, and observe that g maps X into A. For any $\alpha \in \mathbb{R}$, we note that

$$f^{-1}(\alpha, \infty) = \bigcup_{n=1}^{\infty} f_n^{-1}(\alpha, \infty).$$

As each $f_n^{-1}(\alpha, \infty)$ is an open F_σ subset of X (by continuity), the same is true of $f^{-1}(\alpha, \infty)$. Hence, Lemma 9.1 shows that g is continuous, and so $g \in G$.

Given any $x \in X$, we have $f(x) \leq \sup\{f(y) \mid y \in U\}$ for all $U \in \mathcal{N}(x)$, whence $f(x) \leq g(x)$. Thus $f_n \leq f \leq g$ for all $n \in \mathbb{N}$.

Now consider any $h \in G$ such that all $f_n \leq h$, whence $f \leq h$. Given any $x \in X$ and any positive real number ε, there is some $U \in \mathcal{N}(x)$ such that $h \ll h(x) + \varepsilon$ on U. Hence,

$$g(x) \leq \sup\{f(y) \mid y \in U\} \leq \sup\{h(y) \mid y \in U\} \leq h(x) + \varepsilon.$$

Thus $g \leq h$, which proves that g is the supremum of the f_n in G.

Therefore G is Dedekind σ-complete. □

COROLLARY 9.3. *Let X be a compact Hausdorff space. Then $C(X, \mathbb{R})$ is Dedekind σ-complete if and only if X is basically disconnected.*

PROOF. If X is basically disconnected, then the group $G = C(X, \mathbb{R})$ is Dedekind σ-complete by Theorem 9.2. Conversely, assume that G is Dedekind σ-complete.

Given an open F_σ subset U of X, write U as the union of a sequence $X_1, X_2, \ldots$ of closed subsets. For each $n \in \mathbb{N}$, an application of Urysohn's Lemma provides us with a continuous function $f_n: X \to [0,1]$ such that $f_n = 0$ on $X - U$ and $f_n = 1$ on X_n. As each $f_n \leq 1$, the family $\{f_1, f_2, \ldots\}$ has a supremum f in G, and $0 \leq f \leq 1$. For each n, we have $1 = f_n \leq f \leq 1$ on X_n, and hence $f = 1$ on X_n. Thus $f = 1$ on U, and then, by continuity, $f = 1$ on $\overline{U}$.

For any point x in $X - \overline{U}$, there is (by a final application of Urysohn's Lemma) a continuous function $g: X \to [0,1]$ such that $g(x) = 0$ and $g = 1$ on $\overline{U}$. For each $n \in \mathbb{N}$, we have $f_n = 0 \leq g$ on $X - U$ while $f_n \leq 1 = g$ on U, and so $f_n \leq g$. Hence, $f \leq g$, and consequently $f(x) = 0$.

Thus $\overline{U} = \{x \in X \mid f(x) \neq 0\}$, proving that $\overline{U}$ is open in X. Therefore X is basically disconnected. □

• **Additional Examples.** Fairly general Dedekind σ-complete lattice-ordered abelian groups may be constructed by taking direct products of those given in Theorem 9.2. In particular, given compact, Hausdorff, basically disconnected spaces $X_0, X_1, X_2, \ldots$ the group

$$H = C(X_0, \mathbb{R}) \times \prod_{n=1}^{\infty} C(X_n, (1/n)\mathbb{Z})$$

is Dedekind σ-complete. For an example with order-unit, take the direct product

$$(G, u) = (C(X_0, \mathbb{R}), 1) \times \prod_{n=1}^{\infty} (C(X_n, (1/n)\mathbb{Z}), 1)$$

in the category of partially ordered abelian groups with order-unit. (Since G equals the convex subgroup of H generated by u, the Dedekind σ-completeness of G follows directly from that of H.) Not all Dedekind σ-complete lattice-ordered abelian groups with order-unit are of this form, as the following example shows.

EXAMPLE 9.4. *There exists a Dedekind σ-complete lattice-ordered abelian group (G, u) with order-unit that is not isomorphic to*

$$(C(X_0, \mathbb{R}), 1) \times \prod_{n=1}^{\infty} (C(X_n, (1/n)\mathbb{Z}), 1)$$

(as ordered groups with order-unit) for any compact Hausdorff spaces $X_0, X_1, X_2, \ldots$.

PROOF. Let X be an uncountable set, and let u denote the constant function 1 on X. Let H be the subspace of $\mathbb{R}^X$ consisting of all bounded real-valued functions on X with countable support, and let G be the subgroup $H + \mathbb{Z}u$ in $\mathbb{R}^X$. As the minimum and maximum of any pair of functions from G lie in G, we see that G is a lattice-ordered abelian group. Since all the functions in G are bounded, u is an order-unit in G.

Given a sequence $f_1, f_2, \ldots$ in G that has an upper bound in G, the rule

$$f(x) = \sup\{f_n(x) \mid n \in \mathbb{N}\}$$

defines a function $f \in \mathbb{R}^X$ that is the supremum of the f_n in $\mathbb{R}^X$. For each $n \in \mathbb{N}$, there exist $k_n \in \mathbb{Z}$ and a countable subset $Y_n \subset X$ such that $f_n(x) = k_n$ for all $x \in X - Y_n$. The set $Y = \bigcup Y_n$ is then a countable subset of X, and we have $f(x) = \max\{k_n \mid n \in \mathbb{N}\}$ for all $x \in X - Y$, whence $f \in G$. Thus f is the supremum of the f_n in G, proving that G is Dedekind σ-complete.

Now suppose that there exist compact Hausdorff spaces $X_0, X_1, X_2, \ldots$ such that

$$(G, u) \cong (C(X_0, \mathbb{R}), 1) \times \prod_{n=1}^{\infty} (C(X_n, (1/n)\mathbb{Z}), 1)$$

(as ordered groups with order-unit). Under this isomorphism, the order-unit 1 in $C(X_0, \mathbb{R})$ corresponds to a characteristic element e in (G, u) that is divisible by all positive integers. The only elements of G that are divisible by all positive integers are the elements of H, and so $e \in H$. Thus e has countable support, and $u - e \neq 0$. Observing that

$$(p_{u-e}(G), u - e) \cong \prod_{n=1}^{\infty} (C(X_n, (1/n)\mathbb{Z}), 1)$$

(as ordered groups with order-unit), we see that no nonzero element of $p_{u-e}(G)$ can be divisible by all positive integers. But then $H \cap p_{u-e}(G) = \{0\}$, which is impossible, because $u - e \neq 0$.

Therefore (G, u) is not isomorphic to any such direct product of groups of continuous functions. □

All the examples of Dedekind σ-complete lattice-ordered abelian groups discussed so far are groups of continuous real-valued functions on compact, Hausdorff, basically disconnected spaces. Up to isomorphism, this is true of any Dedekind σ-complete lattice-ordered abelian group with an order-unit, as we prove in Corollary 9.14. Without an order-unit, however, it is not always possible to embed a Dedekind σ-complete lattice-ordered abelian group in a direct product of copies of $\mathbb{R}$ (as ordered groups), since there need not exist any nonzero positive real-valued homomorphisms on such a group. An example will be constructed with the help of the following lemma, in which we use the notation $\mathcal{P}(X)$ for the boolean algebra of all subsets of a set X.

LEMMA 9.5. *Let X be a set of cardinality at most $2^{\aleph_0}$, and let $\mathcal{M}$ be a nonprincipal maximal ideal of $\mathcal{P}(X)$. Then there exist pairwise disjoint sets $X_1, X_2, \ldots$ in $\mathcal{M}$ such that $\bigcup X_n = X$.*

PROOF. Choose a set Y of cardinality exactly $2^{\aleph_0}$ that contains X, and set

$$\mathcal{N} = \{A \in \mathcal{P}(Y) \mid A \cap X \in \mathcal{M}\}.$$

Then $\mathcal{N}$ is a nonprincipal maximal ideal of $\mathcal{P}(Y)$. If there exist pairwise disjoint sets $Y_1, Y_2, \ldots$ in $\mathcal{N}$ such that $\bigcup Y_n = Y$, then the sets $Y_n \cap X$ are pairwise disjoint members of $\mathcal{M}$ whose union is X.

Thus we may assume, without loss of generality, that $\operatorname{card}(X) = 2^{\aleph_0}$. Then, furthermore, we may assume that $X = \{0,1\}^{\mathbb{N}}$ (the set of all sequences indexed by $\mathbb{N}$ with values 0 and 1).

For any $n \in \mathbb{N}$, we may express X as the disjoint union of the subsets

$$\{x \in X \mid x_n = 0\} \quad \text{and} \quad \{x \in X \mid x_n = 1\},$$

and so one of these subsets must belong to $\mathcal{M}$. Hence, there is some z_n in $\{0,1\}$ such that the set

$$Y_n = \{x \in X \mid x_n \neq z_n\}$$

is in $\mathcal{M}$. The sequence $z = (z_1, z_2, \ldots)$ is an element of X such that

$$\bigcup_{n=1}^{\infty} Y_n = X - \{z\}.$$

Since $\mathcal{M}$ is not principal, $\{z\} \in \mathcal{M}$. Hence, the set $X_1 = Y_1 \cup \{z\}$ belongs to $\mathcal{M}$. Set

$$X_n = Y_n - (Y_1 \cup \cdots \cup Y_{n-1})$$

for all $n = 2, 3, \ldots$. Then $X_1, X_2, \ldots$ are pairwise disjoint sets in $\mathcal{M}$ such that

$$\bigcup_{n=1}^{\infty} X_n = \left(\bigcup_{n=1}^{\infty} Y_n \right) \cup \{z\} = X. \quad \square$$

EXAMPLE 9.6. *There exists a nonzero, Dedekind σ-complete, lattice-ordered abelian group G such that there are no nonzero positive homomorphisms from G to $\mathbb{R}$.*

PROOF. Choose a set X of cardinality $2^{\aleph_0}$, and set $H = \mathbb{Z}^X$ (with the pointwise ordering). Set

$$K = \{f \in H \mid f \text{ has countable support}\},$$

and observe that K is an ideal of H. Then set $G = H/K$. As H is a lattice-ordered abelian group, so is G, by Corollary 1.14.

Consider a sequence $\{f_1+K, f_2+K, \ldots\}$ in G that has an upper bound $h+K$ in G. For $n = 1, 2, \ldots$, we have $f_n + K \leq h + K$ and so $f_n \leq h + k_n$ for some $k_n \in K$. Since k_n has countable support, it follows that $f_n(x) \leq h(x)$ for all but countably many $x \in X$. Hence, there exists a countable subset $W \subseteq X$ such that $f_n(x) \leq h(x)$ for all $n \in \mathbb{N}$ and all $x \in X - W$. Define $f \in H$ so that $f(x) = 0$ for all $x \in W$ while $f(x) = \max\{f_n(x)\}$ for all $x \in X - W$. As W is countable, we see that $f_n + K \leq f + K$ for all n. Given a coset $g + K$ in G such that each $f_n + K \leq g + K$, choose a countable subset $V \subseteq X$ such that $f_n(x) \leq g(x)$ for all $n \in \mathbb{N}$ and all $x \in X - V$. Then $f(x) \leq g(x)$ for all x in $X - (V \cup W)$, and hence $f + K \leq g + K$, because $V \cup W$ is countable. Thus $f + K$ is the supremum of the $f_n + K$ in G, proving that G is Dedekind σ-complete.

To prove that there exist no nonzero positive homomorphisms from G to $\mathbb{R}$, it suffices to show, for any nonzero positive homomorphism $\phi\colon H \to \mathbb{R}$, that $K \not\subseteq \ker(\phi)$.

For all $Y \in \mathcal{P}(X)$, set

$$K(Y) = \{f \in H \mid \operatorname{support}(f) \subseteq Y\},$$

and observe that $K(Y)$ is an ideal of H. Then set

$$\mathcal{M} = \{Y \in \mathcal{P}(X) \mid K(Y) \subseteq \ker(\phi)\},$$

and observe that $\mathcal{M}$ is an ideal of $\mathcal{P}(X)$. As ϕ is nonzero, $X \notin \mathcal{M}$. We claim that there exists $Y \in \mathcal{P}(X)$ such that $\mathcal{M} \cap \mathcal{P}(Y)$ is a maximal ideal of $\mathcal{P}(Y)$.

Suppose not. First, $\mathcal{M}$ is a proper nonmaximal ideal of $\mathcal{P}(X)$, and so X is a disjoint union of subsets X_1 and Y_1 that are not in $\mathcal{M}$. Then $\mathcal{M} \cap \mathcal{P}(Y_1)$ is a proper nonmaximal ideal of $\mathcal{P}(Y_1)$, and so Y_1 is a disjoint union of subsets X_2 and Y_2 that are not in $\mathcal{M}$. Continuing in this manner, we obtain sets $X_1, X_2, \ldots$ and $Y_1, Y_2, \ldots$ in $\mathcal{P}(X) - \mathcal{M}$ such that each Y_n is the disjoint union of X_{n+1} and Y_{n+1}. Note that the sets $X_1, X_2, \ldots$ are pairwise disjoint.

For any $n \in \mathbb{N}$, the set X_n is not in $\mathcal{M}$, whence $K(X_n) \not\subseteq \ker(\phi)$. Since $K(X_n)$ is directed, there must be some f_n in $K(X_n)^+$ for which $\phi(f_n) \neq 0$. On multiplying f_n by a suitably large positive integer, we obtain a function g_n in $K(X_n)^+$ such that $\phi(g_n) > 1$. As the sets X_n are pairwise disjoint, we may define a function $g \in H$ so that $g(x) = g_n(x)$ whenever $x \in X_n$ while $g(x) = 0$ otherwise. Note that $g \geq 0$. Since each g_n has support contained in X_n, we see that $g_1 + \cdots + g_n \leq g$ for all n. But then

$$\phi(g) \geq \phi(g_1) + \cdots + \phi(g_n) > n$$

for all n, which is impossible.

Therefore there does exist $Y \in \mathcal{P}(X)$ such that the ideal $\mathcal{N} = \mathcal{M} \cap \mathcal{P}(Y)$ is a maximal ideal of $\mathcal{P}(Y)$. In particular, $Y \notin \mathcal{M}$, whence $K(Y) \not\subseteq \ker(\phi)$. As above, it follows that there is some $f \in K(Y)^+$ for which $\phi(f) > 1$.

Suppose that $\mathcal{N}$ is not principal. Then, by Lemma 9.5, there exist pairwise disjoint sets $Y_1, Y_2, \dots$ in $\mathcal{N}$ such that $\bigcup Y_n = Y$. Define $g \in H$ so that $g(x) = nf(x)$ whenever $x \in Y_n$ while $g(x) = 0$ otherwise. Given any positive integer k, we have $g(x) \geq kf(x)$ for all x in $X - (Y_1 \cup \cdots \cup Y_{k-1})$. Hence, $g - kf = h_1 + h_2$ for some h_1 in $K(Y_1 \cup \cdots \cup Y_{k-1})$ and some $h_2 \in H^+$. Since the sets $Y_1, \dots, Y_{k-1}$ are in $\mathcal{N}$, their union is in $\mathcal{N}$, and hence $h_1 \in \ker(\phi)$. Thus

$$\phi(g) = \phi(kf + h_1 + h_2) = k\phi(f) + \phi(h_2) \geq k\phi(f) > k.$$

As k is an arbitrary positive integer, this is impossible.

Thus $\mathcal{N}$ must be principal, and hence there is some $y \in Y$ such that $\{y\}$ is not in $\mathcal{N}$, whence $K(\{y\}) \not\subseteq \ker(\phi)$. Since $\{y\}$ is countable, $K(\{y\}) \subseteq K$. Therefore $K \not\subseteq \ker(\phi)$, as desired. □

As we shall prove shortly, all Dedekind σ-complete lattice-ordered abelian groups are archimedean (Proposition 9.7). Hence, Example 9.6 also provides an example of a nonzero, archimedean, lattice-ordered abelian group that has no nonzero positive homomorphisms to $\mathbb{R}$.

• General Comparability.

PROPOSITION 9.7. *Any Dedekind σ-complete lattice-ordered abelian group G is archimedean.*

PROOF. Consider any $x, y \in G$ such that $nx \leq y$ for all $n \in \mathbb{N}$. By Dedekind σ-completeness, the set $\{x, 2x, 3x, \dots\}$ has a supremum z in G. Applying Proposition 1.4, we find that

$$z + x = \left(\bigvee_{n=1}^{\infty} nx \right) + x = \bigvee_{n=1}^{\infty} (n+1)x \leq z,$$

whence $x \leq 0$. Therefore G is archimedean. □

LEMMA 9.8. *Let (G, u) be a Dedekind σ-complete lattice-ordered abelian group with order-unit, let $x \in G^+$, and set*

$$e = \bigvee_{n=1}^{\infty} (nx \wedge u).$$

Then e is a characteristic element of (G,u), and $p_e(x) = x$. If f is any characteristic element of (G,u) such that $p_f(x) = x$, then $e \le f$.

PROOF. Obviously $0 \le e \le u$. Set $y_n = nx \wedge (u-e)$ for all $n \in \mathbb{N}$. We claim that $ky_n \le e$ for all $k, n \in \mathbb{N}$. For any $n \in \mathbb{N}$, we have $y_n \le nx \wedge u \le e$ by definition of y_n and e. If $ky_n \le e$ for some $k \in \mathbb{N}$, then

$$(k+1)y_n \le e + y_n \le e + (u-e) = u$$

and also $(k+1)y_n \le (k+1)nx$, whence

$$(k+1)y_n \le (k+1)nx \wedge u \le e,$$

completing the induction step. Thus $ky_n \le e$ for all $k, n \in \mathbb{N}$, as claimed. Since G is archimedean by Proposition 9.7, it follows that each $y_n = 0$. Applying Proposition 1.7, we conclude that

$$e \wedge (u-e) = \bigvee_{n=1}^{\infty} (nx \wedge u \wedge (u-e)) = \bigvee_{n=1}^{\infty} y_n = 0.$$

Thus e is a characteristic element of (G,u).

Now $x \le mu$ for some $m \in \mathbb{N}$. Since G is lattice-ordered, it is an interpolation group, and so $x = x_1 + \cdots + x_m$ for some elements $x_1, \ldots, x_m$ in G^+ such that each $x_i \le u$. Then each $x_i \le x \wedge u \le e$, whence $x \le me$. By Proposition 8.3, $p_e(x) = x \wedge me = x$.

Finally, consider any characteristic element f of (G,u) such that $p_f(x) = x$. According to Proposition 8.3, $p_f(x) = x \wedge mf$, and so $x \le mf$. For all $n \in \mathbb{N}$, we obtain

$$nx \wedge u \le u \wedge nmf = p_f(u) = f,$$

with a third application of Proposition 8.3. Therefore $e \le f$. $\square$

THEOREM 9.9. *If (G,u) is a Dedekind σ-complete lattice-ordered abelian group with order-unit, then (G,u) satisfies general comparability.*

PROOF. Given $x, y \in G$, we must find some $e \in B(G,u)$ such that $p_e(x) \le p_e(y)$ while $p_{u-e}(x) \ge p_{u-e}(y)$. Set $x' = x - (x \wedge y)$ and $y' = y - (x \wedge y)$. Then $x', y' \in G^+$, and $x' \wedge y' = 0$ by Proposition 1.4. Since it suffices to find $e \in B(G,u)$ such that $p_e(x') \le p_e(y')$ while $p_{u-e}(x') \ge p_{u-e}(y')$, there is no loss of generality in assuming that $x, y \in G^+$ and $x \wedge y = 0$.

For any $n \in \mathbb{N}$, we have $0 \le x \wedge ny \le ny$, and so

$$x \wedge ny = y_1 + \cdots + y_n$$

for some $y_1, \ldots, y_n \in G^+$ with each $y_i \le y$. Then each $y_i \le x \wedge y = 0$, whence $x \wedge ny = 0$. Set

$$e = \bigvee_{n=1}^{\infty} (ny \wedge u).$$

By Lemma 9.8, e is a characteristic element of (G,u), and $p_e(y)=y$. In view of Proposition 1.7,

$$x\wedge e=\bigvee_{n=1}^{\infty}(x\wedge ny\wedge u)=0,$$

from which we infer that $x\wedge me=0$ for all $m\in\mathbb{N}$. As $x\le mu$ for some $m\in\mathbb{N}$, Proposition 8.3 then shows that $p_e(x)=0$.

Thus $p_e(x)=0\le y=p_e(y)$, and

$$p_{u-e}(x)=x-p_e(x)=x\ge 0=y-p_e(y)=p_{u-e}(y).\quad\square$$

COROLLARY 9.10. *If (G,u) is a Dedekind σ-complete lattice-ordered abelian group with order-unit, then $\partial_e S(G,u)$ is a compact, Hausdorff, basically disconnected space.*

PROOF. By Theorem 8.7, $B(G,u)$ is a σ-complete boolean algebra. Hence, $\operatorname{Max} B(G,u)$ is a compact, Hausdorff, basically disconnected space. Since (G,u) satisfies general comparability by Theorem 9.9, Theorem 8.14 shows that the space $\partial_e S(G,u)$ is homeomorphic to $\operatorname{Max} B(G,u)$. $\square$

• **Functional Representations.** If (G,u) is a Dedekind σ-complete lattice-ordered abelian group with order-unit, then since G is archimedean (Proposition 9.7), G is naturally isomorphic (as an ordered group) to a subgroup of $C(\partial_e S(G,u),\mathbb{R})$, by Theorem 7.7. As (G,u) also satisfies general comparability (Theorem 9.9), the closure of the image of G in $C(\partial_e S(G,u),\mathbb{R})$ is known (Theorem 8.20). Using Dedekind σ-completeness, we show that this image is in fact closed.

LEMMA 9.11. *Let (G,u) be an interpolation group with order-unit, satisfying general comparability, and let $\psi\colon G\to C(\partial_e S(G,u),\mathbb{R})$ be the natural map. Then the set*

$$\{\psi(x)/2^n \mid x\in G \text{ and } n\in\mathbb{N}\}$$

is a dense subgroup of $C(\partial_e S(G,u),\mathbb{R})$.

PROOF. Set $X=\partial_e S(G,u)$, and recall from Corollary 8.15 that X is a compact, Hausdorff, totally disconnected space. Let $p\in C(X,\mathbb{R})$, let ε be a positive real number, and choose $n\in\mathbb{N}$ such that $1/2^n<\varepsilon$. Set

$$V_k=\{s\in X\colon |p(s)-k/2^n|<\varepsilon\}$$

for all integers k. Since $1/2^n<\varepsilon$, these open sets V_k cover X. As X is compact and totally disconnected, X must be the union of pairwise disjoint clopen sets $U_1,\dots,U_m$ such that each $U_i\subseteq V_{k(i)}$ for some integer $k(i)$.

Let $\theta\colon X\to \operatorname{Max} B(G,u)$ be the homeomorphism given in Theorem 8.14; then $\theta(U_1),\dots,\theta(U_m)$ are pairwise disjoint clopen subsets of $\operatorname{Max} B(G,u)$. Hence, there exist $e(1),\dots,e(m)$ in $B(G,u)$ such that

$$\theta(U_i)=\{M\in \operatorname{Max} B(G,u)\mid e(i)\notin M\}$$

for all i and $e(i) \wedge e(j) = 0$ for all $i \neq j$. Then each

$$U_i = \{s \in X \mid s(e(i)) \neq 0\}.$$

In view of Lemma 8.10, we see that $s(e(i)) = 1$ for all $s \in U_i$. For $i \neq j$, we of course have $s(e(j)) = 0$ for all $s \in U_i$.

Now set $x = k(1)e(1) + \cdots + k(n)e(n)$. Any s in X must lie in some U_i, whence $s(x) = k(i)$. As $U_i \subseteq V_{k(i)}$, it follows that

$$|\psi(x)(s)/2^n - p(s)| = |p(s) - k(i)/2^n| < \varepsilon.$$

Thus $\|\psi(x)/2^n - p\| < \varepsilon$. $\square$

LEMMA 9.12. *Let (G, u) be a Dedekind σ-complete lattice-ordered abelian group with order-unit, and set $H = C(\partial_e S(G, u), \mathbb{R})$. Then H is a Dedekind σ-complete lattice-ordered real vector space, and the natural map $\psi: G \to H$ preserves countable infima and suprema.*

PROOF. Set $X = \partial_e S(G, u)$. As X is a compact, Hausdorff, basically disconnected space (Corollary 9.10), Theorem 9.2 shows that the lattice-ordered real vector space H is Dedekind σ-complete.

Consider a sequence $x_1, x_2, \ldots$ in G that has an infimum x in G. Set

$$y_k = x_1 \wedge \cdots \wedge x_k$$

for all $k \in \mathbb{N}$, and note that $\bigwedge y_k = \bigwedge x_k = x$. Since (G, u) satisfies general comparability (Theorem 9.9), Proposition 8.11 shows that all the extremal states on (G, u) are lattice homomorphisms. As a result,

$$\begin{aligned} \psi(y_k)(s) &= s(x_1 \wedge \cdots \wedge x_k) = \min\{s(x_1), \ldots, s(x_k)\} \\ &= \min\{\psi(x_1)(s), \ldots, \psi(x_k)(s)\} \end{aligned}$$

for all $s \in X$, so that $\psi(y_k) = \psi(x_1) \wedge \cdots \wedge \psi(x_k)$. Hence, $\bigwedge \psi(y_k) = \bigwedge \psi(x_k)$, and so to prove that $\psi(x) = \bigwedge \psi(x_k)$ we need only show that $\psi(x) = \bigwedge \psi(y_k)$.

Thus there is no loss of generality in replacing $x_1, x_2, \ldots$ by $y_1, y_2, \ldots$; that is, we may assume that $x_1 \geq x_2 \geq \cdots$. As a result, Proposition 1.8 shows that $mx = \bigwedge mx_k$ for all $m \in \mathbb{N}$.

Obviously $\psi(x) \leq \psi(x_k)$ for all k. Now consider any $p \in H$ such that $p \leq \psi(x_k)$ for all k. Given any positive real number ε, use Lemma 9.11 to choose $z \in G$ and $n \in \mathbb{N}$ such that

$$p - \varepsilon \ll \psi(z)/2^n \ll p.$$

In particular, $\psi(z)/2^n \ll \psi(x_k)$ for all k, whence $\psi(z) \ll \psi(2^n x_k)$ for all k. As G is archimedean (Proposition 9.7), it follows from Theorem 7.7 that $z \leq 2^n x_k$ for all k. Hence,

$$z \leq \bigwedge_{k=1}^{\infty} 2^n x_k = 2^n x,$$

and so $p - \varepsilon \ll \psi(z)/2^n \leq \psi(x)$. Thus $p \leq \psi(x)$, proving that $\psi(x) = \bigwedge \psi(x_k)$.

Therefore ψ preserves countable infima. Applying the order anti-automorphisms $z \mapsto -z$ in G and H, we conclude that ψ also preserves countable suprema. □

THEOREM 9.13. *Let (G,u) be a Dedekind σ-complete lattice-ordered abelian group with order-unit, and let p and q be functions in $C(\partial_e S(G,u),\mathbb{R})$ such that $p \le q$. Assume that $[p(s),q(s)] \cap s(G)$ is nonempty for all discrete s in $\partial_e S(G,u)$. Then there exists $x \in G$ such that $p(s) \le s(x) \le q(s)$ for all $s \in \partial_e S(G,u)$.*

PROOF. Set $X = \partial_e S(G,u)$ and $H = C(X,\mathbb{R})$, and let $\psi: G \to H$ be the natural map.

Given any $m,n \in \mathbb{N}$ and any discrete $s \in X$, note that the set

$$(p(s) - 1/m, q(s) + 1/n) \cap s(G)$$

contains $[p(s),q(s)] \cap s(G)$ and so is nonempty, by hypothesis. In view of Theorem 9.9 and Proposition 8.19, there exists an element $y_{mn} \in G$ such that

$$p - 1/m \ll \psi(y_{mn}) \ll q + 1/n.$$

As X is compact, p and q are bounded; hence, we may choose a positive integer r such that $-r \ll p - 1$ and $q + 1 \ll r$. Then

$$\psi(-ru) = -r \ll \psi(y_{mn}) \ll r = \psi(ru)$$

for all m,n. According to Proposition 9.7 and Theorem 7.7, it follows that $-ru \le y_{mn} \le ru$ for all m,n, so that the set $\{y_{mn} \mid m,n \in \mathbb{N}\}$ is bounded.

For each $m \in \mathbb{N}$, set

$$z_m = \bigwedge_{n=1}^{\infty} y_{mn}.$$

Using Lemma 9.12, we see that

$$p - \frac{1}{m} \le \bigwedge_{n=1}^{\infty} \psi(y_{mn}) = \psi(z_m) \le \bigwedge_{n=1}^{\infty} \left[q + \frac{1}{n}\right] = q.$$

Now set

$$x = \bigvee_{m=1}^{\infty} z_m.$$

With a second application of Lemma 9.12, we conclude that

$$p = \bigvee_{m=1}^{\infty} \left[p - \frac{1}{m}\right] \le \bigvee_{m=1}^{\infty} \psi(z_m) = \psi(x) \le q.$$

Therefore $p(s) \le \psi(x)(s) = s(x) \le q(s)$ for all $s \in X$. □

COROLLARY 9.14. *Let (G,u) be a Dedekind σ-complete lattice-ordered abelian group with order-unit, and set*

$$B = \{p \in C(\partial_e S(G,u),\mathbb{R}) \mid p(s) \in s(G) \text{ for all discrete } s \in \partial_e S(G,u)\}.$$

Then the natural map $\psi: G \to C(\partial_e S(G,u), \mathbb{R})$ *provides an isomorphism of* (G,u) *onto* $(B,1)$ *(as ordered groups with order-unit).*

PROOF. Obviously ψ maps G into B, and it follows from Theorem 9.13 that $\psi(G) = B$. As G is archimedean by Proposition 9.7, Theorem 7.7 shows that ψ provides an isomorphism of G onto $\psi(G)$ (as ordered groups). Since $\psi(u) = 1$, the proof is complete. □

COROLLARY 9.15. *Let* (G,u) *be a Dedekind* σ*-complete lattice-ordered abelian group with order-unit, and set*

$$A = \{p \in \mathrm{Aff}(S(G,u)) \mid p(s) \in s(G) \text{ for all discrete } s \in \partial_e S(G,u)\}.$$

Then the natural map $\phi: G \to \mathrm{Aff}(S(G,u))$ *provides an isomorphism of* (G,u) *onto* $(A,1)$ *(as ordered groups with order-unit).*

PROOF. In view of Proposition 9.7 and Theorem 7.7, ϕ provides an isomorphism of (G,u) onto $(\phi(G),1)$ (as ordered groups with order-unit). Obviously $\phi(G) \subseteq A$, and so it only remains to show that $A \subseteq \phi(G)$.

Any function p in A restricts to a function q in the set B defined in Corollary 9.14. By that corollary, there exists $x \in G$ such that $\psi(x) = q$, so that $\phi(x)$ and p agree on $\partial_e S(G,u)$. Then Corollary 5.20 shows that $\phi(x) = p$. Therefore $A \subseteq \phi(G)$, as required. □

• **Notes.** Corollary 9.3 was proved by Nakano [**96**, Sätze 2, 4] and Stone [**118**, Theorem 15], who also proved the corresponding result that $C(X,\mathbb{R})$ is Dedekind complete if and only if X is extremally disconnected [**96**, Sätze 1, 3; **118**, Theorem 14]. Examples 9.4 and 9.6 were constructed by the author. An example of a Dedekind complete lattice-ordered abelian group without enough positive real-valued homomorphisms to separate points was constructed by Nakayama [**97**, Theorem]. Proposition 9.7 was first proved in the Dedekind complete case by Kantorovitch [**80**, Satz 15(f)], and the Dedekind σ-complete case appeared in Birkhoff [**11**, Theorem 38]. Theorem 9.9 was proved by the author, Handelman, and Lawrence [**60**, Theorem I.4.3]. The analog of Theorem 9.13 for functions in $\mathrm{Aff}(S(G,u))$ was proved more generally for partially ordered abelian groups with order-unit satisfying the countable interpolation property, by the author, Handelman, and Lawrence, along with the corresponding version of Corollary 9.15 [**60**, Theorems I.9.2 and I.9.4].

CHAPTER 10

Choquet Simplices

Up to this point, our development of the natural affine function representation for an interpolation group (G, u) with order-unit has just relied on general properties of arbitrary compact convex sets. To obtain tighter control of this representation requires some study of the geometry of the state space $S(G, u)$ and the corresponding structure of the partially ordered vector space $\mathrm{Aff}(S(G, u))$. In particular, the interpolation property in $\mathrm{Aff}(S(G, u))$ is needed, so that the representation of G in terms of $\mathrm{Aff}(S(G, u))$ remains within the class of interpolation groups. Interpolation in $\mathrm{Aff}(S(G, u))$ can be obtained because $S(G, u)$ is a rather special kind of compact convex set known as a Choquet simplex (which may be viewed as an infinite-dimensional generalization of classical finite-dimensional simplices). Hence, we investigate the geometry of an arbitrary Choquet simplex K and the structure of $\mathrm{Aff}(K)$.

This chapter and the following one are devoted to a general development of Choquet simplices and the affine continuous real-valued functions on them. The simplex property, by which Choquet simplices are characterized among compact convex subsets of locally convex Hausdorff spaces, is a purely algebraic condition. Namely, a convex set is a simplex if and only if it is affinely isomorphic to a base for the positive cone of a lattice-ordered real vector space. Hence, a major portion of this chapter is concerned with algebraic properties of simplices. For example, the extreme boundary of any simplex is affinely independent, and the convex hull of any union of faces in a simplex is a face. After Choquet simplices are introduced, it is shown that a finite-dimensional compact convex set is a Choquet simplex if and only if it is a classical simplex, and that any inverse limit of classical simplices is a Choquet simplex.

• Simplices.

DEFINITION. A *classical simplex* in a real vector space E is any convex subset of E that is the convex hull of a nonempty finite set of affinely independent points of E. For any nonnegative integer n, the *standard n-dimensional simplex* is the convex hull in $\mathbb{R}^{n+1}$ of the standard basis vectors

$$(1,0,0,\ldots,0), \quad (0,1,0,\ldots,0),\ldots,(0,\ldots,0,1,0), \quad (0,\ldots,0,0,1).$$

In particular, a convex set is a classical simplex if and only if it is affinely isomorphic to one of the standard simplices. Note from Proposition 6.6 that the state space of any nonzero simplicial group with order-unit is a classical simplex.

If K is the standard n-dimensional simplex (for some $n \geq 0$), then K generates the positive cone of $\mathbb{R}^{n+1}$ in the sense that the elements of $(\mathbb{R}^{n+1})^+$ are exactly the nonnegative multiples of points of K. The affine independence of the points generating K is reflected (quite faithfully, as it turns out) by the fact that the ordering in $\mathbb{R}^{n+1}$ (which is induced from the cone generated by K) is a lattice-ordering. With much hindsight, this observation provides the key to developing infinite-dimensional simplices.

DEFINITION. A *convex cone* in a real vector space E is any convex subset of the vector space E that is also a cone in the abelian group E. In other words, a convex cone in E is any subset C of E such that $0 \in C$ and also $\alpha_1 x_1 + \alpha_2 x_2 \in C$ for any $\alpha_1, \alpha_2 \in \mathbb{R}^+$ and any $x_1, x_2 \in C$. A *strict convex cone* is any convex cone C that is a strict cone, i.e., the only point $x \in E$ for which $x, -x \in C$ is the point $x = 0$.

For example, the positive cone of any partially ordered real vector space is a strict convex cone. Conversely, if C is a strict convex cone in a real vector space E, then $(E, \leq_C)$ is a partially ordered real vector space with positive cone C.

DEFINITION. Let C be a convex cone in a real vector space. If C is nonzero, then a *base* for C is any convex subset K of C such that every nonzero point of C may be uniquely expressed in the form αx for some $\alpha \in \mathbb{R}^+$ and some $x \in K$. By convention, the only base for the zero cone is the empty set.

LEMMA 10.1. *Let C be a convex cone in a real vector space. If C has a base K, then $0 \notin K$ and C is a strict cone.*

PROOF. As these properties are trivially satisfied by the zero cone, we may assume that C is nonzero. Choose a nonzero point $c \in C$, write c as a nonnegative multiple of a point $w \in K$, and note that $w \neq 0$. If $0 \in K$, then the point $v = \frac{1}{2}w$, being a convex combination of 0 and w, must lie in K. However, since $v \neq 0$, and since v may be expressed in the form $1v$ as well as $\frac{1}{2}w$, with $v, w \in K$, this contradicts the uniqueness condition in the definition of a base. Thus $0 \notin K$.

If C is not strict, there exists a nonzero point $x \in C$ such that $-x \in C$, and we may write $x = \alpha y$ and $-x = \beta z$ for some $y, z \in K$ and some positive real numbers α, β. Then

$$(\alpha+\beta)^{-1}\alpha y + (\alpha+\beta)^{-1}\beta z = (\alpha+\beta)^{-1}(x - x) = 0,$$

whence 0 is a convex combination of y and z. However, as $0 \notin K$, this is impossible. Therefore C is a strict cone. □

If K is a nonempty convex subset of a real vector space E, then K is quite likely to be a base for a convex cone in E. For, as the following proposition shows, K is a base for the convex cone

$$\{\alpha x \mid \alpha \in \mathbb{R}^+ \text{ and } x \in K\}$$

if and only if K lies in a hyperplane in E that misses the origin. If K does not lie in such a hyperplane, then K can easily be maneuvered into one, as follows. Namely, K is affinely isomorphic to the convex set $K' = K \times \{1\}$ in the vector space $E' = E \times \mathbb{R}$, and K' lies in the hyperplane $E \times \{1\}$ in E', which misses the origin. Hence, K' is a base for a convex cone in E'. To complete the remaining case, note that the empty convex set is a base for the zero cone in any vector space.

Thus any convex set may be realized as a base for a convex cone. As a result, convex sets may be investigated by studying the convex cones in which they are bases.

PROPOSITION 10.2. *Let K be a nonempty convex subset of a real vector space E, and set*

$$C = \{\alpha x \mid \alpha \in \mathbb{R}^+ \text{ and } x \in K\}.$$

Then C is a convex cone in E, and the following conditions are equivalent:

(a) *K is a base for C.*

(b) *Any affine map from K to a real vector space E' extends to a linear map from E to E'.*

(c) *There exists a linear functional f on E such that $f = 1$ on K.*

(d) *K is contained in a hyperplane in E which misses the origin.*

PROOF. There exists a point $x \in K$, and then $0 = 0x \in C$. It is clear that C is closed under multiplication by nonnegative scalars. Given $y_1, y_2 \in C$, write each $y_i = \alpha_i x_i$ for some $\alpha_i \in \mathbb{R}^+$ and some $x_i \in K$, and set $\alpha = \alpha_1 + \alpha_2$. If $\alpha = 0$, then each $\alpha_i = 0$, whence $y_1 + y_2 = 0$ and so $y_1 + y_2 \in C$. If $\alpha > 0$, then the point

$$x = \alpha^{-1}\alpha_1 x_1 + \alpha^{-1}\alpha_2 x_2$$

is a convex combination of x_1 and x_2, so that $x \in K$. Since $y_1 + y_2 = \alpha x$, we obtain $y_1 + y_2 \in C$ in this case also. Thus C is a convex cone in E.

(a) $\Rightarrow$ (b): Let $f: K \to E'$ be an affine map. As each nonzero point of C may be uniquely expressed as a positive multiple of a point of K, we may extend f to a well-defined map $g: C \to E'$ such that $g(\alpha x) = \alpha f(x)$ for all $\alpha \in \mathbb{R}^+$ and all $x \in K$. Note that $g(\beta y) = \beta g(y)$ for all $\beta \in \mathbb{R}^+$ and all $y \in C$. We claim that also $g(z_1 + z_2) = g(z_1) + g(z_2)$ for any $z_1, z_2 \in C$. If either $z_1 = 0$ or $z_2 = 0$, this is clear; hence, we may assume that each $z_i \neq 0$. Then each $z_i = \alpha_i x_i$ for some $x_i \in K$ and some positive real number α_i. Set $\alpha = \alpha_1 + \alpha_2$ and $x = \alpha^{-1}(\alpha_1 x_1 + \alpha_2 x_2)$. Then x is a convex combination of x_1 and x_2, whence $x \in K$. As f is affine, we compute that

$$g(z_1 + z_2) = g(\alpha x) = \alpha f(x) = \alpha[\alpha^{-1}\alpha_1 f(x_1) + \alpha^{-1}\alpha_2 f(x_2)] = g(z_1) + g(z_2),$$

which establishes the claim.

Thus g preserves nonnegative linear combinations. If E_0 is the linear subspace of E spanned by C, then g extends uniquely to a linear map $h: E_0 \to E'$, and h extends (non-uniquely) to a linear map $k: E \to E'$.

(b) $\Rightarrow$ (c): The constant function 1 on K is an affine map from K to $\mathbb{R}$, and by (b) it extends to a linear map $f: E \to \mathbb{R}$.

(c) $\Rightarrow$ (d): For f as in (c), the set $f^{-1}(\{1\})$ is a hyperplane in E that contains K and misses the origin.

(d) $\Rightarrow$ (a): Let H be a hyperplane in E that contains K and misses the origin. If K is not a base for C, then some nonzero point $w \in C$ may be written as $w = \alpha x = \beta y$ where α, β are distinct positive real numbers and x, y are distinct points of K. Then 0 may be expressed as an affine combination

$$0 = (\alpha - \beta)^{-1}\alpha x - (\alpha - \beta)^{-1}\beta y$$

with $x, y \in H$. But then $0 \in H$, which is false. Therefore K is a base for C. □

COROLLARY 10.3. *Let K be a base for a nonzero convex cone in a real vector space. If*

$$\alpha_1 x_1 + \cdots + \alpha_n x_n = 0$$

for some $\alpha_i \in \mathbb{R}$ and some $x_i \in K$, then $\alpha_1 + \cdots + \alpha_n = 0$. □

COROLLARY 10.4. *For $i = 1, 2$, let E_i be a partially ordered real vector space with positive cone C_i, and let K_i be a base for C_i. If K_1 and K_2 are affinely isomorphic, then C_1 and C_2 are isomorphic as partially ordered sets. If, furthermore, E_1 and E_2 are directed, then E_1 and E_2 are isomorphic as ordered vector spaces.*

PROOF. Let $f: K_1 \to K_2$ be an affine isomorphism. For $i = 1, 2$, let E_i' be the linear subspace of E_i spanned by K_i, and note that $C_i \subseteq E_i'$. By Proposition 10.2, f extends to a linear map $g: E_1' \to E_2'$, and f^{-1} extends to a linear map $h: E_2' \to E_1'$. Since hg restricts to the identity map on K_1, we see that hg equals the identity map on E_1', and similarly gh equals the identity map on E_2'. As each C_i consists of the nonnegative multiples of points of K_i, we see that $g(C_1) \subseteq C_2$ and $h(C_2) \subseteq C_1$. Hence, g and h are inverse isomorphisms of ordered vector spaces. In addition, $g(C_1) = C_2$, so that g restricts to an order-isomorphism of C_1 onto C_2. If E_1 and E_2 are directed, then each $E_i = E_i'$, whence E_1 and E_2 are isomorphic as ordered vector spaces. □

DEFINITION. A *lattice cone* in a real vector space E is any strict convex cone C in E such that $(C, \leq_C)$ is a lattice.

In view of Proposition 1.5, a strict convex cone C in a real vector space E is a lattice cone if and only if the linear subspace of E spanned by C is a lattice with respect to $\leq_C$.

DEFINITION. A *simplex* in a real vector space E is any convex subset K of E that is affinely isomorphic to a base for a lattice cone in some real vector space.

Suppose that K is affinely isomorphic to bases K_1 and K_2 for convex cones C_1 and C_2 in real vector spaces E_1 and E_2. Then each E_i may be made into a partially ordered real vector space with positive cone C_i. By Corollary 10.4, C_1 and C_2 are isomorphic as partially ordered sets, and so C_1 is a lattice cone if and only if C_2 is a lattice cone. Hence, we may use whichever cone is convenient when checking to see whether K is a simplex.

For example, any classical simplex K is a simplex. Namely, K is affinely isomorphic to the standard n-dimensional simplex (for some $n \geq 0$), and the standard n-dimensional simplex is a base for the positive cone of $\mathbb{R}^{n+1}$, which is lattice-ordered.

PROPOSITION 10.5. *If (G, u) is an interpolation group with order-unit, then $S(G, u)$ is a base for the positive cone of the Dedekind complete lattice-ordered real vector space B of all relatively bounded homomorphisms from G to $\mathbb{R}$.*

PROOF. That B is a Dedekind complete lattice-ordered real vector space is given in Corollary 2.28. If $G = \{0\}$, then $B = \{0\}$ and $S(G, u)$ is empty, in which case it is clear that $S(G, u)$ is a base for B^+. Now assume that G is nonzero.

Since B^+ is the set of all positive homomorphisms from G to $\mathbb{R}$, Lemma 4.5 shows that

$$B^+ = \{\alpha s \mid \alpha \in \mathbb{R}^+ \text{ and } s \in S(G, u)\}.$$

As $S(G, u)$ lies in the hyperplane $\{f \in B \mid f(u) = 1\}$, which misses the origin, we conclude from Proposition 10.2 that $S(G, u)$ is a base for B^+. □

COROLLARY 10.6. *If (G, u) is an interpolation group with order-unit, then $S(G, u)$ is a simplex.* □

PROPOSITION 10.7. *Let K be a simplex, and let*

$$\alpha_1 x_1 + \cdots + \alpha_n x_n = \beta_1 y_1 + \cdots + \beta_k y_k$$

be positive convex combinations in K. Then there exist $\gamma_{ij} \in \mathbb{R}^+$ and $z_{ij} \in K$ (for $i = 1, \ldots, n$ and $j = 1, \ldots, k$) such that

$$\sum_{j=1}^{k} \gamma_{ij} = \alpha_i \quad \text{and} \quad \sum_{j=1}^{k} \alpha_i^{-1} \gamma_{ij} z_{ij} = x_i \quad (\text{for } i = 1, \ldots, n),$$

$$\sum_{i=1}^{n} \gamma_{ij} = \beta_j \quad \text{and} \quad \sum_{i=1}^{n} \beta_j^{-1} \gamma_{ij} z_{ij} = y_j \quad (\text{for } j = 1, \ldots, k).$$

PROOF. We may assume that K is a base for the positive cone of a lattice-ordered real vector space E. Note that E is an interpolation group.

Since $\alpha_i x_i$ and $\beta_j y_j$ are in E^+ for all i, j, we may apply Riesz decomposition to the given equation, obtaining elements $w_{ij} \in E^+$ (for $i = 1, \ldots, n$ and $j = 1, \ldots, k$) such that

$$\sum_{j=1}^{k} w_{ij} = \alpha_i x_i \quad \text{and} \quad \sum_{i=1}^{n} w_{ij} = \beta_j y_j$$

for all i, j. Each $w_{ij} = \gamma_{ij} z_{ij}$ for some $\gamma_{ij} \in \mathbb{R}^+$ and some $z_{ij} \in K$. The desired expressions for x_i and y_j are immediate from the corresponding expressions for $\alpha_i x_i$ and $\beta_j y_j$. The desired expressions for α_i and β_j follow from Corollary 10.3. □

COROLLARY 10.8. *If K is a simplex in a real vector space E, then $\partial_e K$ is an affinely independent subset of E.*

PROOF. If not, there exist distinct points $x_1, \dots, x_n, y_1, \dots, y_k$ in $\partial_e K$ such that some positive convex combination of the x_i equals some positive convex combination of the y_j, say

$$\alpha_1 x_1 + \cdots + \alpha_n x_n = \beta_1 y_1 + \cdots + \beta_k y_k.$$

Then there exist $\gamma_{ij} \in \mathbb{R}^+$ and $z_{ij} \in K$ as in Proposition 10.7. Since

$$\gamma_{11} + \cdots + \gamma_{1k} = \alpha_1 > 0,$$

there is an index j such that $\gamma_{1j} > 0$. As the extreme point x_1 is a convex combination

$$x_1 = \alpha_1^{-1}\gamma_{11} z_{11} + \cdots + \alpha_1^{-1}\gamma_{1k} z_{1k}$$

of the points $z_{11}, \dots, z_{1k}$ in K, with $\alpha_1^{-1}\gamma_{1j} > 0$, it follows that $x_1 = z_{1j}$. Similarly, the extreme point y_j is a convex combination

$$y_j = \beta_j^{-1}\gamma_{1j} z_{1j} + \cdots + \beta_j^{-1}\gamma_{nj} z_{nj}$$

with $\beta_j^{-1}\gamma_{1j} > 0$, whence $y_j = z_{1j}$. But then $x_1 = y_j$, which is a contradiction.

Therefore $\partial_e K$ is affinely independent. □

Corollary 10.8 immediately shows that many convex sets, for instance discs and convex quadrilaterals in the plane, are *not* simplices.

• Faces.

PROPOSITION 10.9. *Any face of a simplex is a simplex.*

PROOF. Let F be a face of a simplex K. We may assume that K is a base for the positive cone of a lattice-ordered real vector space E and that F is nonempty. Set

$$C = \{\alpha x \mid \alpha \in \mathbb{R}^+ \text{ and } x \in F\},$$

and observe that C is a convex cone with base F. Also, $C \subseteq E^+$.

We claim that whenever $x \in E^+$ and $y \in C$ with $x \le y$, then $x \in C$. This is clear if either $x = 0$ or $x = y$; hence, we may assume that $0 < x < y$. Now $x = \alpha u$ and $y = \beta v$ for some positive real numbers α, β and some points $u \in K$ and $v \in F$. Since $y - x > 0$, we also have $y - x = \gamma w$ for some positive real number γ and some point $w \in K$. Then $\alpha u + \gamma w = \beta v$ and so $\alpha + \gamma = \beta$, by Corollary 10.3. Hence, we obtain a positive convex combination

$$(\beta^{-1}\alpha)u + (\beta^{-1}\gamma)w = v$$

in K. Since v lies in the face F, it follows that $u \in F$, and thus $x \in C$, as claimed.

We first use this claim to see that the relative ordering in C (from E) coincides with $\leq_C$. Given $u, v \in C$ with $u \leq_C v$, we have $v - u \in C$, whence $v - u \in E^+$, and so $u \leq v$. On the other hand, if $u \leq v$, then $v - u$ is a point of E^+ such that $v - u \leq v$, whence $v - u \in C$ by the claim, and so $u \leq_C v$.

Given any $a, b \in C$, the infimum $a \wedge b$ and the supremum $a \vee b$ exist in E. As

$$0 \leq a \wedge b \leq a \quad \text{and} \quad 0 \leq a \vee b \leq a + b,$$

we see from the claim above that $a \wedge b$ and $a \vee b$ are in C. Consequently, $a \wedge b$ and $a \vee b$ are the infimum and supremum of $\{a, b\}$ in C. Therefore C is a lattice cone, whence F is a simplex. □

PROPOSITION 10.10. *In a simplex, the convex hull of any union of faces is a face.*

PROOF. Let K be a simplex. As the convex hull of the union of a family $\{F_i\}$ of faces of K equals the union of the convex hulls of finite unions of the F_i, it suffices to show that the convex hull of any finite union of faces of K is a face of K. By induction, this immediately reduces to the case of two faces. Thus let F_1 and F_2 be faces of K, and let F be the convex hull of $F_1 \cup F_2$.

Given a positive convex combination $\alpha_1 x_1 + \alpha_2 x_2 = x$ with $x_1, x_2 \in K$ and $x \in F$, we must show that $x_1 \in F$. Since $x \in F$, there is a convex combination $x = \beta_1 y_1 + \beta_2 y_2$ with each $y_i \in F_i$. If $\beta_1 = 0$, then $x \in F_2$ and so $x_1 \in F_2$, while if $\beta_2 = 0$, then $x \in F_1$ and so $x_1 \in F_1$. In either of these cases, $x_1 \in F$. Thus we may assume that each $\beta_i > 0$.

By Proposition 10.7, there exist $\gamma_{ij} \in \mathbb{R}^+$ and $z_{ij} \in K$ (for $i, j = 1, 2$) such that

$$\gamma_{i1} + \gamma_{i2} = \alpha_i \quad \text{and} \quad \alpha_i^{-1}\gamma_{i1} z_{i1} + \alpha_i^{-1}\gamma_{i2} z_{i2} = x_i \quad (\text{for } i = 1, 2),$$
$$\gamma_{1j} + \gamma_{2j} = \beta_j \quad \text{and} \quad \beta_j^{-1}\gamma_{1j} z_{1j} + \beta_j^{-1}\gamma_{2j} z_{2j} = y_j \quad (\text{for } j = 1, 2).$$

Suppose that $\gamma_{1j} = 0$ for some $j \in \{1, 2\}$. If $j = 1$, set $k = 2$, while if $j = 2$, set $k = 1$. Then

$$\gamma_{1k} = \gamma_{1j} + \gamma_{1k} = \gamma_{11} + \gamma_{12} = \alpha_1 > 0$$

and so $x_1 = z_{1k}$. Since the convex combination

$$\beta_k^{-1}\gamma_{1k} x_1 + \beta_k^{-1}\gamma_{2k} z_{2k} = \beta_k^{-1}\gamma_{1k} z_{1k} + \beta_k^{-1}\gamma_{2k} z_{2k} = y_k$$

lies in the face F_k, we must have $x_1 \in F_k$. Thus $x_1 \in F$ in this case.

Finally, suppose that $\gamma_{1j} > 0$ for each $j = 1, 2$. As

$$\beta_j^{-1}\gamma_{1j} z_{1j} + \beta_j^{-1}\gamma_{2j} z_{2j} = y_j \in F_j,$$

it follows that $z_{1j} \in F_j$. Then each $z_{1j} \in F$. Since

$$x_1 = \alpha_1^{-1}\gamma_{11} z_{11} + \alpha_1^{-1}\gamma_{12} z_{12},$$

we conclude that $x_1 \in F$. Therefore $x_1 \in F$ in all cases, proving that F is a face of K. □

COROLLARY 10.11. *If K is a simplex, then the convex hull of any subset of $\partial_e K$ is a face of K.* □

Many convex sets fail to satisfy the result of Corollary 10.11. For instance, if K is a disc or a convex quadrilateral in the plane, there exist points $x_1, x_2 \in \partial_e K$ such that the line segment between x_1 and x_2 is not a face of K.

- **Complementary Faces.**

PROPOSITION 10.12. *Let F be a face of a simplex K, and let F' be the union of those faces of K that are disjoint from F. Then F' is a face of K, and it is the largest face of K that is disjoint from F. If a point $x \in K$ can be expressed as convex combinations*

$$x = \alpha_1 x_1 + \alpha_2 x_2 = \beta_1 y_1 + \beta_2 y_2$$

with $x_1, y_1 \in F$ and $x_2, y_2 \in F'$, then $\alpha_i = \beta_i$ for each i, and $x_i = y_i$ for those i such that $\alpha_i > 0$.

PROOF. Obviously F and F' are disjoint. Let F^* be the convex hull of F', and note from Proposition 10.10 that F^* is a face of K. If F^* is not disjoint from F, choose a point $x \in F^* \cap F$, and write x as a convex combination

$$x = \alpha_1 x_1 + \cdots + \alpha_n x_n$$

where each $x_i \in F'$. There is an index j such that $\alpha_j > 0$, and since x lies in the face F, we must have $x_j \in F$. However, as $x_j \in F'$, this contradicts the disjointness of F and F'. Thus F^* must be disjoint from F. Consequently, $F^* \subseteq F'$, and so $F' = F^*$. Therefore F' is a face of K.

Now consider any $x \in K$ for which there are convex combinations

$$x = \alpha_1 x_1 + \alpha_2 x_2 = \beta_1 y_1 + \beta_2 y_2$$

with $x_1, y_1 \in F$ and $x_2, y_2 \in F'$. If $\alpha_1 = 0$, then $x = x_2$ and so $x \in F'$. Since $y_1 \notin F'$, it follows that $\beta_1 = 0$ and $x = y_2$. Hence, $\alpha_2 = \beta_2 = 1$ and $x_2 = y_2$. The same conclusions follow if it is assumed that $\beta_1 = 0$. On the other hand, if either $\alpha_2 = 0$ or $\beta_2 = 0$, we find that $\alpha_1 = \beta_1 = 1$ and $\alpha_2 = \beta_2 = 0$, while also $x_1 = y_1$.

Thus we may assume that each $\alpha_i > 0$ and each $\beta_j > 0$. Then Proposition 10.7 provides us with numbers $\gamma_{ij} \in \mathbb{R}^+$ and points $z_{ij} \in K$ (for $i, j = 1, 2$) such that

$$\gamma_{i1} + \gamma_{i2} = \alpha_i \quad \text{and} \quad \alpha_i^{-1}\gamma_{i1}z_{i1} + \alpha_i^{-1}\gamma_{i2}z_{i2} = x_i \quad (\text{for } i = 1, 2),$$
$$\gamma_{1j} + \gamma_{2j} = \beta_j \quad \text{and} \quad \beta_j^{-1}\gamma_{1j}z_{1j} + \beta_j^{-1}\gamma_{2j}z_{2j} = y_j \quad (\text{for } j = 1, 2).$$

If $\gamma_{12} > 0$, then since

$$\alpha_1^{-1}\gamma_{11}z_{11} + \alpha_1^{-1}\gamma_{12}z_{12} = x_1 \in F \quad \text{and} \quad \beta_2^{-1}\gamma_{12}z_{12} + \beta_2^{-1}\gamma_{22}z_{22} = y_2 \in F',$$

we find that $z_{12} \in F \cap F'$, which is impossible. Thus $\gamma_{12} = 0$, and likewise $\gamma_{21} = 0$. Therefore each $\alpha_i = \gamma_{ii} = \beta_i$ and each $x_i = z_{ii} = y_i$. □

The face F' in Proposition 10.12 comes as close as possible to being a complement for F in the lattice of faces of K, and if K equals the convex hull of $F \cup F'$, then F' is a complement for F. Conversely, if F has a complement in the lattice of faces of K, then that complement must equal F'. Hence, the following terminology is used for F'.

DEFINITION. Let F be a face of a simplex K, and let F' be the union of those faces of K that are disjoint from F. Then F' is called the *complementary face of F in K*.

This terminology is only used with respect to simplices, since in other convex sets F' need not be a face. For example, if K is a square in the plane and $F = \{x\}$ for some vertex x of K, then F' is the union of the two edges of K not containing x, and so F' is not a face of K. Moreover, each of the edges of K not containing x is a complement for F in the lattice of faces of K.

It is easy to check that whenever F is a face of a classical simplex K, then K is the direct convex sum of F and its complementary face. This is also true of any closed face of a compact simplex in a locally convex Hausdorff space, as we shall prove in Theorem 11.28. However, the following example shows that simplices (even compact simplices) may have faces whose complementary faces are too small to be truly complementary.

EXAMPLE 10.13. *There exists a lattice-ordered abelian group (G, u) with order-unit such that $S(G, u)$ contains a proper face whose complementary face is empty.*

PROOF. Let $X = \mathbb{N} \cup \{\infty\}$ be the one-point compactification of $\mathbb{N}$, set $G = C(X, \mathbb{R})$, and let u be the constant function 1 in G. Then G is a lattice-ordered abelian group, u is an order-unit in G, and $S(G, u) = M_1^+(X)$ by Proposition 6.8. As G has interpolation, $S(G, u)$ is a simplex, by Corollary 10.6.

Now let F be the convex hull of $\partial_e S(G, u)$. By Corollary 10.11, F is a face of $S(G, u)$. As

$$\partial_e S(G, u) = \{\varepsilon_1, \varepsilon_2, \dots, \varepsilon_\infty\},$$

we see that all points of F, when viewed as measures in $M_1^+(X)$, have finite support. Since the probability measure

$$\sum_{n=1}^{\infty} \frac{1}{2^n} \varepsilon_n$$

on X does not have finite support, $F \neq S(G, u)$. Let F' be the complementary face of F in $S(G, u)$.

Suppose there exists a measure $\mu \in F'$. Since X is countable, μ may be written as a σ-convex combination of point masses:

$$\mu = \sum_{x \in X} \alpha_x \varepsilon_x.$$

Choose $y \in X$ such that $\alpha_y > 0$. Then μ may be written as a convex combination

$$\mu = \alpha_y \varepsilon_y + (1 - \alpha_y)\mu'$$

for some $\mu' \in M_1^+(X)$. As μ lies in the face F', it follows that $\varepsilon_y \in F'$, which contradicts the fact that F and F' are disjoint. Therefore F' is empty. □

DEFINITION. Let C be a lattice cone in a real vector space E, and let E_0 be the linear subspace of E spanned by C. We shall say that C is *Dedekind complete* if and only if the lattice ordered real vector space $(E_0, \leq_C)$ is Dedekind complete.

For example, Proposition 10.5 shows that the state space of any interpolation group with order-unit is a base for a Dedekind complete lattice cone. In simplices that are bases for Dedekind complete lattice cones, faces corresponding to complete subcones have complements in the lattice of faces, as follows.

PROPOSITION 10.14. *Let K be a base for the positive cone of a Dedekind complete lattice-ordered real vector space E, let F be a nonempty face of K, and let*

$$C = \{\alpha x \mid \alpha \in \mathbb{R}^+ \text{ and } x \in F\}.$$

Assume that C contains the supremum of any nonempty subset of C that is bounded above in E. Then K equals the direct convex sum of F and its complementary face.

PROOF. Let F' denote the complementary face of F. In view of Proposition 10.12, we need only prove that K equals the convex hull of $F \cup F'$.

Let x be any point of $K - (F \cup F')$, and set

$$Y = \{c \in C \mid c \leq x\}.$$

Then Y is a nonempty subset of C which is bounded above in E. Hence, Y has a supremum y in E, and $y \in C$ by hypothesis. Then $y = \alpha_1 x_1$ for some $\alpha_1 \in \mathbb{R}^+$ and some $x_1 \in F$. In addition, $y \leq x$, and so $x - y = \alpha_2 x_2$ for some $\alpha_2 \in \mathbb{R}^+$ and some $x_2 \in K$. Then $x = \alpha_1 x_1 + \alpha_2 x_2$, and $\alpha_1 + \alpha_2 = 1$ by Corollary 10.3. Thus x is a convex combination of the point $x_1 \in F$ and the point $x_2 \in K$, and we claim that $x_2 \in F'$. Note that $\alpha_2 > 0$, because $x \notin F$.

If $x_2 \notin F'$, then the face generated by x_2 in K must intersect F. Choose a point y_1 in F that lies in the face generated by x_2. By Corollary 5.8, there exists a positive convex combination $\beta_1 y_1 + \beta_2 y_2 = x_2$, for some $y_2 \in K$. Now $\beta_1 y_1 \leq x_2$, and hence $\alpha_2 \beta_1 y_1 \leq \alpha_2 x_2$. Consequently,

$$y + \alpha_2 \beta_1 y_1 = \alpha_1 x_1 + \alpha_2 \beta_1 y_1 \leq \alpha_1 x_1 + \alpha_2 x_2 = x.$$

As $y \in C$ and $y_1 \in F$, we have $y + \alpha_2 \beta_1 y_1 \in C$. Then $y + \alpha_2 \beta_1 y_1 \leq y$, by definition of y, whence $\alpha_2 \beta_1 y_1 = 0$. Since $\alpha_2 \beta_1 > 0$ and $y_1 \in K$, this is impossible.

Thus $x_2 \in F'$, so that x lies in the convex hull of $F \cup F'$. Therefore K equals the convex hull of $F \cup F'$, as required. □

In particular, Proposition 10.14 applies to faces of state spaces arising from ideals in interpolation groups, as in the following theorem. Later, we shall extend this result to all closed faces of compact simplices in locally convex Hausdorff spaces (Theorem 11.28). In particular, the result will then hold for all closed faces of state spaces of interpolation groups.

THEOREM 10.15. *Let (G, u) be an interpolation group with order-unit, let H be an ideal of G, and set*

$$F = \{s \in S(G, u) \mid H \subseteq \ker(s)\}.$$

Then F is a closed face of $S(G, u)$, and $S(G, u)$ is the direct convex sum of F and its complementary face.

PROOF. By Proposition 6.12, F is a closed face of $S(G, u)$. If F is empty, then its complementary face is $S(G, u)$, in which case it is clear that $S(G, u)$ is the direct convex sum of F and its complementary face. Now assume that F is nonempty.

Let B be the Dedekind complete lattice-ordered real vector space of all relatively bounded homomorphisms from G to $\mathbb{R}$. Then $S(G, u)$ is a base for B^+, by Proposition 10.5. To apply Proposition 10.14, we must show that the cone

$$C = \{\alpha s \mid \alpha \in \mathbb{R}^+ \text{ and } s \in F\}$$

contains the supremum of any nonempty subset of C that is bounded above in B.

Given a nonempty subset $\{f_i \mid i \in I\} \subseteq C$ which is bounded above in B, let $f = \bigvee f_i$, and set $d(x) = \sup\{f_i(x) \mid i \in I\}$ for all $x \in G^+$. According to Theorem 2.27,

$$f(x) = \sup\{d(x_1) + \cdots + d(x_n) \mid x = x_1 + \cdots + x_n \text{ and all } x_j \in G^+\}$$

for all $x \in G^+$. If $x \in H^+$ and $x = x_1 + \cdots + x_n$ for some elements $x_j \in G^+$, then each $x_j \in H$. Then $f_i(x_j) = 0$ for all i, j (because each $f_i \in C$), whence $d(x_j) = 0$ for all j. Thus $f(x) = 0$ for all $x \in H^+$. As H is directed, it follows that $H \subseteq \ker(f)$, from which we conclude that $f \in C$.

Therefore C does contain the supremum of any nonempty subset of C that is bounded above in B. By Proposition 10.14, $S(G, u)$ is the direct convex sum of F and its complementary face. □

• Choquet Simplices.

DEFINITION. A *Choquet simplex* is any compact simplex in a locally convex Hausdorff space.

For example, any classical simplex (in a locally convex Hausdorff space) is a Choquet simplex. On the other hand, all nonempty finite-dimensional Choquet simplices are classical simplices, as follows.

THEOREM 10.16. *Let K be a nonempty compact convex subset of a finite-dimensional locally convex Hausdorff space. Then K is a Choquet simplex if and only if K is a classical simplex.*

PROOF. If K is a Choquet simplex, then $\partial_e K$ is an affinely independent set by Corollary 10.8, and K equals the closure of the convex hull of $\partial_e K$ by the

Krein-Mil'man Theorem. As K lies in a finite-dimensional space, $\partial_e K$ must be finite. Then the convex hull of $\partial_e K$ is compact (Proposition 5.2), whence K equals the convex hull of $\partial_e K$. Therefore $\partial_e K$ is a classical simplex. □

THEOREM 10.17. *If (G, u) is an interpolation group with order-unit, then $S(G, u)$ is a Choquet simplex.*

PROOF. Corollary 10.6 and Proposition 6.2. □

COROLLARY 10.18. *If X is a compact Hausdorff space, then $M_1^+(X)$ is a Choquet simplex.*

PROOF. Proposition 6.8 and Theorem 10.17. □

Not all Choquet simplices have the form $M_1^+(X)$ for compact Hausdorff spaces X. On the one hand, $\partial_e M_1^+(X)$ is homeomorphic to X and so is compact. On the other hand, Example 6.10 provides an example of a dimension group (G, u) with order-unit for which $\partial_e S(G, u)$ is not compact, and $S(G, u)$ is a Choquet simplex by Theorem 10.17. We shall prove in Theorem 11.21 that the Choquet simplices of the form $M_1^+(X)$ are exactly (up to affine homeomorphism) the Choquet simplices with compact extreme boundaries.

• Categorical Properties.

PROPOSITION 10.19. *Any coproduct of simplices (in the category of convex sets) is a simplex.*

PROOF. Let $\{K_i \mid i \in I\}$ be a nonempty family of simplices. We may assume that each K_i is a base for the positive cone of a lattice-ordered real vector space E_i. Set $E = \bigoplus E_i$, and for each $i \in I$ let $q_i \colon E_i \to E$ be the injection map. Let K be the convex hull of $\bigcup q_i(K_i)$, and observe that K is the direct convex sum of the $q_i(K_i)$. By Proposition 5.4, K (together with the restricted injection maps $q_i \colon K_i \to K$) is a coproduct for the K_i. Finally, observe that K is a base for E^+. Since E is lattice-ordered, K is a simplex. □

In the category of convex sets, products of simplices are usually not simplices. For example, the product of a closed line segment with itself is affinely isomorphic to a square in the plane, which is not a simplex.

PROPOSITION 10.20. *If $K_1, \dots, K_n$ are Choquet simplices, then their coproduct (in the category of compact convex sets) is a Choquet simplex.*

PROOF. If K is a coproduct of the K_i in the category of compact convex sets, then K is a compact convex subset of a locally convex Hausdorff space. In view of Propositions 5.4 and 5.6, K is also a coproduct of the K_i in the category of convex sets. Then K is a simplex, by Proposition 10.19. Therefore K is a Choquet simplex. □

PROPOSITION 10.21. *Any inverse limit of classical simplices (in the category of compact convex sets) is a Choquet simplex.*

PROOF. Let K be an inverse limit of an inverse system $\{K_i, f_{ij}\}$ of classical simplices and affine continuous maps, indexed by a directed set I. We may assume that each K_i is a compact convex subset of a euclidean space E_i. Then K is a compact convex subset of the locally convex Hausdorff space $\prod E_i$.

Set $A_i = \mathrm{Aff}(K_i)$ for all $i \in I$. Whenever $i \leq j$ in I, the map f_{ij} induces a map $f_{ij}^*: A_i \to A_j$ such that $f_{ij}^*(p) = pf_{ij}$ for all $p \in A_i$. Note that f_{ij}^* is a normalized positive homomorphism from $(A_i, 1)$ to $(A_j, 1)$. We also observe that whenever $i \leq j \leq k$ in I, then $f_{jk}^* f_{ij}^* = f_{ik}^*$. Hence, the objects $(A_i, 1)$ and morphisms f_{ij}^* form a direct system in the category of partially ordered abelian groups with order-unit. Let (A, u) be a direct limit of this direct system.

Applying the state space functor, we obtain an inverse system of compact convex sets $S(A_i, 1)$ and affine continuous maps $S(f_{ij}^*)$. By Proposition 6.14, the inverse limit of this inverse system is affinely homeomorphic to $S(A, u)$. For each $i \in I$, let ψ_i be the natural affine homeomorphism of K_i onto $S(A_i, 1)$. We claim that whenever $i \leq j$ in I, then $S(f_{ij}^*)\psi_j = \psi_i f_{ij}$. Given any $x \in K_j$ and any $p \in A_i$, we compute that

$$(S(f_{ij}^*)\psi_j(x))(p) = (\psi_j(x) f_{ij}^*)(p) = \psi_j(x)(pf_{ij}) = pf_{ij}(x) = (\psi_i f_{ij}(x))(p).$$

Thus $S(f_{ij}^*)\psi_j = \psi_i f_{ij}$, as claimed. Consequently, K is affinely homeomorphic to $S(A, u)$.

Each K_i is the convex hull of an affinely independent set of $n(i)$ points, for some positive integer $n(i)$. Hence, $A_i \cong \mathbb{R}^{n(i)}$ (as ordered vector spaces), and so A_i is lattice-ordered. In particular, each A_i has interpolation, whence A has interpolation. By Theorem 10.17, $S(A, u)$ is a Choquet simplex, and therefore K is a Choquet simplex. □

There is also a converse to Proposition 10.21, as we prove in Theorem 11.6: any Choquet simplex is affinely homeomorphic to an inverse limit of classical simplices.

• **Notes.** Corollary 10.6 is a preliminary form of Theorem 10.17. Proposition 10.7 and Corollary 10.8 are folklore. Proposition 10.9 is due to Goullet de Rugy [**62**, Proposition 2.8]. Proposition 10.10 was first proved for the case of two faces of a Choquet simplex, by Alfsen [**2**, Proposition 3]. The general case was proved by Goullet de Rugy [**62**, Corollaire 2.5]. Proposition 10.12 was proved for Choquet simplices by Alfsen [**2**, Proposition 2 and Theorem 1]. Theorem 10.16 was observed by Choquet [**19**, p. 139-13]. Theorem 10.17 was proved by the author, Handelman, and Lawrence [**60**, Theorem I.2.5], and, for dimension groups, by Effros, Handelman, and Shen [**30**, Proposition 1.7]. Proposition 10.20 is due to Semadeni [**113**, Theorem 8].

CHAPTER 11

Affine Continuous Functions on Choquet Simplices

As the state space of any interpolation group (G, u) with order-unit is a Choquet simplex, the natural affine representation of (G, u) is thus a representation in terms of affine continuous functions on a Choquet simplex. Just as linear representations of groups are useful because groups of matrices are more well-behaved than arbitrary groups, these affine function representations of interpolation groups are useful because affine continuous function spaces on Choquet simplices exhibit better behavior than arbitrary interpolation groups. First of all, for a Choquet simplex K, the space $\mathrm{Aff}(K)$ is an interpolation group that is also an archimedean partially ordered real vector space. Second, the connections between $\mathrm{Aff}(K)$ and its state space (which is naturally affinely homeomorphic to K) are quite tight. For example, not only does any ideal H of $\mathrm{Aff}(K)$ give rise to a closed face of K (which is affinely homeomorphic to the state space of $\mathrm{Aff}(K)/H$), but also any closed face F of K arises from an ideal H of $\mathrm{Aff}(K)$ such that $\mathrm{Aff}(K)/H$ is naturally isomorphic to $\mathrm{Aff}(F)$ (as ordered Banach spaces). Similarly, any compact subset X of $\partial_e K$ arises from an ideal H of $\mathrm{Aff}(K)$ (so that X is homeomorphic to the extreme boundary of the state space of $\mathrm{Aff}(K)/H$), and $\mathrm{Aff}(K)/H$ is naturally isomorphic to $C(X, \mathbb{R})$ (as ordered Banach spaces).

In this chapter, we develop some of the structure theory for $\mathrm{Aff}(K)$, where K is an arbitrary Choquet simplex, and study the connections between the structure of $\mathrm{Aff}(K)$ and the geometry of K, starting with the fundamental result that $\mathrm{Aff}(K)$ is an interpolation group. Hence, we may characterize Choquet simplices as exactly those compact convex subsets of locally convex Hausdorff spaces that appear as state spaces of interpolation groups. Using the representation of dimension groups as direct limits of simplicial groups, it follows that all Choquet simplices may be represented as inverse limits of classical simplices.

In order to develop the connections between the structures of $\mathrm{Aff}(K)$ and K, we require some machinery with which to construct functions in $\mathrm{Aff}(K)$ with specific properties. This machinery is built in terms of semicontinuous convex and concave functions on K. Namely, given an upper semicontinuous convex function f on K and a lower semicontinuous concave function h on K, such that $f \leq h$, there exists an affine continuous function g in $\mathrm{Aff}(K)$ lying between f

and h. As it is relatively easy to construct semicontinuous convex and concave functions on K, this machinery provides an efficient means of constructing affine continuous functions on K with desirable properties. We use this machinery to derive a number of connections between $\mathrm{Aff}(K)$ and closed faces of K, and between $\mathrm{Aff}(K)$ and compact subsets of $\partial_e K$.

• **Interpolation.** Our first major goal is to prove, for any Choquet simplex K, that $\mathrm{Aff}(K)$ is an interpolation group. The route we follow toward this result involves showing that the dual space $\mathrm{Aff}(K)^*$ is lattice-ordered, which in turn requires a pair of technical lemmas.

LEMMA 11.1. *Let A be a partially ordered real vector space, and let v be an order-unit in A. Let B be the space of all linear functionals on A, and let $p \in B$. If*

$$\alpha = \inf\{q(v) \mid q \in B^+ \text{ and } q \geq^+ p\} \quad \text{and} \quad \beta = \sup\{p(x) \mid x \in A^+ \text{ and } x \leq v\},$$

then $\alpha = \beta$.

PROOF. If $A = \{0\}$, then $\alpha = 0 = \beta$. Hence, we may assume that A is nonzero and, consequently, that $v > 0$.

Note that $\beta \geq p(0) = 0$. Given $q \in B^+$ and $x \in A^+$ such that $q \geq^+ p$ and $x \leq v$, we have $p(x) \leq q(x) \leq q(v)$. Thus $\beta \leq \alpha$. Hence, if $\beta = \infty$, we are done. Now assume that β is finite.

Let ε be an arbitrary positive real number, and choose a positive integer m such that $m\varepsilon \geq \beta - p(v)$. Define linear subspaces

$$C = \{(x, -x) \mid x \in A\} \quad \text{and} \quad D = \mathbb{R}(mv, v) + C$$

in $A \times A$. As $v > 0$, we see that $(mv, v) \notin C$. Hence, we may define a linear functional f on D so that

$$f(\gamma(mv, v) + (x, -x)) = \gamma m(\beta + \varepsilon) + p(x)$$

for all $\gamma \in \mathbb{R}$ and $x \in A$.

We claim that f is positive. Given any $w \in D^+$, write

$$w = \gamma(mv, v) + (x, -x)$$

for some $\gamma \in \mathbb{R}$ and $x \in A$. Then $\gamma mv + x \geq 0$ and $\gamma v - x \geq 0$. Adding these inequalities, we find that $\gamma(m+1)v \geq 0$, and hence $\gamma \geq 0$, because $(m+1)v > 0$. In addition,

$$0 \leq \gamma v - x \leq \gamma v + \gamma mv = \gamma(m+1)v,$$

whence $\gamma^{-1}(m+1)^{-1}(\gamma v - x) \leq v$, and so

$$\gamma^{-1}(m+1)^{-1} p(\gamma v - x) \leq \beta,$$

by definition of β. Then $\gamma p(v) - p(x) \leq \gamma(m+1)\beta$, and consequently

$$-p(x) \leq \gamma m\beta + \gamma(\beta - p(v)) \leq \gamma m\beta + \gamma m\varepsilon,$$

whence $f(w) = \gamma m(\beta+\varepsilon) + p(x) \geq 0$. Thus f is positive, as claimed.

Since v is an order-unit in A, the element (mv, v) is an order-unit in $A \times A$, and hence every element of $A \times A$ is bounded above by an element of D. By Proposition 4.2, f extends to a positive homomorphism $g\colon A\times A \to \mathbb{R}$. Of course g is linear, by Lemma 6.7.

Composing g with the injection of A into the left-hand factor of $A \times A$, we obtain a positive linear functional q on A such that $q(x) = g(x, 0)$ for all $x \in A$. For $x \in A^+$, we have $(0, x) \geq 0$ and $g(0, x) \geq 0$, whence

$$q(x) = g(x,0) \geq g(x,0) - g(0,x) = g(x,-x) = f(x,-x) = p(x).$$

Thus $q \geq^+ p$, and hence $\alpha \leq q(v)$. Also,

$$mq(v) = g(mv,0) \leq g(mv,0) + g(0,v) = f(mv,v) = m(\beta+\varepsilon),$$

so that $\alpha \leq q(v) \leq \beta + \varepsilon$.

Therefore $\alpha = \beta$. □

We shall apply Lemma 11.1 to linear subspaces A of $C(X, \mathbb{R})$ (for a compact Hausdorff space X) such that $1 \in A$. Note that the ordering in B is the same whether A is equipped with the pointwise ordering or the strict ordering. [If $p \in B$ with $p \geq^+ 0$, then obviously $p \gg^+ 0$. Conversely, suppose that $p \gg^+ 0$. If $f \in A$ with $f \geq 0$, then $f + \varepsilon \gg 0$ for all positive real numbers ε, whence $p(f) + \varepsilon p(1) \geq 0$ for all such ε, and so $p(f) \geq 0$. Thus $p \geq^+ 0$.]

LEMMA 11.2. *Let X be a compact Hausdorff space, let A be a linear subspace of $C(X, \mathbb{R})$ with $1 \in A$, and equip A with the strict ordering. Let $p \in A^*$, and set*

$$r(f) = \sup\{p(g) \mid g \in A \text{ and } 0 \ll g \ll f\}$$

for all $f \in A^+$. If the supremum $p^+ = p \vee 0$ exists in A^*, then the restriction of p^+ to A^+ coincides with r, so that r is an additive function from A^+ to $\mathbb{R}$.

PROOF. Given any $f \in A^+$, we must show that $p^+(f) = r(f)$, which is clear in case $f = 0$. Now assume that $f \neq 0$. Then $f \gg 0$, and hence f is an order-unit in A. Let B denote the space of all linear functionals on A, and set

$$\alpha = \inf\{q(f) \mid q \in B^+ \text{ and } q \geq^+ p\}.$$

As $p^+ \in B^+$ and $p^+ \geq^+ p$, we have $\alpha \leq p^+(f)$. In addition, $\alpha = r(f)$, by Lemma 11.1. Thus $r(f) \leq p^+(f)$.

Now consider any $q \in B^+$ such that $q \geq^+ p$. Any function g in the unit ball of A satisfies $-1 \leq g \leq 1$, hence $-q(1) \leq q(g) \leq q(1)$ (because q is positive), and so $|q(g)| \leq q(1)$. Hence, q is bounded, so that $q \in A^*$. Since $q \geq^+ 0$ and $q \geq^+ p$, we must have $q \geq^+ p^+$, whence $p^+(f) \leq q(f)$. Thus $p^+(f) \leq \alpha = r(f)$.

Therefore $p^+(f) = r(f)$. □

THEOREM 11.3. *Let X be a compact Hausdorff space, and let A be a linear subspace of $C(X, \mathbb{R})$ such that $1 \in A$. If $(A^*, \leq^+)$ is a lattice, then $(A, \ll)$ is an interpolation group.*

PROOF. Equip A with the strict ordering. To show that A has interpolation, it suffices to prove that the Riesz decomposition property holds in A. Given any $f_1, f_2 \in A^+$, set

$$F_i = \{f \in A \mid 0 \lll f \lll f_i\} \quad (\text{for } i = 1, 2),$$
$$F = \{f \in A \mid 0 \lll f \lll f_1 + f_2\};$$

we must show that $F = F_1 + F_2$. This is clear if either $f_1 = 0$ or $f_2 = 0$, and hence we may assume that each $f_i \gg 0$.

We first claim that $F_1 + F_2$ is dense in F. If not, choose h in F which does not lie in the closure of $F_1 + F_2$. Note that $F_1 + F_2$ is a convex subset of A, and so its closure is convex. As A is a normed linear space, it is locally convex and Hausdorff. Consequently, by Theorem 5.14, there exists a continuous linear functional $p \in A^*$ such that

$$\sup\{p(f) \mid f \in \overline{F_1 + F_2}\} < p(h).$$

For all $f \in A^+$, define

$$r(f) = \sup\{p(g) \mid g \in A \text{ and } 0 \lll g \lll f\}.$$

Since A^* is assumed to be a lattice, the supremum $p \vee 0$ exists in A^*, and hence r is an additive function from A^+ to $\mathbb{R}$, by Lemma 11.2. However

$$\begin{aligned} r(f_1) + r(f_2) &= \sup(p(F_1)) + \sup(p(F_2)) = \sup(p(F_1 + F_2)) \\ &< p(h) \leq \sup(p(F)) = r(f_1 + f_2), \end{aligned}$$

which contradicts the additivity of r.

Therefore $F_1 + F_2$ is dense in F, as claimed.

Now consider any $f \in F$. If either $f = 0$ or $f = f_1 + f_2$, then obviously f lies in $F_1 + F_2$. Thus we may assume that

$$0 \ll f \ll f_1 + f_2.$$

Choose a positive real number ε such that each $f_i \gg \varepsilon$ and also

$$\varepsilon \ll f \ll f_1 + f_2 - \varepsilon.$$

Then choose a positive real number $\delta < \frac{1}{2}$ such that

$$2\delta(f_1 + f_2) \ll \varepsilon.$$

As a result,

$$2\delta\varepsilon \ll \varepsilon \ll (1 + 2\delta)f \ll (1 + 2\delta)(f_1 + f_2 - \varepsilon) \ll f_1 + f_2,$$

and consequently

$$0 \ll (1 + 2\delta)f - 2\delta\varepsilon \ll f_1 + f_2,$$

so that $(1 + 2\delta)f - 2\delta\varepsilon$ lies in F.

Since $F_1 + F_2$ is dense in F, there must exist $g_1 \in F_1$ and $g_2 \in F_2$ such that

$$\|g_1 + g_2 - (1 + 2\delta)f + 2\delta\varepsilon\| < 2\delta\varepsilon.$$

Set $h = \frac{1}{2}[(1+2\delta)f - 2\delta\varepsilon - g_1 - g_2]$, so that

$$(1+2\delta)f = g_1 + g_2 + 2\delta\varepsilon + 2h$$

and $\|h\| < \delta\varepsilon$. Then set

$$h_i = (1+2\delta)^{-1}(g_i + h + \delta\varepsilon)$$

for each $i = 1, 2$, and note that $h_1 + h_2 = f$. Since $\|h\| < \delta\varepsilon$, we have

$$g_i + h + \delta\varepsilon \gg g_i \gtrsim 0,$$

and hence $h_i \gg 0$. In addition,

$$(1+2\delta)h_i = g_i + h + \delta\varepsilon \lesssim f_i + h + \delta\varepsilon \ll f_i + 2\delta\varepsilon \ll (1+2\delta)f_i,$$

whence $h_i \ll f_i$. Thus each $h_i \in F_i$, proving that f lies in $F_1 + F_2$.

Therefore $F = F_1 + F_2$, as required. □

THEOREM 11.4. *Let K be a compact convex subset of a locally convex Hausdorff space. Then the following conditions are equivalent*:

(a) *K is a Choquet simplex.*

(b) *$(\mathrm{Aff}(K), \leq)$ is an interpolation group.*

(c) *$(\mathrm{Aff}(K), \lesssim)$ is an interpolation group.*

PROOF. Set $A = \mathrm{Aff}(K)$.

(b) ⇒ (a): By Theorem 10.17, $S(A, 1)$ is a Choquet simplex. As K is affinely homeomorphic to $S(A, 1)$ by Theorem 7.1, K must be a Choquet simplex.

(a) ⇒ (c): Since all states on $(A, 1)$ are bounded (Proposition 7.18), $S(A, 1)$ is contained in A^*. Using Lemma 4.5, it follows that $S(A, 1)$ is a base for the positive cone of $(A^*, \leq^+)$. As K is a simplex, $(A^*)^+$ is thus a lattice cone. Hence, the linear subspace of A^* spanned by $(A^*)^+$ is lattice-ordered. By Corollary 7.3, A^* is directed, and thus A^* is a lattice. Then Theorem 11.3 shows that $(A, \lesssim)$ is an interpolation group.

(c) ⇒ (b): We first use (c) to see that A satisfies strict interpolation with respect to $\ll$. Given $p_1, \ldots, p_n$ and $r_1, \ldots, r_k$ in A such that $p_i \ll r_j$ for all i, j, choose a positive real number ε such that $p_i + \varepsilon \ll r_j - \varepsilon$ for all i, j. By (c), there exists $q \in A$ such that $p_i + \varepsilon \lesssim q \lesssim r_j - \varepsilon$ for all i, j, whence $p_i \ll q \ll r_j$ for all i, j.

Given f_1, f_2, h_1, h_2 in A such that $f_i \leq h_j$ for all i, j, we construct functions $g_0, g_1, g_2, \ldots$ in A such that

$$\begin{matrix} f_1 - 1/2^n \\ f_2 - 1/2^n \end{matrix} \ll g_n \ll \begin{matrix} h_1 + 1/2^n \\ h_2 + 1/2^n \end{matrix}$$

for all n, while also

$$g_{n-1} - 1/2^n \ll g_n \ll g_{n-1} + 1/2^n$$

for all $n > 0$. Since $f_i - 1 \ll h_j + 1$ for all i, j, the existence of g_0 is clear.

Suppose that $g_0, \ldots, g_{n-1}$ have been constructed, for some $n > 0$. Note that

$$f_i - 1/2^n \ll h_j + 1/2^n$$

for all i, j, while also $g_{n-1} - 1/2^n \ll g_{n-1} + 1/2^n$. For all i, j, we have

$$f_i - 1/2^{n-1} \ll g_{n-1} \ll h_j + 1/2^{n-1},$$

and consequently

$$f_i - 1/2^n \ll g_{n-1} + 1/2^n \quad \text{and} \quad g_{n-1} - 1/2^n \ll h_j + 1/2^n.$$

Applying strict interpolation again, we obtain $g_n \in A$ such that

$$\begin{matrix} f_1 - 1/2^n & & & & h_1 + 1/2^n \\ f_2 - 1/2^n & \ll & g_n & \ll & h_2 + 1/2^n \\ g_{n-1} - 1/2^n & & & & g_{n-1} + 1/2^n, \end{matrix}$$

completing the induction step.

Now $\|g_n - g_{n-1}\| < 1/2^n$ for all $n = 1, 2, \ldots$, and so $\{g_0, g_1, g_2, \ldots\}$ is a Cauchy sequence in A. This sequence converges to some $g \in A$. Since

$$f_i - 1/2^n \ll g_n \ll h_j + 1/2^n$$

for all i, j, n, we conclude that $f_i \leq g \leq h_j$ for all i, j.

Therefore $(A, \leq)$ is an interpolation group. □

COROLLARY 11.5. *Any Choquet simplex is affinely homeomorphic to the state space of an archimedean dimension group with order-unit.* □

In particular, Theorem 10.17 and Corollary 11.5 show that the "duality" between partially ordered abelian groups with order-unit and compact convex sets restricts to a "duality" between interpolation groups with order-unit and Choquet simplices.

• **Inverse Limits.** The results of the previous section allow us to apply any general result about state spaces of interpolation groups to all Choquet simplices. In particular, this method yields the following properties of inverse limits of Choquet simplices.

THEOREM 11.6. *Let K be a compact convex subset of a locally convex Hausdorff space. Then K is a Choquet simplex if and only if K is affinely homeomorphic to an inverse limit of classical simplices (in the category of compact convex sets).*

PROOF. Sufficiency is given by Proposition 10.21. Conversely, if K is a Choquet simplex, then Corollary 11.5 shows that K is affinely homeomorphic to $S(G, u)$ for some dimension group (G, u) with order-unit. By Corollary 3.20, (G, u) is isomorphic to a direct limit of simplicial groups (G_i, u_i) with order-unit (in the category of partially ordered abelian groups with order-unit). Applying the state space functor, we find that $S(G, u)$ is affinely homeomorphic to $\varprojlim S(G_i, u_i)$, by Proposition 6.14. As each $S(G_i, u_i)$ is a classical simplex (Proposition 6.6), the proof is complete. □

THEOREM 11.7. *Any inverse limit of Choquet simplices (in the category of compact convex sets) is a Choquet simplex.*

PROOF. Let K be an inverse limit of Choquet simplices K_i, and observe that K is a compact convex subset of a locally convex Hausdorff space. Set $A_i = \mathrm{Aff}(K_i)$ for all i. As in Proposition 10.21, there is a partially ordered abelian group (A, u) with order-unit such that (A, u) is a direct limit of the $(A_i, 1)$ and $S(A, u)$ is affinely homeomorphic to K. By Theorem 11.4, each A_i is an interpolation group, whence A is an interpolation group. Therefore Theorem 10.17 shows that K is a Choquet simplex. □

• **Semicontinuous Functions.** Given a Choquet simplex K, we have as yet no means of ensuring a large supply of functions in $\mathrm{Aff}(K)$, beyond the interpolation properties (Theorem 11.4) and the existence of enough functions in $\mathrm{Aff}(K)$ to separate the points of K (Corollary 5.15). In the particular case that $K = M_1^+(X)$ for some compact Hausdorff space X, the partially ordered real Banach space $\mathrm{Aff}(K)$ is isometrically isomorphic to $C(X, \mathbb{R})$ (see Corollary 11.20), so that classical results such as the Tietze Extension Theorem can be used to obtain functions in $\mathrm{Aff}(K)$ with useful properties. A similar result is that if f_1 is an upper semicontinuous function on X and f_2 is a lower semicontinuous function on X, such that $f_1 \leq f_2$, then there exists a function $g \in C(X, \mathbb{R})$ such that $f_1 \leq g \leq f_2$. Such results motivated the search for analogous theorems concerning $\mathrm{Aff}(K)$, for an arbitrary Choquet simplex K.

For instance, by analogy with the Tietze Extension Theorem, we would like to be able to extend any affine continuous real-valued function f on a closed face F of K to an affine continuous real-valued function on K. As a means of rephrasing the problem, define extended real-valued functions f_1 and f_2 on K so that $f_1 = f_2 = f$ on F, while $f_1 = -\infty$ and $f_2 = \infty$ on $K - F$. Then $f_1 \leq f_2$, and we seek a function $g \in \mathrm{Aff}(K)$ such that $f_1 \leq g \leq f_2$. Since F is closed in K, the function f_1 turns out to be upper semicontinuous, while f_2 turns out to be lower semicontinuous. Since F is a face of K, it turns out that f_1 is a convex function, while f_2 is a concave function. Hence, our problem is now one of fitting an affine continuous function between an upper semicontinuous convex function and a lower semicontinuous concave function. After a reminder of the relevant definitions, we develop the machinery to resolve this problem.

DEFINITION. Let X be a topological space, and let

$$f\colon X \to \{-\infty\} \cup \mathbb{R} \quad \text{and} \quad g\colon X \to \mathbb{R} \cup \{\infty\}$$

be extended real-valued functions on X. Then f is *upper semicontinuous* if and only if $f^{-1}[\alpha, \infty)$ is closed in X (or, equivalently, $f^{-1}[-\infty, \alpha)$ is open in X) for every $\alpha \in \mathbb{R}$. Similarly, g is *lower semicontinuous* if and only if $g^{-1}(-\infty, \alpha]$ is closed in X (or, equivalently, $g^{-1}(\alpha, \infty]$ is open in X) for every $\alpha \in \mathbb{R}$.

On a compact space, an upper semicontinuous function attains a maximum value, while a lower semicontinuous function attains a minimum value.

DEFINITION. Let K be a convex subset of a real vector space, and let

$$f\colon K \to \{-\infty\} \cup \mathbb{R} \quad \text{and} \quad g\colon K \to \mathbb{R} \cup \{\infty\}$$

be extended real-valued functions on K. Then f is a *convex function* if and only if

$$f(\alpha_1 x_1 + \alpha_2 x_2) \leq \alpha_1 f(x_1) + \alpha_2 f(x_2)$$

for all positive convex combinations $\alpha_1 x_1 + \alpha_2 x_2$ in K. (If we define $0(-\infty) = 0$, this inequality must hold for all convex combinations in K.) Similarly, g is a *concave function* if and only if

$$g(\alpha_1 x_1 + \alpha_2 x_2) \geq \alpha_1 g(x_1) + \alpha_2 g(x_2)$$

for all positive convex combinations $\alpha_1 x_1 + \alpha_2 x_2$ in K. (If we define $0(\infty) = 0$, this inequality must hold for all convex combinations in K.)

Our goal in this section is to prove that if f is an upper semicontinuous convex function on a Choquet simplex K and h is a lower semicontinuous concave function on K, such that $f \leq h$, then there exists an affine continuous function g on K such that $f \leq g \leq h$. By means of an argument similar to the proof of (c) $\Rightarrow$ (b) in Theorem 11.4, the problem may be reduced to showing that if $f \ll h$, then there exists $g \in \operatorname{Aff}(K)$ such that $f \ll g \ll h$. The construction of such a g requires several intermediate steps, as follows.

First, the value $f(x)$, for any $x \in K$, may be approximated arbitrarily closely by the values $g'(x)$, for continuous convex functions $g'\colon K \to \mathbb{R}$ such that $g' \gg f$. Using the compactness of K, we obtain finitely many such functions $g'_1, \ldots, g'_m$ such that

$$f(x) < \min\{g'_1(x), \ldots, g'_m(x)\} < h(x)$$

for all $x \in K$. Second, there exists a lower semicontinuous convex function $g''\colon K \to \mathbb{R}$ such that $g'' \gg f$ and each $g'_i \geq g''$, whence $f \ll g'' \ll h$. (So far, we have merely traded the upper semicontinuity of f for the lower semicontinuity of g''.) Third, the value $g''(x)$, for any $x \in K$, may be approximated arbitrarily closely by values $f'(x)$, for functions $f' \in \operatorname{Aff}(K)$ such that $f' \ll g''$. Using compactness again, we obtain finitely many functions $f'_1, \ldots, f'_n$ in $\operatorname{Aff}(K)$ such that

$$f(x) < \max\{f'_1(x), \ldots, f'_n(x)\} < g''(x)$$

for all $x \in K$. Then the pointwise maximum of $f'_1, \ldots, f'_n$ is a continuous convex function $f'\colon K \to \mathbb{R}$ such that $f \ll f' \ll h$. Fourth, applying the procedure above to the functions $-h \ll -f'$, we obtain functions $h'_1, \ldots, h'_k$ in $\operatorname{Aff}(K)$ whose pointwise minimum h' satisfies $f' \ll h' \ll h$. Finally, $f'_i \ll h'_j$ for all i, j, and interpolation in $(\operatorname{Aff}(K), \lll)$ provides us with $g \in \operatorname{Aff}(K)$ such that $f'_i \lll g \lll h'_j$ for all i, j, whence $f \ll g \ll h$.

Only the last step of this process (the interpolation) requires K to be a Choquet simplex. The other steps are valid for any compact convex subset of a locally convex Hausdorff space.

PROPOSITION 11.8. *Let K be a compact convex subset of a locally convex Hausdorff space E, and let $g\colon K \to \mathbb{R}$ be a lower semicontinuous convex function. Then*

$$g(x) = \sup\{f(x) \mid f \in \mathrm{Aff}(K) \text{ and } f \ll g\}$$

for all $x \in K$.

PROOF. We first show that the set

$$X = \{(x,a) \in K \times \mathbb{R} \mid g(x) \le a\}$$

is a closed convex subset of $E \times \mathbb{R}$. Given a convex combination

$$(x,a) = \alpha_1(x_1,a_1) + \alpha_2(x_2,a_2)$$

with each $(x_i,a_i) \in X$, we have

$$g(x) = g(\alpha_1 x_1 + \alpha_2 x_2) \le \alpha_1 g(x_1) + \alpha_2 g(x_2) \le \alpha_1 a_1 + \alpha_2 a_2 = a$$

(because g is convex), whence $(x,a) \in X$. Thus X is a convex set.

To show that X is closed in $E \times \mathbb{R}$, it suffices to show that X is closed in $K \times \mathbb{R}$. Given any point (x,a) in $(K \times \mathbb{R}) - X$, we have $g(x) > a$. Choose a real number b such that $g(x) > b > a$. As g is lower semicontinuous, the set

$$U = g^{-1}(b,\infty) \times (-\infty, b)$$

is open in $K \times \mathbb{R}$, and we observe that

$$(x,a) \in U \subseteq (K \times \mathbb{R}) - X.$$

Thus X is closed in $K \times \mathbb{R}$ and hence closed in $E \times \mathbb{R}$.

Given any $y \in K$ and any positive real number ε, we must find some f in $\mathrm{Aff}(K)$ such that $f \ll g$ and $f(y) > g(y) - \varepsilon$. Set $b = g(y) - \varepsilon$, and note that $(y,b) \notin X$. Since $E \times \mathbb{R}$ is a locally convex Hausdorff space, Theorem 5.14 shows that there exist a real number s and a continuous linear functional h on $E \times \mathbb{R}$ such that

$$\sup\{h(z) \mid z \in X\} < s < h(y,b).$$

As $(y, g(y)) \in X$, we have $h(y, g(y)) < h(y,b)$, and so

$$h(0, g(y) - b) < 0.$$

In addition, $g(y) - b = \varepsilon > 0$, whence $h(0,1) < 0$.

Set $c = -h(0,1) > 0$, and define

$$f(x) = c^{-1}(h(x,0) - s)$$

for all $x \in K$. As h is continuous and linear, we see that f is continuous and affine, so that $f \in \mathrm{Aff}(K)$. Given any $x \in K$, the point $(x, g(x))$ lies in X and so

$$s > h(x, g(x)) = h(x,0) + h(0, g(x)) = h(x,0) - cg(x),$$

whence $f(x) < g(x)$. Thus $f \ll g$. On the other hand,

$$s < h(y,b) = h(y,0) - cb,$$

and therefore $f(y) > b = g(y) - \varepsilon$. $\square$

LEMMA 11.9. *Let K be a compact convex subset of a locally convex Hausdorff space E, and let $f\colon K \to \{-\infty\} \cup \mathbb{R}$ be an upper semicontinuous function. Let C be the set of all continuous convex functions from K to $\mathbb{R}$. Then*

$$f(x) = \inf\{g(x) \mid g \in C \text{ and } g \gg f\}$$

for any $x \in K$.

PROOF. Since f is upper semicontinuous, it attains a maximum value m on K.

Given any real number $a > f(x)$, it suffices to find a continuous convex function $g\colon E \to \mathbb{R}$ such that $g|_K \gg f$ and $g(x) < a$. If $a > m$, we may use the constant function $(a+m)/2$ for g. Hence, we may assume that $a \leq m$.

Choose a real number b such that $f(x) < b < a$. By upper semicontinuity, $f^{-1}[b,\infty)$ is a closed subset of K and hence is compact. Consequently, there exists an open neighborhood U of x in E such that the closure of U is disjoint from $f^{-1}[b,\infty)$. Choose an open convex neighborhood V of x which is contained in U. Then $f \ll b$ on $\overline{V}$.

Now $V = A + x$ for some open convex neighborhood A of 0. By Proposition 5.9, there exists a nonnegative, continuous, sublinear functional p on E such that $p^{-1}(-\infty,1] = \overline{A}$. Define

$$g(y) = b + (m-b)p(y-x)$$

for all $y \in E$, and note that g is a continuous function from E to $\mathbb{R}$. Given a convex combination $y = \alpha_1 y_1 + \alpha_2 y_2$ in E, it follows from the sublinearity of p that

$$p(y-x) = p(\alpha_1(y_1-x) + \alpha_2(y_2-x)) \leq \alpha_1 p(y_1-x) + \alpha_2 p(y_2-x).$$

Thus the function $y \mapsto p(y-x)$ is convex. Since $b < a \leq m$, it follows that g is a convex function.

Obviously $g(x) = b < a$. For $y \in K \cap \overline{V}$, we have

$$f(y) < b \leq b + (m-b)p(y-x) = g(y),$$

because $m - b > 0$ and $p(y-x) \geq 0$. On the other hand, for $y \in K - \overline{V}$ we have $y - x \notin \overline{A}$ and so $p(y-x) > 1$, whence

$$g(y) = b + (m-b)p(y-x) > m \geq f(y).$$

Therefore $g|_K \gg f$. □

LEMMA 11.10. *Let K be a compact convex subset of a Hausdorff linear topological space, let $f\colon K \to \{-\infty\} \cup \mathbb{R}$ be a convex function, and let $g_1, \dots, g_n$ be continuous convex functions from K to $\mathbb{R}$ such that each $g_i \gg f$. Then there exists a lower semicontinuous convex function $g\colon K \to \mathbb{R}$ such that $g \gg f$ and each $g_i \geq g$.*

PROOF. Choose $m, M \in \mathbb{R}$ such that $m \leq g_i \leq M$ for all $i = 1, \dots, n$, and set

$$X_i = \{(x,a) \in K \times \mathbb{R} \mid g_i(x) \leq a \leq M\}$$

for each i. As in Proposition 11.8, we observe that X_i is a closed convex subset of $K \times \mathbb{R}$. Since X_i is contained in $K \times [m, M]$, it follows that X_i is compact. Now let X be the convex hull of $X_1 \cup \cdots \cup X_n$, and recall from Proposition 5.2 that X is compact.

Given any $x \in K$, the set

$$I_x = \{a \in \mathbb{R} \mid (x, a) \in X\}$$

is a closed convex subset of $[m, M]$, and I_x is nonempty because $M \in I_x$. Thus we may define

$$g(x) = \inf(I_x)$$

for all $x \in K$, obtaining a function $g\colon K \to [m, M]$. Note that $I_x = [g(x), M]$ for each $x \in K$.

If $x \in K$ and $i \in \{1, \ldots, n\}$, then $(x, g_i(x)) \in X_i \subseteq X$, whence $g_i(x) \in I_x$, and so $g_i(x) \geq g(x)$. Thus each $g_i \geq g$. Since $g(x) \in I_x$, we have $(x, g(x)) \in X$, and so there exists a convex combination

$$(x, g(x)) = \alpha_1(x_1, a_1) + \cdots + \alpha_n(x_n, a_n)$$

with each $(x_i, a_i) \in X_i$. Hence,

$$g(x) = \sum \alpha_i a_i \geq \sum \alpha_i g_i(x_i) > \sum \alpha_i f(x_i) \geq f\left(\sum \alpha_i x_i\right) = f(x),$$

because each $g_i \gg f$ and f is convex. Therefore $g \gg f$.

For any point $(x, a) \in X$, we have $a \in I_x \subseteq [m, M]$, whence $g(x) \leq a \leq M$. Conversely, consider a point $(y, b) \in K \times \mathbb{R}$ such that $g(y) \leq b \leq M$. As I_y equals $[g(y), M]$, it must contain b, and hence $(y, b) \in X$. Thus

$$X = \{(x, a) \in K \times \mathbb{R} \mid g(x) \leq a \leq M\}.$$

Consider a convex combination $y = \alpha_1 y_1 + \alpha_2 y_2$ in K. Each of the points $(y_i, g(y_i))$ lies in the convex set X, whence X contains the point

$$\alpha_1(y_1, g(y_1)) + \alpha_2(y_2, g(y_2)) = (y, \alpha_1 g(y_1) + \alpha_2 g(y_2)),$$

and so $g(y) \leq \alpha_1 g(y_1) + \alpha_2 g(y_2)$. Thus g is a convex function.

Finally, consider any $a \in \mathbb{R}$ and any $x \in g^{-1}(a, \infty)$. Then $a < g(x) \leq M$ and $(x, a) \notin X$. As X is closed in $K \times \mathbb{R}$, there exist neighborhoods U of x and V of a such that

$$(x, a) \in U \times V \subseteq (K \times \mathbb{R}) - X.$$

For any $y \in U$, we then have $(y, a) \notin X$, and so $g(y) > a$. Thus U is a neighborhood of x on which $g \gg a$, proving that the set $g^{-1}(a, \infty)$ is open in K. Therefore g is lower semicontinuous. $\square$

With the help of Proposition 11.8 and Lemmas 11.9 and 11.10, we can now build pointwise maxima and minima of affine continuous real-valued functions on a compact convex set K with which to separate an upper semicontinuous convex function on K from a lower semicontinuous concave function on K.

PROPOSITION 11.11. *Let K be a compact convex subset of a locally convex Hausdorff space, let $f\colon K \to \{-\infty\} \cup \mathbb{R}$ be an upper semicontinuous convex function, and let $h\colon K \to \mathbb{R} \cup \{\infty\}$ be a lower semicontinuous concave function. If $f \ll h$, there exist functions $f', h'\colon K \to \mathbb{R}$ such that $f \ll f' \ll h' \ll h$, where f' is a pointwise maximum of finitely many functions from* Aff(K), *and h' is a pointwise minimum of finitely many functions from* Aff(K).

PROOF. For each $x \in K$, there exists by Lemma 11.9 a continuous convex function $g_x\colon K \to \mathbb{R}$ such that $g_x \gg f$ and $g_x(x) < h(x)$. As $h - g_x$ is lower semicontinuous, the set

$$V_x = \{y \in K \mid h(y) > g_x(y)\}$$

is open. By compactness, $K = V_{x(1)} \cup \cdots \cup V_{x(n)}$ for some points $x(1), \ldots, x(n)$ in K. According to Lemma 11.10, there exists a lower semicontinuous convex function $g\colon K \to \mathbb{R}$ such that $g \gg f$ and each $g_{x(i)} \geq g$. Any point $y \in K$ lies in some $V_{x(i)}$, whence $h(y) > g_{x(i)}(y) \geq g(y)$. Thus $h \gg g$.

By Proposition 11.8, for each $z \in K$ there exists a function f_z in Aff(K) such that $f_z \ll g$ and $f_z(z) > f(z)$. Since $f - f_z$ is upper semicontinuous, the set

$$W_z = \{y \in K \mid f(y) < f_z(y)\}$$

is open. By compactness, $K = W_{z(1)} \cup \cdots \cup W_{z(k)}$ for some points $z(1), \ldots, z(k)$ in K. Set f' equal to the pointwise maximum of $f_{z(1)}, \ldots, f_{z(k)}$, that is,

$$f'(y) = \max\{f_{z(1)}(y), \ldots, f_{z(k)}(y)\}$$

for all $y \in K$. As each $f_{z(j)} \ll g$, we have $f' \ll g \ll h$. Any point $y \in K$ lies in some $W_{z(j)}$, whence $f(y) < f_{z(j)}(y) \leq f'(y)$. Thus $f \ll f'$.

As f' is a pointwise maximum of finitely many continuous affine functions, f' is a continuous convex function. Now $-h \ll -f'$, where $-h$ is an upper semicontinuous convex function from K to $\{-\infty\} \cup \mathbb{R}$, and $-f'$ is a continuous concave function from K to $\mathbb{R}$. By the argument above, there exists a function $k\colon K \to \mathbb{R}$ such that $-h \ll k \ll -f'$ and k is a pointwise maximum of finitely many functions from Aff(K). Then the function $h' = -k$ is a pointwise minimum of finitely many functions from Aff(K), and $f' \ll h' \ll h$. □

THEOREM 11.12. *Let K be a Choquet simplex, let $f\colon K \to \{-\infty\} \cup \mathbb{R}$ be an upper semicontinuous convex function, and let $h\colon K \to \mathbb{R} \cup \{\infty\}$ be a lower semicontinuous concave function. If $f \ll h$, there exists $g \in$ Aff(K) such that $f \ll g \ll h$.*

PROOF. By Proposition 11.11, there exist functions $f', h'\colon K \to \mathbb{R}$ such that $f \ll f' \ll h' \ll h$, where f' is the pointwise maximum of functions $f_1, \ldots, f_n$ from Aff(K), and h' is the pointwise minimum of functions $h_1, \ldots, h_k$ from Aff(K). Then $f_i \leq f' \ll h' \leq h_j$ for all i, j. According to Theorem 11.4, (Aff$(K), \lll$) is an interpolation group, and hence there exists $g \in$ Aff(K) such that $f_i \lll g \lll h_j$ for all i, j. Therefore

$$f \ll f' \lll g \lll h' \ll h. \quad \square$$

Much more useful than Theorem 11.12 is the corresponding result in which $\ll$ has been replaced by $\leq$. We derive this from Theorem 11.12 by the same technique utilized in proving the implication (c) $\Rightarrow$ (b) of Theorem 11.4.

THEOREM 11.13. *Let K be a Choquet simplex, let $f\colon K \to \{-\infty\} \cup \mathbb{R}$ be an upper semicontinuous convex function, and let $h\colon K \to \mathbb{R} \cup \{\infty\}$ be a lower semicontinuous concave function. If $f \leq h$, there exists $g \in \mathrm{Aff}(K)$ such that $f \leq g \leq h$.*

PROOF. We construct functions $g_0, g_1, g_2, \ldots$ in $\mathrm{Aff}(K)$ such that

$$f - 1/2^n \ll g_n \ll h + 1/2^n$$

for all n, while also

$$g_{n-1} - 1/2^n \ll g_n \ll g_{n-1} + 1/2^n$$

for all $n > 0$. Since $f - 1 \ll h + 1$, the existence of g_0 is immediate from Theorem 11.12.

Now assume that $g_0, \ldots, g_{n-1}$ have been constructed, for some $n > 0$. Then

$$f - 1/2^n \ll h + 1/2^n \quad \text{and} \quad g_{n-1} - 1/2^n \ll g_{n-1} + 1/2^n.$$

Since $f - 1/2^{n-1} \ll g_{n-1} \ll h + 1/2^{n-1}$, we also have

$$f - 1/2^n \ll g_{n-1} + 1/2^n \quad \text{and} \quad g_{n-1} - 1/2^n \ll h + 1/2^n.$$

Hence, if we define

$$f'(x) = \max\{f(x), g_{n-1}(x)\} - 1/2^n,$$
$$h'(x) = \min\{h(x), g_{n-1}(x)\} + 1/2^n$$

for all $x \in K$, we obtain $f' \ll h'$. Since f and g_{n-1} are upper semicontinuous convex functions, so is f'. Similarly, since h and g_{n-1} are lower semicontinuous concave functions, so is h'. Thus by Theorem 11.12 there exists $g_n \in \mathrm{Aff}(K)$ such that $f' \ll g_n \ll h'$, which completes the induction step.

As $\|g_n - g_{n-1}\| < 1/2^n$ for all n, the functions g_n form a Cauchy sequence in $\mathrm{Aff}(K)$, and so they converge to some $g \in \mathrm{Aff}(K)$. It is clear that $f \leq g \leq h$. □

• **Compact Sets of Extreme Points.** Our first applications of Theorem 11.13 involve extending continuous real-valued functions on a compact set X of extreme points of a Choquet simplex K to affine continuous real-valued functions on K. In particular, this allows us to present $C(X, \mathbb{R})$ as a natural quotient space of $\mathrm{Aff}(K)$.

THEOREM 11.14. *Let K be a Choquet simplex, let X be a compact subset of $\partial_e K$, and let $g_0 \in C(X, \mathbb{R})$. Let $f : K \to \{-\infty\} \cup \mathbb{R}$ be an upper semicontinuous convex function, and let $h\colon K \to \mathbb{R} \cup \{\infty\}$ be a lower semicontinuous concave function. Assume that $f \leq h$ and that $f|_X \leq g_0 \leq h|_X$. Then there exists $g \in \mathrm{Aff}(K)$ such that $f \leq g \leq h$ and $g|_X = g_0$.*

PROOF. Define functions $f_1\colon K \to \{-\infty\} \cup \mathbb{R}$ and $h_1\colon K \to \mathbb{R} \cup \{\infty\}$ so that $f_1 = h_1 = g_0$ on X, while $f_1 = f$ and $h_1 = h$ on $K - X$, and note that

$f \leq f_1 \leq h_1 \leq h$. We claim that f_1 is upper semicontinuous and convex and that h_1 is lower semicontinuous and concave.

For any $a \in \mathbb{R}$, the set $f^{-1}[a, \infty)$ is closed in K because f is upper semicontinuous, and the set $g_0^{-1}[a, \infty)$ is closed in X because g_0 is continuous. As X is compact, $g_0^{-1}[a, \infty)$ is also closed in K. Since $f|_X \leq g_0$, we see that

$$f_1^{-1}[a, \infty) = f^{-1}[a, \infty) \cup g_0^{-1}[a, \infty),$$

and hence $f_1^{-1}[a, \infty)$ is closed in K. Thus f_1 is upper semicontinuous. Similarly, h_1 is lower semicontinuous.

Now consider a positive convex combination $x = \alpha_1 x_1 + \alpha_2 x_2$ in K. If $x \in X$, then $x \in \partial_e K$, and so $x_1 = x_2 = x$, whence

$$f_1(x) = g_0(x) = \alpha_1 g_0(x) + \alpha_2 g_0(x) = \alpha_1 f_1(x_1) + \alpha_2 f_1(x_2).$$

If $x \notin X$, then since f is convex and $f \leq f_1$ we obtain

$$f_1(x) = f(x) \leq \alpha_1 f(x_1) + \alpha_2 f(x_2) \leq \alpha_1 f_1(x_1) + \alpha_2 f_1(x_2).$$

Thus f_1 is a convex function. Similarly, h_1 is a concave function.

By Theorem 11.13, there exists $g \in \mathrm{Aff}(K)$ for which $f_1 \leq g \leq h_1$. Then $f \leq f_1 \leq g$ and $g \leq h_1 \leq h$. Since $f_1 = h_1 = g_0$ on X, we also have $g|_X = g_0$. □

COROLLARY 11.15. *Let K be a Choquet simplex, let X be a compact subset of $\partial_e K$, and let $g_0 \in C(X, \mathbb{R})$. Then there exists $g \in \mathrm{Aff}(K)$ such that $g|_X = g_0$ and $\|g\| = \|g_0\|$.*

PROOF. Set $M = \|g_0\|$. Considered as constant functions on K, we have $-M \leq M$ and $-M|_X \leq g_0 \leq M|_X$. By Theorem 11.14, there is some $g \in \mathrm{Aff}(K)$ for which $-M \leq g \leq M$ and $g|_X = g_0$. Then $\|g\| = M = \|g_0\|$. □

COROLLARY 11.16. *Let K be a Choquet simplex, let X be a compact subset of $\partial_e K$, and let $f, g \in \mathrm{Aff}(K)$. Then there exist $h, k \in \mathrm{Aff}(K)$ such that*

$$h \leq \begin{matrix} f \\ g \end{matrix} \leq k$$

while also

$$h(x) = \min\{f(x), g(x)\} \quad \text{and} \quad k(x) = \max\{f(x), g(x)\}$$

for all $x \in X$.

PROOF. Set $h'(x) = \min\{f(x), g(x)\}$ for all $x \in K$, and let h_0 be the restriction of h' to X. Then h' is a continuous concave function from X to $\mathbb{R}$, and $h_0 \in C(X, \mathbb{R})$. Choose any function $e \in \mathrm{Aff}(K)$ such that $e \leq h'$, and note that $e|_X \leq h_0 = h'|_X$. By Theorem 11.14, there is some $h \in \mathrm{Aff}(K)$ for which $e \leq h \leq h'$ and $h|_X = h_0$. Then $h \leq h' \leq f$ and $h \leq h' \leq g$, and

$$h(x) = h_0(x) = \min\{f(x), g(x)\}$$

for all $x \in X$.

The function k is constructed in the same manner. □

As an application of Corollary 11.16, we obtain the following criterion for a point of a Choquet simplex to be an extreme point.

THEOREM 11.17. *Let K be a Choquet simplex, and let $x \in K$. Then x is an extreme point of K if and only if for any $f, g \in \mathrm{Aff}(K)$ there exists $h \in \mathrm{Aff}(K)$ such that*

$$h \leq \begin{matrix} f \\ g \end{matrix} \quad \textit{and} \quad h(x) = \min\{f(x), g(x)\}.$$

PROOF. If x is an extreme point, then $\{x\}$ is a compact subset of $\partial_e K$, and the desired condition holds by Corollary 11.16.

If $x \notin \partial_e K$, there exists a positive convex combination $x = \alpha_1 x_1 + \alpha_2 x_2$ where x_1 and x_2 are distinct points of K. Choose a function $f \in \mathrm{Aff}(K)$ such that $f(x_1) \neq f(x_2)$. After multiplying f by a suitable real number, and then adding a suitable constant function, we may assume that $f(x_1) = -1$ and $f(x_2) = 1$. Set $g = -f$, so that $g(x_1) = 1$ and $g(x_2) = -1$. Then

$$f(x) = \alpha_2 - \alpha_1 > -1 \quad \text{and} \quad g(x) = \alpha_1 - \alpha_2 > -1.$$

Given any $h \in \mathrm{Aff}(K)$ satisfying $h \leq f$ and $h \leq g$, we have

$$h(x) = \alpha_1 h(x_1) + \alpha_2 h(x_2) \leq \alpha_1 f(x_1) + \alpha_2 g(x_2) = -1 < \min\{f(x), g(x)\}.$$

Thus there does not exist any $h \in \mathrm{Aff}(K)$ such that $h \leq f$ and $h \leq g$ while $h(x) = \min\{f(x), g(x)\}$. □

THEOREM 11.18. *Let K be a Choquet simplex, let X be a compact subset of $\partial_e K$, and set*

$$H = \{f \in \mathrm{Aff}(K) \mid f|_X = 0\}.$$

Then H is an ideal of $\mathrm{Aff}(K)$, and

$$X = \{x \in \partial_e K \mid f(x) = 0 \quad \textit{for all } f \in H\}.$$

Moreover, the induced map $\mathrm{Aff}(K)/H \to C(X, \mathbb{R})$ is an isometric isomorphism of ordered Banach spaces.

PROOF. Obviously H is a linear subspace and a convex subgroup of $\mathrm{Aff}(K)$. Given any $f, g \in H$, Corollary 11.16 provides us with a function k in $\mathrm{Aff}(K)$ such that $f \leq k$ and $g \leq k$, while also

$$k(x) = \max\{f(x), g(x)\} = 0$$

for all $x \in X$. Then $k \in H$, proving that H is directed. Thus H is an ideal of $\mathrm{Aff}(K)$.

Now consider any x in $\partial_e K - X$. Then $X \cup \{x\}$ is a compact subset of $\partial_e K$, and we may define a continuous real-valued function g_0 on $X \cup \{x\}$ so that $g_0 = 0$ on X while $g_0(x) = 1$. According to Corollary 11.15, g_0 extends to a function $g \in \mathrm{Aff}(K)$. Then g is a function in H for which $g(x) \neq 0$. Therefore X equals the set of those points in $\partial_e K$ at which all functions in H vanish.

If ρ: $\mathrm{Aff}(K) \to C(X, \mathbb{R})$ is the restriction map, then ρ is surjective by Corollary 11.15, and $\ker(\rho) = H$ by definition of H. Hence, ρ induces a linear isomorphism

$$\overline{\rho}\colon \mathrm{Aff}(K)/H \to C(X, \mathbb{R}).$$

Of course $\overline{\rho}$ is a positive map. Conversely, consider any cosets $f+H$ and $g+H$ in $\mathrm{Aff}(K)/H$ for which

$$\overline{\rho}(f+H) \leq \overline{\rho}(g+H),$$

that is, $f|_X \leq g|_X$. By Corollary 11.16, there exists $h \in \mathrm{Aff}(K)$ such that $h \leq f$ and $h \leq g$ while also

$$h(x) = \min\{f(x), g(x)\} = f(x)$$

for all $x \in X$. Then $f - h \in H$, so that $f + H = h + H \leq g + H$. Thus $\overline{\rho}$ is an isomorphism of ordered vector spaces.

Given a coset $f+H$ in $\mathrm{Aff}(K)/H$, observe that

$$\|\overline{\rho}(f+H)\| = \|\overline{\rho}(g+H)\| = \|g|_X\| \leq \|g\|$$

for any $g \in f+H$, whence $\|\overline{\rho}(f+H)\| \leq \|f+H\|$. If $g_0 = f|_X$, then by Corollary 11.15 there exists $g \in \mathrm{Aff}(K)$ such that $g|_X = g_0$ and $\|g\| = \|g_0\|$. Then $f - g \in H$, and hence

$$\|f+H\| \leq \|g\| = \|g_0\| = \|\rho(f)\| = \|\overline{\rho}(f+H)\|.$$

Therefore $\|\overline{\rho}(f+H)\| = \|f+H\|$, proving that $\overline{\rho}$ is an isometry. □

COROLLARY 11.19. *Let K be a Choquet simplex, let X be a compact subset of $\partial_e K$, and let F be the closure of the convex hull of X. Then F is a closed face of K, and $\partial_e F = X$.*

PROOF. Set $H = \{f \in \mathrm{Aff}(K) \mid f|_X = 0\}$ and

$$G = \{x \in K \mid f(x) = 0 \text{ for all } f \in H\},$$

and observe that G is a closed convex subset of K. We claim that G is a face of K. Thus consider any positive convex combination $\alpha_1 x_1 + \alpha_2 x_2 = x$ where $x_1, x_2 \in K$ and $x \in G$. For any $f \in H^+$, we have

$$\alpha_1 f(x_1) + \alpha_2 f(x_2) = f(x) = 0$$

with each $\alpha_i > 0$ and each $f(x_i) \geq 0$, whence each $f(x_i) = 0$. As H is an ideal of $\mathrm{Aff}(K)$ (Theorem 11.18), it is directed, and so all functions in H vanish on x_1 and x_2. Thus $x_1, x_2 \in G$, proving that G is a face.

Since G is a face of K, we have $\partial_e G = G \cap \partial_e K$. On the other hand, Theorem 11.18 shows that $G \cap \partial_e K = X$, and so $\partial_e G = X$. By the Krein-Mil'man Theorem, G equals the closure of the convex hull of X. Therefore $G = F$, whence F is a closed face of K and $\partial_e F = X$. □

In a Choquet simplex K, the closure of the convex hull of a noncompact subset of $\partial_e K$ need not be a face of K. For example, let K be the state space $S(G,u)$ in Example 6.10, which is a Choquet simplex by Theorem 10.17. The face F constructed in that example is the convex hull of a subset of $\partial_e K$, and the closure of F is not a face of K.

COROLLARY 11.20. *Let K be a Choquet simplex. If $\partial_e K$ is compact, then the restriction map ρ: $\mathrm{Aff}(K) \to C(\partial_e K, \mathbb{R})$ is an isometric isomorphism of ordered Banach spaces, and K is affinely homeomorphic to $M_1^+(\partial_e K)$.*

PROOF. In view of Corollary 5.20, the only function in $\mathrm{Aff}(K)$ that vanishes on $\partial_e K$ is the zero function. Hence, ρ is an isometric isomorphism of ordered Banach spaces, by Theorem 11.18. As $\rho(1) = 1$, we see that ρ induces an affine homeomorphism of $S(C(\partial_e K, \mathbb{R}), 1)$ onto $S(\mathrm{Aff}(K), 1)$. However,

$$S(C(\partial_e K, \mathbb{R}), 1) = M_1^+(\partial_e K)$$

by Proposition 6.8, and K is affinely homeomorphic to $S(\mathrm{Aff}(K), 1)$ by Theorem 7.1. Therefore K is affinely homeomorphic to $M_1^+(\partial_e K)$. □

Corollary 11.20 allows us to characterize those Choquet simplices that have compact extreme boundaries, as follows.

THEOREM 11.21. *For a Choquet simplex K, the following conditions are equivalent*:

(a) *$\partial_e K$ is compact.*

(b) *$\mathrm{Aff}(K)$ is a lattice-ordered real vector space.*

(c) *K is affinely homeomorphic to $M_1^+(X)$ for some compact Hausdorff space X.*

PROOF. The implications (a) ⇒ (b) and (a) ⇒ (c) are clear from Corollary 11.20.

(b) ⇒ (a): In view of Theorem 11.17, we see that

$$\partial_e K = \{x \in K \mid (f \wedge g)(x) = \min\{f(x), g(x)\} \text{ for all } f, g \in \mathrm{Aff}(K)\}.$$

Thus $\partial_e K$ is closed in K and hence is compact.

(c) ⇒ (a): In this case $\partial_e K$ is homeomorphic to X, whence $\partial_e K$ is compact. □

For an example of a Choquet simplex whose extreme boundary is not compact, see Example 6.10.

• **Closed Faces.** Most of the results of the previous section have analogs for a closed face F of a Choquet simplex K. In case $\partial_e F$ is compact, these results follow directly from the earlier ones. In general, however, $\partial_e F$ need not be compact, as Example 6.10 shows. (Take $F = K = S(G, u)$.) The results of this section may thus be viewed as proper generalizations of those of the previous section, since any compact subset of $\partial_e K$ equals the extreme boundary of a closed face of K (Corollary 11.19). These results, relating closed faces of K to ideals of $\mathrm{Aff}(K)$, may be viewed as the most desirable prototypes for relationships between closed faces of a state space and ideals of an interpolation group with order-unit.

THEOREM 11.22. *Let K be a Choquet simplex, let $F_1, \dots, F_n$ be pairwise disjoint closed faces of K, and let $g_i \in \operatorname{Aff}(F_i)$ for each $i = 1, \dots, n$. Let $f\colon K \to \{-\infty\} \cup \mathbb{R}$ be an upper semicontinuous convex function, and let $h\colon K \to \mathbb{R} \cup \{\infty\}$ be a lower semicontinuous concave function. Assume that $f \le h$ and that*

$$f|_{F_i} \le g_i \le h|_{F_i}$$

for each i. Then there exists $g \in \operatorname{Aff}(K)$ such that $f \le g \le h$ and $g|_{F_i} = g_i$ for each i.

PROOF. Define functions $f_1\colon K \to \{-\infty\} \cup \mathbb{R}$ and $h_1\colon K \to \mathbb{R} \cup \{\infty\}$ so that $f_1 = h_1 = g_i$ on each F_i, while $f_1 = f$ and $h_1 = h$ on $K - (\bigcup F_i)$, and note that $f \le f_1 \le h_1 \le h$. Since $\bigcup F_i$ is compact, it follows as in Theorem 11.14 that f_1 is upper semicontinuous and that h_1 is lower semicontinuous.

Now consider a positive convex combination $x = \alpha_1 x_1 + \alpha_2 x_2$ in K. If $x \in F_i$ for some i, then $x_1, x_2 \in F_i$, and so

$$f_1(x) = g_i(x) = \alpha_1 g_i(x_1) + \alpha_2 g_i(x_2) = \alpha_1 f_1(x_1) + \alpha_2 f_1(x_2).$$

If $x \notin \bigcup F_i$, then since f is convex and $f \le f_1$, we obtain

$$f_1(x) = f(x) \le \alpha_1 f(x_1) + \alpha_2 f(x_2) \le \alpha_1 f_1(x_1) + \alpha_2 f_1(x_2).$$

Thus f_1 is a convex function. Similarly, h_1 is a concave function.

By Theorem 11.13, there exists $g \in \operatorname{Aff}(K)$ such that $f_1 \le g \le h_1$. The required properties of g are clear. □

COROLLARY 11.23. *Let K be a Choquet simplex, let $F_1, \dots, F_n$ be pairwise disjoint closed faces of K, and let $g_i \in \operatorname{Aff}(F_i)$ for each $i = 1, \dots, n$. Then there exists $g \in \operatorname{Aff}(K)$ such that $g|_{F_i} = g_i$ for each i and*

$$\|g\| = \max\{\|g_1\|, \dots, \|g_n\|\}.$$

PROOF. Set $M = \max\{\|g_1\|, \dots, \|g_n\|\}$. Then $-M \le M$ and $-M \le g_i \le M$ for all i. By Theorem 11.22, there is some $g \in \operatorname{Aff}(K)$ for which $-M \le g \le M$ and $g|_{F_i} = g_i$ for all i. In addition, $\|g\| = M$. □

COROLLARY 11.24. *Let K be a Choquet simplex, let $F_1, \dots, F_n$ be pairwise disjoint closed faces of K, and let $f, g \in \operatorname{Aff}(K)$. For each $i = 1, \dots, n$, assume that either $f \le g$ on F_i or $g \le f$ on F_i. Then there exist $h, k \in \operatorname{Aff}(K)$ such that*

$$h \le \begin{matrix} f \\ g \end{matrix} \le k$$

while also

$$h(x) = \min\{f(x), g(x)\} \quad \text{and} \quad k(x) = \max\{f(x), g(x)\}$$

for all $x \in F_1 \cup \cdots \cup F_n$.

PROOF. Set $h'(x) = \min\{f(x), g(x)\}$ for all $x \in K$, and note that h' is a continuous concave function from K to $\mathbb{R}$. For each $i = 1, \dots, n$, let g_i be the restriction of h' to F_i. By hypothesis, g_i equals the restriction of either f or g to F_i, and hence $g_i \in \mathrm{Aff}(F_i)$. Choose any function $e \in \mathrm{Aff}(K)$ such that $e \leq h'$, and note that $e \leq g_i = h'$ on any F_i. By Theorem 11.22, there is some $h \in \mathrm{Aff}(K)$ for which $e \leq h \leq h'$ and $h|_{F_i} = g_i$ for each i. The required properties of h are clear.

The function k is constructed in the same manner. □

The comparability conditions on the restrictions of f and g to the F_i in Corollary 11.24 are necessary because the pointwise minimum and maximum of f and g usually do not restrict to affine functions on the F_i.

By analogy with Theorem 11.17, we derive the following criterion for a closed convex subset of a Choquet simplex to be a face.

THEOREM 11.25. *Let K be a Choquet simplex, and let F be a closed convex subset of K. Then F is a face of K if and only if for any $f, g \in \mathrm{Aff}(K)$ such that $f|_F \leq g|_F$, there exists $h \in \mathrm{Aff}(K)$ such that*

$$h \quad \leq \quad \begin{matrix} f \\ g \end{matrix} \quad \text{and} \quad h|_F = f|_F.$$

PROOF. If F is a face, the desired condition holds by Corollary 11.24.

If F is not a face, there exists a positive convex combination $\alpha_1 x_1 + \alpha_2 x_2 = x$ where $x_1, x_2 \in K$ and $x \in F$ but $x_1 \notin F$. By Theorem 5.14, there exist $g \in \mathrm{Aff}(K)$ and $b \in \mathbb{R}$ such that

$$g(x_1) < b < \inf\{g(y) \mid y \in F\}.$$

Let $f \in \mathrm{Aff}(K)$ be the constant function b, and note that $f|_F \ll g|_F$. Suppose that there exists $h \in \mathrm{Aff}(K)$ such that $h \leq f$ and $h \leq g$ while also $h|_F = f|_F = b$. Then b is the maximum value of h, and hence $h^{-1}(\{b\})$ is a face of K, by Lemma 5.16. Since

$$\alpha_1 x_1 + \alpha_2 x_2 = x \in F \subseteq h^{-1}(\{b\}),$$

it follows that $x_1 \in h^{-1}(\{b\})$. However, this contradicts the fact that

$$h(x_1) \leq g(x_1) < b.$$

Therefore there does not exist any $h \in \mathrm{Aff}(K)$ such that $h \leq f$ and $h \leq g$ while $h|_F = f|_F$. □

THEOREM 11.26. *Let K be a Choquet simplex, let $F_1, \ldots, F_n$ be pairwise disjoint closed faces of K, and set*

$$H = \{f \in \mathrm{Aff}(K) \mid f|_{F_i} = 0 \textit{ for each } i = 1, \ldots, n\}.$$

Then H is an ideal of Aff(K)*, and the set*

$$F = \{x \in K \mid f(x) = 0 \textit{ for all } f \in H\}$$

equals the convex hull of $F_1 \cup \cdots \cup F_n$. Moreover, the induced map

$$\mathrm{Aff}(K)/H \to \mathrm{Aff}(F_1) \times \cdots \times \mathrm{Aff}(F_n)$$

is an isometric isomorphism of ordered Banach spaces (where the product space $\prod \mathrm{Aff}(F_i)$ *has been equipped with the supremum norm).*

PROOF. Obviously H is a linear subspace and a convex subgroup of $\mathrm{Aff}(K)$. Given any $f, g \in H$, we have $f = g = 0$ on each F_i, and so Corollary 11.24 provides us with a function $k \in \mathrm{Aff}(K)$ such that $f \leq k$ and $g \leq k$, while also

$$k(x) = \max\{f(x), g(x)\} = 0$$

for all $x \in \bigcup F_i$. Then $k \in H$, proving that H is directed. Thus H is an ideal of $\mathrm{Aff}(K)$.

Since H is directed, we infer as in Corollary 11.19 that F is a closed face of K. On the other hand, if G is the convex hull of $F_1 \cup \cdots \cup F_n$, then G is a closed face of K by Propositions 5.2 and 10.10. As $\bigcup F_i \subseteq F$, we must have $G \subseteq F$. If $F \neq G$, then $\partial_e F \not\subseteq G$, because of the Krein-Mil′man Theorem. Choose a point x in $\partial_e F - G$, and note that $x \in \partial_e K$, because F is a face of K. Then $\{x\}$ is a closed face of K that is disjoint from each F_i. Using Corollary 11.23, we obtain a function $g \in \mathrm{Aff}(K)$ such that $g = 0$ on each F_i while $g(x) = 1$. But then $g \in H$ and $g(x) \neq 0$, contradicting the assumption that $x \in F$. Therefore $F = G$.

If $\rho\colon \mathrm{Aff}(K) \to \prod \mathrm{Aff}(F_i)$ is the restriction map, then ρ is surjective by Corollary 11.23, and $\ker(\rho) = H$ by definition of H. Hence, ρ induces a linear isomorphism

$$\overline{\rho}\colon \mathrm{Aff}(K)/H \to \prod \mathrm{Aff}(F_i).$$

Obviously $\overline{\rho}$ is a positive map. Conversely, consider any cosets $f + H$ and $g + H$ in $\mathrm{Aff}(K)/H$ for which

$$\overline{\rho}(f + H) \leq \overline{\rho}(g + H).$$

Then $f \leq g$ on each F_i. By Corollary 11.24, there exists $h \in \mathrm{Aff}(K)$ such that $h \leq f$ and $h \leq g$ while also

$$h(x) = \min\{f(x), g(x)\} = f(x)$$

for all $x \in \bigcup F_i$. Then $f - h \in H$, so that $f + H = h + H \leq g + H$. Thus $\overline{\rho}$ is an isomorphism of ordered vector spaces.

Given a coset $f + H$ in $\mathrm{Aff}(K)/H$, observe that

$$\|\overline{\rho}(f + H)\| = \|\overline{\rho}(g + H)\| = \max\{\|g|_{F_1}\|, \dots, \|g|_{F_n}\|\} \leq \|g\|$$

for any $g \in f + H$, whence $\|\overline{\rho}(f + H)\| \leq \|f + H\|$. For each $i = 1, \dots, n$, let g_i be the restriction of f to F_i. By Corollary 11.23, there exists $g \in \mathrm{Aff}(K)$ such that $g = g_i$ on each F_i, and

$$\|g\| = \max\{\|g_1\|, \dots, \|g_n\|\}.$$

Then $f - g \in H$, and hence

$$\|f + H\| \leq \|g\| = \max\{\|g_i\|\} = \max\{\|f|_{F_i}\|\} = \|\overline{\rho}(f + H)\|.$$

Therefore $\|\overline{\rho}(f + H)\| = \|f + H\|$, proving that $\overline{\rho}$ is an isometry. □

COROLLARY 11.27. *Let K be a Choquet simplex, and let $F_1, \ldots, F_n$ be pairwise disjoint closed faces of K. If K equals the convex hull of $F_1 \cup \cdots \cup F_n$, then the restriction map*

$$\mathrm{Aff}(K) \to \mathrm{Aff}(F_1) \times \cdots \times \mathrm{Aff}(F_n)$$

is an isometric isomorphism of ordered Banach spaces (where the product space $\prod \mathrm{Aff}(F_i)$ has been equipped with the supremum norm). □

• **Complementary Faces.** We conclude this chapter by proving that any Choquet simplex K may be expressed as the direct convex sum of any closed face F of K and the complementary face of F. To prove this result, we use Theorem 11.26 to reduce the general problem to the special case proved in Theorem 10.15.

THEOREM 11.28. *If F is a closed face of a Choquet simplex K, then K equals the direct convex sum of F and its complementary face.*

PROOF. Set $A = \mathrm{Aff}(K)$, and let ψ be the natural affine homeomorphism of K onto $S(A, 1)$. Then $\psi(F)$ is a closed face of $S(A, 1)$. Set

$$H = \{f \in A \mid f|_F = 0\}.$$

According to Theorem 11.26, H is an ideal of A, and

$$F = \{x \in K \mid f(x) = 0 \text{ for all } f \in H\}.$$

As a result, $\psi(F) = \{s \in S(A, 1) \mid H \subseteq \ker(s)\}$.

Now A is an interpolation group, by Theorem 11.4. Hence, Theorem 10.15 shows that $S(A, 1)$ is the direct convex sum of $\psi(F)$ and its complementary face. The complementary face of $\psi(F)$ is just $\psi(F')$, where F' is the complementary face of F in K. Therefore K is the direct convex sum of F and F'. □

If F is a closed face of a Choquet simplex K, and the complementary face F' of F happens to be closed, then it follows from Theorem 11.28 and Corollary 11.27 that $\mathrm{Aff}(K)$ is naturally isometrically isomorphic to $\mathrm{Aff}(F) \times \mathrm{Aff}(F')$. In general, however, F' need not be closed. To construct an example, pick a compact Hausdorff space X with a closed subset $Y \subseteq X$ that is not open. Then set $K = M_1^+(X)$, which is a Choquet simplex by Corollary 10.18, and let F be the closure of the convex hull of $\{\varepsilon_y \mid y \in Y\}$, which by Corollary 11.19 is a closed face of K such that $\partial_e F = \{\varepsilon_y \mid y \in Y\}$. For any $x \in X - Y$, the set $\{\varepsilon_x\}$ is a face of K that is disjoint from F, and so $\varepsilon_x \in F'$. Since Y is not open in X, there is a point z in $Y \cap \overline{(X - Y)}$. Then ε_z lies in F and also in the closure of F', whence F' is not closed.

• **Notes.** Theorem 11.3 and its proof are patterned after a result of Krein concerning a partially ordered real Banach space E such that E^+ has nonempty interior and the values $\|x + y\|$, for all positive points x and y in the unit ball of E, are bounded away from zero. If a new ordering $\leq^\circ$ is defined in E such that $a \leq^\circ b$ if and only if either $a = b$ or $b - a$ lies in the interior of E^+, then $(E, \leq^\circ)$ has interpolation if and only if $(E^*, \leq^+)$ is lattice-ordered [**86**, Théorème]. Theorem 11.3 is also a special case of a result of Andô, who proved that a partially ordered real Banach space E has interpolation if and only if $(E^*, \leq^+)$ has interpolation [**4**, Theorem 2]. The equivalence of conditions (a) and (b) in Theorem 11.4 was proved independently by Edwards [**24**, Théorème] and Semadeni [**113**, Theorem 5].

A strong version of the metrizable case of Theorem 11.6 was proved by Lazar and Lindenstrauss [**90**, Corollary to Theorem 5.2]: any metrizable Choquet simplex is affinely homeomorphic to an inverse limit of a sequence of classical simplices and affine continuous surjections. That all Choquet simplices are affinely homeomorphic to inverse limits of classical simplices was observed by the author and Handelman (unpublished) as a direct corollary of Shen's representation of divisible dimension groups as direct limits of simplicial groups [**115**, Theorem 3.5]. Theorem 11.7 was proved independently by Davies and Vincent-Smith [**23**, Theorem 13] and Jellett [**75**, Theorem 2].

Theorem 11.12 may be viewed as either a preliminary form or an easy corollary of Theorem 11.13; the case in which f and h are continuous and real-valued was proved by Boboc and Cornea [**14**, Théorème 3]. Theorem 11.13 is due to Edwards [**24**, Théorème; **25**, Theorem 2]. Theorems 11.14, 11.17, 11.18 and Corollaries 11.15 and 11.16 are standard kinds of applications of Theorem 11.13. That every compact subset of the extreme boundary of a Choquet simplex equals the extreme boundary of a closed face (Corollary 11.19) was proved by Effros [**28**, Corollary 3.5]. The surjectivity of the restriction map in Corollary 11.20 was proved by Bauer [**10**, Satz 1], who also proved the equivalence of conditions (a) and (b) in Theorem 11.21 [**9**, Satz 13; **10**, Satz 1].

The case of Theorem 11.22 concerning a single closed face was proved by Edwards [**25**, Corollary to Theorem 2]. Corollaries 11.23 and 11.24 and Theorem 11.25 are standard kinds of applications of Theorem 11.13. In the case of Theorem 11.26 concerning a single closed face F, the conclusions that H is an ideal of $\mathrm{Aff}(K)$ and that F equals the set of common zeroes of H are special cases of a result of Effros [**27**, Theorem 3.1]. These results, and the isometric isomorphism of $\mathrm{Aff}(K)/H$ onto $\mathrm{Aff}(F)$, were observed by Edwards and Vincent-Smith to be easy consequences of Theorem 11.13 [**26**, p. 271]. Theorem 11.28 is due to Alfsen [**2**, Corollary to Theorem 1].

CHAPTER 12

Metric Completions

As an instrument for close examination of an arbitrary state s on an interpolation group (G, u) with order-unit, we form a completion of G with respect to a pseudo-metric derived from s. This completion $\overline{G}$ is a Dedekind complete lattice-ordered abelian group, and so computations are much easier in $\overline{G}$ than in G. The convex subgroup G_0 of $\overline{G}$ generated by the image $\overline{u}$ of u is also Dedekind complete, and hence $(G_0, \overline{u})$ satisfies general comparability (Theorem 9.9). Consequently, the construction of elements of G_0 having desirable properties with respect to states on $(G_0, \overline{u})$ is fairly manageable, as seen in the last section of Chapter 8. Approximation of elements of G_0 by elements of G thus provides a mechanism for constructing elements of G that have useful properties relative to s and "nearby" states. [In this context, "nearby" states are those lying in the closure of the face generated by s, since such states on (G, u) are exactly those which are restrictions of states on $(G_0, \overline{u})$.] An initial application of $\overline{G}$ results in a criterion for s to be extremal. Namely, s is extremal if and only if for any $x, y \in G^+$, the minimum of $s(x)$ and $s(y)$ equals the supremum of the values $s(z)$ as z ranges over those elements of G^+ lying below both x and y. The use of this completion is also a key step in approximating affine continuous real-valued functions on $S(G, u)$ by evaluations at elements of G, which is accomplished in the following chapter.

Initially, this completion procedure requires only a positive real-valued homomorphism f on a directed abelian group G. We construct a pseudo-norm on G from f, and we develop the basic properties of the corresponding f-completion of G, which completion is itself a directed abelian group. The fundamental structural result is that if G is an interpolation group, then the f-completion of G is a Dedekind complete lattice-ordered abelian group. We then restrict attention to the case that G is an interpolation group with an order-unit u and investigate completions with respect to a state s on (G, u). In particular, s is extremal if and only if the s-completion of G is totally ordered, from which the extremal state criterion is derived. Roughly speaking, this criterion is a version "to within epsilon" of the criterion for extremality of points in a Choquet simplex proved in Theorem 11.17. In addition, we prove that several other results from the latter part of Chapter 11 have analogs in $S(G,u)$ that hold up to arbitrarily close approximation.

• Completions with Respect to Positive Homomorphisms.

DEFINITION. Let G be a directed abelian group, and let $f\colon G \to \mathbb{R}$ be a positive homomorphism. For any $x \in G$, define

$$|x|_f = \inf\{f(a+b) \mid a,b \in G^+ \text{ and } x = a-b\}.$$

Since G is directed and f is positive, this infimum is taken over a nonempty subset of $\mathbb{R}^+$. Thus $|x|_f \in \mathbb{R}^+$.

LEMMA 12.1. *Let G be a directed abelian group, let $f\colon G \to \mathbb{R}$ be a positive homomorphism, and let $x, y \in G$.*

(a) $|mx|_f \le |m| \cdot |x|_f$ *for all* $m \in \mathbb{Z}$. *In particular,* $|-x|_f = |x|_f$ *and* $|0|_f = 0$.

(b) $|x \pm y|_f \le |x|_f + |y|_f$.

(c) $|f(x) - f(y)| \le |x-y|_f$.

(d) *If* $x \in G^+$, *then* $|x|_f = f(x)$.

(e) *If* $y \in G^+$ *and* $-y \le x \le y$, *then* $|x|_f \le 3|y|_f$. *If, furthermore, G has interpolation, then* $|x|_f \le |y|_f$.

PROOF. (a) If $x = a-b$ with $a, b \in G^+$, then $mx = ma - mb$ and $mx = (-mb) - (-ma)$ with either ma and mb in G^+ or $-mb$ and $-ma$ in G^+. Hence, either

$$|mx|_f \le f(ma+mb) = mf(a+b) \quad \text{or} \quad |mx|_f \le f(-mb-ma) = -mf(a+b).$$

In either case, $|mx|_f \le |m| f(a+b)$. Thus $|mx|_f \le |m| \cdot |x|_f$.

Taking $m = 0$, we obtain $|0|_f = 0$. Taking $m = -1$, we find that $|-x|_f \le |x|_f$ and $|x|_f = |-(-x)|_f \le |-x|_f$, whence $|-x|_f = |x|_f$.

(b) Given a positive real number ε, choose a, b, c, d in G^+ such that $x = a-b$ and $y = c-d$ while also

$$f(a+b) < |x|_f + \varepsilon/2 \quad \text{and} \quad f(c+d) < |y|_f + \varepsilon/2.$$

Then $x + y = (a+c) - (b+d)$ with $a+c$ and $b+d$ in G^+, and so

$$|x+y|_f \le f(a+c+b+d) = f(a+b) + f(c+d) < |x|_f + |y|_f + \varepsilon.$$

Thus $|x+y|_f \le |x|_f + |y|_f$. Consequently, $|x-y|_f \le |x|_f + |-y|_f = |x|_f + |y|_f$, because of (a).

(c) We need $|f(z)| \le |z|_f$, where $z = x-y$. Given $z = a-b$ with $a, b \in G^+$, we observe that

$$-(a+b) \le -b \le z \le a \le a+b.$$

Hence, $-f(a+b) \le f(z) \le f(a+b)$, so that $|f(z)| \le f(a+b)$. Thus $|f(z)| \le |z|_f$, as required.

(d) Since $x = x - 0$, we have $|x|_f \le f(x+0) = f(x)$. On the other hand, $f(x) = |f(x)| \le |x|_f$ by (c). Thus $|x|_f = f(x)$.

(e) Note that $x + 2y \le 3y$. Since $x = (x + y) - y$ with $x + y$ and y in G^+, we conclude that

$$|x|_f \le f(x + 2y) \le f(3y) = 3f(y) = 3|y|_f,$$

because of (d).

Now assume that G has interpolation. As $0 \le x + y \le 2y$, there must exist elements y_1, y_2 in G^+ such that $x + y = y_1 + y_2$ and each $y_i \le y$. Then

$$x = x + y - y = y_1 - (y - y_2) = y_2 - (y - y_1)$$

with each y_i and each $y - y_i$ in G^+, whence

$$|x|_f \le f(y_1 + y - y_2) \quad \text{and} \quad |x|_f \le f(y_2 + y - y_1).$$

Since either $f(y_1 - y_2) \le 0$ or $f(y_2 - y_1) \le 0$, we must have $|x|_f \le f(y) = |y|_f$. □

DEFINITION. Let G be a directed abelian group, and let $f\colon G \to \mathbb{R}$ be a positive homomorphism. In view of parts (a) and (b) of Lemma 12.1, we see that the rule $d(x, y) = |x - y|_f$ defines a pseudo-metric d on G. By way of abbreviation, we refer to d as *the f-metric on G*, although d is not always a metric. *The f-completion of G* is the (Hausdorff) completion of G with respect to d.

Because of the triangle inequality (Lemma 12.1(b)), addition and subtraction in G are uniformly continuous with respect to d. Hence, the f-completion $\overline{G}$ of G is a topological abelian group, and the natural map $\phi\colon G \to \overline{G}$ is a continuous group homomorphism. Note that $\ker(\phi)$ is just the set of those $x \in G$ for which $|x|_f = 0$.

DEFINITION. Let G be a directed abelian group, and let $f\colon G \to \mathbb{R}$ be a positive homomorphism. Let $\overline{G}$ denote the f-completion of G, and let $\phi\colon G \to \overline{G}$ be the natural map. We define a relation $\le$ on $\overline{G}$ so that for any $x, y \in \overline{G}$, we have $x \le y$ if and only if $y - x$ lies in the closure of $\phi(G^+)$.

As the following proposition shows, this relation is a translation-invariant partial order on $\overline{G}$. We shall always assume that $\overline{G}$ has been made into a partially ordered abelian group using this relation.

PROPOSITION 12.2. *Let G be a directed abelian group, let $f\colon G \to \mathbb{R}$ be a positive homomorphism, and let d_f denote the f-metric on G. Let $\overline{G}$ denote the f-completion of G, let $\phi\colon G \to \overline{G}$ be the natural map, and let $\overline{d}_f$ denote the induced metric on $\overline{G}$.*

(a) *$\overline{G}$ is a directed abelian group with positive cone equal to the closure of $\phi(G^+)$.*

(b) *There is a unique continuous map $\overline{f}\colon \overline{G} \to \mathbb{R}$ such that $\overline{f}\phi = f$, and $\overline{f}$ is a positive homomorphism.*

(c) *The $\overline{f}$-metric $d_{\overline{f}}$ on $\overline{G}$ coincides with $\overline{d}_f$.*

(d) *$|\phi(x)|_{\overline{f}} = |x|_f$ for all $x \in G$.*

PROOF. (a) Let C be the closure of $\phi(G^+)$ in $\overline{G}$. Since $\phi(0) = 0$, we have $0 \in C$. As $\phi(G^+)$ is closed under addition, it follows from the continuity of addition in $\overline{G}$ that C is closed under addition. Thus C is a cone in $\overline{G}$.

Now consider any $x \in C$ for which $-x \in C$. Choose sequences $\{x_n\}$ and $\{y_n\}$ in G^+ such that $\phi(x_n) \to x$ and $\phi(y_n) \to -x$. Then $\phi(x_n + y_n) \to 0$. Since

$$\overline{d}_f(\phi(x_n + y_n), 0) = d_f(x_n + y_n, 0) = |x_n + y_n|_f = f(x_n + y_n)$$

for all n, we find that $f(x_n + y_n) \to 0$. Now

$$0 \leq |x_n|_f = f(x_n) \leq f(x_n + y_n)$$

for all n, whence $|x_n|_f \to 0$. Hence, $\overline{d}_f(\phi(x_n), 0) \to 0$, and so $x = 0$.

Thus C is a strict cone, and hence $\overline{G}$ becomes a partially ordered abelian group with positive cone C.

Given any $x \in \overline{G}$, we may choose a sequence $\{x_1, x_2, \ldots\}$ in G such that $\phi(x_n) \to x$ and

$$|x_{n+1} - x_n|_f < 1/2^n$$

for all n. Then each $x_{n+1} - x_n = a_n - b_n$ for some $a_n, b_n \in G^+$ satisfying

$$f(a_n + b_n) < 1/2^n.$$

Since $|a_n|_f = f(a_n) \leq f(a_n + b_n) < 1/2^n$ for all n, the partial sums of the series $\sum a_n$ form a Cauchy sequence with respect to d_f. Consequently, the series $\sum \phi(a_n)$ converges to an element $a \in \overline{G}$. As the partial sums of this series all lie in $\phi(G^+)$, we see that $a \in \overline{G}^+$. Similarly, the series $\sum \phi(b_n)$ converges to an element $b \in \overline{G}^+$. We compute that

$$\begin{aligned} a - b &= \lim_{k\to\infty} \sum_{n=1}^{k} \phi(a_n - b_n) = \lim_{k\to\infty} \sum_{n=1}^{k} \phi(x_{n+1} - x_n) \\ &= \lim_{k\to\infty} [\phi(x_{k+1}) - \phi(x_1)] = x - \phi(x_1). \end{aligned}$$

Since $x_1 = c - d$ for some $c, d \in G^+$, we find that

$$x = (a + \phi(c)) - (b + \phi(d))$$

with $a + \phi(c)$ and $b + \phi(d)$ in $\overline{G}^+$.

Therefore $\overline{G}$ is directed.

(b) By Lemma 12.1(c), f is uniformly continuous with respect to d_f. Hence, there is a unique continuous map $\overline{f}\colon \overline{G} \to \mathsf{R}$ such that $\overline{f}\phi = f$. Since $\overline{f}$ is additive on $\phi(G)$, it follows from the continuity of addition in $\overline{G}$ that $\overline{f}$ is a homomorphism. As $\overline{f}\phi(G^+) \subseteq \mathsf{R}^+$ and $\phi(G^+)$ is dense in $\overline{G}^+$, we must have $\overline{f}(\overline{G}^+) \subseteq \mathsf{R}^+$. Thus $\overline{f}$ is a positive homomorphism.

(c) We must show that $\overline{d}_f(x, y) = d_{\overline{f}}(x, y) = |x - y|_{\overline{f}}$ for any $x, y \in \overline{G}$. Choose sequences $\{x_n\}$ and $\{y_n\}$ in G such that $\phi(x_n) \to x$ and $\phi(y_n) \to y$. Then $d_f(x_n, y_n) \to \overline{d}_f(x, y)$. As

$$d_f(x_n, y_n) = |x_n - y_n|_f = d_f(x_n - y_n, 0)$$

for all n, it follows that $\overline{d}_f(x, y) = \overline{d}_f(x - y, 0)$. Hence, it suffices to show that $\overline{d}_f(z, 0) = |z|_{\overline{f}}$, where $z = x - y$.

First let $z = a - b$ for some $a, b \in \overline{G}^+$, and choose sequences $\{a_n\}$ and $\{b_n\}$ in G^+ such that $\phi(a_n) \to a$ and $\phi(b_n) \to b$. Then

$$d_f(a_n - b_n, 0) \to \overline{d}_f(z, 0) \quad \text{and} \quad \overline{f}\phi(a_n + b_n) \to \overline{f}(a + b).$$

Since

$$d_f(a_n - b_n, 0) = |a_n - b_n|_f \leq f(a_n + b_n) = \overline{f}\phi(a_n + b_n)$$

for all n, it follows that $\overline{d}_f(z, 0) \leq \overline{f}(a + b)$. Thus $\overline{d}_f(z, 0) \leq |z|_{\overline{f}}$.

Given a positive real number ε, choose a sequence $\{z_1, z_2, \ldots\}$ in G such that $\phi(z_n) \to z$, while $\overline{d}_f(\phi(z_1), z) < \varepsilon/2$ and

$$|z_{n+1} - z_n|_f < \varepsilon/2^{n+1}$$

for all n. Then each $z_{n+1} - z_n = a_n - b_n$ for some $a_n, b_n \in G^+$ with

$$f(a_n + b_n) < \varepsilon/2^{n+1}.$$

As in the proof of (a), the series $\sum \phi(a_n)$ converges to an element $a \in \overline{G}^+$, the series $\sum \phi(b_n)$ converges to an element $b \in \overline{G}^+$, and

$$a - b = z - \phi(z_1).$$

Observe that

$$\overline{f}(a + b) = \sum_{n=1}^{\infty} \overline{f}\phi(a_n + b_n) = \sum_{n=1}^{\infty} f(a_n + b_n) < \sum_{n=1}^{\infty} \frac{\varepsilon}{2^{n+1}} = \frac{\varepsilon}{2}.$$

In addition,

$$|z_1|_f = d_f(z_1, 0) = \overline{d}_f(\phi(z_1), 0) \leq \overline{d}_f(\phi(z_1), z) + \overline{d}_f(z, 0) < \overline{d}_f(z, 0) + \varepsilon/2,$$

and hence $z_1 = c - d$ for some $c, d \in G^+$ satisfying

$$f(c + d) < \overline{d}_f(z, 0) + \varepsilon/2.$$

We now have elements $a + \phi(c)$ and $b + \phi(d)$ in $\overline{G}^+$ such that

$$z = a - b + \phi(z_1) = (a + \phi(c)) - (b + \phi(d)),$$

and consequently

$$|z|_{\overline{f}} \leq \overline{f}(a + \phi(c) + b + \phi(d)) = f(c + d) + \overline{f}(a + b) < \overline{d}_f(z, 0) + \varepsilon.$$

Therefore $\overline{d}_f(z, 0) = |z|_{\overline{f}}$.

(d) Using (c), we conclude that

$$|\phi(x)|_{\overline{f}} = d_{\overline{f}}(\phi(x), 0) = \overline{d}_f(\phi(x), 0) = d_f(x, 0) = |x|_f. \quad \square$$

DEFINITION. In the situation of Proposition 12.2, the map $\overline{f}$ is called *the natural extension of f to $\overline{G}$.*

Since $\overline{d}_f = d_{\overline{f}}$, there is no need to refer explicitly to the metric on $\overline{G}$; rather, we may work with $|\cdot|_{\overline{f}}$.

LEMMA 12.3. *Let G be a directed abelian group, let $f\colon G \to \mathbb{R}$ be a positive homomorphism, and let $\overline{G}$ denote the f-completion of G. Let $x, y \in \overline{G}$, let I be a directed set, and let $\{x_i\}$ and $\{y_i\}$ be nets on I in $\overline{G}$ such that $x_i \to x$ and $y_i \to y$. If $x_i \leq y_i$ for all $i \in I$, then $x \leq y$.*

PROOF. The differences $y_i - x_i$ form a net on I in $\overline{G}^+$ which converges to $y - x$. As $\overline{G}^+$ is closed in $\overline{G}$, we conclude that $y - x$ lies in $\overline{G}^+$. □

• **Dedekind Completeness.** The completion of a directed abelian group G with respect to a positive homomorphism $f\colon G \to \mathbb{R}$ is not only metrically complete, but also enjoys certain lattice completeness properties. The most useful result occurs when G has interpolation, in which case we shall prove that the f-completion of G is a Dedekind complete lattice-ordered abelian group.

PROPOSITION 12.4. *Let G be a directed abelian group, and let $f\colon G \to \mathbb{R}$ be a positive homomorphism. Let $\overline{G}$ denote the f-completion of G, and let X be a nonempty upward directed subset of $\overline{G}$ which is bounded above. If X is considered as a net (indexed by itself), then X converges to an element x^* in $\overline{G}$, and x^* is the supremum of X in $\overline{G}$.*

PROOF. Let $\overline{f}$ be the natural extension of f to $\overline{G}$. As $\overline{f}$ is positive, its restriction to X is a nondecreasing net in $\mathbb{R}$ which is bounded above. The net $\overline{f}|_X$ must converge in $\mathbb{R}$.

Now given any positive real number ε, there exists $x \in X$ such that

$$|\overline{f}(x_1) - \overline{f}(x_2)| < \varepsilon/2$$

for all $x_1, x_2 \in X$ such that each $x_i \geq x$. In particular, $|\overline{f}(x_i) - \overline{f}(x)| < \varepsilon/2$ for such x_i. Since

$$x_1 - x_2 = (x_1 - x) - (x_2 - x)$$

with each $x_i - x \in \overline{G}^+$, we have

$$|x_1 - x_2|_{\overline{f}} \leq \overline{f}(x_1 - x + x_2 - x) = |\overline{f}(x_1) - \overline{f}(x)| + |\overline{f}(x_2) - \overline{f}(x)| < \varepsilon.$$

This proves that X is a Cauchy net in $\overline{G}$. Hence, X must converge to some x^* in $\overline{G}$.

For any $x \in X$, the subnet $\{y \in X \mid y \geq x\}$ also converges to x^*, whence $x^* \geq x$ by Lemma 12.3. Thus x^* is an upper bound for X. On the other hand, if z is any upper bound for X in $\overline{G}$, then as $x \leq z$ for all $x \in X$ we conclude from Lemma 12.3 that $x^* \leq z$. Therefore x^* is the supremum of X in $\overline{G}$. □

We may view Proposition 12.4 as providing a partial lattice completeness in f-completions. However, this is not sufficient for $\overline{G}$ to be a lattice in general, as the following example shows.

EXAMPLE 12.5. *There exists a directed abelian group G with a positive homomorphism $f\colon G \to \mathbb{R}$ such that the f-completion of G is not a lattice, nor even an interpolation group.*

PROOF. Let G be the abelian group $\mathbb{Z}^2$ equipped with the strict ordering, and note that G is directed (in fact, G has an order-unit). Now

$$\begin{matrix}(0,1)\\(1,0)\end{matrix} \ll \begin{matrix}(2,3)\\(3,2)\end{matrix}$$

in G, yet there is no element $(x, y) \in G$ for which

$$\begin{matrix}(0,1)\\(1,0)\end{matrix} \leqq (x,y) \leqq \begin{matrix}(2,3)\\(3,2).\end{matrix}$$

Thus G is not an interpolation group and hence not a lattice.

The rule $f(x, y) = x$ defines a positive homomorphism $f \colon G \to \mathbb{R}$. Observing that $f(a, b) \geq 1$ for all nonzero elements (a, b) in G^+, we see that $|(x, y)|_f \geq 1$ for all nonzero elements (x, y) in G. Hence, the f-metric on G is actually a metric and is discrete. Therefore the f-completion of G is just G itself. □

In fact, the lack of interpolation in the f-completion of Example 12.5 is all that stands in the way of Dedekind completeness, as the following lemma shows.

LEMMA 12.6. *Let G be a directed abelian group, let $f \colon G \to \mathbb{R}$ be a positive homomorphism, and let $\overline{G}$ denote the f-completion of G. If $\overline{G}$ has interpolation, then it is Dedekind complete and lattice-ordered.*

PROOF. First consider any two elements $x, y \in \overline{G}$, and let W be the set of lower bounds for $\{x, y\}$ in $\overline{G}$. Obviously W is bounded above, and W is nonempty because $\overline{G}$ is directed. As $\overline{G}$ has interpolation, W is also upward directed. By Proposition 12.4, W has a supremum w in $\overline{G}$, and clearly $w = x \wedge y$. Similarly, $(-x) \wedge (-y)$ exists in $\overline{G}$, and its negative equals $x \vee y$. Thus $\overline{G}$ is a lattice.

Now consider any nonempty subset X of $\overline{G}$ that is bounded above, and set

$$Y = \{x_1 \vee \cdots \vee x_n \mid x_1, \ldots, x_n \in X\}.$$

Then Y is a nonempty upward-directed subset of $\overline{G}$ which is bounded above. According to Proposition 12.4, Y has a supremum y in $\overline{G}$, and y is the supremum of X as well. Therefore $\overline{G}$ is Dedekind complete. □

THEOREM 12.7. *Let G be a directed abelian group, let $f \colon G \to \mathbb{R}$ be a positive homomorphism, and let $\overline{G}$ denote the f-completion of G. If G is an interpolation group, then $\overline{G}$ is a Dedekind complete lattice-ordered abelian group.*

PROOF. Let $\phi \colon G \to \overline{G}$ be the natural map, and let $\overline{f}$ denote the natural extension of f to $\overline{G}$.

Because of Lemma 12.6, it suffices to prove that $\overline{G}$ has interpolation. Thus let x_1, x_2, y_1, y_2 be any elements of $\overline{G}$ satisfying $x_i \leq y_j$ for all i, j. Choose sequences

$$\{x_{11}, x_{12}, \ldots\}, \quad \{x_{21}, x_{22}, \ldots\}, \quad \{y_{11}, y_{12}, \ldots\}, \quad \{y_{21}, y_{22}, \ldots\}$$

in G such that

$$|\phi(x_{in}) - x_i|_{\overline{f}} < 1/2^{n+5} \quad \text{and} \quad |\phi(y_{jn}) - y_j|_{\overline{f}} < 1/2^{n+5}$$

for all i, j, n. For all i, n, we observe that

$$|x_{i,n+1} - x_{in}|_f = |\phi(x_{i,n+1}) - \phi(x_{in})|_{\overline{f}} \leq |\phi(x_{i,n+1}) - x_i|_{\overline{f}} + |\phi(x_{in}) - x_i|_{\overline{f}}$$
$$< 1/2^{n+6} + 1/2^{n+5} < 1/2^{n+4}.$$

Similarly, $|y_{j,n+1} - y_{jn}|_f < 1/2^{n+4}$ for all j, n. We shall construct Cauchy sequences $\{b_n\}$ and $\{z_n\}$ in G such that $b_n \to 0$ and $x_{in} \leq z_n \leq y_{jn} + b_n$ for all i, j, n. The limit of the sequence $\{\phi(z_n)\}$ will then provide an element of $\overline{G}$ to interpolate between x_1, x_2 and y_1, y_2.

We first construct elements $a_1, a_2, \ldots$ in G^+ such that $f(a_n) < 1/2^{n+2}$ for all n, while also

$$x_{in} - a_n \leq x_{i,n+1} \leq x_{in} + a_n \quad \text{and} \quad y_{jn} - a_n \leq y_{j,n+1} \leq y_{jn} + a_n$$

for all i, j, n.

For each i, n, we have $|x_{i,n+1} - x_{in}|_f < 1/2^{n+4}$, whence

$$x_{i,n+1} - x_{in} = p_{in} - q_{in}$$

for some $p_{in}, q_{in} \in G^+$ satisfying $f(p_{in} + q_{in}) < 1/2^{n+4}$. Similarly, each

$$y_{j,n+1} - y_{jn} = r_{jn} - s_{jn}$$

for some $r_{jn}, s_{jn} \in G^+$ satisfying $f(r_{jn} + s_{jn}) < 1/2^{n+4}$. Set

$$a_n = p_{1n} + q_{1n} + p_{2n} + q_{2n} + r_{1n} + s_{1n} + r_{2n} + s_{2n}$$

for all n, and note that $f(a_n) < 4/2^{n+4} = 1/2^{n+2}$. In addition,

$$x_{in} - a_n \leq x_{in} - q_{in} = x_{i,n+1} - p_{in} \leq x_{i,n+1}$$
$$\leq x_{i,n+1} + q_{in} = x_{in} + p_{in} \leq x_{in} + a_n,$$

$$y_{jn} - a_n \leq y_{jn} - s_{jn} = y_{j,n+1} - r_{jn} \leq y_{j,n+1}$$
$$\leq y_{j,n+1} + s_{jn} = y_{jn} + r_{jn} \leq y_{jn} + a_n$$

for all i, j, n.

Next, we construct elements $b_1, b_2, \ldots$ in G^+ such that $f(b_n) < 1/2^{n+1}$ for all n, while also $x_{in} \leq y_{jn} + b_n$ for all i, j, n.

Fix n for a while. Since each $y_j - x_i$ lies in $\overline{G}^+$, we must have

$$|\phi(t_{ij}) - (y_j - x_i)|_{\overline{f}} < 1/2^{n+4}$$

for some $t_{ij} \in G^+$. Then

$$|t_{ij} - y_{jn} + x_{in}|_f = |\phi(t_{ij}) - \phi(y_{jn}) + \phi(x_{in})|_{\overline{f}}$$
$$\leq |\phi(t_{ij}) - y_j + x_i|_{\overline{f}} + |y_j - \phi(y_{jn})|_{\overline{f}} + |\phi(x_{in}) - x_i|_{\overline{f}}$$
$$< 1/2^{n+4} + 1/2^{n+5} + 1/2^{n+5} = 1/2^{n+3},$$

and consequently

$$t_{ij} - y_{jn} + x_{in} = u_{ij} - v_{ij}$$

for some $u_{ij}, v_{ij} \in G^+$ satisfying $f(u_{ij}+v_{ij}) < 1/2^{n+3}$. Set

$$b_n = u_{11} + u_{12} + u_{21} + u_{22},$$

and observe that $f(b_n) \leq \sum f(u_{ij}+v_{ij}) < 4/2^{n+3} = 1/2^{n+1}$. In addition,

$$x_{in} \leq x_{in} + t_{ij} = y_{jn} + u_{ij} - v_{ij} \leq y_{jn} + u_{ij} \leq y_{jn} + b_n$$

for all i, j.

Finally, we construct elements $z_1, z_2, \ldots$ in G such that

$$x_{in} \leq z_n \leq y_{jn} + b_n$$

for all i, j, n, while also $|z_{n+1} - z_n|_f < 1/2^n$ for all n.

As $x_{i1} \leq y_{j1} + b_1$ for all i, j, interpolation in G immediately provides us with an appropriate element z_1. Now suppose that $z_1, \ldots, z_n$ have been constructed, for some n. Then

$$x_{i,n+1} \leq y_{j,n+1} + b_{n+1}, \qquad x_{i,n+1} \leq x_{in} + a_n \leq z_n + a_n,$$
$$z_n - b_n - a_n \leq y_{jn} - a_n \leq y_{j,n+1} \leq y_{j,n+1} + b_{n+1}$$

for all i, j. Hence, there exists $z_{n+1} \in G$ such that

$$\begin{matrix} x_{1,n+1} & & & & y_{1,n+1} + b_{n+1} \\ x_{2,n+1} & \leq & z_{n+1} & \leq & y_{2,n+1} + b_{n+1} \\ z_n - b_n - a_n & & & & z_n + a_n. \end{matrix}$$

Since $-(a_n+b_n) \leq z_{n+1} - z_n \leq a_n \leq a_n + b_n$, we conclude using Lemma 12.1(e) that

$$|z_{n+1} - z_n|_f \leq |a_n + b_n|_f = f(a_n) + f(b_n) < 1/2^{n+2} + 1/2^{n+1} < 1/2^n.$$

This completes the induction step.

The z_n form a Cauchy sequence in G, and hence there exists $z \in \overline{G}$ such that $\phi(z_n) \to z$. As

$$|\phi(b_n)|_{\overline{f}} = |b_n|_f = f(b_n) < 1/2^{n+1}$$

for all n, we also have $\phi(b_n) \to 0$. Since

$$\phi(x_{in}) \leq \phi(z_n) \leq \phi(y_{jn}) + \phi(b_n)$$

for all i, j, n, we conclude that $x_i \leq z \leq y_j$ for all i, j.

Therefore $\overline{G}$ has interpolation, as desired. □

• **Completions with Respect to Extremal States.** We now consider completions with respect to states on an interpolation group (G, u) with order-unit, aiming to prove that a state s on (G, u) is extremal if and only if the s-completion of G is totally ordered. For this purpose, we need to be able to extend to the s-completion any states s_1, s_2 on (G, u) appearing in a positive convex combination $s = \alpha_1 s_1 + \alpha_2 s_2$. Suitable extensions exist because of the following kind of continuity.

LEMMA 12.8. *Let G be a directed abelian group, and let $f, g\colon G \to \mathbb{R}$ be positive homomorphisms. Then the following conditions are equivalent:*

(a) *The map $g\colon (G, |\cdot|_f) \to \mathbb{R}$ is (uniformly) continuous.*

(b) *The set-theoretic identity map $(G, |\cdot|_f) \to (G, |\cdot|_g)$ is (uniformly) continuous.*

(c) *Given any positive real number ε, there exists a positive real number δ such that all $x \in G^+$ satisfying $f(x) < \delta$ also satisfy $g(x) < \varepsilon$.*

PROOF. (a)$\Rightarrow$(c): Just assume that g is continuous at 0. Then, given any positive real number ε, there exists a positive real number δ such that all $y \in G$ satisfying $|y|_f < \delta$ also satisfy $|g(y)| < \varepsilon$. Since $|x|_f = f(x)$ for any $x \in G^+$, condition (c) follows.

(c)$\Rightarrow$(a): Given a positive real number ε, there exists a positive real number δ as in (c). For any $x, y \in G$ with $|x-y|_f < \delta$, we have $x - y = a - b$ for some $a, b \in G^+$ such that $f(a+b) < \delta$, whence $g(a+b) < \varepsilon$, and so

$$|g(x) - g(y)| = |g(a) - g(b)| \leq |g(a)| + |g(b)| = g(a+b) < \varepsilon.$$

Thus g is uniformly continuous with respect to $|\cdot|_f$.

(b)$\Rightarrow$(c): Just assume continuity at 0. Then, given any positive real number ε, there exists a positive real number δ such that all $y \in G$ satisfying $|y|_f < \delta$ also satisfy $|y|_g < \varepsilon$. Condition (c) follows.

(c)$\Rightarrow$(b): Given a positive real number ε, there exists a positive real number δ as in (c). For any $x, y \in G$ with $|x-y|_f < \delta$, we have $x - y = a - b$ for some $a, b \in G^+$ such that $f(a+b) < \delta$, whence $g(a+b) < \varepsilon$, and so $|x-y|_g < \varepsilon$. Thus the identity map from $(G, |\cdot|_f)$ to $(G, |\cdot|_g)$ is uniformly continuous. $\square$

DEFINITION. Let G be a directed abelian group, and let $f, g\colon G \to \mathbb{R}$ be positive homomorphisms. Then g is *absolutely continuous with respect to* f if and only if the equivalent conditions of Lemma 12.8 hold.

For instance, if $g \leq^+ \alpha f$ for some positive real number α, then g is absolutely continuous with respect to f.

LEMMA 12.9. *Let G be a directed abelian group, and let $f, g\colon G \to \mathbb{R}$ be positive homomorphisms. Let $\overline{G}$ denote the f-completion of G, and let $\phi\colon G \to \overline{G}$ be the natural map. If g is absolutely continuous with respect to f, then g extends uniquely to a continuous map $\overline{g}\colon \overline{G} \to \mathbb{R}$ such that $\overline{g}\phi = g$, and $\overline{g}$ is a positive homomorphism.*

PROOF. The existence and uniqueness of $\overline{g}$ follows from the uniform continuity of g with respect to $|\cdot|_f$. Since $\overline{g}$ is additive on $\phi(G)$, it follows from the continuity of addition in $\overline{G}$ that $\overline{g}$ is a homomorphism. In addition, $\overline{g}\phi(G^+) \subseteq \mathbb{R}^+$ and $\phi(G^+)$ is dense in $\overline{G}^+$, whence $\overline{g}(\overline{G}^+) \subseteq \mathbb{R}^+$. Thus $\overline{g}$ is a positive homomorphism. $\square$

DEFINITION. In the situation of Lemma 12.9, the map $\overline{g}$ is called *the continuous extension of g to $\overline{G}$.*

THEOREM 12.10. *Let (G,u) be an interpolation group with order-unit, let s be a state on (G,u), and let $\overline{G}$ denote the s-completion of G. Let $\phi\colon G \to \overline{G}$ be the natural map, and let G_0 be the convex subgroup of $\overline{G}$ generated by $\phi(u)$. Then the following conditions are equivalent*:

(a) *$\overline{G}$ is totally ordered.*

(b) *G_0 is totally ordered.*

(c) *s is an extremal state.*

PROOF. By Theorem 12.7, $\overline{G}$ is Dedekind complete, from which it follows that G_0 is Dedekind complete. Note that $\phi(u)$ is an order-unit in G_0 and that $\phi(G) \subseteq G_0$. Let $\overline{s}$ be the natural extension of s to $\overline{G}$, and let s_0 be the restriction of $\overline{s}$ to G_0. Then s_0 is a state on $(G_0, \phi(u))$.

(a)$\Rightarrow$(b): This is automatic.

(b)$\Rightarrow$(a): Since $\phi(G) \subseteq G_0$, we see that G_0 is dense in $\overline{G}$. Hence, any element $x \in \overline{G}$ may be expressed as the limit of a sequence $\{x_n\} \subseteq G_0$. For each n, we have either $x_n \geq 0$ or $x_n \leq 0$. After passing to a subsequence, we may assume that either all $x_n \geq 0$ or all $x_n \leq 0$. Consequently, either $x \geq 0$ or $x \leq 0$, because of Lemma 12.3. Thus $\overline{G}$ is totally ordered.

(b)$\Rightarrow$(c): In view of Corollary 4.17, s_0 is the only state on $(G_0, \phi(u))$. Consider any positive convex combination $s = \alpha t + (1-\alpha)t'$ in $S(G,u)$. Then $t \leq^+ \alpha^{-1}s$, from which we see that t is absolutely continuous with respect to s. Let $\overline{t}$ be the continuous extension of t to $\overline{G}$, and let t_0 be the restriction of $\overline{t}$ to G_0. Then t_0 is a state on $(G_0, \phi(u))$, whence $t_0 = s_0$, and hence $t = t_0\phi = s_0\phi = s$. Similarly, $t' = s$. Thus s is extremal.

(c)$\Rightarrow$(b): We first show that s_0 is an extremal state of $(G_0, \phi(u))$. Thus consider any positive convex combination $s_0 = \alpha t_0 + (1-\alpha)t_0'$ in $S(G_0, \phi(u))$. Then we obtain a positive convex combination

$$s = s_0\phi = \alpha(t_0\phi) + (1-\alpha)(t_0'\phi)$$

in $S(G,u)$, whence $t_0\phi = t_0'\phi = s = s_0\phi$. Since $t_0 \leq^+ \alpha^{-1}s_0$, we see that t_0 is absolutely continuous with respect to s_0. Now t_0 and s_0 are continuous maps from G_0 to $\mathbb{R}$ that agree on the dense subgroup $\phi(G)$, whence $t_0 = s_0$. Similarly, $t_0' = s_0$. Therefore s_0 is extremal, as claimed.

Consequently, Theorem 9.9 and Proposition 8.11 show that $\ker(s_0)$ is an ideal of G_0, and that $G_0/\ker(s_0)$ is totally ordered.

Because the $\overline{s}$-metric on $\overline{G}$ is actually a metric, any nonzero element x in G_0^+ satisfies

$$s_0(x) = \overline{s}(x) = |x|_{\overline{s}} \neq 0.$$

Hence, $\ker(s_0)^+ = \{0\}$. As $\ker(s_0)$ is an ideal of G_0, it is directed, and so we conclude that $\ker(s_0) = \{0\}$. Therefore G_0 is totally ordered. $\square$

COROLLARY 12.11. *Let (G,u) be an interpolation group with order-unit, let s be an extremal state on (G,u), and let $\overline{G}$ denote the s-completion of G. Let $\phi\colon G \to \overline{G}$ be the natural map, and let $\overline{s}$ be the natural extension of s to $\overline{G}$. Then*

$\phi(u)$ is an order-unit in $\overline{G}$, and $\overline{s}$ provides an isomorphism of $\overline{G}$ onto $\overline{s}(\overline{G})$ (as ordered groups). If s is discrete, then $\overline{s}(\overline{G}) = s(G)$, while if s is indiscrete, then $\overline{s}(\overline{G}) = \mathbb{R}$.

PROOF. By Theorems 12.7 and 12.10, $\overline{G}$ is Dedekind complete and totally ordered.

Given any $x \in \ker(\overline{s})$, we have either $x \geq 0$ or $-x \geq 0$, whence either

$$|x|_{\overline{s}} = \overline{s}(x) = 0 \quad \text{or} \quad |x|_{\overline{s}} = |-x|_{\overline{s}} = \overline{s}(-x) = 0.$$

Since the $\overline{s}$-metric on $\overline{G}$ is a metric, $x = 0$. Thus $\overline{s}$ is injective.

Now $\overline{s}$ gives a group isomorphism of $\overline{G}$ onto $\overline{s}(\overline{G})$. If $x, y \in \overline{G}$ with $x \leq y$, then of course $\overline{s}(x) \leq \overline{s}(y)$, because $\overline{s}$ is positive. If $x \not\leq y$, then $x > y$, whence $\overline{s}(x) > \overline{s}(y)$ (because $\overline{s}$ is positive and injective), and so $\overline{s}(x) \not\leq \overline{s}(y)$. Thus $\overline{s}$ also gives an order-isomorphism of $\overline{G}$ onto $\overline{s}(\overline{G})$.

As $\overline{s}\phi(u) = 1$, we see that $\overline{s}\phi(u)$ is an order-unit in $\overline{s}(\overline{G})$. Hence, $\phi(u)$ must be an order-unit in $\overline{G}$.

If s is discrete, then $s(G) = (1/n)\mathbb{Z}$ for some $n \in \mathbb{N}$, in which case the only possible values for $|\cdot|_s$ are $0, \frac{1}{n}, \frac{2}{n}, \ldots$. Consequently, the s-metric on G is discrete, and $\overline{G} = \phi(G)$. Thus $\overline{s}(\overline{G}) = \overline{s}\phi(G) = s(G)$ in this case.

If s is indiscrete, then $s(G)$ is dense in $\mathbb{R}$. As $s(G) \subseteq \overline{s}(\overline{G})$, it follows that $\overline{s}(\overline{G})$ is dense in $\mathbb{R}$. Since $\overline{s}(\overline{G})$ is isomorphic to $\overline{G}$ (as ordered groups), $\overline{s}(\overline{G})$ is Dedekind complete. Therefore $\overline{s}(\overline{G})$ must equal $\mathbb{R}$. □

• **Criterion for Extremal States.** As an application of Theorem 12.10, we derive a criterion for extremality of states on interpolation groups. This criterion may be regarded as an approximate version of the criterion for extremality in Choquet simplices proved in Theorem 11.17. (Recall that any Choquet simplex K may be identified with the state space of the interpolation group $(\mathrm{Aff}(K), 1)$.) The sufficiency of the criterion does not require interpolation, as follows.

LEMMA 12.12. *Let (G, u) be a partially ordered abelian group with order-unit, and let s be a state on (G, u). Assume that*

$$\min\{s(x), s(y)\} = \sup\{s(z) \mid z \in G^+;\ z \leq x;\ z \leq y\}$$

for all $x, y \in G^+$. Then s is an extremal state.

PROOF. If not, there exists a positive convex combination $s = \alpha_1 s_1 + \alpha_2 s_2$ where s_1 and s_2 are distinct states on (G, u). As G is directed, there must be an element $a \in G^+$ for which $s_1(a) \neq s_2(a)$. After renumbering s_1 and s_2, if necessary, we may assume that $0 \leq s_1(a) < s_2(a)$. Choose $m, n \in \mathbb{N}$ such that $s_1(a) < m/n < s_2(a)$, and set $x = na$ and $y = mu$. Then x and y are elements of G^+ for which $s_1(x) < s_1(y)$ and $s_2(x) > s_2(y)$.

For any $z \in G^+$ with $z \leq x$ and $z \leq y$, we have

$$s(z) = \alpha_1 s_1(z) + \alpha_2 s_2(z) \leq \alpha_1 s_1(x) + \alpha_2 s_2(y).$$

As a result,

$$\min\{s(x), s(y)\} = \sup\{s(z) \mid z \in G^+;\ z \le x;\ z \le y\} \le \alpha_1 s_1(x) + \alpha_2 s_2(y).$$

Since $\alpha_1 > 0$ and $s_1(x) < s_1(y)$, we have

$$\alpha_1 s_1(x) + \alpha_2 s_2(y) < \alpha_1 s_1(y) + \alpha_2 s_2(y) = s(y).$$

Similarly, $\alpha_1 s_1(x) + \alpha_2 s_2(y) < s(x)$, which is a contradiction.

Therefore s must be extremal. □

In general, extremal states need not satisfy the condition given in Lemma 12.12, as the following example shows.

EXAMPLE 12.13. *There exist a partially ordered real vector space (G, u) with order-unit and an extremal state s on (G, u) such that*

$$\min\{s(x), s(y)\} > \sup\{s(z) \mid z \in G^+;\ z \le x;\ z \le y\}$$

for some $x, y \in G^+$.

PROOF. Let K denote the unit square in $\mathbb{R}^2$, that is,

$$K = \{(\alpha, \beta) \in \mathbb{R}^2 \mid 0 \le \alpha \le 1 \text{ and } 0 \le \beta \le 1\}.$$

The compact convex set K has exactly four extreme points, which we label as follows:

$$e_1 = (0,0), \quad e_2 = (1,1), \quad e_3 = (1,0), \quad e_4 = (0,1).$$

Note that $\frac{1}{2}e_1 + \frac{1}{2}e_2 = \frac{1}{2}e_3 + \frac{1}{2}e_4$. Now set $G = \mathrm{Aff}(K)$ and let u be the constant function 1 in G. If s is the state on (G, u) defined by evaluation at e_2, then in view of Theorem 7.1, we see that s is extremal.

Let $x, y \colon K \to \mathbb{R}$ be the coordinate projections, so that $x(\alpha, \beta) = \alpha$ and $y(\alpha, \beta) = \beta$ for all $(\alpha, \beta) \in K$. Then $x, y \in G^+$ and $s(x) = s(y) = 1$. Given any $z \in G^+$ for which $z \le x$ and $z \le y$, we have

$$0 \le z(e_1) \le x(e_1) = 0, \quad 0 \le z(e_3) \le y(e_3) = 0, \quad 0 \le z(e_4) \le x(e_4) = 0,$$

and so $z(e_1) = z(e_3) = z(e_4) = 0$. Since

$$\tfrac{1}{2}z(e_1) + \tfrac{1}{2}z(e_2) = \tfrac{1}{2}z(e_3) + \tfrac{1}{2}z(e_4),$$

it follows that $z(e_2) = 0$, that is, $s(z) = 0$. Therefore

$$\min\{s(x), s(y)\} = 1 > 0 = \sup\{s(z) \mid z \in G^+;\ z \le x;\ z \le y\}. \quad \square$$

An alternate example may be obtained by using the partially ordered real vector space G constructed in Example 3.2, with the order-unit $u = (1,1,1)$. There are states s and t on (G, u) such that

$$s(a, b, c) = (a + b)/2 \quad \text{and} \quad t(a, b, c) = c$$

for all $(a, b, c) \in G$, and it may be checked that $S(G, u)$ equals the convex hull of $\{s, t\}$. Hence, s and t are extremal states on (G, u). Although t satisfies the condition given in Lemma 12.12, s does not.

THEOREM 12.14. *Let (G,u) be an interpolation group with order-unit, and let s be a state on (G,u). Then s is extremal if and only if*

$$\min\{s(x), s(y)\} = \sup\{s(z) \mid z \in G^+;\ z \le x;\ z \le y\}$$

for all $x, y \in G^+$.

PROOF. Sufficiency is given in Lemma 12.12. Conversely, assume that s is extremal. Let $x, y \in G^+$, and let ε be a positive real number. Either $s(x) \le s(y)$ or $s(y) \le s(x)$; say $s(x) \le s(y)$.

Let $\overline{G}$ denote the s-completion of G, let $\phi\colon G \to \overline{G}$ be the natural map, and let $\overline{s}$ be the natural extension of s to $\overline{G}$. As s is extremal, $\overline{s}$ provides an order-isomorphism of $\overline{G}$ onto a subgroup of $\mathbb{R}$, by Corollary 12.11. Since

$$\overline{s}\phi(x) = s(x) \le s(y) = \overline{s}\phi(y),$$

we obtain $\phi(x) \le \phi(y)$, so that $\phi(y - x)$ lies in $\overline{G}^+$. Consequently, there is some $w \in G^+$ for which

$$|\phi(w) - \phi(y - x)|_{\overline{s}} < \varepsilon.$$

Then $|w - y + x|_s < \varepsilon$, and hence $w - y + x = a - b$ for some $a, b \in G^+$ satisfying $s(a + b) < \varepsilon$. Note that

$$x \le w + x = y + a - b \le y + a.$$

By Riesz decomposition, $x = z + c$ for some $z, c \in G^+$ such that $z \le y$ and $c \le a$. Now $z \le x$ as well, and

$$s(c) \le s(a) \le s(a + b) < \varepsilon,$$

whence $s(z) = s(x) - s(c) > s(x) - \varepsilon$. Therefore $\min\{s(x), s(y)\}$ equals the required supremum. □

For a dimension group (G,u) with order-unit, the extremal state criterion of Theorem 12.14 may be derived from the extreme point criterion for Choquet simplices (Theorem 11.17), without the use of s-completions, as follows. Let s be an extremal state on (G,u), let $x, y \in G^+$, and let ε be a positive real number. If ϕ denotes the natural map from G to $\mathrm{Aff}(S(G,u))$, then by Theorem 11.17 there exists a function h in $\mathrm{Aff}(S(G,u))$ such that $h \le \phi(x)$ and $h \le \phi(y)$, while also $h(s) = \min\{s(x), s(y)\}$. Using Theorem 7.9, we obtain $v \in G$ and $n \in \mathbb{N}$ such that

$$h - \varepsilon \ll \phi(v)/2^n \ll h.$$

Then $\phi(v) \ll \phi(2^n x)$ and $\phi(v) \ll \phi(2^n y)$, whence $v \le 2^n x$ and $v \le 2^n y$, by Theorem 7.8. According to Corollary 2.22, there exists $w \in G$ such that $w \le x$ and $w \le y$, while also $v \le 2^n w$. Hence, $\phi(w) \ge \phi(v)/2^n \gg h - \varepsilon$, and so

$$s(w) > h(s) - \varepsilon = \min\{s(x), s(y)\} - \varepsilon.$$

Finally, interpolation provides an element $z \in G$ such that

$$\begin{matrix} w \\ 0 \end{matrix} \le z \le \begin{matrix} x \\ y. \end{matrix}$$

Then $z \in G^+$ and $s(z) \ge s(w) > \min\{s(x), s(y)\} - \varepsilon$.

One application of Theorem 12.14 is to derive the following simple description of a neighborhood base for an extreme point in the state space of an interpolation group.

COROLLARY 12.15. *Let (G, u) be an interpolation group with order-unit, let t be an extremal state on (G, u), and let V be a neighborhood of t in $S(G, u)$. Then there exist $x \in G^+$ and $k \in \mathbb{N}$ such that*

$$t \in \{s \in S(G,u) \mid s(x) < k\} \subseteq V.$$

PROOF. In view of Proposition 6.1, t has a neighborhood of the form

$$V_1 = \{s \in S(G,u) \mid s(y_i) < k_i \text{ for each } i = 1, \dots, n\},$$

where $y_1, \dots, y_n \in G^+$ and $k_1, \dots, k_n \in \mathbb{N}$, such that $V_1 \subseteq V$. Let k be the maximum of $k_1, \dots, k_n$, and set $x_i = y_i + (k - k_i)u$ for each i. Then $x_1, \dots, x_n \in G^+$ and

$$V_1 = \{s \in S(G,u) \mid s(x_i) < k \text{ for each } i = 1, \dots, n\}.$$

Choose $m \in \mathbb{N}$ such that each $x_i \leq mu$. Then each $mu - x_i$ is an element of G^+ such that $t(mu - x_i) > m - k$. As t is extremal, Theorem 12.14 says that there exists $z \in G^+$ such that $t(z) > m - k$ and $z \leq mu - x_i$ for each i. Since $z \leq mu - x_1 \leq mu$, the element $x = mu - z$ lies in G^+. Note that $t(x) < k$ and that each $x_i \leq x$. Now the set

$$V_2 = \{s \in S(G,u) \mid s(x) < k\}$$

is a neighborhood of t. Given any $s \in V_2$, we have $s(x_i) \leq s(x) < k$ for all $i = 1, \dots, n$, whence $s \in V_1$. Therefore $t \in V_2 \subseteq V_1 \subseteq V$. □

Theorem 12.14 may be extended to compact sets of extremal states as follows.

COROLLARY 12.16. *Let (G, u) be an interpolation group with order-unit, and let X be a compact subset of $\partial_e S(G, u)$. Let $x, y \in G^+$, and let ε be a positive real number. Then there exists $z \in G^+$ such that $z \leq x$ and $z \leq y$, while also*

$$s(z) > \min\{s(x), s(y)\} - \varepsilon$$

for all $s \in X$.

PROOF. Since G has interpolation, the set

$$W = \{w \in G^+ \mid w \leq x \text{ and } w \leq y\}$$

is upward directed. For each $w \in W$, the set

$$V(w) = \{s \in S(G,u) \mid s(w) > \min\{s(x), s(y)\} - \varepsilon\}$$

is an open subset of $S(G, u)$, and Theorem 12.14 shows that these open sets cover $\partial_e S(G, u)$. As X is compact, we obtain

$$X \subseteq V(w_1) \cup \cdots \cup V(w_k)$$

for some $w_1, \dots, w_k$ in W. There exists $z \in W$ such that $w_i \leq z$ for all $i = 1, \dots, k$, and z has the desired properties. □

COROLLARY 12.17. *Let (G,u) be an interpolation group with order-unit and let X be a compact subset of $\partial_e S(G,u)$. Let $x,y \in G^+$ and $m \in \mathbb{N}$ such that $s(mx) \le s(y)$ for all $s \in X$, and let ε be a positive real number. Then there exists $z \in G^+$ such that $z \le x$ and $mz \le y$, while also $s(z) > s(x) - \varepsilon$ for all $x \in X$.*

PROOF. According to Corollary 12.16, there is some $w \in G^+$ for which $w \le mx$ and $w \le y$, while also $s(w) > s(mx) - \varepsilon$ for all $s \in X$. Then $w = w_1 + \cdots + w_m$ for some $w_i \in G^+$ such that each $w_i \le x$. Each

$$w - w_i = \sum_{j \ne i} w_j \le (m-1)x,$$

and so $w - (m-1)x \le w_i$. We also have $0 \le w_i$ for each i. Then, by interpolation, there exists $z \in G$ such that

$$\begin{matrix} w-(m-1)x \\ 0 \end{matrix} \quad \le \quad z \quad \le \quad w_i$$

for all i. Now $z \in G^+$, while also $z \le w_1 \le x$ and

$$mz \le w_1 + \cdots + w_m = w \le y.$$

Since $w - (m-1)x \le z$, we obtain $x - z \le mx - w$. For any $s \in X$, it follows that

$$s(x) - s(z) \le s(mx) - s(w) < \varepsilon,$$

and therefore $s(z) > s(x) - \varepsilon$. □

In a lattice-ordered abelian group with order-unit, we may sharpen the extremal state criterion as follows.

THEOREM 12.18. *Let (G,u) be a lattice-ordered abelian group with order-unit, and let s be a state on (G,u). Then the following conditions are equivalent:*

(a) *s is extremal.*

(b) *s is a lattice homomorphism.*

(c) *$s(x \wedge y) = \min\{s(x), s(y)\}$ for all $x, y \in G^+$.*

PROOF. (a)⇔(c): This is clear from Theorem 12.14.

(b)⇒(c): This is automatic.

(c)⇒(b): Given $a, b \in G$, choose $c \in G^+$ such that

$$-c \quad \le \quad \begin{matrix} a \\ b \end{matrix} \quad \le \quad c,$$

so that the elements $a+c$, $b+c$, $c-a$, $c-b$ all lie in G^+. Using Proposition 1.4, we infer from (c) that

$$\begin{aligned} s(a \wedge b) &= s([(a+c) \wedge (b+c)] - c) \\ &= \min\{s(a+c), s(b+c)\} - s(c) = \min\{s(a), s(b)\}, \end{aligned}$$

and, similarly,

$$\begin{aligned} s(a \vee b) &= s(c - [(c-a) \wedge (c-b)]) \\ &= s(c) - \min\{s(c-a), s(c-b)\} = \max\{s(a), s(b)\}. \end{aligned}$$

Therefore s is a lattice homomorphism. □

COROLLARY 12.19. *If (G,u) is a lattice-ordered abelian group with order-unit, then $\partial_e S(G,u)$ is compact.*

PROOF. In view of Theorem 12.18, we see that $\partial_e S(G,u)$ is closed in $S(G,u)$ and hence is compact. $\square$

COROLLARY 12.20. *If (G,u) is a lattice-ordered abelian group with order-unit and X is a subset of $\partial_e S(G,u)$, then* $\ker(X)$ *is an ideal of G.*

PROOF. Obviously $\ker(X)$ is a convex subgroup of G. Given any x, y in $\ker(X)$, Theorem 12.18 shows that

$$s(x \vee y) = \max\{s(x), s(y)\} = 0$$

for all $s \in X$. Hence, $x \vee y \in \ker(X)$, proving that $\ker(X)$ is directed. Therefore $\ker(X)$ is an ideal of G. $\square$

COROLLARY 12.21. *If (G,u) is a lattice-ordered abelian group with order-unit and F is a closed face of $S(G,u)$, then* $\ker(F)$ *is an ideal of G.*

PROOF. In view of Corollary 5.20, we see that $\ker(F) = \ker(\partial_e F)$. As F is a face of $S(G,u)$, the set $\partial_e F$ is contained in $\partial_e S(G,u)$. Hence, $\ker(\partial_e F)$ is an ideal of G by Corollary 12.20. $\square$

• **Closed Faces.** For use in state spaces with noncompact extreme boundary, we derive versions of Corollaries 12.16 and 12.17 for closed faces of state spaces of interpolation groups. Since we must then work with non-extremal states, we first develop a further property of states on completions.

PROPOSITION 12.22. *Let (G,u) be an interpolation group with order-unit, let s be a state on (G,u), and let $\overline{G}$ denote the s-completion of G. Let $\phi\colon G \to \overline{G}$ be the natural map, and let G_0 denote the convex subgroup of $\overline{G}$ generated by $\phi(u)$. Then the set*

$$K = \{t\phi \mid t \in S(G_0, \phi(u))\}$$

equals the closure of the face generated by s in $S(G,u)$.

PROOF. Since ϕ induces an affine continuous map from $S(G_0, \phi(u))$ to $S(G,u)$ whose image is K, we see that K is a compact convex subset of $S(G,u)$.

Let F be the face generated by s in $S(G,u)$. Given any $s' \in F$, Proposition 6.15 shows that $s' \leq^+ \alpha s$ for some positive real number α, from which we see that s' is absolutely continuous with respect to s. Hence, the continuous extension of s' to $\overline{G}$ restricts to a state t' in $S(G_0, \phi(u))$ such that $t'\phi = s'$, so that $s' \in K$. Thus $F \subseteq K$, and so $\overline{F} \subseteq K$.

Let $\overline{s}$ be the natural extension of s to $\overline{G}$, and let s_0 be the restriction of $\overline{s}$ to G_0, so that s_0 is a state on $(G_0, \phi(u))$. As the $\overline{s}$-metric on $\overline{G}$ is a metric, we see that

$$s_0(x) = \overline{s}(x) = |x|_{\overline{s}} \neq 0$$

for all nonzero $x \in G_0^+$, whence $\ker(s_0)^+ = \{0\}$. Because of Theorem 12.7, $\overline{G}$ is Dedekind complete, whence G_0 is Dedekind complete. Then $(G_0, \phi(u))$ satisfies general comparability (Theorem 9.9), and hence Corollary 8.17 shows that $S(G_0, \phi(u))$ equals the closure of the face generated by s_0.

Now given any $t \in S(G_0, \phi(u))$, there exists a net $\{t_i\}$, in the face generated by s_0 in $S(G_0, \phi(u))$, which converges to t. According to Proposition 6.15, each $t_i \leq^+ \alpha_i s_0$ for some positive real number α_i. Then each $t_i\phi \leq^+ \alpha_i(s_0\phi) = \alpha_i s$, and so each $t_i\phi \in F$, by that same proposition. Observing that $t_i\phi \to t\phi$, we conclude that $t\phi \in \overline{F}$, which proves that $K \subseteq \overline{F}$.

Therefore $K = \overline{F}$. □

THEOREM 12.23. *Let (G, u) be an interpolation group with order-unit, and let F be a closed face of $S(G, u)$. Let $x, y \in G^+$ such that $s(x) \leq s(y)$ for all $s \in F$, and let ε be a positive real number. Then there exists $z \in G^+$ such that $z \leq x$ and $z \leq y$, while also $s(z) > s(x) - \varepsilon$ for all $s \in F$.*

PROOF. Since G has interpolation, the set

$$W = \{w \in G^+ \mid w \leq x \text{ and } w \leq y\}$$

is upward directed. For each $w \in W$, the set

$$V(w) = \{s \in S(G, u) \mid s(w) > s(x) - \varepsilon\}$$

is an open subset of $S(G, u)$, and we claim that these open sets cover F.

Now consider any $s \in F$. Let $\overline{G}$ denote the s-completion of G, let $\overline{s}$ denote the natural extension of s to $\overline{G}$, and let $\phi \colon G \to \overline{G}$ be the natural map. Recall from Theorem 12.7 that $\overline{G}$ is a Dedekind complete lattice-ordered abelian group, whence $\overline{G}$ is archimedean, by Proposition 9.7. Let G_0 be the convex subgroup of $\overline{G}$ generated by $\phi(u)$. Then G_0 is archimedean.

As F is a closed face of $S(G, u)$, it contains the closure of the face generated by s. Consequently, Proposition 12.22 shows that $t\phi \in F$ for any t in $S(G_0, \phi(u))$, and hence $t\phi(x) \leq t\phi(y)$ for all t in $S(G_0, \phi(u))$. Since G_0 is archimedean, it follows that $\phi(x) \leq \phi(y)$, by Theorem 4.14.

Thus $\phi(y - x) \in \overline{G}^+$, and so there is some $v \in G^+$ for which

$$|\phi(v) - \phi(y - x)|_{\overline{s}} < \varepsilon.$$

Then $|v - y + x|_s < \varepsilon$, and hence $v - y + x = a - b$ for some $a, b \in G^+$ satisfying $s(a + b) < \varepsilon$. Note that

$$x \leq v + x = y + a - b \leq y + a.$$

By Riesz decomposition, $x = w + c$ for some $w, c \in G^+$ such that $w \leq y$ and $c \leq a$. Note that $w \leq x$, so that $w \in W$. In addition,

$$s(c) \leq s(a) \leq s(a + b) < \varepsilon,$$

whence $s(w) = s(x) - s(c) > s(x) - \varepsilon$, and so $s \in V(w)$.

Therefore the $V(w)$ cover F, as claimed. Since F is compact,

$$F \subseteq V(w_1) \cup \cdots \cup V(w_n)$$

for some $w_i \in W$. As W is upward directed, there exists $z \in W$ such that all $w_i \leq z$, and z has the desired properties. □

COROLLARY 12.24. *Let (G, u) be an interpolation group with order-unit, and let F be a closed face of $S(G, u)$. Let $x, y \in G^+$ and $m \in \mathbb{N}$ such that $s(mx) \leq s(y)$ for all $s \in F$, and let ε be a positive real number. Then there exists $z \in G^+$ such that $z \leq x$ and $mz \leq y$, while also $s(z) > s(x) - \varepsilon$ for all $s \in F$.*

PROOF. This follows from Theorem 12.23 by the same argument used in deriving Corollary 12.17 from Corollary 12.16. □

• **Notes.** The results of this chapter are due to the author and Handelman [**57**]. In particular, Theorems 12.7, 12.10, and 12.14 are [**57**, Theorems 1.6, 2.3, 3.1], Theorem 12.18 and Corollary 12.19 are [**57**, Corollaries 3.2, 3.3], and Theorem 12.23 is [**57**, Theorem 3.4].

CHAPTER 13

Affine Continuous Functions on State Spaces

This chapter represents the culmination of our program for approximating affine continuous real-valued functions on a state space $S(G,u)$, where (G,u) is an interpolation group with order-unit, by evaluations at elements of G. A function p in $\mathrm{Aff}(S(G,u))$ is eligible for such approximation if $p(s)$ lies in $s(G)$ for every discrete extremal state s in $S(G,u)$. Provided G is 2-unperforated, p can be approximated arbitrarily closely by functions from $\phi(G)$, where ϕ is the natural map from G to $\mathrm{Aff}(S(G,u))$. Moreover, if G is unperforated and $p \geq 0$, then p can be approximated arbitrarily closely by functions from $\phi(G^+)$. Corresponding approximations are obtained for suitable continuous real-valued functions on compact subsets of $\partial_e S(G,u)$ and for suitable affine continuous real-valued functions on closed faces of $S(G,u)$.

• **Approximations.** Our main goal in this section is the approximation of affine continuous real-valued functions on the state space of an interpolation group (G,u) with order-unit by evaluations at elements of G. If there exist discrete extremal states s on (G,u), then a function r in $\mathrm{Aff}(S(G,u))$ which is to be approximated arbitrarily closely by evaluations must necessarily satisfy $r(s) \in s(G)$ for any such s. We prove that as long as G is 2-unperforated, no other restrictions on r are necessary.

Given a positive real number ε, we wish to find an element $x \in G$ such that evaluation at x lies between $r - \varepsilon$ and $r + \varepsilon$. Having specified ε, the hypothesis on r may be relaxed to the requirement that for any discrete extremal state s on (G,u), some number in the image $s(G)$ lies in the open interval $(r(s)-\varepsilon, r(s)+\varepsilon)$. Thus the problem may now be stated entirely in terms of the functions $r - \varepsilon$ and $r + \varepsilon$. Namely, if $p = r - \varepsilon$ and $q = r + \varepsilon$, then p and q are functions in $\mathrm{Aff}(S(G,u))$ such that $p \ll q$ and $s(G)$ intersects the open interval $(p(s), q(s))$ for each discrete extremal state s on (G,u), and we want evaluation at some element of G to lie between p and q.

As $S(G,u)$ is a Choquet simplex (Theorem 10.17), we already have a somewhat parallel result allowing us to fit an affine continuous function on $S(G,u)$ between an upper semicontinuous convex function and a lower semicontinuous concave function (Theorem 11.12). Recalling the applications of Theorems 11.12 and 11.13 to compact sets of extreme points and to closed faces (e.g., Theorems 11.18 and 11.26), it would appear advantageous for potential applications to compact sets of extremal states and to closed faces of state spaces that we extend our problem by allowing p and q to be semicontinuous convex and concave functions. Thus we shall assume only that p is an upper semicontinuous convex function on $S(G,u)$ and that q is a lower semicontinuous concave function on $S(G,u)$. Of course we shall first work toward reducing the problem to the case that p and q are affine and continuous.

LEMMA 13.1. *Let K be a Choquet simplex, let $q\colon K \to \mathbb{R} \cup \{\infty\}$ be a lower semicontinuous concave function, and set*

$$R = \{r \in \text{Aff}(K) \mid r \ll q\}.$$

Then R is upward directed with respect to $\ll$, and

$$q(x) = \sup\{r(x) \mid r \in R\}$$

for all extreme points x of K.

PROOF. Given $r_1, r_2 \in R$, let r' be the pointwise maximum of r_1 and r_2. Then r' is a continuous convex function from K to $\mathbb{R}$, and $r' \ll q$. According to Theorem 11.12, there exists $r \in \text{Aff}(K)$ such that $r' \ll r \ll q$. Then $r \in R$ and each $r_i \leq r' \ll r$, proving that $(R, \ll)$ is upward directed.

Now let $x \in \partial_e K$. We must show that for any real number $a < q(x)$, there is some $r \in R$ for which $r(x) > a$. Define a function $p\colon K \to \{-\infty\} \cup \mathbb{R}$ so that $p(x) = a$ while $p = -\infty$ on $K - \{x\}$. It is clear that p is upper semicontinuous, and we claim that p is also convex. Hence, consider any positive convex combination $y = \alpha_1 y_1 + \alpha_2 y_2$ in K. If $y = x$, then each $y_i = x$, and so

$$p(y) = a = \alpha_1 a + \alpha_2 a = \alpha_1 p(y_1) + \alpha_2 p(y_2).$$

If $y \neq x$, then $p(y) = -\infty \leq \alpha_1 p(y_1) + \alpha_2 p(y_2)$. Thus p is convex, as claimed. Finally, observe that $p \ll q$. By Theorem 11.12, there exists $r \in \text{Aff}(K)$ such that $p \ll r \ll q$. Then $r \in R$ and $r(x) > p(x) = a$, as desired. $\square$

LEMMA 13.2. *Let (G,u) be a nonzero interpolation group with order-unit, and set $S = S(G,u)$. Let $p\colon S \to \{-\infty\} \cup \mathbb{R}$ be an upper semicontinuous convex function, and let $q\colon S \to \mathbb{R} \cup \{\infty\}$ be a lower semicontinuous concave function. Assume that $p \ll q$ and that $(p(s), q(s)) \cap s(G)$ is nonempty for every discrete extremal state $s \in S$. Then there exist $p', q' \in \text{Aff}(S)$ such that*

$$p \ll p' \ll q' \ll q$$

and $(p'(s), q'(s)) \cap s(G)$ is nonempty for all $s \in S$.

PROOF. Set $R = \{r \in \text{Aff}(S) \mid r \ll q\}$. As S is a Choquet simplex (Theorem 10.17), $(R, \ll)$ is upward directed, by Lemma 13.1. Set

$$W(r,x) = \{s \in S \mid p(s) < s(x) < r(s)\}$$

for all $r \in R$ and $x \in G$. Since p is upper semicontinuous, while r and evaluation at x are continuous, $W(r,x)$ is an open subset of S. We claim that these open sets cover S.

First consider any discrete extremal state $s \in S$. By assumption, the open interval $(p(s), q(s))$ intersects $s(G)$, and so there is some $x \in G$ for which $p(s) < s(x) < q(s)$. In view of Lemma 13.1, there exists $r \in R$ such that $r(s) > s(x)$. Thus $s \in W(r,x)$.

Next, consider an arbitrary discrete state $s \in S$. By Proposition 6.22, s is a convex combination of discrete extremal states. Hence, we may express s as a positive convex combination of distinct discrete extremal states, say

$$s = \alpha_1 s_1 + \cdots + \alpha_n s_n.$$

Each $s_i \in W(r_i, x_i)$ for some $r_i \in R$ and $x_i \in G$, that is,

$$p(s_i) < s_i(x_i) < r_i(s_i).$$

Since $(R, \ll)$ is upward directed, there exists $r \in R$ such that each $r_i \ll r$, and we note that $s_i(x_i) < r_i(s_i) < r(s_i)$ for each i. By Corollary 6.20, the n-tuple $(s_1, \ldots, s_n)$ induces an isomorphism

$$G/(\ker(s_1) \cap \cdots \cap \ker(s_n)) \to s_1(G) \times \cdots \times s_n(G).$$

Hence, there exists $x \in G$ such that $s_i(x) = s_i(x_i)$ for all i, so that

$$p(s_i) < s_i(x) < r(s_i)$$

for all i. As p is convex and r is affine, we find that

$$p(s) \leq \sum \alpha_i p(s_i) < \sum \alpha_i s_i(x) = s(x) < \sum \alpha_i r(s_i) = r(s),$$

whence $s \in W(r,x)$.

Finally, consider an indiscrete state $s \in S$. Since $p \ll q$, Theorem 11.12 provides us with a function $r \in \mathrm{Aff}(S)$ such that $p \ll r \ll q$, so that $r \in R$ and $p(s) < r(s)$. As s is indiscrete, $s(G)$ is dense in $\mathbb{R}$. Consequently, there is some $x \in G$ for which $p(s) < s(x) < r(s)$, and hence $s \in W(r,x)$.

Thus S is covered by the open sets $W(r,x)$, as claimed. By compactness,

$$S = W(r_1, x_1) \cup \cdots \cup W(r_k, x_k)$$

for some $r_i \in R$ and $x_i \in G$. Since $(R, \ll)$ is upward directed, there exists $q' \in R$ such that each $r_i \ll q'$. Then $q' \in \mathrm{Aff}(S)$ and $q' \ll q$. Any s in S lies in some $W(r_i, x_i)$, so that

$$p(s) < s(x_i) < r_i(s) < q'(s),$$

and hence $s(x_i) \in (p(s), q'(s))$. Therefore $p \ll q' \ll q$, and $(p(s), q'(s)) \cap s(G)$ is nonempty for all $s \in S$.

Now $-q' \in \mathrm{Aff}(S)$, and $-p \colon S \to \mathbb{R} \cup \{\infty\}$ is a lower semicontinuous concave function, such that $-q' \ll -p$ and $(-q'(s), -p(s)) \cap s(G)$ is nonempty for all $s \in S$. By the argument above, there exists $p'' \in \mathrm{Aff}(S)$ such that $-q' \ll p'' \ll -p$ and $(-q'(s), p''(s)) \cap s(G)$ is nonempty for all $s \in S$. Set $p' = -p''$. $\square$

Lemma 13.2 brings our problem to a consideration of affine continuous real-valued functions p and q on the state space of an interpolation group (G, u) with order-unit, such that $p \ll q$ and $(p(s), q(s)) \cap s(G)$ is nonempty for all $s \in S(G, u)$. Thus $S(G, u)$ is covered by open sets of the form

$$\{s \in S(G, u) \mid p(s) < s(x) < q(s)\}$$

(for $x \in G$), and by compactness finitely many of these sets cover $S(G, u)$. However, in order to cover $S(G, u)$ with a single such set, we must restrict attention to elements $x \in G$ for which $\phi(x) \ll q$, where ϕ is the natural map from G to $\mathrm{Aff}(S(G, u))$. Thus, given any $s \in S(G, u)$, we need to find an element $x \in G$ such that $p(s) < s(x) < q(s)$ and also $\phi(x) \ll q$. Such elements can be found by working with the s-completion $\overline{G}$ of G. For if G_0 is the convex subgroup of $\overline{G}$ generated by the image $\overline{u}$ of u, then G_0 is Dedekind complete (using Theorem 12.7), and so $(G_0, \overline{u})$ satisfies general comparability (Theorem 9.9). Thus we may use p and q to induce affine continuous functions on $S(G_0, \overline{u})$, apply Proposition 8.19 to generate an element $\overline{x} \in G_0$, and then approximate $\overline{x}$ by an element of G.

For technical reasons, we approximate q from below by $\phi(z)/2^k$ for some $z \in G$ and $k \in \mathbb{N}$, and we seek elements $x \in G$ for which $2^k x \leq z$. In order to obtain such elements as approximations of elements of s-completions of G, we require a pair of lemmas.

LEMMA 13.3. *Let G be a directed abelian group, and let $s\colon G \to \mathbb{R}$ be a positive homomorphism. Let $v \in G$, let m be an integer greater than 1, and let ε be a positive real number. Assume that $|mw - v|_s < \varepsilon$ for some $w \in G$. Then there exists $x \in G$ such that $mx \leq v$ and*

$$|mx - v|_s < (m-1)\varepsilon.$$

PROOF. As $|mw - v|_s < \varepsilon$, there exist $a, b \in G^+$ such that

$$mw - v = a - b$$

and $s(a + b) < \varepsilon$. Set $x = w - a$. Since $a \geq 0$, we see that

$$mx = mw - ma \leq mw - a = v - b \leq v.$$

In addition,

$$v - mx = v - mw + ma = b - a + ma \leq (m-1)(a+b).$$

As $v - mx \in G^+$, we conclude that

$$|mx - v|_s = |v - mx|_s = s(v - mx) \leq (m-1)s(a+b) < (m-1)\varepsilon. \quad \square$$

LEMMA 13.4. *Let G be a directed abelian group, let $s\colon G \to \mathbb{R}$ be a positive homomorphism, and let $\overline{G}$ denote the s-completion of G. Let $\psi\colon G \to \overline{G}$ be the natural map, and let $\overline{s}$ be the natural extension of s to $\overline{G}$. Let $y \in \overline{G}$ and*

$z \in G$, let m be an integer greater than 1, and let ε be a positive real number. If $my \leq \psi(z)$, then there exists $x \in G$ such that $mx \leq z$ and $s(x) > \bar{s}(y) - \varepsilon$.

PROOF. Choose $w \in G$ satisfying $|\psi(w) - y|_{\bar{s}} < \varepsilon/2m$, and note that

$$|\psi(mw) - my|_{\bar{s}} \leq m|\psi(w) - y|_{\bar{s}} < \varepsilon/2.$$

Since $\psi(z) - my$ lies in $\overline{G}^{+}$, there is some $a \in G^{+}$ for which

$$|\psi(a) - \psi(z) + my|_{\bar{s}} < \varepsilon/2.$$

As a result,

$$|s(z-a) - m\bar{s}(y)| = |\bar{s}(\psi(z) - \psi(a) - my)| \leq |\psi(z) - \psi(a) - my|_{\bar{s}} < \varepsilon/2,$$

and hence $s(z-a) > m\bar{s}(y) - \varepsilon/2$. Now observe that

$$\begin{aligned} |mw - (z-a)|_s &= |\psi(mw) - \psi(z) + \psi(a)|_{\bar{s}} \\ &\leq |\psi(mw) - my|_{\bar{s}} + |my - \psi(z) + \psi(a)|_{\bar{s}} < \varepsilon. \end{aligned}$$

By Lemma 13.3, there exists $x \in G$ such that $mx \leq z - a$ and

$$|mx - (z-a)|_s < (m-1)\varepsilon.$$

As $a \geq 0$, we have $mx \leq z$. Since $z - a - mx$ lies in G^{+}, we see that

$$s(z - a - mx) = |z - a - mx|_s < (m-1)\varepsilon,$$

and consequently

$$ms(x) > s(z-a) - (m-1)\varepsilon > m\bar{s}(y) - \varepsilon/2 - (m-1)\varepsilon > m\bar{s}(y) - m\varepsilon.$$

Therefore $s(x) > \bar{s}(y) - \varepsilon$. □

THEOREM 13.5. *Let (G, u) be a nonzero interpolation group with order-unit, set $S = S(G, u)$, and let $\phi\colon G \to \mathrm{Aff}(S)$ be the natural map. Suppose that $p\colon S \to \{-\infty\} \cup \mathbb{R}$ is an upper semicontinuous convex function, and suppose that $q\colon S \to \mathbb{R} \cup \{\infty\}$ is a lower semicontinuous concave function. Assume that $p \ll q$ and that $(p(s), q(s)) \cap s(G)$ is nonempty for all discrete extremal states $s \in S$. If G is 2-unperforated, there exists $x \in G$ such that $p \ll \phi(x) \ll q$. If G is unperforated and $q \gg 0$, there exists $x \in G^{+}$ such that $p \ll \phi(x) \ll q$.*

PROOF. First assume that G is 2-unperforated. Because of Lemma 13.2, we may assume, without loss of generality, that p and q are in $\mathrm{Aff}(S)$ and that $(p(s), q(s)) \cap s(G)$ is nonempty for all $s \in S$. With a second application of Lemma 13.2, we obtain a function $q' \in \mathrm{Aff}(S)$ such that $p \ll q' \ll q$ and $(p(s), q'(s)) \cap s(G)$ is nonempty for all $s \in S$. Choose a positive real number ε such that $q' + 2\varepsilon \leq q$. By Theorem 7.9, there exist $z \in G$ and $k \in \mathbb{N}$ such that

$$\|\phi(z)/2^k - (q' + \varepsilon)\| < \varepsilon,$$

whence $q' \ll \phi(z)/2^k \ll q' + 2\varepsilon \leq q$.

Set $Y = \{y \in G \mid 2^k y \leq z\}$, and recall from Corollary 2.22 that Y is upward directed. For each $y \in Y$, set

$$W(y) = \{s \in S \mid s(y) > p(s)\},$$

and note that $W(y)$ is an open subset of S. We claim that these open sets cover S.

Fix $s \in S$ for a while. Let $\overline{G}$ denote the s-completion of G, and let $\bar{s}$ be the natural extension of s to $\overline{G}$. Let $\psi\colon G \to \overline{G}$ be the natural map, and let G_0 be the convex subgroup of $\overline{G}$ generated by $\psi(u)$. According to Theorem 12.7, $\overline{G}$ is Dedekind complete and lattice-ordered, from which it follows that G_0 is Dedekind complete and lattice-ordered. Hence, G_0 is archimedean, by Proposition 9.7, and $(G_0, \psi(u))$ satisfies general comparability, by Theorem 9.9.

Set $S_0 = S(G_0, \psi(u))$, let $\phi_0\colon G_0 \to \mathrm{Aff}(S_0)$ be the natural map, and let s_0 be the restriction of $\bar{s}$ to G_0, so that $s_0 \in S_0$. Viewing ψ as a normalized positive homomorphism from (G, u) to $(G_0, \psi(u))$, we see that ψ induces an affine continuous map from S_0 to S. Composing this map with p and q', we obtain maps p_0, q_0 in $\mathrm{Aff}(S_0)$ such that $p_0(t) = p(t\psi)$ and $q_0(t) = q'(t\psi)$ for all $t \in S_0$. Since $p \ll q' \ll \phi(z)/2^k$, we obtain

$$p_0 \ll q_0 \ll \phi_0\psi(z)/2^k.$$

Because $(p(s_1), q'(s_1)) \cap s_1(G)$ is nonempty for every $s_1 \in S$, it follows that $(p_0(t), q_0(t)) \cap t(G_0)$ is nonempty for all $t \in S_0$.

According to Proposition 8.19, there is some $w \in G_0$ such that $p_0(t) < t(w) < q_0(t)$ for all t in $\partial_e S_0$. Hence,

$$p_0 \ll \phi_0(w) \ll q_0 \ll \phi_0\psi(z)/2^k,$$

because of Corollary 5.20. Now $\phi_0(2^k w) \ll \phi_0\psi(z)$. As G_0 is archimedean, Theorem 7.7 shows that $2^k w \leq \psi(z)$. Note that

$$\bar{s}(w) = s_0(w) = \phi_0(w)(s_0) > p_0(s_0) = p(s_0\psi) = p(s).$$

By Lemma 13.4, there exists $y \in G$ such that $2^k y \leq z$ and $s(y) > p(s)$. Thus $y \in Y$ and $s \in W(y)$.

Therefore the open sets $W(y)$ cover S, as claimed. By compactness,

$$S = W(y_1) \cup \cdots \cup W(y_n)$$

for some $y_1, \ldots, y_n$ in Y. Since Y is upward directed, there exists $x \in Y$ such that each $y_i \leq x$. Then $2^k x \leq z$, and hence

$$\phi(x) \leq \phi(z)/2^k \ll q.$$

Any s in S lies in some $W(y_i)$, whence $s(x) \geq s(y_i) > p(s)$. Therefore $\phi(x) \gg p$.

Finally, suppose that G is unperforated and that $q \gg 0$. Using Lemma 13.2 again, we obtain $q' \in \mathrm{Aff}(S)$ such that $p \ll q' \ll q$ and $(p(s), q'(s)) \cap s(G)$ is nonempty for all $s \in S$. Because of the result just proved, there exists $y \in G$ satisfying $p \ll \phi(y) \ll q'$. Now q and $q - q'$ are strictly positive lower semicontinuous functions, and so they attain strictly positive minimum values on S. Hence, there is a positive real number ε such that $q \gg \varepsilon$ and $q \gg q' + \varepsilon$. By Theorem 7.9, there exist $z \in G$ and $n \in \mathbb{N}$ such that

$$\|\phi(z)/2^n - (q - \varepsilon/2)\| < \varepsilon/2,$$

so that $q - \varepsilon \ll \phi(z)/2^n \ll q$. Then

$$0 \ll q - \varepsilon \ll \phi(z)/2^n \quad \text{and} \quad \phi(y) \ll q' \ll q - \varepsilon \ll \phi(z)/2^n,$$

whence $\phi(z) \gg 0$ and $\phi(2^n y) \ll \phi(z)$. As G is unperforated, Theorem 7.8 shows that $z \geq 0$ and $2^n y \leq z$. Note that $2^n 0 \leq z$ as well. By Corollary 2.22, there exists $x \in G$ such that $2^n x \leq z$, while also $y \leq x$ and $0 \leq x$. Therefore $x \in G^+$ and

$$p \ll \phi(y) \leq \phi(x) \leq \phi(z)/2^n \ll q. \quad \square$$

COROLLARY 13.6. *Let (G, u) be a nonzero interpolation group with order-unit, let $\phi\colon G \to \mathrm{Aff}(S(G,u))$ be the natural map, and set*

$$A = \{p \in \mathrm{Aff}(S(G,u)) \mid p(s) \in s(G) \text{ for all discrete } s \in \partial_e S(G,u)\}.$$

If G is 2-unperforated, then $\phi(G)$ is a dense subgroup of A. If G is unperforated, then $\phi(G^+)$ is dense in A^+.

PROOF. Obviously $\phi(G) \subseteq A$. Assume that G is 2-unperforated, and consider any $p \in A$ and any positive real number ε. Then $p - \varepsilon$ and $p + \varepsilon$ are affine continuous functions from $S(G,u)$ to $\mathbb{R}$ such that $p - \varepsilon \ll p + \varepsilon$ and

$$((p-\varepsilon)(s), (p+\varepsilon)(s)) \cap s(G)$$

is nonempty for all discrete extremal states s in $S(G,u)$. By Theorem 13.5, there exists $x \in G$ such that $p - \varepsilon \ll \phi(x) \ll p + \varepsilon$, whence $\|\phi(x) - p\| < \varepsilon$. Thus $\phi(G)$ is dense in A.

Now assume that G is unperforated, and consider any $p \in A^+$ and any positive real number ε. Since $p + \varepsilon \gg 0$, Theorem 13.5 provides us with an element $x \in G^+$ for which $p - \varepsilon \ll \phi(x) \ll p + \varepsilon$, and hence $\|\phi(x) - p\| < \varepsilon$. Therefore $\phi(G^+)$ is dense in A^+. $\square$

COROLLARY 13.7. *Let (G, u) be a nonzero interpolation group with order-unit, let $\phi\colon G \to \mathrm{Aff}(S(G,u))$ be the natural map, and assume that there are no discrete extremal states in $S(G,u)$. If G is 2-unperforated, then $\phi(G)$ is a dense subgroup of $\mathrm{Aff}(S(G,u))$. If G is unperforated, then $\phi(G^+)$ is dense in $\mathrm{Aff}(S(G,u))^+$.* $\square$

COROLLARY 13.8. *Let (G, u) be a nonzero interpolation group with order-unit, and let $\phi\colon G \to \mathrm{Aff}(S(G,u))$ be the natural map. Let $p \in \mathrm{Aff}(S(G,u))$, and let $\varepsilon_1, \varepsilon_2 \in \mathbb{R}^+$ such that $\varepsilon_1 + \varepsilon_2 > 1$. If G is 2-unperforated, there exists $x \in G$ such that*

$$p - \varepsilon_1 \ll \phi(x) \ll p + \varepsilon_2.$$

If G is unperforated and $p + \varepsilon_2 \gg 0$, then there exists such an x in G^+.

PROOF. The functions $p-\varepsilon_1$ and $p+\varepsilon_2$ are affine continuous functions from $S(G,u)$ to $\mathbb{R}$ such that $p-\varepsilon_1 \ll p+\varepsilon_2$. For any s in $S(G,u)$ we have

$$(p(s)+\varepsilon_2)-(p(s)-\varepsilon_1)=\varepsilon_1+\varepsilon_2>1.$$

Since $\mathbb{Z} \subseteq s(G)$, the set $(p(s)-\varepsilon_1, p(s)+\varepsilon_2) \cap s(G)$ must be nonempty. Thus the desired element of G exists by Theorem 13.5. □

COROLLARY 13.9. *Let (G,u) be a nonzero interpolation group with order-unit, and let $\phi: G \to \operatorname{Aff}(S(G,u))$ be the natural map. Let $p \in \operatorname{Aff}(S(G,u))$, and let ε be a positive real number. If G is 2-unperforated, there exists $x \in G$ such that*

$$\|\phi(x)-p\| < \tfrac{1}{2}+\varepsilon.$$

If G is unperforated and $p \geq 0$, there exists such an x in G^+ □

• **Compact Sets of Extremal States.** In order to obtain Corollaries 13.6–13.9 from Theorem 13.5, only the case of that theorem in which p and q are affine continuous real-valued functions is needed. One advantage of proving the theorem in terms of semicontinuous convex and concave functions is that it is applicable to functions defined on compact sets of extremal states or on closed faces of the state space, as in the following results. The method by which Theorem 13.5 is made to apply in these situations is of course analogous to the derivation of Theorems 11.14 and 11.22 from Theorem 11.13.

THEOREM 13.10. *Let (G,u) be a nonzero interpolation group with order-unit, let X be a nonempty compact subset of $\partial_e S(G,u)$, and let $\psi: G \to C(X,\mathbb{R})$ be the evaluation map. Let $p: X \to \{-\infty\} \cup \mathbb{R}$ be an upper semicontinuous function, and let $q: X \to \mathbb{R} \cup \{\infty\}$ be a lower semicontinuous function. Assume that $p \ll q$ and that $(p(s),q(s)) \cap s(G)$ is nonempty for all discrete states $s \in X$. If G is 2-unperforated, there exists $x \in G$ such that $p \ll \psi(x) \ll q$. If G is unperforated and $q \gg 0$, then there exists such an x in G^+.*

PROOF. Set $S=S(G,u)$, and let $\phi: G \to \operatorname{Aff}(S)$ be the natural map. Define functions

$$p': S \to \{-\infty\} \cup \mathbb{R} \quad \text{and} \quad q': S \to \mathbb{R} \cup \{\infty\}$$

so that $p'=p$ and $q'=q$ on X, while $p'=-\infty$ and $q'=\infty$ on $S-X$. As X is compact, we see that p' is upper semicontinuous and that q' is lower semicontinuous. We claim that p' is convex and that q' is concave. Thus consider any positive convex combination $s=\alpha_1 s_1+\alpha_2 s_2$ in S. If $s \in X$, then $s \in \partial_e S$ and so $s_1=s_2=s$, whence

$$p'(s)=p(s)=\alpha_1 p(s)+\alpha_2 p(s)=\alpha_1 p'(s_1)+\alpha_2 p'(s_2).$$

On the other hand, if $s \notin X$, then

$$p'(s)=-\infty \leq \alpha_1 p'(s_1)+\alpha_2 p'(s_2).$$

Thus p' is convex, and similarly q' is concave.

Obviously $p' \ll q'$. Now consider any discrete extremal state s in S. If $s \in X$, then

$$(p'(s), q'(s)) \cap s(G) = (p(s), q(s)) \cap s(G),$$

which is nonempty by hypothesis. If $s \notin X$, then

$$(p'(s), q'(s)) \cap s(G) = \mathbb{R} \cap s(G),$$

which is nonempty because $0 \in s(G)$. Thus $(p'(s), q'(s)) \cap s(G)$ is nonempty for all discrete extremal states $s \in S$. If G is 2-unperforated, then according to Theorem 13.5 there exists $x \in G$ such that $p' \ll \phi(x) \ll q'$. Restricting to X, we conclude that $p \ll \psi(x) \ll q$. If G is unperforated and $q \gg 0$, then we have $q' \gg 0$, in which case Theorem 13.5 provides such an x in G^+. □

COROLLARY 13.11. *Let (G, u) be a nonzero interpolation group with order-unit, let X be a nonempty compact subset of $\partial_e S(G, u)$, and let $\psi: G \to C(X, \mathbb{R})$ be the evaluation map. Set*

$$B = \{p \in C(X, \mathbb{R}) \mid p(s) \in s(G) \text{ for all discrete } s \in X\}.$$

If G is 2-unperforated, then $\psi(G)$ is a dense subgroup of B. If G is unperforated, then $\psi(G^+)$ is dense in B^+. □

COROLLARY 13.12. *Let (G, u) be a nonzero interpolation group with order-unit, let X be a nonempty compact subset of $\partial_e S(G, u)$, and let $\psi: G \to C(X, \mathbb{R})$ be the evaluation map. Assume that there are no discrete states in X. If G is 2-unperforated, then $\psi(G)$ is a dense subgroup of $C(X, \mathbb{R})$. If G is unperforated, then $\psi(G^+)$ is dense in $C(X, \mathbb{R})^+$.* □

• **Closed Faces.** We conclude the chapter with analogs of Theorem 13.10 and its corollaries for finite families of pairwise disjoint nonempty closed faces in state spaces of interpolation groups. Since the proofs are essentially the same as those of the previous section, we leave the details to the reader.

THEOREM 13.13. *Let (G, u) be a nonzero interpolation group with order-unit, let $F_1, \dots, F_n$ be pairwise disjoint nonempty closed faces of $S(G, u)$, and for each $i = 1, \dots, n$ let $\phi_i: G \to \mathrm{Aff}(F_i)$ be the evaluation map. For each i, let $p_i: F_i \to \{-\infty\} \cup \mathbb{R}$ be an upper semicontinuous convex function, and let $q_i: F_i \to \mathbb{R} \cup \{\infty\}$ be a lower semicontinuous concave function. Assume that $p_i \ll q_i$ and that $(p_i(s), q_i(s)) \cap s(G)$ is nonempty for all discrete extremal states $s \in F_i$. If G is 2-unperforated, there exists $x \in G$ such that $p_i \ll \phi_i(x) \ll q_i$ for all i. If G is unperforated and each $q_i \gg 0$, then there exists such an x in G^+.* □

COROLLARY 13.14. *Let (G, u) be a nonzero interpolation group with order-unit, and let $F_1, \dots, F_n$ be pairwise disjoint nonempty closed faces of $S(G, u)$. For each $i = 1, \dots, n$, set*

$$A_i = \{p \in \mathrm{Aff}(F_i) \mid p(s) \in s(G) \text{ for all discrete } s \in \partial_e F_i\},$$

and let $\phi_i \colon G \to \mathrm{Aff}(F_i)$ *be the evaluation map. Let* ϕ° *denote the map*

$$(\phi_1, \dots, \phi_n) \colon G \to \mathrm{Aff}(F_1) \times \cdots \times \mathrm{Aff}(F_n).$$

If G *is 2-unperforated, then* $\phi^\circ(G)$ *is a dense subgroup of* $\prod A_i$*. If* G *is unperforated, then* $\phi^\circ(G^+)$ *is dense in* $\prod A_i^+$*.* □

COROLLARY 13.15. *Let* (G, u) *be a nonzero interpolation group with order-unit, let* $F_1, \dots, F_n$ *be pairwise disjoint nonempty closed faces of* $S(G, u)$*, and for each* $i = 1, \dots, n$ *let* $\phi_i \colon G \to \mathrm{Aff}(F_i)$ *be the evaluation map. Let* ϕ° *denote the map*

$$(\phi_1, \dots, \phi_n) \colon G \to \mathrm{Aff}(F_1) \times \cdots \times \mathrm{Aff}(F_n).$$

Assume that there are no discrete extremal states in $F_1 \cup \cdots \cup F_n$*. If* G *is 2-unperforated, then* $\phi^\circ(G)$ *is a dense subgroup of* $\prod \mathrm{Aff}(F_i)$*. If* G *is unperforated, then* $\phi^\circ(G^+)$ *is dense in* $\prod \mathrm{Aff}(F_i)^+$*.* □

• **Notes.** Theorem 13.5 and Corollary 13.6 are due to the author, Handelman, and Lawrence [**60**, **57**, **58**]. A version of Theorem 13.5 was first proved in the case that G satisfies the countable interpolation property; in this case, if p and q are functions in $\mathrm{Aff}(S(G,u))$ such that $p \leq q$ and $[p(s), q(s)] \cap s(G)$ is nonempty for all s in $\partial_e S(G,u)$, there exists $x \in G$ such that $p \leq \phi(x) \leq q$ [**60**, Theorem I.9.2]. A corresponding version of Corollary 13.6 was proved at the same time: if G is archimedean and satisfies the countable interpolation property, then ϕ provides an isomorphism of G onto A (as ordered groups) [**60**, Theorem I.9.4]. The present form of Corollary 13.6 was proved in [**57**, Theorem 4.8], and the first conclusion of Theorem 13.5 was proved in [**58**, Theorem 2.12].

Corollaries 13.11 and 13.14 are applications of Theorem 13.5 patterned after the applications of [**60**, Theorem I.9.2] obtained in [**60**]. In the case of Corollary 13.11 in which G satisfies the countable interpolation property, $\psi(G) = B$ and $\psi(G^+) = B^+$ [**60**, Corollary I.9.6]. In the case of Corollary 13.14 in which $n = 1$ and G satisfies the countable interpolation property, $\phi_1(G) = A_1$ and $\phi_1(G^+) = A_1^+$ [**60**, Corollary I.9.5]. All of these consequences of the countable interpolation property will be derived in Chapter 16.

CHAPTER 14

Simple Dimension Groups

By analogy with rings, algebras, and other algebraic systems, a dimension group G is said to be "simple" if it has no nontrivial ideals, i.e., if the only ideals of G are $\{0\}$ and G. The class of simple dimension groups is described and classified in this chapter, in terms of data arising from maps of such groups into the affine continuous function spaces on their state spaces, with the help of the approximation results obtained in the previous chapter. The state spaces arising in this classification are essentially arbitrary, for every nonempty Choquet simplex is affinely homeomorphic to a state space of a simple dimension group. Also, every nonempty metrizable Choquet simplex is affinely homeomorphic to a state space of a countable simple dimension group.

A simple dimension group G has plenty of order-units, for any nonzero element $u \in G^+$ is an order-unit in G. If there exist any discrete states on (G, u), then G is isomorphic to $\mathbb{Z}$ (as an ordered group). Otherwise, the image of the natural map from G to $\mathrm{Aff}(S(G,u))$ is a dense subgroup of $\mathrm{Aff}(S(G,u))$ (Corollary 13.7), and such maps provide a means of classifying simple dimension groups. Namely, if H is a torsion-free abelian group, K is a nonempty Choquet simplex, and ψ is a group homomorphism of H onto a dense subgroup of $\mathrm{Aff}(K)$, then H may be made into a simple dimension group by using ψ to pull back the strict ordering from $\mathrm{Aff}(K)$ to H; in other words,

$$H^+ = \{0\} \cup \{x \in H \mid \psi(x) \gg 0\}.$$

Conversely, any simple dimension group is isomorphic (as an ordered group) either to $\mathbb{Z}$ or to one constructed as above, and the isomorphism types in this construction can be classified in terms of isomorphisms of the groups H and isomorphisms of the ordered vector spaces $\mathrm{Aff}(K)$.

A cleaner version of the classification is provided for simple dimension groups having finite-dimensional state spaces. One source of such examples is the result that a simple dimension group which is a free abelian group of finite rank $n \geq 2$ has a state space of dimension at most $n-2$. By way of illustration, the classification of the rank 2 simple dimension groups is reduced to irrational numerical invariants.

• **Simplicity.**

DEFINITION. A partially ordered abelian group G is *simple* provided G is nonzero and directed (so that G is a nonzero ideal of itself) and the only ideals of G are $\{0\}$ and G.

For example, any nonzero subgroup of $\mathbb{R}$ is simple as a partially ordered abelian group. Any nonzero partially ordered abelian group (G, u) with order-unit has a quotient group that is simple. Namely, since ideals of G are proper if and only if they do not contain u, Zorn's Lemma provides a maximal proper ideal H in G, and then G/H is a simple partially ordered abelian group.

LEMMA 14.1. *Let G be a nonzero directed abelian group. Then G is simple if and only if every nonzero element of G^+ is an order-unit in G.*

PROOF. First assume that G is simple. If u is any nonzero element of G^+, then the convex subgroup of G generated by u, that is, the subgroup

$$H = \{x \in G \mid -nu \leq x \leq nu \text{ for some } n \in \mathbb{N}\},$$

is a nonzero ideal of G. By simplicity, $H = G$, whence u is an order-unit in G.

Conversely, assume that every nonzero element of G^+ is an order-unit in G. Given a nonzero ideal H in G, the cone H^+ is nonzero because H is directed, and hence we may choose a nonzero element $u \in H^+$. Then u is an order-unit in G, and so the convex subgroup of G generated by u equals G, whence $H = G$. Therefore G is simple. $\square$

For instance, if X is a nonempty compact Hausdorff space, then $C(X, \mathbb{R})$ is a simple dimension group under the strict ordering, since every strictly positive function in $C(X, \mathbb{R})$ is an order-unit. Similarly, any nonzero directed subgroup of $C(X, \mathbb{R})$ is a simple partially ordered abelian group under the strict ordering. In particular, for each positive integer n, the groups $\mathbb{Z}^n$, $\mathbb{Q}^n$, and $\mathbb{R}^n$ are all simple under the strict ordering. (The latter two are dimension groups as well.) On the other hand, using the pointwise ordering, $\mathbb{Z}^n$, $\mathbb{Q}^n$, and $\mathbb{R}^n$ are not simple unless $n = 1$.

LEMMA 14.2. *Let G be a simple interpolation group, and suppose that G^+ contains an atom x. Then $G = \mathbb{Z}x$, and $G \cong \mathbb{Z}$ as ordered groups.*

PROOF. By Proposition 3.12, the subgroup $\mathbb{Z}x$ is an ideal of G, and $\mathbb{Z}x$ is a simplicial group with simplicial basis $\{x\}$. Hence, $\mathbb{Z}x \cong \mathbb{Z}$ as ordered groups. Since G is simple, $G = \mathbb{Z}x$. $\square$

PROPOSITION 14.3. *Let (G, u) be a simple interpolation group with order-unit. Then the following conditions are equivalent:*

(a) *G is cyclic (as an abelian group).*

(b) *There exists a discrete state on (G, u).*

(c) *There exists an atom in G^+.*

(d) *$(G, u) \cong (\mathbb{Z}, m)$ (as ordered groups with order-unit) for some $m \in \mathbb{N}$.*

PROOF. (a) $\Rightarrow$ (b): As G is simple, it is nonzero, and so there exists a state s on (G, u), by Corollary 4.4. Since G is cyclic, so is $s(G)$, whence s is a discrete state.

(b) $\Rightarrow$ (c): Let s be a discrete state on (G, u), and let H be the subgroup of G generated by $\ker(s)^+$. According to Proposition 4.22, H is an ideal of G, and G/H is a simplicial group. By simplicity, $H = \{0\}$, and hence G is simplicial. As a result, there exist atoms in G^+.

(c) $\Rightarrow$ (d): This is clear from Lemma 14.2.

(d) $\Rightarrow$ (a): This is automatic. $\square$

In particular, since every simple interpolation group contains order-units, Proposition 14.3 shows that there exists only one cyclic simple interpolation group (up to isomorphism of ordered groups), namely $\mathbb{Z}$. Hence, most of our study of simple dimension groups is devoted to the noncyclic case. (Although our first few results have not required the interpolation groups to be unperforated, this condition is essential for most of the remaining results.) To illustrate the dichotomy between the cyclic and noncyclic cases, we next prove that noncyclic simple interpolation groups enjoy the strict interpolation property (Proposition 14.6).

LEMMA 14.4. *Let G be a simple interpolation group, and let $x_1, \ldots, x_n$ be elements of G. If each $x_i > 0$, then there exists an element $x \in G$ such that $x > 0$ and each $x_i \geq x$.*

PROOF. By induction, it suffices to prove the case that $n = 2$. Since $x_1 > 0$, it is an order-unit in G (by Lemma 14.1), and so $x_2 \leq kx_1$ for some $k \in \mathbb{N}$. By Riesz decomposition, $x_2 = y_1 + \cdots + y_k$ for some $y_i \in G^+$ such that each $y_i \leq x_1$. Note that each $y_i \leq x_2$ as well. Since $x_2 > 0$, some $y_j > 0$, and we set $x = y_j$. $\square$

LEMMA 14.5. *Let G be a simple interpolation group, let $x \in G$, and let $n \in \mathbb{N}$. If $x > 0$ and G is noncyclic, then there exists $y \in G$ such that $y > 0$ and $ny < x$.*

PROOF. Since G^+ contains no atoms (by Proposition 14.3), there exists a chain of elements

$$x = x_0 > x_1 > x_2 > \cdots > x_n > 0$$

in G^+. Then $x_{i-1} - x_i > 0$ for each $i = 1, \ldots, n$. By Lemma 14.4, there is some $y \in G$ for which $y > 0$ and each $x_{i-1} - x_i \geq y$. Thus

$$ny \leq (x_0 - x_1) + (x_1 - x_2) + \cdots + (x_{n-1} - x_n) = x - x_n < x. \quad \square$$

PROPOSITION 14.6. *If G is a noncyclic simple interpolation group, then G satisfies the strict interpolation property.*

PROOF. If $x_1, x_2, y_1, y_2 \in G$ with $x_i < y_j$ for all i, j, then $y_j - x_i > 0$ for all i, j. According to Lemma 14.4, there exists $a \in G$ such that $a > 0$ and

$y_j - x_i \geq a$ for all i, j. Then Lemma 14.5 provides us with an element $b \in G$ such that $b > 0$ and $2b < a$. Now $y_j - x_i > 2b$ for all i, j, and so $x_i + b < y_j - b$ for all i, j. Interpolating, we obtain $z \in G$ such that

$$x_i + b \leq z \leq y_j - b$$

for all i, j, and hence $x_i < z < y_j$ for all i, j. Therefore G satisfies strict interpolation. □

We close this section with an example to show that the class of simple interpolation groups is strictly larger than the class of simple dimension groups.

EXAMPLE 14.7. *There exist simple interpolation groups which are perforated.*

PROOF. Choose an abelian group A which is not torsion-free, and make A into a partially ordered abelian group using the discrete ordering. Then A is a perforated interpolation group. Let G be the lexicographic direct product of $\mathbb{Q}$ with A. As $\mathbb{Q}$ satisfies strict interpolation, Proposition 2.10 shows that G is an interpolation group, and G is perforated because A is perforated.

It is clear that G is directed. Since $A^+ = \{0\}$, any nonzero element $u \in G^+$ has the form $u = (q, a)$ with $q > 0$, whence u is an order-unit in G. Therefore G is simple, by Lemma 14.1. □

• **State Spaces.** A large class of simple dimension groups can be constructed directly from Choquet simplices, as follows.

THEOREM 14.8. *If K is a nonempty Choquet simplex, then* $(\mathrm{Aff}(K), \ll)$ *is a simple dimension group.*

PROOF. Let A denote $\mathrm{Aff}(K)$ equipped with the strict ordering. It is clear that A is directed and unperforated, and Theorem 11.4 shows that A has interpolation. Thus A is a dimension group. Any nonzero element $f \in A^+$ is strictly positive, whence f is bounded away from zero (by compactness), and hence f is an order-unit in A. By Lemma 14.1, A is simple. □

In contrast, if K is a Choquet simplex with more than one point, then $\mathrm{Aff}(K)$ with the usual ordering is not simple. For if x is any extreme point of K, then by Theorem 11.18 the set

$$H = \{f \in \mathrm{Aff}(K) \mid f(x) = 0\}$$

is an ideal of $\mathrm{Aff}(K)$, and $\mathrm{Aff}(K)/H$ is one-dimensional. As $\mathrm{Aff}(K)$ separates the points of K, it cannot be one-dimensional, and so $H \neq \{0\}$. Thus H is a nontrivial ideal of $\mathrm{Aff}(K)$.

COROLLARY 14.9. *Given any nonempty Choquet simplex K, there exists a simple dimension group (G, u) with order-unit such that $S(G, u)$ is affinely homeomorphic to K.*

PROOF. Theorem 14.8 and Corollary 7.2. □

Thus all nonempty Choquet simplices appear as state spaces of simple dimension groups. In the metrizable case, we may reduce to countable simple dimension groups, by means of the following results.

PROPOSITION 14.10. *If K is a metrizable compact convex subset of a linear topological space, then* Aff(K) *is separable.*

PROOF. As K is a compact metrizable space, Proposition 5.23 shows that $C(K, \mathbb{R})$ is separable. Since all subspaces of separable metric spaces are separable, we conclude that Aff(K) is separable. $\square$

LEMMA 14.11. *Let G be an interpolation group, and let X be a countable subset of G. Then there exists a countable subgroup H of G such that $X \subseteq H$ and H has interpolation.*

PROOF. Let H_1 be the subgroup of G generated by X, and observe that H_1 is countable. Set

$$W_1 = \{(x_1, x_2, y_1, y_2) \in H_1^4 \mid x_i \leq y_j \text{ for all } i, j\}.$$

Since G has interpolation, there is a map $\lambda: W_1 \to G$ such that

$$\begin{matrix} x_1 \\ x_2 \end{matrix} \quad \leq \quad \lambda(x_1, x_2, y_1, y_2) \quad \leq \quad \begin{matrix} y_1 \\ y_2 \end{matrix}$$

for all $(x_1, x_2, y_1, y_2) \in W_1$. Set H_2 equal to the subgroup of G generated by $H_1 \cup \lambda(W_1)$. As H_1 and W_1 are countable, so is H_2.

Continuing in this manner, we obtain countable subgroups $H_1 \subseteq H_2 \subseteq \cdots$ in G such that for any x_1, x_2, y_1, y_2 in any H_n with $x_i \leq y_j$ for all i, j, there exists $z \in H_{n+1}$ such that $x_i \leq z \leq y_j$ for all i, j. Set $H = \bigcup H_n$. $\square$

THEOREM 14.12. *Given any nonempty metrizable Choquet simplex K, there exists a countable simple dimension group (G, u) with order-unit such that $S(G, u)$ is affinely homeomorphic to K.*

PROOF. Let A denote Aff(K) equipped with the strict ordering, and let u denote the constant function 1 in A. Then A is a simple dimension group, by Theorem 14.8. According to Proposition 14.10, there is a countable dense subset $X \subseteq A$. In view of Lemma 14.11, there exists a countable subgroup G of A such that $X \cup \{u\} \subseteq G$ and G has interpolation. Since G contains the order-unit u, it is nonzero and directed. In addition, G is unperforated because A is unperforated, and hence G is a dimension group. Any nonzero element of G^+ is an order-unit in A (because A is simple) and so is an order-unit in G. Thus G is simple.

By Proposition 7.18, all states on (A, u) are bounded. Since G contains X and so is dense in A, it follows that the restriction map

$$\rho: S(A, u) \to S(G, u)$$

is injective. On the other hand, ρ is surjective by Corollary 4.3, whence ρ is a homeomorphism. Therefore $S(G, u)$ is affinely homeomorphic to $S(A, u)$. By Corollary 7.2, $S(A, u)$ is affinely homeomorphic to K. $\square$

COROLLARY 14.13. *Given any nonempty compact metric space X, there exists a countable simple dimension group (G,u) with order-unit such that $S(G,u)$ is affinely homeomorphic to $M_1^+(X)$.*

PROOF. Proposition 5.23, Corollary 10.18, and Theorem 14.12. □

• Classification.

THEOREM 14.14. *Let G be a noncyclic simple dimension group. Choose an order-unit $u \in G$, and let $\phi: G \to \mathrm{Aff}(S(G,u))$ be the natural map. Then $\phi(G)$ is a dense subgroup of $\mathrm{Aff}(S(G,u))$, and $\phi(G^+)$ is dense in $\mathrm{Aff}(S(G,u))^+$. Moreover,*

$$G^+ = \{0\} \cup \{x \in G | \phi(x) \gg 0\}.$$

PROOF. Since G is not cyclic, Proposition 14.3 shows that there are no discrete states on (G,u). Hence, the density results follow from Corollary 13.7. Using Lemma 14.1 and Theorem 7.8, we conclude that a nonzero element $x \in G$ lies in G^+ if and only if x is an order-unit in G, if and only if $\phi(x) \gg 0$. □

Theorem 14.14 provides a means of recovering the ordered group structure of a noncyclic simple dimension group G from three pieces of data: the group structure of G, the Choquet simplex $S(G,u)$, and the map ϕ from G onto a dense subgroup of $\mathrm{Aff}(S(G,u))$. Conversely, a simple noncyclic dimension group may be constructed from such data, as follows.

PROPOSITION 14.15. *Let G be a torsion-free abelian group, let K be a nonempty Choquet simplex, and let ψ be a group homomorphism of G onto a dense subgroup of $\mathrm{Aff}(K)$. Then G can be made into a noncyclic simple dimension group with positive cone*

$$G^+ = \{0\} \cup \{x \in G \mid \psi(x) \gg 0\}.$$

PROOF. The set $\{0\} \cup \{x \in G \mid \psi(x) \gg 0\}$ is clearly a strict cone in G, and hence it is the positive cone of a partially ordered abelian group structure on G. Any nonzero element $x \in G^+$ satisfies $\psi(x) \gg 0$, whence $\psi(x)$ is bounded away from zero. For any $y \in G$, the function $\psi(y)$ is bounded above, and so $\psi(y) \ll n\psi(x)$ for some $n \in \mathbb{N}$, whence $y \leq nx$. Thus every nonzero element of G^+ is an order-unit in G. Since $\psi(G)$ is dense in $\mathrm{Aff}(K)$, there exists $u \in G$ such that $\|\psi(u) - 1\| < 1$. Then $\psi(u) \gg 0$, and hence $u \in G^+$. Now u is an order-unit in G, which shows that G is directed. In addition, u must be nonzero, because K is nonempty. Therefore G is a simple partially ordered abelian group, by Lemma 14.1.

If $nx \geq 0$ for some $n \in \mathbb{N}$ and $x \in G$, then either $nx = 0$ or $\psi(nx) \gg 0$. In the first case, $x = 0$ because G is torsion-free, while in the second case, $\psi(x) \gg 0$ and so $x > 0$. Thus $x \geq 0$ in either case, proving that G is unperforated.

Now consider any $x_1, x_2, y_1, y_2 \in G$ such that $x_i \leq y_j$ for all i, j. If $x_m = y_n$ for some m, n, then $x_i \leq x_m \leq y_j$ for all i, j. Hence, we may assume that $x_i < y_j$

for all i, j, and so $\psi(x_i) \ll \psi(y_j)$ for all i, j. Choose a positive real number ε such that $\psi(x_i) + \varepsilon \leq \psi(y_j) - \varepsilon$ for all i, j. Since $\mathrm{Aff}(K)$ is an interpolation group (Theorem 11.4), there exists $f \in \mathrm{Aff}(K)$ such that

$$\psi(x_i) + \varepsilon \leq f \leq \psi(y_j) - \varepsilon$$

for all i, j. Then there exists $z \in G$ with $\|\psi(z) - f\| < \varepsilon$, and so

$$\psi(x_i) \leq f - \varepsilon \ll \psi(z) \ll f + \varepsilon \leq \psi(y_j)$$

for all i, j. Consequently, $x_i \leq z \leq y_j$ for all i, j, proving that G has interpolation. Therefore G is a dimension group.

As K is nonempty, $\mathrm{Aff}(K)$ is nonzero, and hence no cyclic subgroup of $\mathrm{Aff}(K)$ can be dense in $\mathrm{Aff}(K)$. Thus G cannot be cyclic. □

We shall classify noncyclic simple dimension groups in terms of the data used in Proposition 14.15, that is, triples (G, K, ψ) where G is a torsion-free abelian group, K is a nonempty Choquet simplex, and ψ is a group homomorphism of G onto a dense subgroup of $\mathrm{Aff}(K)$. Given such data, let us write $G[\psi]$ to stand for the dimension group constructed in Proposition 14.15. Thus as a group, $G[\psi] = G$, and

$$G[\psi]^+ = \{0\} \cup \{x \in G \mid \psi(x) \gg 0\}.$$

THEOREM 14.16. *Let Γ denote the family of all triples (G, K, ψ), where G is a torsion-free abelian group, K is a nonempty Choquet simplex, and ψ is a group homomorphism of G onto a dense subgroup of* $\mathrm{Aff}(K)$.

(a) *If $(G, K, \psi) \in \Gamma$, then $G[\psi]$ is a noncyclic simple dimension group.*

(b) *If G is a noncyclic simple dimension group, then there exist K, ψ such that $(G, K, \psi) \in \Gamma$ and $G[\psi] = G$.*

(c) *Let $(G_i, K_i, \psi_i) \in \Gamma$ for $i = 1, 2$. Then $G_1[\psi_1] \cong G_2[\psi_2]$ as ordered groups if and only if there exist a group isomorphism $\sigma: G_1 \to G_2$ and an isomorphism*

$$\rho: (\mathrm{Aff}(K_1), \lll) \to (\mathrm{Aff}(K_2), \lll)$$

of ordered vector spaces such that $\rho\psi_1 = \psi_2\sigma$.

PROOF. (a) This is Proposition 14.15.

(b) As G is unperforated, it is torsion-free. Choose an order-unit $u \in G$, set $K = S(G, u)$, and let $\psi: G \to \mathrm{Aff}(K)$ be the natural map. Since G is nonzero, K is nonempty by Corollary 4.4, and Theorem 10.17 shows that K is a Choquet simplex. That $\psi(G)$ is dense in $\mathrm{Aff}(K)$ and that $G[\psi] = G$ are given by Theorem 14.14.

(c) First assume that there exist σ, ρ as described. For any nonzero $x \in G_1$, we have $x > 0$ in $G_1[\psi_1]$ if and only if $\psi_1(x) \gg 0$, if and only if $\rho\psi_1(x) \gg 0$, if and only if $\psi_2\sigma(x) \gg 0$, if and only if $\sigma(x) > 0$ in $G_2[\psi_2]$. Thus σ provides an isomorphism of $G_1[\psi_1]$ onto $G_2[\psi_2]$ as ordered groups.

Conversely, assume that there exists an isomorphism

$$\sigma: G_1[\psi_1] \to G_2[\psi_2]$$

in the category of partially ordered abelian groups. In particular, σ is a group isomorphism of G_1 onto G_2. Pick an element $u_1 \in G_1$ such that $\|\psi_1(u_1)-2\| < 1$, and set $u_2 = \sigma(u_1)$. Then $1 \ll \psi_1(u_1) \ll 3$, and so, in particular, $u_1 > 0$. Consequently, $u_2 > 0$, whence $\psi_2(u_2) \gg 0$ and $\|\psi_2(u_2)\| > 0$.

We first show that $\sigma(\ker(\psi_1)) \subseteq \ker(\psi_2)$. Given any x in $\ker(\psi_1)$, we have $\psi_1(x) = 0$ and so

$$\psi_1(-u_1) \ll -1 \ll \psi_1(mx) \ll 1 \ll \psi_1(u_1)$$

for all $m \in \mathbb{N}$. Then $-u_1 \leq mx \leq u_1$ for all $m \in \mathbb{N}$, whence

$$-u_2 \leq m\sigma(x) \leq u_2$$

for all $m \in \mathbb{N}$. Consequently, $-\psi_2(u_2) \leq m\psi_2\sigma(x) \leq \psi_2(u_2)$ for all $m \in \mathbb{N}$, and hence $\psi_2\sigma(x) = 0$. Then $\sigma(x) \in \ker(\psi_2)$, as desired.

As a result, σ induces a group homomorphism

$$\overline{\sigma}\colon \psi_1(G_1) \to \psi_2(G_2)$$

such that $\overline{\sigma}\psi_1 = \psi_2\sigma$. We claim that $\overline{\sigma}$ is uniformly continuous. Since $\overline{\sigma}$ is a group homomorphism, it suffices to prove that given any positive real number ε, there exists a positive real number δ such that any element $x \in G_1$ satisfying $\|\psi_1(x)\| < \delta$ also satisfies $\|\psi_2\sigma(x)\| < \varepsilon$. In fact, we shall prove this for $\delta = \varepsilon/\|\psi_2(u_2)\|$.

Given $x \in G_1$ satisfying $\|\psi_1(x)\| < \varepsilon/\|\psi_2(u_2)\|$, choose $k, n \in \mathbb{N}$ such that

$$\|\psi_1(x)\| < k/n < \varepsilon/\|\psi_2(u_2)\|.$$

Then $\|\psi_1(nx)\| < k$, and so

$$\psi_1(-ku_1) \ll -k \ll \psi_1(nx) \ll k \ll \psi_1(ku_1),$$

whence $-ku_1 \leq nx \leq ku_1$. Consequently, $-ku_2 \leq n\sigma(x) \leq ku_2$, and hence

$$-k\psi_2(u_2) \leq n\psi_2\sigma(x) \leq k\psi_2(u_2).$$

Thus $\|\psi_2\sigma(x)\| \leq k\|\psi_2(u_2)\|/n < \varepsilon$, as desired.

Therefore $\overline{\sigma}$ is uniformly continuous, as claimed. Since $\psi_1(G_1)$ is dense in $\mathrm{Aff}(K_1)$, and since $\mathrm{Aff}(K_2)$ is complete, $\overline{\sigma}$ extends uniquely to a uniformly continuous map

$$\rho\colon \mathrm{Aff}(K_1) \to \mathrm{Aff}(K_2).$$

As the linear operations in $\mathrm{Aff}(K_1)$ and $\mathrm{Aff}(K_2)$ are uniformly continuous, we infer from the additivity of $\overline{\sigma}$ and the uniform continuity of ρ that ρ is a linear map.

Given any $f \in \mathrm{Aff}(K_1)$ with $f \gg 0$, choose a positive integer m such that $f \gg \frac{1}{m}$, and then choose a sequence $\{x_1, x_2, \dots\}$ in G_1 such that $\psi_1(x_i) \to f$ and $\psi_1(x_i) \gg \frac{1}{m}$ for all i. Since $\psi_1(u_1) \ll 3$, we see that $\psi_1(3mx_i) \gg \psi_1(u_1)$ for all i, so that $3mx_i > u_1$ for all i. Hence, $3m\sigma(x_i) > u_2$ for all i, and so $3m\psi_2\sigma(x_i) \gg \psi_2(u_2)$ for all i. As a result,

$$\overline{\sigma}\psi_1(x_i) = \psi_2\sigma(x_i) \gg \psi_2(u_2)/3m$$

for all i. Since $\overline{\sigma}\psi_1(x_i) \to \rho(f)$, we obtain

$$\rho(f) \geq \psi_2(u_2)/3m \gg 0.$$

Therefore ρ is a positive linear map from $(\text{Aff}(K_1), \ll)$ to $(\text{Aff}(K_2), \ll)$ such that $\rho\psi_1 = \psi_2\sigma$. By symmetry, there exists a positive linear map

$$\rho': (\text{Aff}(K_2), \ll) \to (\text{Aff}(K_1), \ll)$$

such that $\rho'\psi_2 = \psi_1\sigma^{-1}$. Then $\rho'\rho\psi_1 = \rho'\psi_2\sigma = \psi_1$, and hence $\rho'\rho$ must be the identity map on $\text{Aff}(K_1)$, because $\psi_1(G_1)$ is dense in $\text{Aff}(K_1)$. Similarly, $\rho\rho'$ is the identity map on $\text{Aff}(K_2)$. Thus ρ is an isomorphism of ordered vector spaces. □

Theorem 14.16 and Proposition 14.3 together provide a complete classification of all simple dimension groups.

In the situation of Theorem 14.16(c), it is possible for $G_1[\psi_1]$ and $G_2[\psi_2]$ to be isomorphic as ordered groups even though K_1 and K_2 are not affinely homeomorphic. (An example may be constructed by using the strict ordering in place of the product ordering in Example 6.18.) Such problems do not occur if we modify Theorem 14.16 to classify noncyclic simple dimension groups with specified order-units, as follows.

THEOREM 14.17. *Let Γ' denote the family of all quadruples (G, K, ψ, u) where G is a torsion-free abelian group, K is a nonempty Choquet simplex, ψ is a group homomorphism of G onto a dense subgroup of* Aff(K), *and u is an element of G such that $\psi(u) = 1$.*

(a) *If $(G, K, \psi, u) \in \Gamma'$, then $G[\psi]$ is a noncyclic simple dimension group and u is an order-unit in $G[\psi]$. Moreover, there is an affine homeomorphism*

$$\alpha: K \to S(G[\psi], u)$$

such that $\alpha(x)(y) = \psi(y)(x)$ for all $x \in K$ and $y \in G$.

(b) *If (G, u) is a noncyclic simple dimension group with order-unit, then there exist K, ψ such that $(G, K, \psi, u) \in \Gamma'$ and $G[\psi] = G$.*

(c) *Let $(G_i, K_i, \psi_i, u_i) \in \Gamma'$ for $i = 1, 2$. Then*

$$(G_1[\psi_1], u_1) \cong (G_2[\psi_2], u_2)$$

as ordered groups with order-unit if and only if there exist a group isomorphism $\sigma: G_1 \to G_2$ and an affine homeomorphism $\tau: K_2 \to K_1$ such that $\sigma(u_1) = u_2$ and $\tau^*\psi_1 = \psi_2\sigma$ (where τ^* denotes the isomorphism of Aff(K_1) *onto* Aff(K_2) *induced by τ).*

PROOF. (a) That $G[\psi]$ is a noncyclic simple dimension group is given by Proposition 14.15. Since $\psi(u) = 1$, it is clear that u is an order-unit in $G[\psi]$.

Let A denote $\text{Aff}(K)$ equipped with the strict ordering. Then ψ is a normalized positive homomorphism from $(G[\psi], u)$ to $(A, 1)$, and so it induces an affine continuous map

$$S(\psi): S(A, 1) \to S(G[\psi], u).$$

Since $\psi(G)$ is dense in A, and since all states on $(A,1)$ are bounded (by Proposition 7.18), $S(\psi)$ must be injective.

Now consider any state s in $S(G[\psi],u)$. Given any x in $\ker(\psi)$, we have

$$\psi(-u) = -1 \ll 0 = \psi(nx) \ll 1 = \psi(u)$$

for all $n \in \mathbb{N}$, and so $-u \leq nx \leq u$ for all $n \in \mathbb{N}$. Then $-1 \leq ns(x) \leq 1$ for all $n \in \mathbb{N}$, whence $s(x) = 0$. Thus s vanishes on $\ker(\psi)$. As a result, s induces a group homomorphism $\overline{s}: \psi(G) \to \mathbb{R}$ such that $\overline{s}\psi = s$. Observing that $\psi(G)^+ = \psi(G^+)$, we see that $\overline{s}$ is a state on $(\psi(G),1)$. By Corollary 4.3, $\overline{s}$ extends to a state t on $(A,1)$, and $S(\psi)(t) = t\psi = s$, proving that $S(\psi)$ is surjective.

Thus $S(\psi)$ is an affine continuous bijection. By compactness, $S(\psi)$ is a homeomorphism. If $\phi: K \to S(A,1)$ is the natural affine homeomorphism (Corollary 7.2), then the composite map $\alpha = S(\psi)\phi$ is an affine homeomorphism of K onto $S(G[\psi],u)$. For all $x \in K$ and $y \in G$, we check that

$$\alpha(x)(y) = (S(\psi)\phi(x))(y) = \phi(x)(\psi(y)) = \psi(y)(x).$$

(b) Proceed as in Theorem 14.16(b).

(c) First assume that there exist σ, τ as described. For any nonzero $x \in G_1$, we have $x > 0$ in $G_1[\psi_1]$ if and only if $\psi_1(x) \gg 0$, if and only if $\psi_1(x)\tau \gg 0$, if and only if $\psi_2\sigma(x) \gg 0$ (because $\psi_2\sigma(x) = \tau^*\psi_1(x) = \psi_1(x)\tau$), if and only if $\sigma(x) > 0$ in $G_2[\psi_2]$. Thus σ provides an isomorphism of $(G_1[\psi_1],u_1)$ onto $(G_2[\psi_2],u_2)$ as ordered groups with order-unit.

Conversely, assume that there exists an isomorphism

$$\sigma: (G_1[\psi_1],u_1) \to (G_2[\psi_2],u_2)$$

in the category of partially ordered abelian groups with order-unit. In particular, σ is a group isomorphism of G_1 onto G_2, and $\sigma(u_1) = u_2$. Also, σ induces an affine homeomorphism

$$S(\sigma): S(G_2[\psi_2],u_2) \to S(G_1[\psi_1],u_1).$$

For $i = 1,2$, there exists by (a) an affine homeomorphism

$$\alpha_i: K_i \to S(G_i[\psi_i],u_i)$$

such that $\alpha_i(x)(y) = \psi_i(y)(x)$ for all $x \in K_i$ and $y \in G_i$. Then the composite map

$$\tau = \alpha_1^{-1}S(\sigma)\alpha_2$$

is an affine homeomorphism of K_2 onto K_1. For all $x \in K_2$ and $y \in G_1$, we compute that

$$\begin{aligned}(\tau^*\psi_1(y))(x) &= \psi_1(y)(\tau(x)) = (\alpha_1\tau(x))(y) = (S(\sigma)\alpha_2(x))(y)\\ &= \alpha_2(x)(\sigma(y)) = (\psi_2\sigma(y))(x).\end{aligned}$$

Therefore $\tau^*\psi_1 = \psi_2\sigma$. □

• **Finite-Dimensional State Spaces.** The classification of noncyclic simple dimension groups in Theorems 14.16 and 14.17 becomes somewhat neater when applied to simple dimension groups with finite-dimensional state spaces, for then the Choquet simplices K that appear are classical simplices and the corresponding partially ordered real vector spaces $\mathrm{Aff}(K)$ are isomorphic to products of copies of $\mathbb{R}$. In this case, typical data include a torsion-free abelian group G and a group homomorphism ψ from G onto a dense subgroup of some $\mathbb{R}^n$. As in the general case, we shall write $G[\psi]$ for the dimension group constructed from G with positive cone

$$G[\psi]^+ = \{0\} \cup \{x \in G \mid \psi(x) \gg 0\}.$$

THEOREM 14.18. *Let Γ° denote the family of all triples (G, n, ψ), where G is a torsion-free abelian group, n is a positive integer, and ψ is a group homomorphism of G onto a dense subgroup of $\mathbb{R}^n$.*

(a) *If $(G, n, \psi) \in \Gamma^\circ$, then $G[\psi]$ is a noncyclic simple dimension group, and for any order-unit $u \in G[\psi]$, the state space $S(G[\psi], u)$ is finite-dimensional.*

(b) *Let G be a noncyclic simple dimension group, and assume that G has an order-unit u such that $S(G[\psi], u)$ is finite-dimensional. Then there exist n, ψ such that $(G, n, \psi) \in \Gamma^\circ$ and $G[\psi] = G$.*

(c) *Let $(G_i, n(i), \psi_i) \in \Gamma^\circ$ for $i = 1, 2$. Then $G_1[\psi_1] \cong G_2[\psi_2]$ as ordered groups if and only if $n(1) = n(2)$ and there exist a group isomorphism $\sigma: G_1 \to G_2$ and an isomorphism*

$$\rho: (\mathbb{R}^{n(1)}, \ll) \to (\mathbb{R}^{n(2)}, \ll)$$

of ordered vector spaces such that $\rho\psi_1 = \psi_2\sigma$.

PROOF. (a) If K is the standard $(n-1)$-dimensional simplex, then

$$(\mathrm{Aff}(K), \ll) \cong (\mathbb{R}^n, \ll)$$

as ordered vector spaces. Hence, it follows from Proposition 14.15 that $G[\psi]$ is a noncyclic simple dimension group.

Let A denote $\mathbb{R}^n$ equipped with the strict ordering. Since $u > 0$ in $G[\psi]$, we have $\psi(u) \gg 0$, and so $\psi(u)$ is an order-unit in A. Thus ψ is a normalized positive homomorphism from $(G[\psi], u)$ to $(A, \psi(u))$. As in the proof of Theorem 14.17(a), the induced map

$$S(\psi): S(A, \psi(u)) \to S(G[\psi], u)$$

is an affine homeomorphism. Therefore $S(G[\psi], u)$ must be finite-dimensional.

(b) According to Theorem 14.17(b), there exist a nonempty Choquet simplex K and a group homomorphism ψ' of G onto a dense subgroup of $\mathrm{Aff}(K)$ such that $\psi'(u) = 1$ and $G[\psi'] = G$. Then Theorem 14.17(a) shows that K is affinely homeomorphic to $S(G, u)$, whence K is finite-dimensional. Consequently, K is a classical simplex, by Theorem 10.16, and so there exists an isomorphism

$$q: (\mathrm{Aff}(K), \ll) \to (\mathbb{R}^n, \ll)$$

of ordered vector spaces, for some $n \in \mathbb{N}$. Now the map $\psi = q\psi'$ is a group homomorphism of G onto a dense subgroup of $\mathbb{R}^n$, so that $(G, n, \psi) \in \Gamma^\circ$, and $G[\psi] = G[\psi'] = G$.

(c) If $n(1) = n(2)$ and there exist σ, ρ as described, then it is clear that $G_1[\psi_1] \cong G_2[\psi_2]$ as ordered groups. Conversely, assume that $G_1[\psi_1]$ and $G_2[\psi_2]$ are isomorphic ordered groups.

For $i = 1, 2$, let K_i denote the standard $(n(i) - 1)$-dimensional simplex, and choose an isomorphism

$$q_i \colon (\mathbb{R}^{n(i)}, \ll) \to (\mathrm{Aff}(K_i), \ll)$$

of ordered vector spaces. Then $q_i\psi_i$ is a group homomorphism of G_i onto a dense subgroup of $\mathrm{Aff}(K_i)$, and $G_i[q_i\psi_i] = G_i[\psi_i]$. Now Theorem 14.16(c) shows that there exist a group isomorphism $\sigma \colon G_1 \to G_2$ and an isomorphism

$$\rho' \colon (\mathrm{Aff}(K_1), \ll) \to (\mathrm{Aff}(K_2), \ll)$$

of ordered vector spaces such that $\rho' q_1 \psi_1 = q_2 \psi_2 \sigma$. The composite map $\rho = q_2^{-1} \rho' q_1$ is then an isomorphism of $(\mathbb{R}^{n(1)}, \ll)$ onto $(\mathbb{R}^{n(2)}, \ll)$ as ordered vector spaces, and $\rho\psi_1 = \psi_2\sigma$. In particular, since ρ is a vector space isomorphism we conclude that $n(1) = n(2)$. $\square$

DEFINITION. Let $n \in \mathbb{N}$. A *permutation automorphism of* $\mathbb{R}^n$ is a real vector space automorphism ρ of $\mathbb{R}^n$ which permutes the entries of all n-tuples in $\mathbb{R}^n$, that is, there exists a permutation π of $\{1, 2, \ldots, n\}$ such that

$$\rho(x_1, x_2, \ldots, x_n) = (x_{\pi(1)}, x_{\pi(2)}, \ldots, x_{\pi(n)})$$

for all $(x_1, x_2, \ldots, x_n) \in \mathbb{R}^n$.

Of course, the only permutation automorphism of $\mathbb{R}^1$ is the identity map.

THEOREM 14.19. *Let $n \in \mathbb{N}$, and let Γ_n denote the family of all triples (G, ψ, u) where G is a torsion-free abelian group, ψ is a group homomorphism of G onto a dense subgroup of $\mathbb{R}^n$, and u is an element of G such that $\psi(u) = (1, 1, \ldots, 1)$.*

(a) *If $(G, \psi, u) \in \Gamma_n$, then $G[\psi]$ is a noncyclic simple dimension group, u is an order-unit in $G[\psi]$, and $S(G[\psi], u)$ is $(n-1)$-dimensional.*

(b) *If (G, u) is a noncyclic simple dimension group with order-unit, and $S(G, u)$ is $(n-1)$-dimensional, then there exists ψ such that $(G, \psi, u) \in \Gamma_n$ and $G[\psi] = G$.*

(c) *Let $(G_i, \psi_i, u_i) \in \Gamma_n$ for $i = 1, 2$. Then*

$$(G_1[\psi_1], u_1) \cong (G_2[\psi_2], u_2)$$

as ordered groups with order-unit if and only if there exist a group isomorphism $\sigma \colon G_1 \to G_2$ and a permutation automorphism ρ of $\mathbb{R}^n$ such that $\sigma(u_1) = u_2$ and $\rho\psi_1 = \psi_2\sigma$.

PROOF. Let v denote the n-tuple $(1, 1, \ldots, 1)$ in $\mathbb{R}^n$. Let K be the standard $(n-1)$-dimensional simplex, and let $e_1, \ldots, e_n$ be the extreme points of K. There is an isometric isomorphism of ordered Banach spaces

$$q\colon (\mathrm{Aff}(K), \leqslant) \to (\mathbb{R}^n, \leqslant)$$

such that $q(f) = (f(e_1), \ldots, f(e_n))$ for all $f \in \mathrm{Aff}(K)$. Note that $q(1) = v$.

(a) The composite map $q^{-1}\psi$ is a homomorphism of G onto a dense subgroup of $\mathrm{Aff}(K)$, and $q^{-1}\psi(u) = 1$. In the notation of Theorem 14.17, the quadruple $(G, K, q^{-1}\psi, u)$ lies in Γ'. Hence, $G[q^{-1}\psi]$ is a noncyclic simple dimension group, u is an order-unit in $G[q^{-1}\psi]$, and $S(G[q^{-1}\psi], u)$ is affinely homeomorphic to K. In particular, $S(G[q^{-1}\psi], u)$ is $(n-1)$-dimensional. As q and q^{-1} preserve the strict ordering, $G[q^{-1}\psi] = G[\psi]$.

(b) According to Theorem 14.17(b), there exist a nonempty Choquet simplex L and a homomorphism ϕ of G onto a dense subgroup of $\mathrm{Aff}(L)$ such that $\phi(u) = 1$ and $G[\phi] = G$. Then $S(G, u)$ is affinely homeomorphic to L, by Theorem 14.17(a), whence L is $(n-1)$-dimensional. Hence, L is affinely homeomorphic to K (Theorem 10.18), and so there is no loss of generality in assuming that $L = K$. The composite map $\psi = q\phi$ is then a homomorphism of G onto a dense subgroup of $\mathbb{R}^n$, and $\psi(u) = v$. Thus $(G, \psi, u) \in \Gamma_n$, and $G[\psi] = G[\phi] = G$.

(c) As in (a), we observe that the quadruples $(G_i, K, q^{-1}\psi_i, u_i)$ lie in Γ', and each $G_i[q^{-1}\psi_i] = G_i[\psi_i]$. Consequently, Theorem 14.17(c) shows that $(G_1[\psi_1], u_1)$ and $(G_2[\psi_2], u_2)$ are isomorphic (as ordered groups with order-unit) if and only if there exist a group isomorphism $\sigma\colon G_1 \to G_2$ and an affine homeomorphism $\tau\colon K \to K$ such that $\sigma(u_1) = u_2$ and $\tau^* q^{-1}\psi_1 = q^{-1}\psi_2\sigma$.

Given such σ, τ, observe that τ permutes the extreme points of K. Hence, there is a permutation π of $\{1, 2, \ldots, n\}$ such that $\tau(e_i) = e_{\pi(i)}$ for all $i = 1, \ldots, n$, and we compute that

$$(q\tau^* q^{-1})(x_1, \ldots, x_n) = (x_{\pi(1)}, \ldots, x_{\pi(n)})$$

for all $(x_1, \ldots, x_n) \in \mathbb{R}^n$. Thus the map $\rho = q\tau^* q^{-1}$ is a permutation automorphism of $\mathbb{R}^n$. Since $\tau^* q^{-1}\psi_1 = q^{-1}\psi_2\sigma$, we also have $\rho\psi_1 = \psi_2\sigma$.

Conversely, assume that there exist a group isomorphism $\sigma\colon G_1 \to G_2$ and a permutation automorphism ρ of $\mathbb{R}^n$ such that $\sigma(u_1) = u_2$ and $\rho\psi_1 = \psi_2\sigma$. Then there is a permutation π of $\{1, 2, \ldots, n\}$ such that

$$\rho(x_1, \ldots, x_n) = (x_{\pi(1)}, \ldots, x_{\pi(n)})$$

for all $(x_1, \ldots, x_n) \in \mathbb{R}^n$. Using π to permute the extreme points of K, we obtain an affine homeomorphism $\tau\colon K \to K$ such that $\tau(e_i) = e_{\pi(i)}$ for all $i = 1, \ldots, n$, and we observe that $q\tau^* q^{-1} = \rho$. Therefore $\tau^* q^{-1}\psi_1 = q^{-1}\psi_2\sigma$. $\square$

• **Finite Rank.** If (G, u) is a simple dimension group with order-unit, then finite-dimensionality of the state space $S(G, u)$ can be obtained from group-theoretic finiteness conditions on G. For instance, if G has finite rank n when viewed as a torsion-free abelian group, then $S(G, u)$ has dimension at most $n-1$. A stronger result is obtained by assuming that G is a free abelian group of finite rank n, with $n \geq 2$, in which case we shall prove that $S(G, u)$ has dimension at most $n-2$. This latter result leads to a quite explicit description and classification of those simple dimension groups that are free abelian groups of rank 2.

DEFINITION. The *torsion-free rank* of an abelian group G is the dimension of the rational vector space $G \otimes_{\mathbb{Z}} \mathbb{Q}$.

PROPOSITION 14.20. *Let (G, u) be a partially ordered abelian group with order-unit, and let n be the torsion-free rank of G. If n is finite, then the dimension of $S(G, u)$ is at most $n - 1$.*

PROOF. Set $H = G \otimes_{\mathbb{Z}} \mathbb{Q}$, and let $f: G \to H$ be the natural homomorphism given by the rule $f(x) = x \otimes 1$. We observe that the induced map

$$f^*: \operatorname{Hom}_{\mathbb{Q}}(H, \mathbb{R}) \to \operatorname{Hom}_{\mathbb{Z}}(G, \mathbb{R})$$

is an isomorphism of real vector spaces. Since H is an n-dimensional rational vector space, $\operatorname{Hom}_{\mathbb{Q}}(H, \mathbb{R})$ must be an n-dimensional real vector space. Thus $\operatorname{Hom}_{\mathbb{Z}}(G, \mathbb{R})$ is an n-dimensional real vector space. As $S(G, u)$ lies in the hyperplane

$$\{g \in \operatorname{Hom}_{\mathbb{Z}}(G, \mathbb{R}) \mid g(u) = 1\},$$

which misses the origin, we see that $S(G, u)$ is at most $(n - 1)$-dimensional. □

PROPOSITION 14.21. *Let (G, u) be a simple interpolation group with order-unit, and assume that G is a free abelian group of finite rank n. If $n \geq 2$, then the dimension of $S(G, u)$ is at most $n - 2$.*

PROOF. Let $\{x_1, \ldots, x_n\}$ be a basis for G as a free abelian group, and let $p_1, \ldots, p_n$ be the corresponding coordinate projections. That is, the p_i are group homomorphisms from G to $\mathbb{Z}$, and $p_i(x_i) = 1$ for all i, while $p_i(x_j) = 0$ for all $i \neq j$. Then $\{p_1, \ldots, p_n\}$ is a basis for the real vector space $\operatorname{Hom}_{\mathbb{Z}}(G, \mathbb{R})$.

Suppose that the dimension of $S(G, u)$ is $n - 1$ or greater. Then $S(G, u)$ contains n affinely independent states $s_1, \ldots, s_n$. Since $S(G, u)$ lies in a hyperplane in $\operatorname{Hom}_{\mathbb{Z}}(G, \mathbb{R})$ that misses the origin, $s_1, \ldots, s_n$ must be linearly independent. As $\operatorname{Hom}_{\mathbb{Z}}(G, \mathbb{R})$ is n-dimensional, $\{s_1, \ldots, s_n\}$ must span it. Hence, there exist $\alpha_{ij} \in \mathbb{R}$ (for $i, j = 1, \ldots, n$) such that

$$p_i = \alpha_{i1} s_1 + \cdots + \alpha_{in} s_n$$

for each $i = 1, \ldots, n$.

Choose $m \in \mathbb{N}$ such that $m > n|\alpha_{ij}|$ for all i, j. Since G is free of rank at least 2, it is not cyclic. Hence, by Lemma 14.5 there exists an element $y \in G$ such that $y > 0$ and $my < u$. Then $0 \leq s_j(y) \leq \frac{1}{m}$ for all j, whence

$$|p_i(y)| \leq \sum_{j=1}^{n} |\alpha_{ij}| \cdot |s_j(y)| \leq \sum_{j=1}^{n} \frac{|\alpha_{ij}|}{m} < \sum_{j=1}^{n} \frac{1}{n} = 1$$

for all i. As $p_i(G) = \mathbb{Z}$ for each i, we conclude that each $p_i(y) = 0$. But then $y = 0$, which is a contradiction.

Therefore $S(G, u)$ is at most $(n - 2)$-dimensional. □

COROLLARY 14.22. *Let G be a simple dimension group. If G is a free abelian group of rank 2, then G is isomorphic (as an ordered group) to a dense subgroup of* $\mathbb{R}$.

PROOF. Choose an order-unit $u \in G$, and recall from Corollary 4.4 that $S(G, u)$ is nonempty. According to Proposition 14.21, $S(G, u)$ is zero-dimensional, and so $S(G, u)$ consists of a single state s. As $s(G)$ is a nonzero torsion-free homomorphic image of G, it is a free abelian group of rank 1 or 2. If $s(G)$ is free of rank 1, then it is a cyclic group and so s is a discrete state. But then Proposition 14.3 shows that G is cyclic, which contradicts the assumption that G is free of rank 2. Thus $s(G)$ must be free of rank 2. As a result, $s(G)$ is a dense subgroup of $\mathbb{R}$, and s gives a group isomorphism of G onto $s(G)$.

Certainly s is a positive homomorphism. Conversely, consider any $x, y \in G$ for which $s(x) \leq s(y)$. If $s(x) = s(y)$, then $x = y$ because s is injective, while if $s(x) < s(y)$, then $x < y$ by Corollary 4.13. Therefore s also provides an order-isomorphism of G onto $s(G)$. □

In the rank 2 case, Theorems 14.18 and 14.19 reduce to the following classifications in terms of irrational numbers.

THEOREM 14.23. *Let A denote the set of all positive irrational real numbers.*

(a) *For $\alpha \in A$, let G_α denote the subgroup $\mathbb{Z} + \alpha\mathbb{Z}$ of $\mathbb{R}$. Then G_α is a simple dimension group and a free abelian group of rank 2.*

(b) *If G is a simple dimension group, and G is also a free abelian group of rank 2, then there exists $\alpha \in A$ such that $G \cong G_\alpha$ (as ordered groups).*

(c) *Let $\alpha, \beta \in A$. Then $G_\alpha \cong G_\beta$ (as ordered groups) if and only if there exist integers a, b, c, d such that*

(i) $ad - bc = \pm 1$;

(ii) $a + b\beta > 0$;

(iii) $(c + d\beta)/(a + b\beta) = \alpha$.

PROOF. (a) This is clear.

(b) By Corollary 14.22, there exists an isomorphism h of G onto a subgroup of $\mathbb{R}$ (as ordered groups). Then $h(G) = \beta\mathbb{Z} + \gamma\mathbb{Z}$ for some nonzero $\beta, \gamma \in \mathbb{R}$, and there is no loss of generality in assuming that β and γ are each positive. Then the map $\beta^{-1}h$ (that is, the composition of h with multiplication by β^{-1}) provides an ordered group isomorphism of G onto $\mathbb{Z} + \alpha\mathbb{Z}$, where $\alpha = \beta^{-1}\gamma$. As G is not cyclic, α is not rational, and hence $\alpha \in A$.

(c) In the notation of Theorem 14.18, G_α arises from the triple $(G_1, 1, \psi_1)$, where G_1 denotes the abelian group G_α and ψ_1 denotes the inclusion map $G_1 \to \mathbb{R}$. Similarly, G_β arises from $(G_2, 1, \psi_2)$, where G_2 denotes the abelian group G_β and ψ_2 denotes the inclusion map $G_2 \to \mathbb{R}$. According to Theorem 14.18(c), $G_\alpha \cong G_\beta$ as ordered groups if and only if there exist a group isomorphism $\sigma: G_\alpha \to G_\beta$ and an isomorphism

$$\rho: (\mathbb{R}, \ll) \to (\mathbb{R}, \ll)$$

of ordered vector spaces such that $\rho\psi_1 = \psi_2\sigma$. The only ordered vector space automorphisms of $(\mathbb{R}, \leqslant)$ are the maps given by multiplication by positive real numbers. Thus $G_\alpha \cong G_\beta$ as ordered groups if and only if there exists a positive real number γ such that $\gamma G_\alpha = G_\beta$.

Given such a γ, we must have $\gamma, \gamma\alpha \in G_\beta$, and so there exist integers a, b, c, d such that $\gamma = a + b\beta$ and $\gamma\alpha = c + d\beta$. Properties (ii) and (iii) are immediate. Multiplication by γ provides a group isomorphism of G_α onto G_β, and these groups are free with bases $\{1, \alpha\}$ and $\{1, \beta\}$. The matrix of the isomorphism induced by γ, with respect to these bases, is either $\left(\begin{smallmatrix} a & b \\ c & d \end{smallmatrix}\right)$ or $\left(\begin{smallmatrix} a & c \\ b & d \end{smallmatrix}\right)$, depending on whether the coordinates of elements of G_α and G_β are written as row or column vectors. This matrix must be invertible as a matrix over $\mathbb{Z}$, whence its determinant must be invertible in $\mathbb{Z}$, and thus property (i) holds.

Conversely, assume that there exist integers a, b, c, d satisfying properties (i), (ii), (iii). Set $\gamma = a + b\beta$. Then γ is a positive real number lying in G_β, and the product $\gamma\alpha = c + d\beta$ is also in G_β, whence $\gamma G_\alpha \subseteq G_\beta$. Thus multiplication by γ induces a group homomorphism of G_α into G_β. With respect to the bases $\{1, \alpha\}$ and $\{1, \beta\}$, the matrix of this homomorphism has determinant ± 1 because of property (i), and so this homomorphism is an isomorphism. Therefore $\gamma G_\alpha = G_\beta$. $\square$

THEOREM 14.24. *Let A denote the set of all positive irrational real numbers.*

(a) *For $m \in \mathbb{N}$ and $\alpha \in A$, let $G_{m,\alpha}$ denote the subgroup $(1/m)\mathbb{Z} + \alpha\mathbb{Z}$ of $\mathbb{R}$. Then $G_{m,\alpha}$ is a simple dimension group and a free abelian group of rank 2, and 1 is an order-unit in $G_{m,\alpha}$.*

(b) *If (G, u) is a simple dimension group with order-unit, and G is a free abelian group of rank 2, then there exist $m \in \mathbb{N}$ and $\alpha \in A$ such that $(G, u) \cong (G_{m,\alpha}, 1)$ (as ordered groups with order-unit).*

(c) *Let $m, n \in \mathbb{N}$ and $\alpha, \beta \in A$. Then $(G_{m,\alpha}, 1) \cong (G_{n,\beta}, 1)$ (as ordered groups with order-unit) if and only if $m = n$ and either $\alpha + \beta \in (1/m)\mathbb{Z}$ or $\alpha - \beta \in (1/m)\mathbb{Z}$.*

PROOF. (a) This is clear.

(b) By Corollary 14.22, there exists an isomorphism h of G onto a dense subgroup H of $\mathbb{R}$ (as ordered groups). Since h may be multiplied by a positive real number if necessary, there is no loss of generality in assuming that $h(u) = 1$. Thus h provides an isomorphism of (G, u) onto $(H, 1)$.

Now H is a finitely generated abelian group, and so all its subgroups are finitely generated. In particular, $H \cap \mathbb{Q}$ is a finitely generated subgroup of $\mathbb{Q}$, whence $H \cap \mathbb{Q}$ is cyclic. As $1 \in H \cap \mathbb{Q}$, we thus find that $H \cap \mathbb{Q} = (1/m)\mathbb{Z}$ for some $m \in \mathbb{N}$. The quotient group $H/(H \cap \mathbb{Q})$ embeds in the rational vector space $\mathbb{R}/\mathbb{Q}$ and so is torsion-free. Since H is free of rank 2 and $H \cap \mathbb{Q}$ is free of rank 1, it follows that $H/(H \cap \mathbb{Q})$ must be free of rank 1, and hence $H/(H \cap \mathbb{Q})$ is cyclic. Consequently,

$$H = (H \cap \mathbb{Q}) + \alpha\mathbb{Z} = (1/m)\mathbb{Z} + \alpha\mathbb{Z}$$

for some $\alpha \in \mathbb{R}$. There is no loss of generality in assuming that $\alpha > 0$. As $H \not\subseteq \mathbb{Q}$, we also have $\alpha \notin \mathbb{Q}$, so that $\alpha \in A$. Now $H = G_{m,\alpha}$, and therefore h provides an isomorphism of (G, u) onto $(G_{m,\alpha}, 1)$ (as ordered groups with order-unit).

(c) If $m = n$ and either $\alpha + \beta$ or $\alpha - \beta$ is in $(1/m)\mathbb{Z}$, then $G_{m,\alpha} = G_{n,\beta}$, whence $(G_{m,\alpha}, 1)$ and $(G_{n,\beta}, 1)$ are identical ordered groups with order-unit.

Conversely, let

$$f: (G_{m,\alpha}, 1) \to (G_{n,\beta}, 1)$$

be an isomorphism in the category of partially ordered abelian groups with order-unit. There is a unique state s on $(G_{m,\alpha}, 1)$, namely, the inclusion map $G_{m,\alpha} \to \mathbb{R}$, and there is a unique state t on $(G_{n,\beta}, 1)$, namely, the inclusion map $G_{n,\beta} \to \mathbb{R}$. As tf is a state on $(G_{m,\alpha}, 1)$, we must have $tf = s$. Therefore

$$G_{m,\alpha} = s(G_{m,\alpha}) = tf(G_{m,\alpha}) = t(G_{n,\beta}) = G_{n,\beta}.$$

Now $1/m = p/n + \beta q$ for some $p, q \in \mathbb{Z}$, and $q = 0$ because β is irrational. Then $n = mp$, and so $m|n$. By symmetry, $n|m$, whence $m = n$.

Also, $\alpha = a/m + \beta b$ and $\beta = c/m + \alpha d$, for some $a, b, c, d \in \mathbb{Z}$. Then

$$\alpha = a/m + (c/m + \alpha d)b,$$

and so $\alpha(1 - bd) = (a + bc)/m$. Since α is irrational, we must have $1 - bd = 0$, and hence $b = \pm 1$. As $\alpha - \beta b = a/m$, it follows that either $\alpha + \beta$ or $\alpha - \beta$ is in $(1/m)\mathbb{Z}$. □

We conclude this chapter by proving that, up to isomorphism of ordered groups, there are only two nonsimple dimension groups that are free abelian groups of rank 2. Together with Theorem 14.23, this provides a complete classification of all dimension groups that are free of rank 2.

THEOREM 14.25. *Let G be a dimension group, and assume that G is a free abelian group of rank 2. If G is not simple, then G is isomorphic (as an ordered group) either to $\mathbb{Z}^2$ or to the lexicographic direct product of $\mathbb{Z}$ with itself.*

PROOF. By assumption, G contains a nontrivial ideal H. Since G/H is a dimension group, it is torsion-free. As G is free of rank 2, it follows that H and G/H must each be free of rank 1, and hence cyclic. If H' is any ideal of H, then H/H' is torsion-free, whence either $H' = \{0\}$ or $H' = H$. Thus H is a simple dimension group, and, similarly, G/H is a simple dimension group. Since these groups are cyclic, Proposition 14.3 shows that $H \cong G/H \cong \mathbb{Z}$ (as ordered groups).

First suppose that G contains an ideal K distinct from $\{0\}$, H, G. As above, $K \cong \mathbb{Z}$ (as ordered groups). By Proposition 2.4, $H \cap K$ and $H + K$ are ideals of G. Since H and K are simple, we must have $H \cap K = \{0\}$. On the other hand, $(H+K)/H$ is a nonzero ideal of the simple dimension group G/H, whence $(H + K)/H = G/H$, and so $H + K = G$. Thus $G = H \oplus K$ as an abelian group. By Proposition 2.6, $G = H \oplus K$ as an ordered group, and therefore $G \cong \mathbb{Z}^2$ as ordered groups.

Now assume that $\{0\}$, H, G are the only ideals of G. Given any element x in $G^+ - H$, observe that the convex subgroup of G generated by x is an ideal of G not contained in H. This ideal must equal G, and so x is an order-unit in G. Thus all elements of $G^+ - H$ are order-units in G.

We claim that if $x \in G^+ - H$ and $y \in H$, then $y \leq x$. There is an atom $e \in H^+$ such that $H = \mathbb{Z}e$, and we observe that e is also an atom in G^+. Now $y = ke$ for some $k \in \mathbb{Z}$, and if $k \leq 0$, then $y \leq 0 \leq x$. Thus we may assume that $k > 0$.

If $k = 1$, then $y = e$. Since x is an order-unit in G, there is some $n \in \mathbb{N}$ such that $y \leq nx$. By Riesz decomposition, $y = y_1 + \cdots + y_n$ for some $y_i \in G^+$ with each $y_i \leq x$. Since y is an atom in G^+, each $y_i \in \{0, y\}$. Then some $y_i = y$, whence $y \leq x$. Now suppose that $k > 1$, and that $(k-1)e \leq x$. Then $x-(k-1)e$ lies in $G^+ - H$, whence $e \leq x - (k-1)e$ by the first stage of the induction, and so $y = ke \leq x$. Thus the induction works, proving the claim.

There is an atom $u \in (G/H)^+$ such that $G/H = \mathbb{Z}u$ and $(G/H)^+ = \mathbb{Z}^+u$. Choose an element $v \in G^+$ such that $v + H = u$. Since $H = \mathbb{Z}e$, we see that $\{v, e\}$ is a basis for G (as a free abelian group). We claim that

$$G^+ = \mathbb{Z}^+e \cup (\mathbb{N}v + \mathbb{Z}e).$$

Given any $x \in G^+$, write $x = av + be$ for some $a, b \in \mathbb{Z}$. Since $x + H$ is an element of $(G/H)^+$, and since $x + H = au$, we must have $a \geq 0$. If $a > 0$, then $x \in \mathbb{N}v + \mathbb{Z}e$. If $a = 0$, then $x = be$ and so $be \in H^+$, whence $b \geq 0$ and $x \in \mathbb{Z}^+e$. Thus

$$G^+ \subseteq \mathbb{Z}^+e \cup (\mathbb{N}v + \mathbb{Z}e).$$

As $e \in G^+$, we obviously have $\mathbb{Z}^+e \subseteq G^+$. Given any z in $\mathbb{N}v + \mathbb{Z}e$, write $z = av + be$ for some $a \in \mathbb{N}$ and some $b \in \mathbb{Z}$. Then $av \in G^+ - H$ and $-be \in H$. By the claim above, $-be \leq av$, whence $z \in G^+$. Therefore

$$G^+ = \mathbb{Z}^+e \cup (\mathbb{N}v + \mathbb{Z}e),$$

as claimed.

The rule $(a, b) \mapsto av + be$ defines a group isomorphism of $\mathbb{Z}^2$ onto G. In view of the description of G^+ just obtained, we conclude that this map is an isomorphism from the lexicographic direct product of $\mathbb{Z}$ with itself onto G (as ordered groups). □

Note that $\mathbb{Z}^2$ and the lexicographic direct product of $\mathbb{Z}$ with itself are non-isomorphic dimension groups: for instance, the lexicographic product is totally ordered while $\mathbb{Z}^2$ is not.

• **Notes.** Proposition 14.6 is contained in a result of Effros, Handelman, and Shen [**30**, Corollary 1.2]. A more pathological example than Example 14.7—namely, a perforated simple interpolation group which is also torsion-free—was constructed by Lawrence [**89**]. Theorem 14.8 may be considered folklore, and Theorem 14.12 was observed by the author and Handelman (unpublished). The description of G^+ in Theorem 14.14 was obtained by Effros, Handelman, and Shen [**30**, Corollary 1.5], who also proved the density of $\phi(G)$ in $\mathrm{Aff}(S(G,u))$ in the case that $\partial_e S(G,u)$ is finite [**30**, Theorem 3.5]. The general density result was obtained by the author and Handelman [**57**, Corollary 4.10]. Proposition 14.15 is due to Effros, Handelman, and Shen [**30**, Lemmas 3.1, 3.2]. Theorems 14.16–14.19 were proved by the author.

Proposition 14.21 was proved by Handelman (unpublished). Corollary 14.22 and the first two parts of Theorem 14.23 are contained in results of Effros and Shen [**32**, Theorem 2.1; **115**, Theorem 4.4, Proposition 4.5]. Part (c) of Theorem 14.23 was proved by Shen [**115**, Lemma 4.7], along with Theorem 14.25 [**115**, Theorem 4.3]. Theorem 14.24 is a revised version of Theorem 14.23.

CHAPTER 15

Norm-Completeness

We have already seen that lattice completeness properties in a partially ordered abelian group (G, u) with order-unit, in particular Dedekind σ-completeness, can lead to a powerful structure theory for G. Norm-completeness properties (with respect to the order-unit norm) can also lead to a strong structure theory. In this chapter, we derive such a structure theory for dimension groups (G, u) with order-unit that are archimedean and norm-complete. The archimedean assumption may be viewed as an additional completeness condition, since G is archimedean if and only if G^+ is norm-closed in G (Proposition 7.17). Thus once G is norm-complete, G is archimedean if and only if G^+ is norm-complete. One source of examples is the result that an archimedean norm-complete dimension group may be obtained from any dimension group (G, u) with order-unit by completing G with respect to its order-unit norm.

Now let (G, u) be any archimedean norm-complete dimension group with order-unit. The ideal theory in G is related to the state space $S(G, u)$ by the result that an ideal H of G is norm-closed in G if and only if H equals the kernel of some subset of $S(G, u)$. Moreover, H is norm-closed if and only if G/H is norm-complete, if and only if G/H is archimedean. Later in the chapter, it is proved that the kernel of any compact set of extremal states in $S(G, u)$ is an ideal of G and that the kernel of any closed face of $S(G, u)$ is an ideal of G. The latter result leads to an order anti-isomorphism between the family of closed faces of $S(G, u)$ and the family of norm-closed ideals of G. These results are of course parallel to the ideal theory of affine continuous function spaces on Choquet simplices developed in Chapter 11 (e.g., Theorems 11.18 and 11.26).

Because of the completeness assumptions on G, the natural affine continuous function representation of G is completely determined. Namely, (G, u) is naturally isomorphic (as an ordered group with order-unit) to $(A, 1)$, where

$$A = \{p \in \mathrm{Aff}(S(G, u)) \mid p(s) \in s(G) \text{ for all discrete } s \in \partial_e S(G, u)\}.$$

Furthermore, if $\partial_e S(G, u)$ is compact, then (G, u) is naturally isomorphic to $(B, 1)$, where

$$B = \{p \in C(\partial_e S(G, u), \mathbb{R}) \mid p(s) \in s(G) \text{ for all discrete } s \in \partial_e S(G, u)\},$$

and G is lattice-ordered. From this a complete description of all archimedean norm-complete lattice-ordered abelian groups with order-unit is obtained. Corresponding representations are obtained for quotient groups $G/\ker(X)$ when X is either a compact subset of $\partial_e S(G,u)$ or a closed face of $S(G,u)$.

Finally, the collection $\operatorname{MaxSpec}(G)$ of all maximal (proper) ideals of G is investigated. This collection is topologized by defining its closed subsets to be the subsets of the form

$$\{M \in \operatorname{MaxSpec}(G) \mid M \supseteq H\},$$

where H is any ideal of G. The rule $s \mapsto \ker(s)$ then defines a continuous bijection of $\partial_e S(G,u)$ onto $\operatorname{MaxSpec}(G)$. This map is a homeomorphism if and only if $\partial_e S(G,u)$ is compact, if and only if $\operatorname{MaxSpec}(G)$ is Hausdorff.

• Norm-Completeness.

DEFINITION. Let (G,u) be a partially ordered abelian group with order-unit. For all $x,y \in G$, set $d_u(x,y) = \|x-y\|_u$. In view of Proposition 7.12, we see that d_u is a pseudo-metric on G. By way of abbreviation, we refer to d_u as *the order-unit metric* on G (relative to u), although d_u is not always a metric. If d_u is actually a metric (equivalently, if the norm-topology on G is Hausdorff), and if G is complete with respect to d_u, then we say that G is *norm-complete*.

If v is another order-unit in G, then there exists $m \in \mathbb{N}$ such that $u \le mv$ and $v \le mu$, whence

$$\|\cdot\|_u \le m\|\cdot\|_v \quad \text{and} \quad \|\cdot\|_v \le m\|\cdot\|_u.$$

As a result, the order-unit metrics d_u and d_v are equivalent, and so G is complete with respect to d_u if and only if G is complete with respect to d_v. Thus the property of norm-completeness is independent of the choice of the order-unit.

For example, let K be a compact convex subset of a linear topological space, and equip $\operatorname{Aff}(K)$ with either the pointwise ordering or the strict ordering. In either case, the order-unit norm in $\operatorname{Aff}(K)$ (with respect to the order-unit 1) coincides with the usual supremum norm, and hence $\operatorname{Aff}(K)$ is norm-complete. Similarly, it follows from Corollary 9.15 that all Dedekind σ-complete lattice-ordered abelian groups with order-unit are norm-complete.

In dimension groups, norm-completeness may be tested by investigating the convergence of certain infinite series, as follows.

LEMMA 15.1. *Let (G,u) be a dimension group with order-unit, and assume that the norm-topology on G is Hausdorff. Then G is norm-complete if and only if given any sequence $\{x_1, x_2, \dots\}$ in G^+ such that $2^n x_n \le u$ for all n, the series $\sum x_n$ converges in norm to some element of G.*

PROOF. First assume that G is norm-complete, and consider any $x_1, x_2, \dots$ in G^+ satisfying $2^n x_n \le u$ for all n. Then $\|x_n\| \le 1/2^n$ for all n, and hence the partial sums of the series $\sum x_n$ form a Cauchy sequence. Thus $\sum x_n$ must converge in G.

Conversely, assume that any such series is convergent, and consider any Cauchy sequence $\{y_1, y_2, \ldots\}$ in G. By passing to a subsequence, we may assume that

$$\|y_{n+1} - y_n\| < 1/2^n$$

for all n. Then each $y_{n+1} - y_n = z_n - w_n$ for some $z_n, w_n \in G^+$ such that $2^n z_n \leq u$ and $2^n w_n \leq u$, by Lemma 7.14. The series $\sum z_n$ converges to some element $z \in G$, and the series $\sum w_n$ converges to some element $w \in G$. Since

$$\begin{aligned} y_n &= y_1 + (y_2 - y_1) + (y_3 - y_2) + \cdots + (y_n - y_{n-1}) \\ &= y_1 + (z_1 + z_2 + \cdots + z_{n-1}) - (w_1 + w_2 + \cdots + w_{n-1}) \end{aligned}$$

for all n, we conclude that $y_n \to y_1 + z - w$. Therefore G is norm-complete. □

• **Norm-Completions.** As a source of examples, we briefly investigate norm-completions of dimension groups. We verify that such norm-completions are norm-complete (in particular, the norm-completion does have an order-unit, and the order-unit metric on the norm-completion coincides with the metric obtained from the completion process) and that such norm-completions are dimension groups.

DEFINITION. Let (G, u) be a partially ordered abelian group with order-unit. The *norm-completion of* G is the (Hausdorff) completion of G with respect to its order-unit norm.

As the group operation in G is uniformly continuous with respect to the order-unit norm, the norm-completion of G is an abelian group $\overline{G}$. If $\psi \colon G \to \overline{G}$ is the natural map, then $\psi(G^+)$ is a cone in $\overline{G}$, and so the closure of $\psi(G^+)$ is a cone also. We make $\overline{G}$ into a pre-ordered abelian group with positive cone $\overline{G}^+$ equal to the closure of $\psi(G^+)$ in $\overline{G}$, and we always assume that norm-completions have been made into pre-ordered abelian groups in this manner.

Several obvious questions arise: Is $\overline{G}$ partially ordered, is $\psi(u)$ an order-unit in $\overline{G}$, and if so, is $\overline{G}$ norm-complete? In case G is a dimension group, these questions can be answered using the following concrete representation of $\overline{G}$.

THEOREM 15.2. *Let (G, u) be a nonzero dimension group with order-unit, let $\overline{G}$ be the norm-completion of G, and let $\psi \colon G \to \overline{G}$ be the natural map. Set $S = S(G, u)$, let $\phi \colon G \to \mathrm{Aff}(S)$ be the natural map, and set*

$$A = \{p \in \mathrm{Aff}(S) \mid p(s) \in s(G) \text{ for all discrete } s \in \partial_e S\}.$$

Then there exists an isomorphism $f \colon \overline{G} \to A$ (as ordered groups) such that $f\psi = \phi$.

PROOF. Since $\|\phi(x)\| = \|x\|$ for all $x \in G$ (by Proposition 7.12), ϕ is uniformly continuous. Consequently, there is a unique uniformly continuous map $f \colon \overline{G} \to \mathrm{Aff}(S)$ such that $f\psi = \phi$, and we observe that f is a group homomorphism. As $\psi(G)$ is dense in $\overline{G}$ and $f\psi(G) = \phi(G) \subseteq A$, we see that $f(\overline{G}) \subseteq A$. Similarly, since $\psi(G^+)$ is dense in $\overline{G}^+$ and $f\psi(G^+) \subseteq A^+$, we have $f(\overline{G}^+) \subseteq A^+$. Thus f is actually a positive homomorphism from $\overline{G}$ to A.

Given any $x \in \ker(f)$, choose a sequence $\{x_1, x_2, \ldots\}$ in G such that $\psi(x_i) \to x$. Then $\phi(x_i) \to 0$ in Aff(S). As $\|x_i\| = \|\phi(x_i)\|$ for all i, we obtain $\|x_i\| \to 0$, whence $x = 0$. Thus f is injective.

By Corollary 13.6, $\phi(G^+)$ is dense in A^+. Hence, given any $p \in A^+$ there exists a sequence $\{x_1, x_2, \ldots\}$ in G^+ such that $\phi(x_i) \to p$. Since

$$\|x_i - x_j\| = \|\phi(x_i) - \phi(x_j)\|$$

for all i, j, the x_i form a Cauchy sequence in G^+. Then there exists $x \in \overline{G}^+$ such that $\psi(x_i) \to x$, and $f(x) = p$ by the continuity of f. Thus $f(\overline{G}^+) = A^+$.

As A contains an order-unit (e.g., 1), it is directed. Hence, from $f(\overline{G}^+) = A^+$ we conclude that $f(\overline{G}) = A$. Therefore f is an isomorphism of $\overline{G}$ onto A, as ordered groups. □

In the situation of Theorem 15.2, we may now identify $\overline{G}$ with A and ψ with ϕ. Certainly $\overline{G}$ is partially ordered and $\psi(u) = 1$ is an order-unit in $\overline{G}$. Given $p \in \overline{G}$ and $k, n \in \mathbb{N}$, note that $-k \cdot 1 \leq np \leq k \cdot 1$ if and only if $\|p\|_{\text{sup}} \leq k/n$. Hence, the order-unit norm on $\overline{G}$ (with respect to the order-unit 1) coincides with the supremum norm. Therefore $\overline{G}$ is norm-complete. Note also that the order-unit metric in $\overline{G}$ coincides with the metric obtained from the completion process; namely, given elements x and y in $\overline{G}$, expressed as limits of sequences $\{\psi(x_i)\}$ and $\{\psi(y_i)\}$ with $x_i, y_i \in G$, the value $\|x - y\|$ is the limit of the values $\|x_i - y_i\|$.

If there are no discrete states in S, then $\overline{G} = \text{Aff}(S)$. In this case, Theorems 10.17 and 11.4 show that $\overline{G}$ is a dimension group. In general, the norm-completion of any dimension group with order-unit is a dimension group, as we now prove, following a version of the method used in Theorem 12.7.

THEOREM 15.3. *If (G, u) is a dimension group with order-unit, then the norm-completion of G is an archimedean norm-complete dimension group.*

PROOF. There is no loss of generality in assuming that G is nonzero. Let $\overline{G}$ denote the norm-completion of G, and let $\phi\colon G \to \overline{G}$ be the natural map. We have already observed that $\overline{G}$ is norm-complete. Since $\overline{G}$ is clearly archimedean, unperforated, and directed, it only remains to show that $\overline{G}$ has interpolation.

Thus let f_1, f_2, g_1, g_2 by any elements of $\overline{G}$ satisfying $f_i \leq g_j$ for all i, j. Choose sequences

$$\{x_{11}, x_{12}, \ldots\}, \quad \{x_{21}, x_{22}, \ldots\}, \quad \{y_{11}, y_{12}, \ldots\}, \quad \{y_{21}, y_{22}, \ldots\}$$

in G such that

$$\|\phi(x_{in}) - f_i\| < 1/2^{n+2} \quad \text{and} \quad \|\phi(y_{jn}) - g_j\| < 1/2^{n+2}$$

for all i, j, n. For all i, n, we observe that

$$\|x_{i,n+1} - x_{in}\| = \|\phi(x_{i,n+1}) - \phi(x_{in})\| \leq \|\phi(x_{i,n+1}) - f_i\| + \|\phi(x_{in}) - f_i\| < 1/2^{n+3} + 1/2^{n+2} < 1/2^{n+1}.$$

Similarly, $\|y_{j,n+1} - y_{jn}\| < 1/2^{n+1}$ for all j, n. We shall construct Cauchy sequences $\{b_n\}$ and $\{z_n\}$ in G such that $b_n \to 0$ and $x_{in} \leq z_n \leq y_{jn} + b_n$ for all i, j, n. The limit of the sequence $\{\phi(z_n)\}$ will then provide an element of $\overline{G}$ to interpolate between f_1, f_2 and g_1, g_2.

We first construct elements $a_1, a_2, \ldots$ in G^+ such that $\|a_n\| \leq 1/2^{n+1}$ for all n, while also

$$x_{in} - a_n \leq x_{i,n+1} \leq x_{in} + a_n \quad \text{and} \quad y_{jn} - a_n \leq y_{j,n+1} \leq y_{jn} + a_n$$

for all i, j, n.

For each i, n, we have $\|x_{i,n+1} - x_{in}\| < 1/2^{n+1}$. By Lemma 7.14,

$$x_{i,n+1} - x_{in} = p_{in} - q_{in}$$

for some $p_{in}, q_{in} \in G^+$ satisfying $2^{n+1}p_{in} \leq u$ and $2^{n+1}q_{in} \leq u$. Similarly, each

$$y_{j,n+1} - y_{jn} = r_{jn} - s_{jn}$$

for some $r_{jn}, s_{jn} \in G^+$ satisfying $2^{n+1}r_{jn} \leq u$ and $2^{n+1}s_{jn} \leq u$. According to Proposition 2.21, there exists $a_n \in G^+$ such that $2^{n+1}a_n \leq u$ and a_n is an upper bound for the set

$$\{p_{1n}, p_{2n}, q_{1n}, q_{2n}, r_{1n}, r_{2n}, s_{1n}, s_{2n}\}.$$

The required properties of a_n are clear.

Next, we construct elements $b_1, b_2, \ldots$ in G^+ such that $\|b_n\| \leq 1/2^{n+1}$ for all n, while also $x_{in} \leq y_{jn} + b_n$ for all i, j, n.

Fix n for a while. For each i, j, we have

$$\phi(x_{in}) - 1/2^{n+2} \ll f_i \leq g_j \ll \phi(y_{jn}) + 1/2^{n+2},$$

whence $\phi(x_{in}) \ll \phi(y_{jn}) + 1/2^{n+1}$, and so

$$\phi(2^{n+1}x_{in}) \ll \phi(2^{n+1}y_{jn}) + 1 = \phi(2^{n+1}y_{jn} + u).$$

Consequently, $2^{n+1}x_{in} \leq 2^{n+1}y_{jn} + u$, by Theorem 7.8. According to Proposition 2.19, $x_{in} = p_{ij} + q_{ij}$ for some $p_{ij}, q_{ij} \in G^+$ such that $p_{ij} \leq y_{jn}$ and $2^{n+1}q_{ij} \leq u$. Using Proposition 2.21 again, we obtain an element $b_n \in G^+$ such that $2^{n+1}b_n \leq u$ and $q_{ij} \leq b_n$ for all i, j. Then $\|b_n\| \leq 1/2^{n+1}$ and

$$x_{in} = p_{ij} + q_{ij} \leq y_{jn} + b_n$$

for all i, j.

Finally, we construct elements $z_1, z_2, \ldots$ in G such that

$$x_{in} \leq z_n \leq y_{jn} + b_n$$

for all i, j, n, while also $\|z_{n+1} - z_n\| \leq 1/2^n$ for all n.

As $x_{i1} \leq y_{j1} + b_1$ for all i, j, interpolation in G immediately provides us with an appropriate element z_1. Now suppose that $z_1, \ldots, z_n$ have been constructed, for some n. Then

$$x_{i,n+1} \leq y_{j,n+1} + b_{n+1},$$

$$x_{i,n+1} \leq x_{in} + a_n \leq z_n + a_n,$$

$$z_n - b_n - a_n \leq y_{jn} - a_n \leq y_{j,n+1} \leq y_{j,n+1} + b_{n+1}$$

for all i, j. Hence, there exists $z_{n+1} \in G$ such that

$$\begin{matrix} x_{1,n+1} \\ x_{2,n+1} \\ z_n - b_n - a_n \end{matrix} \quad \leq \quad z_{n+1} \quad \leq \quad \begin{matrix} y_{1,n+1} + b_{n+1} \\ y_{2,n+1} + b_{n+1} \\ z_n + a_n. \end{matrix}$$

Since $-(a_n + b_n) \leq z_{n+1} - z_n \leq a_n \leq a_n + b_n$, we conclude that

$$\|z_{n+1} - z_n\| \leq \|a_n + b_n\| \leq \|a_n\| + \|b_n\| \leq 1/2^n.$$

This completes the induction step.

The z_n form a Cauchy sequence in G, and hence there exists $h \in \overline{G}$ such that $\phi(z_n) \to h$. Note also that $\phi(b_n) \to 0$. Since

$$\phi(x_{in}) \leq \phi(z_n) \leq \phi(y_{jn}) + \phi(b_n)$$

for all i, j, n, we conclude that $f_i \leq h \leq g_j$ for all i, j.

Therefore $\overline{G}$ has interpolation, as desired. □

Although norm-completions are always archimedean, norm-completeness alone is insufficient to imply the archimedean property, because a partially ordered abelian group (G, u) with order-unit that is norm-complete need only be isomorphic as a group to its norm-completion. For example, if K is any Choquet simplex, then Aff(K) when equipped with the strict ordering is a norm-complete dimension group, but $(\mathrm{Aff}(K), \ll)$ is not archimedean if K contains more than one point. Note that the norm-completion of $(\mathrm{Aff}(K), \ll)$ is isomorphic (as an ordered group) to $(\mathrm{Aff}(K), \leq)$.

• Quotient Groups.

THEOREM 15.4. *Let (G, u) be a norm-complete dimension group with order-unit, and let H be an ideal of G. Then G/H is norm-complete if and only if H is norm-closed in G.*

PROOF. First assume that G/H is norm-complete. In particular, the norm-topology on G/H is Hausdorff. Consider a sequence $\{x_1, x_2, \ldots\}$ in H that converges to an element $x \in G$. For all n, we have $x - x_n \in x + H$, and so $\|x + H\| \leq \|x - x_n\|$, by Proposition 7.15. As a result, $\|x + H\| = 0$, whence $x + H = 0$ and $x \in H$. Thus H is norm-closed.

Conversely, assume that H is norm-closed. Given a coset $x + H$ in G/H with $\|x + H\| = 0$, Proposition 7.15 shows that there exist elements $x_1, x_2, \ldots$ in $x + H$ such that $\|x_n\| < 1/n$ for all n. Then $x_n \to 0$ and $x - x_n \to x$. Since $x - x_n \in H$ for all n, we obtain $x \in H$, whence $x + H = 0$. Thus the norm-topology on G/H is Hausdorff.

Now consider a Cauchy sequence $\{y_1 + H, y_2 + H, \ldots\}$ in G/H. By passing to a subsequence, we may assume that

$$\|(y_{n+1} + H) - (y_n + H)\| < 1/2^n$$

for all n. Set $z_1 = y_1$. Since $\|(y_2 - z_1) + H\| < \frac{1}{2}$, Proposition 7.15 says that there is an element w in $(y_2 - z_1) + H$ such that $\|w\| < \frac{1}{2}$. Set $z_2 = z_1 + w$; then $z_2 + H = y_2 + H$ and $\|z_2 - z_1\| < \frac{1}{2}$. Continuing in this manner, we obtain a sequence $\{z_1, z_2, \ldots\}$ in G such that $z_n + H = y_n + H$ and $\|z_{n+1} - z_n\| < 1/2^n$ for all n. The z_n form a Cauchy sequence in G, which must converge to some element $z \in G$. As

$$\|(y_n + H) - (z + H)\| \leq \|z_n - z\|$$

for all n, we conclude that $y_n + H \to z + H$ in G/H. Therefore G/H is norm-complete. □

For an archimedean norm-complete dimension group with order-unit, we shall show that quotient groups modulo norm-closed ideals are archimedean as well as norm-complete. One lemma is needed first.

LEMMA 15.5. *Let (G, u) be a dimension group with order-unit, and let H be an ideal of G. Let x, y be elements of G^+ such that*

$$\|(x + H) - (y + H)\| < 1/2^n$$

for some $n \in \mathbb{N}$. Then there exists $z \in G^+$ such that $z + H = x + H$ and

$$-u \leq 2^n(z - y) \leq u.$$

PROOF. Set $x_1 = x$ and $y_1 = y$. According to Proposition 7.15, there is an element w in $(x - y) + H$ for which $\|w\| < 1/2^n$. Write

$$w = x_1 - y_1 + x_2 - y_2$$

for some $x_2, y_2 \in H^+$. By Lemma 7.14, $w = y_3 - x_3$ for some $y_3, x_3 \in G^+$ such that $2^n y_3 \leq u$ and $2^n x_3 \leq u$. Now $x_1 + x_2 + x_3 = y_1 + y_2 + y_3$. Hence, there exist elements $z_{ij} \in G^+$ (for $i, j = 1, 2, 3$) such that

$$z_{i1} + z_{i2} + z_{i3} = x_i \quad \text{and} \quad z_{1j} + z_{2j} + z_{3j} = y_j$$

for all i, j. For each i, we have $0 \leq z_{i2} \leq y_2$, and so $z_{i2} \in H$. Similarly, each $z_{2j} \in H$. Also, $2^n z_{i3} \leq 2^n y_3 \leq u$ for each i and $2^n z_{3j} \leq 2^n x_3 \leq u$ for each j.

Set $z = z_{11} + z_{21} + z_{13}$. Then $z - x = z_{21} - z_{12}$, whence $z - x \in H$ and so $z + H = x + H$. As $z - y = z_{13} - z_{31}$, we conclude that

$$-u \leq -2^n z_{31} \leq 2^n(z - y) \leq 2^n z_{13} \leq u. \quad \square$$

THEOREM 15.6. *Let (G, u) be an archimedean norm-complete dimension group with order-unit, and let H be an ideal of G. Then the following conditions are equivalent:*

(a) *G/H is archimedean.*

(b) *G/H is norm-complete.*

(c) *H is norm-closed in G.*

(d) *$H = \ker(X)$ for some $X \subseteq S(G, u)$.*

PROOF. As these results are trivial in case G is zero, we may assume that G is nonzero.

(a) ⇒ (d): In view of Theorem 7.7, we see that

$$\ker(S(G/H, u+H)) = \{0\}.$$

Consequently, $H = \ker(X)$, where

$$X = \{s \in S(G,u) \mid H \subseteq \ker(s)\}.$$

(d) ⇒ (c): Since all states in $S(G,u)$ are continuous with respect to the norm-topology on G, the set $\ker(X)$ must be norm-closed in G.

(c) ⇒ (b): Theorem 15.4.

(b) ⇒ (a): In particular, the norm-topology on G/H is Hausdorff. In view of Proposition 7.17, we see that G^+ is norm-closed in G, and it suffices to prove that $(G/H)^+$ is norm-closed in G/H.

Thus consider any sequence $\{a_1, a_2, \ldots\}$ in $(G/H)^+$ that converges to an element $a \in G/H$. After passing to some subsequence, we may assume that $\|a_n - a_{n-1}\| < 1/2^n$ for all $n > 1$.

Choose elements $x_1, x_2, \ldots$ in G^+ such that $x_n + H = a_n$ for all n. We shall construct elements $y_1, y_2, \ldots$ in G^+ such that $y_n + H = a_n$ for all n and $\|y_n - y_{n-1}\| \leq 1/2^n$ for all $n > 1$. To start, set $y_1 = x_1$. Now let $n > 1$, and assume that $y_1, \ldots, y_{n-1}$ have been constructed. Then

$$\|(x_n + H) - (y_{n-1} + H)\| = \|a_n - a_{n-1}\| < 1/2^n.$$

By Lemma 15.5, there exists $y_n \in G^+$ such that $y_n + H = x_n + H$ and

$$-u \leq 2^n(y_n - y_{n-1}) \leq u,$$

whence $\|y_n - y_{n-1}\| \leq 1/2^n$. Thus the induction works.

Now $\{y_1, y_2, \ldots\}$ is a Cauchy sequence in G, and so there is an element $y \in G$ such that $y_n \to y$. Since G^+ is norm-closed in G and each $y_n \in G^+$, we have $y \in G^+$. As $a_n = y_n + H$ for all n, we also have $a_n \to y + H$. Thus $a = y + H$, because the norm-topology on G/H is Hausdorff, and hence $a \in (G/H)^+$.

Therefore $(G/H)^+$ is norm-closed in G/H, as desired. □

Part (d) of Theorem 15.6 should not be read as saying that the kernel of any subset of $S(G,u)$ is a norm-closed ideal of G, since kernels of subsets of $S(G,u)$ need not be ideals of G. However, kernels of compact subsets of $\partial_e S(G,u)$ are ideals of G (Corollary 15.15), and kernels of closed faces of $S(G,u)$ are ideals of G (Corollary 15.22).

• **Functional Representations.** We have already noted that Corollary 13.6 provides a precise description of the norm-completion of any nonzero dimension group with order-unit (Theorem 15.2). From this result, we obtain a complete affine continuous function representation of any nonzero archimedean norm-complete dimension group, as follows.

THEOREM 15.7. *Let (G,u) be a nonzero, archimedean, norm-complete dimension group with order-unit, and set*

$$A = \{p \in \text{Aff}(S(G,u)) \mid p(s) \in s(G) \textit{ for all discrete } s \in \partial_e S(G,u)\}.$$

Then the natural map $\phi\colon G \to \mathrm{Aff}(S(G,u))$ *provides an isomorphism of* (G,u) *onto* $(A,1)$ *(as ordered groups with order-unit).*

PROOF. In view of Corollary 13.6, $\phi(G)$ is a norm-dense subgroup of A, whence $\phi(G) = A$ by norm-completeness. As G is archimedean, Theorem 7.7 shows that ϕ provides an ordered group isomorphism of G onto A. In addition, $\phi(u) = 1$. □

COROLLARY 15.8. *Let* (G,u) *be a nonzero, archimedean, norm-complete dimension group with order-unit. If there are no discrete extremal states on* (G,u)*, then the natural map*

$$\phi\colon (G,u) \to (\mathrm{Aff}(S(G,u)), 1)$$

is an isomorphism of ordered groups with order-unit. □

THEOREM 15.9. *Let* (G,u) *be a nonzero, archimedean, norm-complete dimension group with order-unit, set* $X = \partial_{\mathrm{e}} S(G,u)$*, and set*

$$B = \{p \in C(X,\mathbb{R}) \mid p(s) \in s(G) \text{ for all discrete } s \in X\}.$$

If X *is compact, then the natural map* $\psi\colon G \to C(X,\mathbb{R})$ *provides an isomorphism of* (G,u) *onto* $(B,1)$ *(as ordered groups with order-unit).*

PROOF. Let $\phi\colon (G,u) \to (A,1)$ be the isomorphism given in Theorem 15.7. Then $\psi = \rho\phi$, where $\rho\colon (A,1) \to (B,1)$ is the restriction map. Since $S(G,u)$ is a Choquet simplex with compact extreme boundary, it follows from Corollary 11.20 that ρ is an isomorphism of ordered groups with order-unit. Therefore ψ is an isomorphism of ordered groups with order-unit as well. □

COROLLARY 15.10. *Let* (G,u) *be an archimedean norm-complete dimension group with order-unit. Then* G *is lattice-ordered if and only if* $\partial_{\mathrm{e}} S(G,u)$ *is compact.*

PROOF. If G is zero, then it is lattice-ordered. In this case, $\partial_{\mathrm{e}} S(G,u)$ is empty and so is compact. Thus we may assume that G is nonzero.

If G is lattice-ordered, then $\partial_{\mathrm{e}} S(G,u)$ is compact by Corollary 12.19. Conversely, if $\partial_{\mathrm{e}} S(G,u)$ is compact, then G is isomorphic (as an ordered group) to the group B defined in Theorem 15.9. Since finite suprema and infima in $C(X,\mathbb{R})$ are given by pointwise maxima and minima, B is closed under finite suprema and infima. Therefore B is lattice-ordered, whence G is lattice-ordered. □

By following the pattern set in Theorem 15.9, we obtain all nonzero, archimedean, norm-complete, lattice-ordered abelian groups with order-unit, as follows.

THEOREM 15.11. *Let* X *be a nonempty compact Hausdorff space. For each* $x \in X$ *let* A_x *be either* $\mathbb{R}$ *or* $(1/n_x)\mathbb{Z}$ *for some* $n_x \in \mathbb{N}$*, and set*

$$B = \{p \in C(X,\mathbb{R}) \mid p(x) \in A_x \text{ for all } x \in X\}.$$

Then $(B,1)$ *is a nonzero, archimedean, norm-complete, lattice-ordered abelian group with order-unit.*

Conversely, any nonzero, archimedean, norm-complete, lattice-ordered abelian group with order-unit is isomorphic (as an ordered group with order-unit) to one of this form.

PROOF. It is clear that $(B, 1)$ is a nonzero, archimedean, lattice-ordered abelian group with order-unit. Note that the order-unit norm on $(B, 1)$ coincides with the supremum norm. Since B is closed in $C(X, \mathbb{R})$, we conclude that $(B, 1)$ is norm-complete.

The converse follows from Corollary 15.10 and Theorem 15.9, because all lattice-ordered abelian groups with order-unit are dimension groups (see Proposition 1.22). □

Using the pattern of Theorem 15.7 in place of the pattern of Theorem 15.9, we might expect to obtain a complete description of all nonzero, archimedean, norm-complete dimension groups with order-unit. However, not all groups constructed in such a manner have interpolation, as the following example shows.

EXAMPLE 15.12. *There exists a Choquet simplex K such that the partially ordered abelian group*

$$A = \{p \in \mathrm{Aff}(K) \mid p(x) \in \mathbb{Z} \text{ for all } x \in \partial_e K\}$$

is not an interpolation group.

PROOF. Set $G = \{x \in \mathbb{R}^{\mathbb{N}} \mid x_n \to (x_1 + x_2)/2\}$, and set $u = (1, 1, 1, \ldots)$. We saw in Example 6.10 that (G, u) is a dimension group with order-unit. It is clear that G is archimedean, and that the order-unit norm on (G, u) is the supremum norm, whence G is norm-complete.

Set $K = S(G, u)$, and observe that there are no discrete states in K. By Theorem 10.17, K is a Choquet simplex, and Corollary 15.8 shows that the natural map

$$\phi\colon (G, u) \to (\mathrm{Aff}(K), 1)$$

is an isomorphism of ordered groups with order-unit. Note that ϕ is also a linear isomorphism.

As shown in Examples 6.10 and 6.18, the extremal states in K are exactly the coordinate projections $s_n\colon G \to \mathbb{R}$. Hence, the group A equals the group

$$\{p \in \mathrm{Aff}(K) \mid p(s_n) \in \mathbb{Z} \text{ for all } n \in \mathbb{N}\},$$

and consequently,

$$\phi^{-1}(A) = \{x \in G \mid s_n(x) \in \mathbb{Z} \text{ for all } n \in \mathbb{N}\} = G \cap \mathbb{Z}^{\mathbb{N}}.$$

We shall show that $\phi^{-1}(A)$ fails to have interpolation.

Set $a = (1, -1, 0, 0, 0, \ldots)$ and $b = (2, 0, 1, 1, 1, \ldots)$ in $\phi^{-1}(A)$, and note that

$$\begin{matrix} a \\ 0 \end{matrix} \le \begin{matrix} b \\ u. \end{matrix}$$

Suppose that there exists $c \in \phi^{-1}(A)$ satisfying

$$\begin{matrix} a & & & & b \\ & \leq & c & \leq & \\ 0 & & & & u. \end{matrix}$$

As $a_1 \leq c_1 \leq u_1$ and $0 \leq c_2 \leq b_2$, we must have $c_1 = 1$ and $c_2 = 0$. But then $c_n \to \frac{1}{2}$, which is impossible, because the c_n are all integers.

Thus $\phi^{-1}(A)$ does not have interpolation, and therefore A does not have interpolation. $\square$

The functional representation for a nonzero, archimedean, norm-complete dimension group (G, u) with order-unit (Theorem 15.7) was derived as a consequence of our ability to fit evaluations at elements of G between certain semicontinuous convex and concave functions on $S(G, u)$ (Theorem 13.5). The analogy between this result and Theorem 11.12 suggests that a version of Theorem 13.5 with the strictness removed from the inequalities should hold in the norm-complete case, by analogy with Theorem 11.13. However, the strongest form of such a result does not hold in general, as the following example shows.

EXAMPLE 15.13. *Let X be the Cantor set in the unit interval, and set $G = C(X, \mathbb{Z})$. Then $(G, 1)$ is an archimedean norm-complete lattice-ordered abelian group with order-unit. Set $S = S(G, 1)$ and let $\phi\colon G \to \mathrm{Aff}(S)$ be the natural map. There exist affine continuous real-valued functions p and q on S such that $p \ll q$ and $[p(s), q(s)] \cap s(G)$ is nonempty for all $s \in S$ yet there are no elements $x \in G$ for which $p \leq \phi(x) \leq q$.*

PROOF. That $(G, 1)$ is an archimedean norm-complete lattice-ordered abelian group with order-unit follows from Theorem 15.11.

Since X is totally disconnected, G separates the points of X. Set $A = C(X, \mathbb{R})$, and let B be the linear subspace of A spanned by G. Then B is a real subalgebra of A that contains the constant functions and separates the points of X. By the Stone-Weierstrass Theorem, B is dense in A.

Let $\rho\colon S(A, 1) \to S$ be the restriction map, and note from Corollary 4.3 that ρ is surjective. Since B is dense in A, and since all states on $(A, 1)$ are bounded linear maps (Proposition 7.18 and Lemma 6.7), ρ is injective. Thus ρ is an affine continuous bijection. By compactness, ρ is a homeomorphism.

Define $f, h \in A$ according to the rules

$$f(x) = x - \tfrac{1}{3} \quad \text{and} \quad h(x) = x + \tfrac{2}{3}.$$

Evaluation at f and at h defines two affine continuous real-valued functions on $S(A, 1)$. Composing these functions with ρ^{-1}, we obtain functions p and q in $\mathrm{Aff}(S)$ such that

$$p(s) = \rho^{-1}(s)(f) \quad \text{and} \quad q(s) = \rho^{-1}(s)(h)$$

for all $s \in S$. As $h = f + 1$, we see that $q = p + 1$. Hence, $p \ll q$. Since the image of every state in S contains $\mathbb{Z}$, we also see that $[p(s), q(s)] \cap s(G)$ is nonempty for all $s \in S$.

Suppose that there is some $g \in G$ for which $p \leq \phi(g) \leq q$. Given any $x \in X$, consider the state $s_x = \rho(\varepsilon_x)$ in S. Then

$$p(s_x) = \rho^{-1}(s_x)(f) = \varepsilon_x(f) = f(x),$$

and, similarly, $q(s_x) = h(x)$. Moreover,

$$\phi(g)(s_x) = s_x(g) = \varepsilon_x(g) = g(x),$$

whence $f(x) \leq g(x) \leq h(x)$.

Thus $f \leq g \leq h$. For all x in $X \cap [0, \frac{1}{3})$, we have

$$-\tfrac{1}{3} \leq f(x) \leq g(x) \leq h(x) < 1,$$

and hence $g(x) = 0$, because g is integer-valued. For x in $X \cap (\frac{1}{3}, 1]$, we have

$$0 < f(x) \leq g(x) \leq h(x) \leq \tfrac{5}{3},$$

and hence $g(x) = 1$. Since $\frac{1}{3}$ is a point of X that lies in the closure of $X \cap [0, \frac{1}{3})$ as well as in the closure of $X \cap (\frac{1}{3}, 1]$, the continuity of g is contradicted.

Therefore no $g \in G$ satisfies $p \leq \phi(g) \leq q$. □

• **Compact Sets of Extremal States.** In this section and the following one, we show that many of the structural results derived for affine continuous function spaces on Choquet simplices in Chapter 11 have analogs for archimedean norm-complete dimension groups.

THEOREM 15.14. *Let (G, u) be an archimedean norm-complete dimension group with order-unit, let X be a compact subset of $\partial_e S(G, u)$, and let $x, y \in G$. Then there exist $z, w \in G$ such that*

$$z \leq \begin{matrix} x \\ y \end{matrix} \leq w,$$

while also

$$s(z) = \min\{s(x), s(y)\} \quad \textit{and} \quad s(w) = \max\{s(x), s(y)\}$$

for all $s \in X$. Moreover, if $x, y \in G^+$, then such elements z, w can be found in G^+.

PROOF. We first construct z in the case that $x, y \in G^+$. Let

$$p \colon S(G, u) \to \mathbb{R}^+$$

be the continuous map given by the rule $p(s) = \min\{s(x), s(y)\}$. We construct elements $z_1, z_2, \ldots$ in G^+ such that $z_n \leq x$ and $z_n \leq y$ for all n, while also

$$s(z_n) > p(s) - 1/2^n \quad \text{and} \quad \|z_{n+1} - z_n\| \leq 1/2^n$$

for all n and all $s \in X$. Using Corollary 12.16, we first obtain elements $w_1, w_2, \ldots$ in G^+ such that $w_n \leq x$ and $w_n \leq y$ for all n, while also $s(w_n) > p(s) - 1/2^n$ for all n and all $s \in X$. Set $z_1 = w_1$.

Now suppose that $z_1, \ldots, z_n$ have been constructed, for some n. Since

$$\begin{matrix} z_n & & x \\ & \leq & \\ w_{n+2} & & y, \end{matrix}$$

there is an element $a \in G$ satisfying

$$\begin{matrix} z_n & & & & x \\ & \leq & a & \leq & \\ w_{n+2} & & & & y. \end{matrix}$$

Note that $s(a) \geq s(w_{n+2}) > p(s) - 1/2^{n+2}$ for all $s \in X$.

The element $a - z_n$ lies in G^+. For all $s \in X$, we have

$$s(z_n) > \min\{s(x), s(y)\} - 1/2^n \geq s(a) - 1/2^n,$$

and hence $s(2^n(a - z_n)) < 1 = s(u)$. According to Corollary 12.17, there is some $b \in G^+$ such that $b \leq a - z_n$ and $2^n b \leq u$, while also

$$s(b) > s(a - z_n) - 1/2^{n+2}$$

for all $s \in X$. Set $z_{n+1} = z_n + b$, and observe that

$$\begin{matrix} & & & & & & x \\ z_n & \leq & z_{n+1} & \leq & a & \leq & \\ & & & & & & y. \end{matrix}$$

As $0 \leq 2^n(z_{n+1} - z_n) = 2^n b \leq u$, we have $\|z_{n+1} - z_n\| \leq 1/2^n$. In addition, for all $s \in X$ we have

$$s(z_{n+1}) = s(z_n) + s(b) > s(a) - 1/2^{n+2} > p(s) - 1/2^{n+1}.$$

This completes the induction step.

The z_n form a Cauchy sequence in G, which must converge to an element $z \in G$. In view of Lemma 7.16, we see that $z \in G^+$ and that $z \leq x$ and $z \leq y$. Also, since $z_k \geq z_n$ whenever $k \geq n$, we obtain $z \geq z_n$ for all n. For any $s \in X$, it follows that

$$\min\{s(x), s(y)\} \geq s(z) \geq s(z_n) > \min\{s(x), s(y)\} - 1/2^n$$

for all n, whence $s(z) = \min\{s(x), s(y)\}$.

Given arbitrary elements $x, y \in G$, choose an element $a \in G$ such that $x + a$ and $y + a$ lie in G^+. Using the result just proved, there exists $b \in G^+$ such that $b \leq x + a$ and $b \leq y + a$, while also

$$s(b) = \min\{s(x + a), s(y + a)\}$$

for all $s \in X$. Set $z = b - a$. Then $z \leq x$ and $z \leq y$, and

$$s(z) = \min\{s(x), s(y)\}$$

for all $s \in X$.

Similarly, there exists an element $c \in G$ such that $c \leq -x$ and $c \leq -y$, while also

$$s(c) = \min\{s(-x), s(-y)\}$$

for all $s \in X$. Set $w = -c$. $\square$

COROLLARY 15.15. *Let (G, u) be an archimedean norm-complete dimension group with order-unit, and let X be a compact subset of $\partial_e S(G, u)$. Then $\ker(X)$ is an ideal of G, and $G/\ker(X)$ is an archimedean norm-complete lattice-ordered abelian group.*

PROOF. The set $H = \ker(X)$ is clearly a convex subgroup of G. Given any $x, y \in H$, we have $s(x) = s(y) = 0$ for all $s \in X$. In view of Theorem 15.14, there is some $w \in G$ for which $w \geq x$ and $w \geq y$, while also

$$s(w) = \max\{s(x), s(y)\} = 0$$

for all $s \in X$. Then $w \in H$, proving that H is directed. Thus H is an ideal of G. By Theorem 15.6, G/H is archimedean and norm-complete.

Given $a, b \in G/H$, write $a = x + H$ and $b = y + H$ for some $x, y \in G$. According to Theorem 15.14, there is some $z \in G$ for which $z \leq x$ and $z \leq y$, while also

$$s(z) = \min\{s(x), s(y)\}$$

for all $s \in X$. Set $c = z + H$, and note that $c \leq a$ and $c \leq b$. We claim that $c = a \wedge b$.

Thus consider any $d \in G/H$ for which $d \leq a$ and $d \leq b$. Write $d = w + H$ for some $w \in G$. Then $w \leq x + h'$ and $w \leq y + h''$ for some $h', h'' \in H$. As H is directed, there exists $h \in H$ such that $h' \leq h$ and $h'' \leq h$, whence $w - h \leq x$ and $w - h \leq y$. By interpolation, there is some $v \in G$ satisfying

$$\begin{matrix} z \\ w-h \end{matrix} \leq v \leq \begin{matrix} x \\ y. \end{matrix}$$

Since $s(z) \leq s(v) \leq \min\{s(x), s(y)\} = s(z)$ for all $s \in X$, the element $v - z$ lies in H. Thus

$$d = w + H = (w - h) + H \leq v + H = z + H = c,$$

which shows that $c = a \wedge b$, as claimed.

Therefore G/H is lattice-ordered. □

The quotient group $G/\ker(X)$ in Corollary 15.15 must be representable in terms of continuous real-valued functions on a compact Hausdorff space, as in Theorem 15.11. The compact Hausdorff space in question is just X, as we shall prove in Theorem 15.20.

COROLLARY 15.16. *Let (G, u) be an archimedean norm-complete dimension group with order-unit, and let s be an extremal state on (G, u). Then $\ker(s)$ is an ideal of G, and s induces an isomorphism of $G/\ker(s)$ onto the subgroup $s(G) \subseteq \mathbb{R}$ (as ordered groups). Moreover, either $s(G) = \mathbb{R}$ or $s(G) = (1/m)\mathbb{Z}$ for some $m \in \mathbb{N}$.*

PROOF. By Corollary 15.15, the subgroup $H = \ker(s)$ is an ideal of G, and the quotient group G/H is archimedean and norm-complete. The induced map $\bar{s}\colon G/H \to s(G)$ is a group isomorphism and a positive homomorphism.

Now let us consider any coset $x + H$ in G/H for which $\overline{s}(x + H) \geq 0$. Then $s(x) \geq 0 = s(0)$. By Theorem 15.14, there is some $z \in G$ such that $z \leq x$ and $z \leq 0$, while also $s(z) = 0$. Since $z \in H$ and $x - z \geq 0$, we conclude that

$$x + H = (x - z) + H \geq 0.$$

Thus $\overline{s}$ is an order-isomorphism.

Therefore $\overline{s}$ is an isomorphism of $(G/H, u + H)$ onto $(s(G), 1)$ (as ordered groups with order-unit). Consequently, $s(G)$ is norm-complete. Since the order-unit norm on $s(G)$ is just absolute value, it follows that $s(G)$ is closed in $\mathbb{R}$. Hence, either $s(G) = \mathbb{R}$ or $s(G)$ is a discrete subgroup of $\mathbb{R}$. In the latter case, s is a discrete state, and $s(G) = (1/m)\mathbb{Z}$ for some $m \in \mathbb{N}$. □

In the absence of norm-completeness, Corollaries 15.15 and 15.16 may fail, as the following examples show.

EXAMPLE 15.17. *There exists an archimedean simple dimension group (G, u) with order-unit possessing an extremal state s such that* $\ker(s) = \{0\}$ *but G is not lattice-ordered.*

PROOF. Let G be the rational linear subspace of $\mathbb{R}^2$ spanned by the vectors $u = (1, 1)$ and $v = (\pi, -\pi)$. It is clear that G is archimedean and that u is an order-unit in G. Note also that u and v are $\mathbb{R}$-linearly independent, whence G is dense in $\mathbb{R}^2$. Since 1 and π are $\mathbb{Q}$-linearly independent, we see that

$$G^+ = \{(0, 0)\} \cup \{(a, b) \in G \mid a > 0 \text{ and } b > 0\}.$$

Thus, the ordering on G is the strict ordering. As G is dense in $\mathbb{R}^2$, Theorem 14.18 shows that G is a simple dimension group.

Since G is dense in $\mathbb{R}^2$, the restriction map

$$\rho \colon S(\mathbb{R}^2, u) \to S(G, u)$$

must be injective. In addition, ρ is surjective by Corollary 4.3, and hence ρ is an affine homeomorphism. Define a map $t \colon \mathbb{R}^2 \to \mathbb{R}$ according to the rule $t(x_1, x_2) = x_1$. Then t is an extremal state on $(\mathbb{R}^2, u)$, and so the restriction $s = \rho(t)$ is an extremal state on (G, u). That $\ker(s) = \{0\}$ is clear.

It remains to show that G is not lattice-ordered, which we do by showing that the set $\{u, v\}$ has no infimum in G. Thus consider any $x \in G$ for which $x \leq u$ and $x \leq v$. Write $x = au + bv$ for some $a, b \in \mathbb{Q}$, and set

$$\alpha = \min\{1 - b\pi, 1 + b\pi, (1 - b)\pi, (b - 1)\pi\}.$$

From the inequalities $x \leq u$ and $x \leq v$, we compute that $a \leq \alpha$. As α is irrational, $a < \alpha$, and we may choose $c \in \mathbb{Q}$ satisfying $a < c < \alpha$. Setting $y = cu + bv$, we conclude that

$$x < y \leq \begin{matrix} u \\ v. \end{matrix}$$

Therefore $\{u, v\}$ has no infimum in G. □

EXAMPLE 15.18. *There exists an archimedean simple dimension group (G,u) with order-unit possessing an extremal state s such that* $\ker(s)$ *is not an ideal of G.*

PROOF. Let G be the rational linear subspace of $\mathbb{R}^3$ spanned by the vectors

$$u = (1,1,1); \quad v = (\pi, -\pi, 0); \quad w = (\pi^2, \pi^2, \pi).$$

It is clear that G is archimedean and that u is an order-unit in G. Note also that u, v, w are $\mathbb{R}$-linearly independent, whence G is dense in $\mathbb{R}^3$.

We claim that if $x = (x_1, x_2, x_3)$ is any nonzero element of G^+, each $x_i > 0$. Write $x = au + bv + cw$ for some a, b, c in $\mathbb{Q}$. If $x_1 = 0$, then $a + b\pi + c\pi^2 = 0$. In this case, $a = b = c = 0$ (because $1, \pi, \pi^2$ are $\mathbb{Q}$-linearly independent), and hence $x = 0$, which is a contradiction. Thus $x_1 > 0$, and similarly $x_2 > 0$. If $x_3 = 0$, then $a + c\pi = 0$, whence $a = c = 0$ and $x = b(\pi, -\pi, 0)$. Since $x \geq 0$, it follows that $b = 0$ and hence $x = 0$, a contradiction. Thus $x_3 > 0$, proving the claim.

In light of this claim, the ordering on G is the strict ordering. Since G is dense in $\mathbb{R}^3$, Theorem 14.18 now shows that G is a simple dimension group.

Define a state $s \in S(G,u)$ by the rule $s(x_1, x_2, x_3) = x_3$. As in Example 15.17, the restriction map from $S(\mathbb{R}^3, u)$ to $S(G,u)$ is an affine isomorphism, from which it follows that s is extremal.

Now $u \notin \ker(s)$ and $v \in \ker(s)$, whence $\ker(s) \neq G$ and $\ker(s) \neq \{0\}$. Since G is simple, $\ker(s)$ thus cannot be an ideal of G. $\square$

LEMMA 15.19. *Let (G,u) be an interpolation group with order-unit, and let X be either a compact subset of $\partial_e S(G,u)$ or a closed face of $S(G,u)$. Let t be an extremal state in $S(G,u)$ that does not lie in X. Then there exists $x \in G^+$ such that $t(x) > 1$ while $s(x) < 1$ for all $s \in X$.*

PROOF. Set $A = \{a \in G^+ \mid t(a) > 1\}$. As t is extremal, Theorem 12.14 shows that A is downward directed. For all $a \in A$, set

$$W(a) = \{s \in S(G,u) \mid s(a) < 1\},$$

which is an open subset of $S(G,u)$. We claim that these open sets cover X.

Thus consider any $s \in X$. Since $\{t\}$ is a face of $S(G,u)$, and since either $\{s\}$ or X is a face of $S(G,u)$, we see that s and t lie in disjoint faces of $S(G,u)$. According to Proposition 6.16, there exist elements $a, b \in G^+$ such that $2u = a+b$ while also $t(2u - a) < 1$ and $s(2u - b) < 1$. Then $t(a) > t(2u) - 1 = 1$, whence $a \in A$. In addition, $s(a) = s(2u - b) < 1$, so that $s \in W(a)$. Therefore the $W(a)$ cover X, as claimed.

By compactness, $X \subseteq W(a_1) \cup \cdots \cup W(a_n)$ for some $a_1, \ldots, a_n$ in A. As A is downward directed, there is some $x \in A$ such that each $a_i \geq x$. Any $s \in X$ lies in $W(a_i)$ for some i, whence $s(x) \leq s(a_i) < 1$. $\square$

THEOREM 15.20. *Let (G,u) be an archimedean norm-complete dimension group with order-unit, let X be a nonempty compact subset of $\partial_e S(G,u)$, and set*

$H = \ker(X)$. *Then*

$$X = \{s \in \partial_e S(G,u) \mid H^+ \subseteq \ker(s)\},$$

and the map $S(G/H, u+H) \to S(G,u)$ *induced by the quotient map* $G \to G/H$ *restricts to a homeomorphism of* $\partial_e S(G/H, u+H)$ *onto* X. *Moreover, if*

$$B = \{p \in C(X,\mathbb{R}) \mid p(s) \in s(G) \textit{ for all discrete } s \in X\},$$

then the evaluation map $G \to C(X,\mathbb{R})$ *induces an isomorphism of* $(G/H, u+H)$ *onto* $(B,1)$ *(as ordered groups with order-unit).*

PROOF. Recall from Corollary 15.15 that H is an ideal of G and that G/H is an archimedean norm-complete lattice-ordered abelian group with order-unit.

Obviously any $s \in X$ satisfies $H^+ \subseteq \ker(s)$. Now consider any t in $\partial_e S(G,u)$ such that $t \notin X$. By Lemma 15.19, there exists $x \in G^+$ such that $t(x) > 1$ while $s(x) < 1$ for all $s \in X$. Applying Theorem 15.14 to the compact subset $X \cup \{t\}$ in $\partial_e S(G,u)$ and the elements x, u in G^+, we obtain an element $w \in G^+$ such that $w \geq x$ and $w \geq u$, while also $t(w) = t(x) > 1$ and $s(w) = s(u) = 1$ for all $s \in X$. Then $w - u$ is an element of H^+ for which $t(w-u) > 0$, proving that $H^+ \nsubseteq \ker(t)$. Therefore

$$X = \{s \in \partial_e S(G,u) \mid H^+ \subseteq \ker(s)\}.$$

Set $Y = \partial_e S(G/H, u+H)$, and let $\pi\colon G \to G/H$ be the quotient map. According to Proposition 6.12, the induced map

$$S(\pi)\colon S(G/H, u+H) \to S(G,u)$$

provides an affine homeomorphism of $S(G/H, u+H)$ onto the set

$$F = \{s \in S(G,u) \mid H \subseteq \ker(s)\},$$

which is a closed face of $S(G,u)$. Hence, $S(\pi)$ restricts to a homeomorphism of Y onto $\partial_e F$. In view of the description of X obtained above, we have

$$\partial_e F = F \cap \partial_e S(G,u) = X.$$

Thus $S(\pi)$ restricts to a homeomorphism τ of Y onto X.

Finally, let $\psi\colon G/H \to C(Y,\mathbb{R})$ be the natural map, and let

$$C = \{p \in C(Y,\mathbb{R}) \mid p(s) \in s(G/H) \text{ for all discrete } s \in Y\}.$$

Theorem 15.9 shows that ψ provides an isomorphism of $(G/H, u+H)$ onto $(C,1)$. The homeomorphism $\tau\colon Y \to X$ induces an isomorphism

$$\tau^*\colon (C(X,\mathbb{R}),1) \to (C(Y,\mathbb{R}),1),$$

and we observe that $\tau^*(B) = C$. Hence, τ^* restricts to an isomorphism

$$f\colon (B,1) \to (C,1).$$

Let $\theta\colon G \to C(X,\mathbb{R})$ be the evaluation map, and observe that $\tau^*\theta = \psi\pi$. Thus the map

$$\bar{\theta}\colon (G/H, u+H) \to (B,1)$$

induced by θ coincides with $f^{-1}\psi$, whence $\overline{\theta}$ is an isomorphism of ordered groups with order-unit. $\square$

• **Closed Faces.** We now derive analogs of the results of the previous section for closed faces in state spaces of archimedean norm-complete dimension groups with order-unit. At the end of the section, we extend the main results to finite families of pairwise disjoint closed faces in such state spaces.

THEOREM 15.21. *Let (G,u) be an archimedean norm-complete dimension group with order-unit, and let F be a closed face of $S(G,u)$. Let $x,y \in G$, and assume that $s(x) \leq s(y)$ for all $s \in F$. Then there exist $z,w \in G$ such that*

$$z \leq \begin{matrix} x \\ y \end{matrix} \leq w,$$

while also $s(z) = s(x)$ and $s(w) = s(y)$ for all $s \in F$. Moreover, if $x,y \in G^+$, then such elements z,w can be found in G^+.

PROOF. The proof of Theorem 15.14 may be applied, using Theorem 12.23 and Corollary 12.24 in place of Corollaries 12.16 and 12.17. $\square$

COROLLARY 15.22. *Let (G,u) be an archimedean norm-complete dimension group with order-unit, and let F be a closed face of $S(G,u)$. Then $\ker(F)$ is an ideal of G, and $G/\ker(F)$ is an archimedean norm-complete dimension group.*

PROOF. The set $H = \ker(F)$ is clearly a convex subgroup of G. Given any $x,y \in H$, we have $s(x) = s(y) = 0$ for all $s \in F$. By Theorem 15.21, there is some $w \in G$ for which $w \geq x$ and $w \geq y$, while also $s(w) = 0$ for all $s \in F$. Then $w \in H$, proving that H is directed. Thus H is an ideal of G. Now Theorem 15.6 shows that G/H is archimedean and norm-complete. $\square$

THEOREM 15.23. *Let (G,u) be an archimedean norm-complete dimension group with order-unit, let F be a nonempty closed face of $S(G,u)$, and set $H = \ker(F)$. Then*

$$F = \{s \in S(G,u) \mid H^+ \subseteq \ker(s)\},$$

and the map $S(G/H, u+H) \to S(G,u)$ induced by the quotient map $G \to G/H$ provides an affine homeomorphism of $S(G/H, u+H)$ onto F. Moreover, if

$$A = \{p \in \mathrm{Aff}(F) \mid p(s) \in s(G) \text{ for all discrete } s \in \partial_e F\},$$

then the evaluation map $G \to \mathrm{Aff}(F)$ induces an isomorphism of $(G/H, u+H)$ onto $(A,1)$ (as ordered groups with order-unit).

PROOF. Set $F' = \{s \in S(G,u) \mid H^+ \subseteq \ker(s)\}$, which by Lemma 6.11 is a closed face of $S(G,u)$. Clearly $F \subseteq F'$. To prove the reverse inclusion, we first show that $\partial_e F' \subseteq F$.

If $\partial_e F' \not\subseteq F$, choose a state $t \in \partial_e F'$ that is not in F. Since F' is a face of $S(G,u)$, we see that t is an extreme point of $S(G,u)$. According to Lemma

15.19, there exists $x \in G^+$ such that $t(x) > 1$ while $s(x) < 1$ for all $s \in F$. By Theorem 15.21, there is an element $w \in G^+$ such that $w \geq x$ and $w \geq u$, while also $s(w) = s(u) = 1$ for all $s \in F$. Then $w - u$ is an element of H^+. As $t(w) \geq t(x) > 1$, we also have $t(w - u) > 0$, contradicting the assumption that $t \in F'$.

Thus $\partial_e F' \subseteq F$. Since F' and F are compact convex subsets of a locally convex Hausdorff space, we conclude from the Krein-Mil′man Theorem that $F' \subseteq F$. Therefore $F = F'$.

In particular, since H is directed (Corollary 15.22), it follows that

$$F = \{s \in S(G,u) \mid H \subseteq \ker(s)\}.$$

Hence, if $S = S(G/H, u+H)$ and $\pi\colon G \to G/H$ is the quotient map, Proposition 6.12 shows that the induced map $S(\pi)$ provides an affine homeomorphism of S onto F.

Finally, let $\phi\colon G/H \to \mathrm{Aff}(S)$ be the natural map, and let

$$B = \{p \in \mathrm{Aff}(S) \mid p(s) \in s(G/H) \text{ for all discrete } s \in \partial_e S\}.$$

Since G/H is archimedean and norm-complete (Corollary 15.22), Theorem 15.7 shows that ϕ provides an isomorphism of $(G/H, u+H)$ onto $(B, 1)$. The affine homeomorphism $S(\pi)\colon S \to F$ induces an isomorphism

$$S(\pi)^*\colon (\mathrm{Aff}(F), 1) \to (\mathrm{Aff}(S), 1),$$

and we observe that $S(\pi)^*(A) = B$. Hence, $S(\pi)^*$ restricts to an isomorphism

$$f\colon (A, 1) \to (B, 1).$$

Let $\theta\colon G \to \mathrm{Aff}(F)$ be the evaluation map, and observe that $S(\pi)^*\theta = \phi\pi$. Thus the map

$$\overline{\theta}\colon (G/H, u+H) \to (A, 1)$$

induced by θ coincides with $f^{-1}\phi$, whence θ is an isomorphism of ordered groups with order-unit. □

With the aid of Theorem 15.23, we construct a bijection between the set $\mathcal{F}$ of closed faces in the state space of an archimedean norm-complete dimension group (G, u) with order-unit and the set $\mathcal{L}$ of norm-closed ideals in G. As $\mathcal{F}$ is closed under arbitrary intersections, $(\mathcal{F}, \subseteq)$ is a complete lattice. The set $\mathcal{L}$ is also a complete lattice under inclusion, since $(\mathcal{L}, \subseteq)$ is order anti-isomorphic to $(\mathcal{F}, \subseteq)$, as we now prove.

THEOREM 15.24. *Let (G, u) be an archimedean norm-complete dimension group with order-unit, let $\mathcal{F}$ be the lattice of closed faces of $S(G, u)$, and let $\mathcal{L}$ be the collection of all norm-closed ideals of G, ordered by inclusion. Then the rule $\sigma(F) = \ker(F)$ defines an order anti-isomorphism σ of $\mathcal{F}$ onto $\mathcal{L}$.*

PROOF. For any $F \in \mathcal{F}$, Corollary 15.22 shows that $\ker(F)$ is an ideal of G. Since all states on (G, u) are norm-continuous, $\ker(F)$ is norm-closed. Thus the rule $\sigma(F) = \ker(F)$ does define a map $\sigma\colon \mathcal{F} \to \mathcal{L}$. Note that σ is inclusion-reversing.

Given any $H \in \mathcal{L}$, set

$$\tau(H) = \{s \in S(G,u) \mid H \subseteq \ker(s)\}.$$

By Proposition 6.12, $\tau(H)$ is a closed face of $S(G,u)$. Thus we obtain a map $\tau\colon \mathcal{L} \to \mathcal{F}$, and we note that τ is inclusion-reversing.

For any nonempty $F \in \mathcal{F}$, Theorem 15.23 shows that

$$F = \{s \in S(G,u) \mid \sigma(F)^+ \subseteq \ker(s)\},$$

from which we obtain $F = \tau\sigma(F)$. If F is the empty face of $S(G,u)$, then $\sigma(F) = G$, and again $\tau\sigma(F) = F$. Thus $\tau\sigma$ is the identity map on $\mathcal{F}$.

Finally, consider any $H \in \mathcal{L}$. Theorem 15.6 says that G/H is archimedean, and hence

$$\ker(S(G/H, u+H)) = \{0\},$$

because of Theorem 7.7. As a result, $H = \ker(\tau(H)) = \sigma\tau(H)$. Thus $\sigma\tau$ is the identity map on $\mathcal{L}$.

Therefore σ and τ are inverse order anti-isomorphisms. $\square$

We conclude this section with the following extension of Theorem 15.23 to finite sets of pairwise disjoint closed faces.

THEOREM 15.25. *Let (G,u) be an archimedean norm-complete dimension group with order-unit, let $F_1, \dots, F_k$ be pairwise disjoint nonempty closed faces of $S(G,u)$, and set*

$$H = \ker(F_1) \cap \cdots \cap \ker(F_k).$$

Then H is an ideal of G, and G/H is an archimedean norm-complete dimension group. Let F be the convex hull of $F_1 \cup \cdots \cup F_k$. Then

$$F = \{s \in S(G,u) \mid H^+ \subseteq \ker(s)\},$$

and the map $S(G/H, u+H) \to S(G,u)$ induced by the quotient map $G \to G/H$ provides an affine homeomorphism of $S(G/H, u+H)$ onto F. Moreover, if

$$A_i = \{p \in \operatorname{Aff}(F_i) \mid p(s) \in s(G) \text{ for all discrete } s \in \partial_e F_i\}$$

for each $i = 1, \dots, k$, then the evaluation map

$$G \to \operatorname{Aff}(F_1) \times \cdots \times \operatorname{Aff}(F_k)$$

induces an isomorphism of $(G/H, u+H)$ onto $(A_1, 1) \times \cdots \times (A_k, 1)$ (as ordered groups with order-unit).

PROOF. In view of Propositions 5.2 and 10.10, F is a nonempty closed face of $S(G,u)$. In addition, $H = \ker(F)$. Hence, Corollary 15.22 shows that H is an ideal of G and that G/H is archimedean and norm-complete. Furthermore, Theorem 15.23 shows that, first,

$$F = \{s \in S(G,u) \mid H^+ \subseteq \ker(s)\},$$

that, second, the quotient map $G \to G/H$ induces an affine homeomorphism of $S(G/H, u+H)$ onto F, and that, third, the evaluation map $G \to \mathrm{Aff}(F)$ induces an isomorphism of $(G/H, u+H)$ onto $(A, 1)$, where

$$A = \{p \in \mathrm{Aff}(F) \mid p(s) \in s(G) \text{ for all discrete } s \in \partial_e F\}.$$

Hence, it only remains to prove that the restriction map

$$\rho\colon (A, 1) \to (A_1, 1) \times \cdots \times (A_k, 1)$$

is an isomorphism.

As F is the convex hull of $F_1 \cup \cdots \cup F_k$, any extreme point of F must lie in some F_i and so must be an extreme point of F_i. On the other hand, any extreme point of any F_i is also an extreme point of $S(G, u)$ and hence is an extreme point of F. Thus

$$\partial_e F = \partial_e F_1 \cup \cdots \cup \partial_e F_k.$$

Because of Theorem 10.17 and Proposition 10.9, F is a Choquet simplex, and we note that $F_1, \dots, F_k$ are pairwise disjoint nonempty closed faces of F. By Corollary 11.27, the restriction map

$$\overline{\rho}\colon \mathrm{Aff}(F) \to \mathrm{Aff}(F_1) \times \cdots \times \mathrm{Aff}(F_k)$$

is an isomorphism of ordered vector spaces. Since $\partial_e F = \bigcup \partial_e F_i$, we conclude that $\overline{\rho}(A) = \prod A_i$. Therefore ρ is an isomorphism, as desired. □

• **Maximal Ideals.** The family of all ideals in a directed abelian group G has only one maximal element, namely G itself. Hence, there is little need to discuss the maximal ideal of G. More interesting are the maximal proper ideals of G, that is, the maximal elements in the family of all proper ideals of G, ordered by inclusion. Following long-standing practice with regard to rings and algebras, we use the term "maximal ideal of G" as an abbreviation for "maximal proper ideal of G".

If (G, u) is a nonzero partially ordered abelian group with order-unit, then an ideal H of G is proper if and only if $u \notin H$. Consequently, by Zorn's Lemma every proper ideal of G is contained in a maximal ideal. On the other hand, a partially ordered abelian group without an order-unit need not have any maximal ideals. For example, if I is a nonempty totally ordered set with no minimal elements, and if $G_i = \mathbb{Z}$ for each $i \in I$, then the lexicographic direct sum of the family $\{G_i \mid i \in I\}$ is a dimension group with no maximal ideals.

THEOREM 15.26. *Let (G, u) be an archimedean norm-complete dimension group with order-unit, and let M be a maximal ideal of G. Then M is norm-closed in G, and G/M is an archimedean norm-complete totally ordered abelian group. Moreover, $M = \ker(s)$ for some $s \in \partial_e S(G, u)$.*

PROOF. As M is proper, G/M is nonzero. Hence, Corollary 4.4 shows that the state space of $(G/M, u+M)$ is nonzero. Consequently, by the Krein-Mil′man

Theorem, there exists an extremal state t on $(G/M, u+M)$. Let $p\colon G \to G/M$ be the quotient map, and set $s = tp$. By Proposition 6.12, s is an extreme point of $S(G,u)$.

According to Corollaries 15.15 and 15.16, $\ker(s)$ is an ideal of G, and $G/\ker(s)$ is an archimedean norm-complete totally ordered abelian group. Moreover, the norm-continuity of s implies that $\ker(s)$ is norm-closed in G. Since $M \subseteq \ker(s)$, we conclude from the maximality of M that $M = \ker(s)$. □

In particular, if (G,u) is a simple archimedean norm-complete dimension group with order-unit, then Theorem 15.26 shows that G is totally ordered. This is not always the case in the absence of norm-completeness, as Examples 15.17 and 15.18 show. Furthermore, this may also fail if the interpolation property is removed, as shown by the following example.

EXAMPLE 15.27. *There exists an archimedean norm-complete partially ordered abelian group (G,u) with order-unit such that G is simple but not totally ordered.*

PROOF. Let G be the subgroup of $\mathbb{R}^2$ generated by the vectors $u = (1,1)$ and $v = (\pi, 2\pi)$. It is clear that G is an archimedean partially ordered abelian group, that u is an order-unit in G, and that G is not totally ordered. As 1 and π are $\mathbb{Z}$-linearly independent, we see that all nonzero positive elements of G are strictly positive and so are order-units in G. Thus G is simple, by Lemma 14.1.

By Corollary 4.3, the restriction map from $S(\mathbb{R}^2, u)$ to $S(G,u)$ is surjective. Hence, all states on (G,u) are convex combinations of the two coordinate projections $G \to \mathbb{R}$, from which we infer that

$$\|(a,b)\| = \max\{|a|, |b|\}$$

for all $(a,b) \in G$.

We claim that if x is any nonzero element of G, then $\|x\| \geq 1$. Write

$$x = au + bv = (a + b\pi, a + 2b\pi)$$

for some $a, b \in \mathbb{Z}$; then $\|x\| = \max\{|a + b\pi|, |a + 2b\pi|\}$. Now

$$|b\pi| = |(a + 2b\pi) - (a + b\pi)| \leq |a + 2b\pi| + |a + b\pi| \leq 2\|x\|,$$

whence $\|x\| \geq (\pi/2)|b|$. Thus if $b \neq 0$, then $\|x\| \geq \pi/2 > 1$. On the other hand, if $b = 0$ then $a \neq 0$ and $\|x\| = |a| \geq 1$. This establishes the claim.

Now the order-unit metric on G is actually a metric, and it is discrete. Therefore G is norm-complete. □

PROPOSITION 15.28. *Let (G,u) be an archimedean norm-complete dimension group with order-unit, let $s \in S(G,u)$, and let H be the subgroup of G generated by $\ker(s)^+$. Then the following conditions are equivalent:*

(a) *s is an extremal state.*

(b) *H is a maximal ideal of G.*

(c) *G/H is totally ordered.*

(d) *There is a unique state on $(G/H, u+H)$.*

PROOF. As $\ker(s)$ is a convex subgroup of G, Proposition 1.9 shows that H is an ideal of G. Let $\overline{s}$ denote the state on $(G/H, u+H)$ induced by s, and note that $\ker(\overline{s})^+ = \{0\}$.

(a) $\Rightarrow$ (c): According to Corollary 15.16, $\ker(s)$ is an ideal of G, and $G/\ker(s)$ is totally ordered. Since $\ker(s)$ is directed, $H = \ker(s)$.

(c) $\Rightarrow$ (d): Corollary 4.17.

(d) $\Rightarrow$ (a): The convex set $S = S(G/H, u+H)$ is a singleton, whence $\overline{s}$ is an extreme point of S. By Proposition 6.12, s must be an extreme point of $S(G,u)$.

(d) $\Rightarrow$ (b): Let K be any proper ideal of G that contains H. By Corollary 4.4, there exists a state on $(G/K, u+K)$. Composing this state with the natural map $G/H \to G/K$, we obtain a state t on $(G/H, u+H)$ such that $K/H \subseteq \ker(t)$. Then $t = \overline{s}$ by uniqueness, whence $\ker(t)^+ = \{0\}$, and so $(K/H)^+ = \{0\}$. As K/H is an ideal of G/H, we find that $K/H = \{0\}$, and hence $K = H$. Thus H is a maximal ideal of G.

(b) $\Rightarrow$ (c): Theorem 15.26. □

COROLLARY 15.29. *If (G,u) is an archimedean norm-complete dimension group with order-unit, then the intersection of the maximal ideals of G equals* $\{0\}$.

PROOF. For each extremal state s on (G,u), Corollary 15.16 shows that $\ker(s)$ is an ideal of G, and then Proposition 15.28 shows that $\ker(s)$ is a maximal ideal. Hence, any element $x \in G$ that lies in all the maximal ideals of G must satisfy $s(x) = 0$ for all s in $\partial_e S(G,u)$. As G is archimedean, we conclude from Theorem 7.7 that $x = 0$. Therefore the intersection of the maximal ideals of G is zero. □

DEFINITION. For any directed abelian group G, let $\operatorname{MaxSpec}(G)$ denote the collection of all maximal ideals of G.

We wish to topologize $\operatorname{MaxSpec}(G)$ so that the closed subsets of $\operatorname{MaxSpec}(G)$ will be those of the form

$$\{M \in \operatorname{MaxSpec}(G) \mid M \supseteq H\},$$

where H is an arbitrary ideal of G. Since the ideals of G are exactly the convex subgroups of G generated by nonempty subsets of G^+ (see Proposition 1.9), the closed subsets of $\operatorname{MaxSpec}(G)$ could also be described as those subsets of the form

$$\{M \in \operatorname{MaxSpec}(G) \mid M \supseteq X\},$$

where X is an arbitrary subset of G^+.

It is clear that the empty set and $\operatorname{MaxSpec}(G)$ itself are closed subsets of $\operatorname{MaxSpec}(G)$ and that any intersection of closed subsets of $\operatorname{MaxSpec}(G)$ is a closed subset. However, to get finite unions of closed subsets of $\operatorname{MaxSpec}(G)$ to be closed, we restrict to the case in which G has interpolation, so that the following lemma may be applied.

LEMMA 15.30. *Let G be a directed interpolation group, and let M be a maximal ideal of G. If H_1 and H_2 are ideals of G such that $H_1 \cap H_2 \subseteq M$, then either $H_1 \subseteq M$ or $H_2 \subseteq M$.*

PROOF. Since $H_1 \cap H_2 \subseteq M$, we compute using Proposition 2.5 that

$$M = M + (H_1 \cap H_2) = (M + H_1) \cap (M + H_2).$$

Hence, $M + H_1$ and $M + H_2$ cannot both equal G, and so at least one $M + H_j$ is a proper subset of G. Since $M + H_j$ is an ideal of G (by Proposition 2.4), we conclude from the maximality of M that $M + H_j = M$. Therefore $H_j \subseteq M$. □

LEMMA 15.31. *If G is a directed interpolation group, then there is a T_1 topology on* MaxSpec(G) *such that the closed subsets of* MaxSpec(G) *are exactly the subsets of the form*

$$\{M \in \operatorname{MaxSpec}(G) \mid M \supseteq H\},$$

where H is any ideal of G. If G has an order-unit, this topology is compact.

PROOF. Set $\mathcal{M} = \operatorname{MaxSpec}(G)$, and let $\mathcal{C}$ be the collection of all subsets of $\mathcal{M}$ of the form described. Since the empty set and $\mathcal{M}$ may be expressed in the forms

$$\{M \in \mathcal{M} \mid M \supseteq G\} \quad \text{and} \quad \{M \in \mathcal{M} \mid M \supseteq \{0\}\},$$

they belong to $\mathcal{C}$.

If $\{X_i \mid i \in I\}$ is a nonempty family of sets in $\mathcal{C}$, then each

$$X_i = \{M \in \mathcal{M} \mid M \supseteq H_i\}$$

for some ideal H_i of G, whence

$$\bigcap_{i \in I} X_i = \left\{M \in \mathcal{M} \mid M \supseteq \sum_{i \in I} H_i\right\}.$$

Since $\sum H_i$ is an ideal of G (Proposition 2.4), $\bigcap X_i \in \mathcal{C}$. Thus $\mathcal{C}$ is closed under arbitrary intersections.

Given $X_1, \dots, X_n$ in $\mathcal{C}$, write each $X_i = \{M \in \mathcal{M} \mid M \supseteq H_i\}$ for some ideal H_i of G. In view of Lemma 15.30, we see that

$$X_1 \cup \cdots \cup X_n = \{M \in \mathcal{M} \mid M \supseteq H_1 \cap \cdots \cap H_n\}.$$

As $\bigcap H_i$ is an ideal of G (Proposition 2.4 again), $\bigcup X_i \in \mathcal{C}$. Thus $\mathcal{C}$ is closed under finite unions.

Therefore $\mathcal{C}$ is the collection of all closed sets for a topology on $\mathcal{M}$. For any $M \in \mathcal{M}$, we have

$$\{M\} = \{N \in \mathcal{M} \mid N \supseteq M\},$$

whence $\{M\}$ is closed. Thus $\mathcal{M}$ is T_1.

Now assume that G has an order-unit u, and let $\{U_i \mid i \in I\}$ be a nonempty family of open sets covering $\mathcal{M}$. Each

$$U_i = \{M \in \mathcal{M} \mid M \not\supseteq H_i\}$$

for some ideal H_i of G. As the U_i cover $\mathcal{M}$, no maximal ideal of G can contain the ideal $\sum H_i$. Hence, $u \in \sum H_i$, and so there exist $i(1), \dots, i(n)$ in I such that u is in $H_{i(1)} + \cdots + H_{i(n)}$. Since $H_{i(1)} + \cdots + H_{i(n)}$ is an ideal of G containing u, it must equal G. Consequently, $U_{i(1)}, \dots, U_{i(n)}$ cover $\mathcal{M}$. Therefore $\mathcal{M}$ is compact. $\square$

Given any directed interpolation group G, we shall assume that $\operatorname{MaxSpec}(G)$ has been equipped with the topology described in Lemma 15.31.

THEOREM 15.32. *Let (G, u) be an archimedean norm-complete dimension group with order-unit.*

(a) *There is a continuous bijection $\theta\colon \partial_e S(G,u) \to \operatorname{MaxSpec}(G)$ given by the rule $\theta(s) = \ker(s)$.*

(b) *θ maps compact subsets of $\partial_e S(G,u)$ onto closed subsets of* $\operatorname{MaxSpec}(G)$.

(c) *Let $X \subseteq \partial_e S(G,u)$. Then $\theta(X)$ is closed in* $\operatorname{MaxSpec}(G)$ *if and only if $X = \partial_e F$ for some closed face F in $S(G,u)$.*

(d) *θ is a homeomorphism if and only if $\partial_e S(G,u)$ is compact, if and only if* $\operatorname{MaxSpec}(G)$ *is Hausdorff.*

PROOF. Set $\nabla = \partial_e S(G,u)$ and $\mathcal{M} = \operatorname{MaxSpec}(G)$.

(a) If $s \in \nabla$, then $\ker(s)$ is an ideal of G, by Corollary 15.16. Hence, $\ker(s)$ is generated by $\ker(s)^+$, and so Proposition 15.28 shows that $\ker(s)$ is a maximal ideal of G. Thus the rule given for θ does define a map from ∇ to $\mathcal{M}$. The injectivity of θ follows from Proposition 15.28(d), while the surjectivity of θ follows from Theorem 15.26. Thus θ is a bijection.

Any closed set X in $\mathcal{M}$ has the form

$$X = \{M \in \mathcal{M} \mid M \supseteq H\}$$

for some ideal H of G. Consequently,

$$\theta^{-1}(X) = \{s \in \nabla \mid s(x) = 0 \text{ for all } x \in H\},$$

which is a closed subset of ∇. Therefore θ is continuous.

(b) Given any compact subset Y of ∇, we obtain

$$Y = \{s \in \nabla \mid \ker(Y)^+ \subseteq \ker(s)\}$$

from Theorem 15.20. As a result,

$$\theta(Y) = \{M \in \mathcal{M} \mid \ker(Y)^+ \subseteq M\},$$

which is closed in $\mathcal{M}$.

(c) If $\theta(X)$ is closed in $\mathcal{M}$, then

$$\theta(X) = \{M \in \mathcal{M} \mid M \supseteq H\}$$

for some ideal H of G. By Proposition 6.12, the set

$$F = \{s \in S(G,u) \mid H \subseteq \ker(s)\}$$

is a closed face of $S(G,u)$, and $X = \nabla \cap F = \partial_e F$.

Conversely, assume that $X = \partial_e F$ for some closed face F in $S(G,u)$. In view of Theorem 15.23, we see that

$$X = \nabla \cap F = \{s \in \nabla \mid \ker(F)^+ \subseteq \ker(s)\}.$$

Then $\theta(X) = \{M \in \mathcal{M} \mid \ker(F)^+ \subseteq M\}$, whence $\theta(X)$ is closed in $\mathcal{M}$.

(d) Note that ∇ is Hausdorff while $\mathcal{M}$ is compact (Lemma 15.31). Thus if θ is a homeomorphism, then $\mathcal{M}$ must be Hausdorff, and ∇ must be compact. On the other hand, if ∇ is compact, then using (b) we see that θ is a closed map, whence θ is a homeomorphism.

Finally, assume that $\mathcal{M}$ is Hausdorff. We shall prove that ∇ is compact, by proving that ∇ is closed in $S(G,u)$.

If not, there exists a state s in $\overline{\nabla} - \nabla$. Let H be the subgroup of G generated by $\ker(s)^+$, and note that $H^+ = \ker(s)^+$. Then H is an ideal of G (by Proposition 1.9), and Proposition 15.28 shows that there exist at least two states on $(G/H, u+H)$. Because of the Krein-Mil′man Theorem, there must exist two distinct extreme points t_1 and t_2 in $S(G/H, u+H)$. Composing t_1 and t_2 with the quotient map $G \to G/H$, we obtain distinct states s_1 and s_2 in $S(G,u)$ such that each $\ker(s_i) \supseteq H$. By Proposition 6.12, each s_i is an extreme point of $S(G,u)$.

Thus each $s_i \in \nabla$, and each $M_i = \theta(s_i)$ is a maximal ideal of G such that $M_i \supseteq H$. As $\mathcal{M}$ is Hausdorff, there exist disjoint open sets V_1 and V_2 in $\mathcal{M}$ such that each $M_i \in V_i$. Each

$$V_i = \{M \in \mathcal{M} \mid M \not\supseteq H_i\}$$

for some ideal H_i of G. Since $M_i \not\supseteq H_i$ and H_i is directed, there exists $x_i \in H_i^+$ such that $x_i \notin M_i$. Note that any M in $\mathcal{M}$ that does not contain x_i must be in V_i. Thus each M in $\mathcal{M}$ must contain either x_1 or x_2.

As each $x_i \notin M_i$, we have $x_i \notin H$, whence $s(x_i) > 0$. Set

$$W = \{t \in S(G,u) \mid t(x_1) > 0 \text{ and } t(x_2) > 0\},$$

which is an open subset of $S(G,u)$ that contains s. Since $s \in \overline{\nabla}$, there must exist a state $r \in W \cap \nabla$. But then $\theta(r) \in \mathcal{M}$ with $x_1 \notin \theta(r)$ and $x_2 \notin \theta(r)$, which is a contradiction.

Thus ∇ is closed in $S(G,u)$ and hence is compact. Therefore θ is a homeomorphism in this case also. □

• **Notes.** Theorem 15.3 is due to Burgess and Handelman [**18**, Corollary 1.16], while Theorems 15.4 and 15.6 were proved by the author. Theorems 15.7, 15.9, and 15.11 are due to the author and Handelman [**57**, Theorems 5.1, 5.3, 5.5], along with Corollary 15.10 [**57**, Corollary 5.4] and Example 15.12 [**57**, pp. 894, 895]. Example 15.13 was constructed by the author.

Most of the remainder of the chapter is due to the author. Theorems 15.14 and 15.21 are [**53**, Theorems 3.7, 3.15], and Corollary 15.16 is [**53**, Theorem 3.10]. In Corollaries 15.15 and 15.22, the results that $\ker(X)$ and $\ker(F)$ are ideals and $G/\ker(X)$ is lattice-ordered are [**53**, Corollaries 3.8, 3.16]. Example 15.17 is a revised version of an example suggested by a referee [**53**, Example 3.9]. The characterizations of X and F obtained in Theorems 15.20 and 15.23 are in [**53**, Theorems 3.12, 3.17].

The material on maximal ideals is an interpolation group analog of a series of results first proved for maximal ideals in regular rings. For example, the analog of Theorem 15.32 was proved for unit-regular rings with countable interpolation by the author, Handelman, and Lawrence in [**60**, Proposition II.14.5 and Theorem II.14.6], and for N^*-complete regular rings by the author in [**53**, Theorems 4.5, 4.9].

CHAPTER 16

Countable Interpolation and Monotone σ-Completeness

In this chapter we investigate the structure of a partially ordered abelian group (G, u) with order-unit which satisfies either the countable interpolation property (a countably infinite analog of the Riesz interpolation property) or the monotone σ-completeness property (a weakening of Dedekind σ-completeness in which only bounded ascending or descending sequences are required to possess suprema or infima). The countable interpolation property is more general, for if G is a monotone σ-complete interpolation group, then G is archimedean and has countable interpolation, and all quotients of G by ideals have countable interpolation. The assumption of countable interpolation in G makes G nearly norm-complete, for if K is the subgroup of G generated by $\ker(S(G,u))^+$, then G/K is an archimedean norm-complete dimension group. Thus the subsequent development of the structure of G provides an illustration and application of the structure theory for archimedean norm-complete dimension groups developed in the previous chapter. For instance, the relationships between compact subsets of $\partial_e S(G,u)$, or closed faces of $S(G,u)$, and quotients of G modulo the kernels of such sets of states are essentially the same as in the norm-complete case. Special to the case of countable interpolation are further restrictions on the structure of $S(G,u)$, such as that all compact subsets of $\partial_e S(G,u)$ are F-spaces and that the closure of the convex hull of any countable union of closed faces in $S(G,u)$ is again a face. The final section of the chapter is concerned with characterizing certain families of quotient groups of G in terms of properties of sets of states from which they arise. For example, a quotient G/H is Dedekind σ-complete if and only if H is the kernel of some compact basically disconnected subset of $\partial_e S(G,u)$.

• Countable Interpolation.

DEFINITION. A partially ordered set G satisfies the *countable interpolation property* provided that for any countable sequences $\{x_1, x_2, \ldots\}$ and $\{y_1, y_2, \ldots\}$ of elements of G such that $x_i \leq y_j$ for all i, j, there exists an element $z \in G$ such that $x_i \leq z \leq y_j$ for all i, j. The sequences $\{x_i\}$ and $\{y_j\}$ are allowed to have repetitions, so that either sequence, or both, might have only finitely many distinct values. Thus, in particular, the countable interpolation property implies the Riesz interpolation property.

For example, any Dedekind σ-complete lattice has countable interpolation. In particular, if X is any compact Hausdorff basically disconnected space, and A is either $\mathbb{R}$ or a discrete subgroup of $\mathbb{R}$, then $C(X, A)$ is Dedekind σ-complete by Theorem 9.2 and so has countable interpolation. That the countable interpolation property is weaker than Dedekind σ-completeness is shown by the following example.

EXAMPLE 16.1. *There exists a dimension group (G, u) with order-unit such that G has countable interpolation but is not lattice-ordered.*

PROOF. Choose an uncountable set X, and choose two distinct elements $a, b \in X$. Let G be the set of those bounded functions $f\colon X \to \mathbb{Z}$ satisfying $f(x) = (f(a) + f(b))/2$ for all but countably many $x \in X$. Certainly G is an unperforated partially ordered abelian group. If u denotes the constant function 1 in G, then u is an order-unit in G. In particular, G is directed.

Now consider any sequences $\{f_1, f_2, \ldots\}$ and $\{g_1, g_2, \ldots\}$ in G such that $f_i \leq g_j$ for all i, j. For any $x \in X$, the set $\{f_1(x), f_2(x), \ldots\}$ of integers is bounded above (by $g_1(x)$, for example), and hence this set has a maximum value, call it m_x. Note that $m_x \leq g_j(x)$ for all j. For each $i = 1, 2, \ldots$, there is a countable set $V_i \subseteq X$ such that

$$f_i(x) = (f_i(a) + f_i(b))/2$$

for all $x \in X - V_i$. Likewise, for each $j = 1, 2, \ldots$, there is a countable set $W_j \subseteq X$ such that

$$g_j(x) = (g_j(a) + g_j(b))/2$$

for all $x \in X - W_j$. We may assume that $a, b \in V_1$. Set

$$Y = (V_1 \cup V_2 \cup \cdots) \cup (W_1 \cup W_2 \cup \cdots),$$

which is a countable subset of X. Define a map $h\colon X \to \mathbb{Z}$ so that $h(y) = m_y$ for all $y \in Y$ while $h(x) = (m_a + m_b)/2$ for all $x \in X - Y$. In particular, $h(a) = m_a$ and $h(b) = m_b$ (because $a, b \in Y$).

For $i \in \mathbb{N}$ and $y \in Y$, we have $f_i(y) \leq m_y = h(y)$. On the other hand, for $i \in \mathbb{N}$ and $x \in X - Y$, we have $x \notin V_i$ and hence

$$f_i(x) = (f_i(a) + f_i(b))/2 \leq (m_a + m_b)/2 = h(x).$$

Thus $f_i \leq h$ for all i. For $j \in \mathbb{N}$ and $y \in Y$, we have $h(y) = m_y \leq g_j(y)$. On the other hand, for $j \in \mathbb{N}$ and $x \in X - Y$, we have $x \notin W_j$ and hence

$$h(x) = (m_a + m_b)/2 \leq (g_j(a) + g_j(b))/2 = g_j(x).$$

Thus $h \leq g_j$ for all j.

In particular, $f_1 \leq h \leq g_1$, whence h is bounded. Since

$$h(x) = (m_a + m_b)/2 = (h(a) + h(b))/2$$

for all $x \in X - Y$, we have $h \in G$. Therefore G has countable interpolation. In particular, as G is directed and unperforated, it follows that G is a dimension group.

Define a map $v \in G$ by setting $v(a) = 0$ and $v(b) = 2$ while $v(x) = 1$ for all x in $X - \{a, b\}$. Consider any $h \in G$ such that $h \leq u$ and $h \leq v$. Then $h(a) \leq v(a) = 0$ and $h(b) \leq u(b) = 1$, and hence $h(x) \leq \frac{1}{2}$ for all but countably many $x \in X$. Choose $z \in X - \{a, b\}$ satisfying $h(z) \leq \frac{1}{2}$, and define a map $h' \colon X \to \mathbb{Z}$ so that $h'(z) = 1$ and $h'(x) = h(x)$ for all x in $X - \{z\}$. Then h' is an element of G for which

$$h \quad < \quad h' \quad \leq \quad \begin{matrix} u \\ v. \end{matrix}$$

Therefore the set $\{u, v\}$ has no infimum in G, whence G is not lattice-ordered. $\square$

LEMMA 16.2. *Let G be a partially ordered set with interpolation. Then G has countable interpolation if and only if whenever*

(a) $x_1 \leq x_2 \leq \cdots$ *in G and $y_1, y_2 \in G$, or*

(b) $x_1, x_2 \in G$ *and* $y_1 \geq y_2 \geq \cdots$ *in G, or*

(c) $x_1 \leq x_2 \leq \cdots$ *and* $y_1 \geq y_2 \geq \cdots$ *in G,*

and $x_i \leq y_j$ for all i, j, there exists $z \in G$ such that $x_i \leq z \leq y_j$ for all i, j.

PROOF. These conditions are obviously necessary. Conversely, assume that countable interpolation always occurs in cases (a), (b), (c).

Let $x_1, x_2, \ldots$ and $y_1, y_2, \ldots$ be elements of G such that $x_i \leq y_j$ for all i, j. Using finite interpolation, there exist elements $u_2, u_3, \ldots$ in G such that

$$\begin{matrix} x_1 \\ x_2 \end{matrix} \quad \leq \quad u_2 \quad \leq \quad \begin{matrix} y_1 \\ y_2 \end{matrix} \quad \text{and} \quad \begin{matrix} u_i \\ x_{i+1} \end{matrix} \quad \leq \quad u_{i+1} \quad \leq \quad \begin{matrix} y_1 \\ y_2 \end{matrix}$$

for all i. Since $u_2 \leq u_3 \leq \cdots$, countable interpolation in case (a) provides an element $v_2 \in G$ such that

$$u_i \quad \leq \quad v_2 \quad \leq \quad \begin{matrix} y_1 \\ y_2 \end{matrix}$$

for all i. Note that each $x_i \leq v_2$.

Repeating this procedure, we obtain elements $v_2, v_3, \ldots$ in G such that

$$x_i \quad \leq \quad v_{j+1} \quad \leq \quad \begin{matrix} v_j \\ y_{j+1} \end{matrix}$$

for all i, j. As $v_2 \geq v_3 \geq \ldots$, repeated countable interpolation in case (b) provides elements $w_2, w_3, \ldots$ in G such that

$$\begin{matrix} x_1 \\ x_2 \end{matrix} \quad \leq \quad w_2 \quad \leq \quad v_j \quad \text{and} \quad \begin{matrix} w_i \\ x_{i+1} \end{matrix} \quad \leq \quad w_{i+1} \quad \leq \quad v_j$$

for all i, j.

Finally, countable interpolation in case (c) yields an element $z \in G$ such that $w_i \le z \le v_j$ for all i, j, whence $x_i \le z \le y_j$ for all i, j. Therefore G satisfies countable interpolation. □

In particular, if G is a directed set with interpolation, Lemma 16.2 shows that G has countable interpolation if and only if all closed intervals in G have countable interpolation. In case G is an interpolation group with an order-unit u, it is only necessary that the closed interval $[0, u]$ have countable interpolation, as follows.

PROPOSITION 16.3. *Let (G, u) be an interpolation group with order-unit. Then G has countable interpolation if and only if $[0, u]$ has countable interpolation.*

PROOF. If G has countable interpolation, then clearly $[0, u]$ does also. Conversely, assume that $[0, u]$ has countable interpolation. For any $v \in G$, the interval $[v, v + u]$ is order-isomorphic to $[0, u]$, whence $[v, v + u]$ has countable interpolation.

Claim I: Given elements $x_1, x_2, \dots$ in $[0, u]$ and $y_1, y_2, \dots$ in $[0, 2u]$ satisfying $x_i \le y_j$ for all i, j, there exists an element $z \in [0, u]$ such that $x_i \le z \le y_j$ for all i, j.

Fix j for a moment. By Riesz decomposition, $y_j = a_j + b_j$ for some a_j, b_j in $[0, u]$. For each $i \in \mathbb{N}$, we have

$$\begin{matrix} x_i & & u \\ & \le & \\ a_j & & y_j, \end{matrix}$$

and hence there is some $v_{ij} \in G$ for which

$$\begin{matrix} x_i & & & & u \\ & \le & v_{ij} & \le & \\ a_j & & & & y_j. \end{matrix}$$

As $y_j = a_j + b_j \le a_j + u$, the elements v_{ij}, u, y_j all lie in the interval $[a_j, a_j + u]$. Since countable interpolation holds in this interval, there exists an element w_j in $[a_j, a_j + u]$ such that

$$\begin{matrix} & & & & u \\ v_{ij} & \le & w_j & \le & \\ & & & & y_j \end{matrix}$$

for all i. Now $w_1, w_2, \dots$ are elements of $[0, u]$ such that $x_i \le w_j$ for all i, j. Using countable interpolation in $[0, u]$, we obtain $z \in [0, u]$ satisfying $x_i \le z \le w_j$ for all i, j, whence $x_i \le z \le y_j$ for all i, j.

Claim II: Given elements $x_1, x_2, \dots$ in $[0, 2u]$ and $y_1, y_2, \dots$ in $[u, 2u]$ satisfying $x_i \le y_j$ for all i, j, there exists an element $z \in [u, 2u]$ such that $x_i \le z \le y_j$ for all i, j.

In this case, we have $2u - y_1, 2u - y_2, \dots$ in $[0, u]$ and $2u - x_1, 2u - x_2, \dots$ in $[0, 2u]$ satisfying $2u - y_j \le 2u - x_i$ for all i, j. Applying Claim I, we obtain $w \in [0, u]$ such that

$$2u - y_j \le w \le 2u - x_i$$

for all i, j. Then $2u - w$ is an element of $[u, 2u]$ such that $x_i \le 2u - w \le y_j$ for all i, j.

Claim III: The interval $[0, 2u]$ satisfies countable interpolation.

Let $x_1, x_2, \ldots$ and $y_1, y_2, \ldots$ be elements of $[0, 2u]$ satisfying $x_i \le y_j$ for all i, j. For each $i \in \mathbb{N}$, Riesz decomposition provides us with some a_i, b_i in $[0, u]$ such that $x_i = a_i + b_i$. Since $a_i \le x_i \le y_j$ for all i, j, Claim I shows that there exists $a \in [0, u]$ such that $a_i \le a \le y_j$ for all i, j. Note that $x_i = a_i + b_i \le a + u$ for all i. For each i, j, we have

$$\begin{matrix} x_i & & a+u \\ & \le & \\ a & & y_j, \end{matrix}$$

and hence there is an element $s_{ij} \in G$ such that

$$\begin{matrix} x_i & & & & a+u \\ & \le & s_{ij} & \le & \\ a & & & & y_j. \end{matrix}$$

Note that $s_{ij} - a \in [0, u]$ for all i, j. With j fixed, we have

$$\begin{matrix} & & u \\ s_{ij} - a & \le & \\ & & y_j - a \end{matrix}$$

for all i. By Claim I, there exists $t_j \in [0, u]$ such that

$$\begin{matrix} & & & & u \\ s_{ij} - a & \le & t_j & \le & \\ & & & & y_j - a \end{matrix}$$

for all i. Note that $t_j + u \in [u, 2u]$ for all j. Since

$$x_i + u - a \le s_{ij} + u - a \le t_j + u$$

for all i, j, Claim II provides us with an element $w \in [u, 2u]$ such that

$$x_i + u - a \le w \le t_j + u$$

for all i, j. Hence, $x_i \le w + a - u \le t_j + a \le y_j$ for all i, j.

Thus countable interpolation in $[0, u]$ implies countable interpolation in $[0, 2u]$. By induction, it follows that $[0, 2^n u]$ has countable interpolation for each $n \in \mathbb{N}$. Given any elements $p \le q$ in G, we have $0 \le q - p \le 2^n u$ for some n, whence the interval $[p, q]$ is order-isomorphic to a subinterval of $[0, 2^n u]$. Consequently, $[p, q]$ must satisfy countable interpolation.

Therefore all closed intervals in G satisfy countable interpolation. As G is directed, we conclude from Lemma 16.2 that G satisfies countable interpolation. □

PROPOSITION 16.4. *Let H be an ideal in a partially ordered abelian group G. If G has countable interpolation, then H and G/H have countable interpolation.*

PROOF. It is clear that H has countable interpolation.

Consider cosets $a_1 + H, a_2 + H, \ldots$ and $b_1 + H, b_2 + H, \ldots$ in G/H such that $a_i + H \le b_j + H$ for all i, j. We construct elements $x_1, y_1, x_2, y_2, \ldots$ in G such that $x_i + H = a_i + H$ for all i and $y_j + H = b_j + H$ for all j, while also $x_i \leq y_j$ for all i, j. To start, set $x_1 = a_1$.

Now assume that $x_1, y_1, \dots, x_{n-1}, y_{n-1}, x_n$ have been constructed, for some n. For each $i = 1, \dots, n$, we have

$$x_i + H = a_i + H \leq b_n + H,$$

and so $x_i \leq b_n + d_i$ for some $d_i \in H$. Since H is directed, there exists $d \in H$ such that each $d_i \leq d$. Setting $y_n = b_n + d$, we obtain $y_n + H = b_n + H$ and $x_i \leq y_n$ for each $i = 1, \dots, n$. A similar argument yields an element $x_{n+1} \in G$ such that $x_{n+1} + H = a_{n+1} + H$ and $x_{n+1} \leq y_j$ for each $j = 1, \dots, n$.

Thus the induction works. By countable interpolation in G, there exists $z \in G$ such that $x_i \leq z \leq y_j$ for all i, j. Therefore

$$a_i + H = x_i + H \leq z + H \leq y_j + H = b_j + H$$

for all i, j. □

In particular, Proposition 16.4 shows that any quotient of a Dedekind σ-complete lattice-ordered abelian group G by an ideal H has countable interpolation, even though G/H need not be Dedekind σ-complete. For example, let G be the direct product of an infinite sequence of copies of $\mathbb{Z}$, and let H be the direct sum of the sequence. Then G/H is a lattice-ordered abelian group with countable interpolation which is not Dedekind σ-complete. (If $X_1, X_2, \dots$ are pairwise disjoint infinite subsets of $\mathbb{N}$, and if f_j denotes the characteristic function of X_j, then the set $\{f_1 + H, f_2 + H, \dots\}$ has no supremum in G/H.)

On the other hand, partially ordered abelian groups with order-unit satisfying countable interpolation have a good supply of Dedekind complete quotient groups, as follows.

PROPOSITION 16.5. *Let (G, u) be a partially ordered abelian group with order-unit, satisfying countable interpolation. Let X be a countable subset of $S(G, u)$, and let H be the subgroup of G generated by* $\ker(X)^+$. *Then G/H is a Dedekind complete lattice-ordered abelian group.*

PROOF. Note from Proposition 1.9 that H is an ideal of G. Since G/H has countable interpolation (Proposition 16.4), there is no loss of generality in assuming that $H = \{0\}$. List X as $\{s_1, s_2, \dots\}$, and set

$$s = \sum_{n=1}^{\infty} \frac{1}{2^n} s_n.$$

Then s is a state in $S(G, u)$ for which

$$\ker(s)^+ = \bigcap_{n=1}^{\infty} \ker(s_n)^+ = \ker(X)^+ = \{0\}.$$

We first prove that G is Dedekind σ-complete. Thus consider any nonempty countable subset $V \subseteq G$ which is bounded above, and let W be the set of upper bounds for V in G. Set

$$\alpha = \inf\{s(w) \mid w \in W\},$$

and choose a countable subset $W_0 \subseteq W$ such that

$$\inf\{s(w) \mid w \in W_0\} = \alpha.$$

By countable interpolation, there exists $x \in G$ such that $v \leq x \leq w$ for all $v \in V$ and $w \in W_0$. Then $x \in W$ and $s(x) = \alpha$.

Given any $w \in W$, countable interpolation provides an element $y \in G$ such that

$$v \leq y \leq \begin{matrix} x \\ w \end{matrix}$$

for all $v \in V$. Then $y \in W$ and $s(y) = \alpha$, whence $x - y$ is an element of $\ker(s)^+$. Consequently, $x - y = 0$, and hence $x \leq w$. Thus x is the supremum of V.

Therefore every nonempty countable subset of G which is bounded above has a supremum in G, and so G is Dedekind σ-complete. As G is directed, it follows that G is a lattice.

Now consider an arbitrary nonempty subset $V \subseteq G$ which is bounded above, and let W be the set of upper bounds for V in G. We again set

$$\alpha = \inf\{s(w) \mid w \in W\},$$

and we choose a countable set $W_0 \subseteq W$ such that

$$\inf\{s(w) \mid w \in W_0\} = \alpha.$$

Since W_0 is nonempty and bounded below, it has an infimum x in G. Then $x \in W$ and $s(x) = \alpha$. Given any $w \in W$, we have $w \wedge x \in W$ and $s(w \wedge x) = \alpha$. Hence, $x - (w \wedge x)$ is an element of $\ker(s)^+$, so that $x = w \wedge x \leq w$. Therefore x is the supremum of V. $\square$

COROLLARY 16.6. *Let (G, u) be a partially ordered abelian group with order-unit, satisfying countable interpolation, and let s be an extremal state on (G, u). Then $\ker(s)$ is an ideal of G, and s induces an isomorphism of $G/\ker(s)$ onto $s(G)$ (as ordered groups). Moreover, either $s(G) = \mathbb{R}$ or $s(G) = (1/m)\mathbb{Z}$ for some $m \in \mathbb{N}$.*

PROOF. If H is the subgroup of G generated by $\ker(s)^+$, then G/H is a Dedekind complete lattice-ordered abelian group, by Proposition 16.5. Let $\bar{s}$ be the state on $(G/H, u + H)$ induced by s, and observe that $\bar{s}$ is extremal. According to Theorem 9.9 and Proposition 8.11, $\ker(\bar{s})$ is an ideal of G/H, and $\bar{s}$ induces an isomorphism of $(G/H)/\ker(\bar{s})$ onto $\bar{s}(G/H)$ (as ordered groups). Since $\ker(\bar{s})^+ = \{0\}$, it follows that $\ker(\bar{s}) = \{0\}$, whence $\ker(s) = H$. Thus $\ker(s)$ is an ideal of G, and $\bar{s}$ provides an isomorphism of $G/\ker(s)$ onto $s(G)$ (as ordered groups). Consequently, $s(G)$ is Dedekind complete, and therefore either $s(G) = \mathbb{R}$ or $s(G) = (1/m)\mathbb{Z}$ for some $m \in \mathbb{N}$. $\square$

• Monotone σ-Completeness.

DEFINITION. A partially ordered set G is *monotone σ-complete* provided that every ascending (descending) sequence $x_1 \leq x_2 \leq \cdots$ ($x_1 \geq x_2 \geq \cdots$) in G which is bounded above (below) in G has a supremum (infimum) in G.

Obviously all Dedekind σ-complete posets are monotone σ-complete. However, the converse fails even for posets with interpolation, as Example 16.8 shows. In addition, there exist monotone σ-complete partially ordered abelian groups that are not interpolation groups, such as $\mathbb{Z}^2$ with the strict ordering. (Monotone σ-completeness follows from the observation that all bounded ascending or descending sequences in $(\mathbb{Z}^2, \ll)$ are eventually constant.) For a directed poset, the existence of finite infima and suprema is the sole difference between Dedekind σ-completeness and monotone σ-completeness, as follows.

LEMMA 16.7. *Let G be a directed monotone σ-complete partially ordered set. Then G is Dedekind σ-complete if and only if G is a lattice.*

PROOF. If G is Dedekind σ-complete, then since G is directed, it must be a lattice.

Conversely, assume that G is a lattice, and consider any countable subset X of G which is bounded below. List X as $\{x_1, x_2, \dots\}$, and set $y_n = x_1 \wedge \cdots \wedge x_n$ for all $n = 1, 2, \dots$. The y_n form a descending sequence in G which is bounded below(by any lower bound for X), and hence the set $\{y_1, y_2, \dots\}$ has an infimum y in G. Then $y = \bigwedge X$. Dually, any countable subset of G which is bounded above has a supremum in G, and thus G is Dedekind σ-complete. □

EXAMPLE 16.8. *There exists a dimension group (G, u) with order-unit such that G is monotone σ-complete but not lattice-ordered.*

PROOF. Define G and u as in Example 16.1. Then (G, u) is a dimension group with order-unit, and G is not lattice-ordered.

Now consider any ascending sequence $f_1 \leq f_2 \leq \cdots$ in G which is bounded above. For any $x \in X$, the sequence $\{f_1(x), f_2(x), \dots\}$ is a bounded ascending sequence of integers and so is eventually constant. Thus we may define a map $f\colon X \to \mathbb{Z}$ by setting

$$f(x) = \max\{f_1(x), f_2(x), \dots\}$$

for all $x \in X$.

Choose a positive integer k such that $f_n(a) = f_k(a)$ and $f_n(b) = f_k(b)$ for all $n \geq k$. For each $n \geq k$, there is a countable set $Y_n \subseteq X$ such that

$$f_n(x) = (f_n(a) + f_n(b))/2 = (f(a) + f(b))/2$$

for all $x \in X - Y_n$. Set $Y = Y_k \cup Y_{k+1} \cup \cdots$. For any $x \in X - Y$, we have

$$f_n(x) = (f(a) + f(b))/2$$

for all $n \geq k$, whence $f(x) = (f(a) + f(b))/2$. As Y is countable, it follows that $f \in G$, and then obviously f is the supremum of the f_n.

Since G is order anti-isomorphic to itself, the existence of infima for bounded descending sequences in G follows. Therefore G is monotone σ-complete. □

PROPOSITION 16.9. *Let (G,u) be an interpolation group with order-unit. Then G is monotone σ-complete if and only if $[0,u]$ is monotone σ-complete.*

PROOF. If G is monotone σ-complete, then clearly $[0,u]$ is also. Conversely, assume that $[0,u]$ is monotone σ-complete. Since $[0,u]$ also satisfies Riesz interpolation, it follows from Lemma 16.2 that $[0,u]$ has countable interpolation. Hence, Proposition 16.3 shows that G has countable interpolation.

Any descending sequence $w_1 \geq w_2 \geq \cdots$ in $[0,u]$ has an infimum w in $[0,u]$, by hypothesis. We claim that w is also the infimum of the w_j in G. Given any $p \in G$ such that $p \leq w_j$ for all j, countable interpolation in G provides us with an element $q \in G$ such that

$$\begin{matrix} p \\ w \end{matrix} \quad \leq \quad q \quad \leq \quad w_j$$

for all j. In particular, $0 \leq w \leq q \leq w_1 \leq u$, so that $q \in [0,u]$. Then $q \leq w$ (by definition of w), whence $p \leq w$. Thus w is the infimum of the w_j in G, as claimed. Hence, we may use the notation $w = \bigwedge w_j$ without ambiguity.

We next show that $[0,2u]$ is monotone σ-complete. Given a descending sequence $x_1 \geq x_2 \geq \cdots$ in $[0,2u]$, we use Riesz decomposition to obtain $x_1 = y_1+z_1$ for some $y_1, z_1 \in [0,u]$. As $x_2 \leq x_1$, a second application of Riesz decomposition yields $x_2 = y_2 + z_2$ for some $y_2 \in [0,y_1]$ and some $z_2 \in [0,z_1]$. Continuing in this manner, we obtain decompositions $x_i = y_i + z_i$ for all i such that each $y_{i+1} \in [0,y_i]$ and each $z_{i+1} \in [0,z_i]$.

Since $y_1 \geq y_2 \geq \cdots$ is a descending chain in $[0,u]$, it has an infimum y, and similarly the descending chain $z_1 \geq z_2 \geq \cdots$ has an infimum z. Using Proposition 1.4, we compute that

$$\begin{aligned} y+z &= \left(\bigwedge_{i=1}^{\infty} y_i\right) + z = \bigwedge_{i=1}^{\infty} (y_i + z) \\ &= \bigwedge_{i=1}^{\infty} \left(y_i + \left(\bigwedge_{j=1}^{\infty} z_j \right) \right) = \bigwedge_{i=1}^{\infty} \bigwedge_{j=1}^{\infty} (y_i + z_j). \end{aligned}$$

As the terms $y_i + z_i$ (for $i \in \mathbb{N}$) form a downward-cofinal subset of the set

$$\{y_i + z_j \mid i,j \in \mathbb{N}\},$$

we find that

$$y+z = \bigwedge_{i=1}^{\infty} (y_i + z_i) = \bigwedge_{i=1}^{\infty} x_i.$$

Since $[0,2u]$ is order anti-isomorphic to itself, it also has suprema for ascending sequences.

Thus monotone σ-completeness in $[0,u]$ implies monotone σ-completeness in $[0,2u]$. By induction, it follows that $[0,2^n u]$ is monotone σ-complete for all $n \in \mathbb{N}$.

Finally, consider any descending sequence $x_1 \geq x_2 \geq \cdots$ in G which has a lower bound y in G. Since $[y, x_1]$ is order-isomorphic to a subinterval of $[0, 2^n u]$ for some n, it is monotone σ-complete, and hence the x_i have an infimum x in $[y, x_1]$. As above, we infer that x is also the infimum of the x_i in G. Therefore G is monotone σ-complete. □

THEOREM 16.10. *If G is a monotone σ-complete interpolation group, then G is archimedean and has countable interpolation.*

PROOF. Countable interpolation follows from Lemma 16.2.

Now consider any $x, y \in G$ such that $nx \leq y$ for all $n \in \mathbb{N}$. Set $z_1 = y$. For any $n \in \mathbb{N}$, we have $(n+1)x \leq z_1$, whence $nx \leq z_1 - x$. By countable interpolation, there exists $z_2 \in G$ such that

$$nx \quad \leq \quad z_2 \quad \leq \quad \begin{matrix} z_1 \\ z_1 - x \end{matrix}$$

for all $n \in \mathbb{N}$. Continuing in this manner, we obtain elements $z_2, z_3, \ldots$ in G such that

$$nx \quad \leq \quad z_{j+1} \quad \leq \quad \begin{matrix} z_j \\ z_j - x \end{matrix}$$

for all $n, j \in \mathbb{N}$. Since the descending sequence $z_1 \geq z_2 \geq \cdots$ is bounded below by x, it has an infimum z in G. For all $j = 1, 2, \ldots$, we have $z \leq z_{j+1} \leq z_j - x$, and so $z + x \leq z_j$. Hence, $z + x \leq z$, and thus $x \leq 0$. Therefore G is archimedean. □

In particular, it follows from Theorem 16.10 and Proposition 1.24 that any directed monotone σ-complete interpolation group is unperforated. From Theorem 16.10 and Proposition 16.4, we see that any quotient of a monotone σ-complete interpolation group G by an ideal H has countable interpolation. However, G/H need not be monotone σ-complete. For example, let G be the direct product of an infinite sequence of copies of $\mathbb{Z}$, and let H be the direct sum of the sequence. (If $X_1, X_2, \ldots$ are pairwise disjoint infinite subsets of $\mathbb{N}$, and if f_j denotes the characteristic function of $X_1 \cup \cdots \cup X_j$, then the bounded ascending sequence $f_1 + H \leq f_2 + H \leq \cdots$ has no supremum in G/H.)

• **Norm-Completeness.** Our most powerful tool for investigating the structure of a partially ordered abelian group (G, u) with order-unit which satisfies countable interpolation is norm-completeness. While G itself need not be norm-complete (for example, consider the quotient of the group of all bounded sequences of integers modulo the ideal of all eventually zero sequences), we shall see that quotients of G by ideals arising from kernels of subsets of the state space $S(G, u)$ are norm-complete.

LEMMA 16.11. *Let G be a 2-unperforated partially ordered abelian group satisfying countable interpolation. Let $n \in \mathbb{N}$, and let $z, x_1, x_2, \ldots$ be elements of G^+ such that $2^n x_i \leq z$ for all i. Then there exists $x \in G^+$ such that $2^n x \leq z$ and each $x_i \leq x$.*

PROOF. First consider the case in which $n = 1$, so that $2x_i \leq z$ for all i. Given any $i, j \in \mathbb{N}$, we have $2(x_i + x_j) \leq 2z$. Then $x_i + x_j \leq z$, and hence $x_i \leq z - x_j$, for all i, j. By countable interpolation, there exists $w \in G$ such that $x_i \leq w \leq z - x_j$ for all i, j. In particular, $x_j \leq z - w$ for all j. Using countable interpolation again, we obtain an element $x \in G$ such that

$$x_i \quad \leq \quad x \quad \leq \quad \begin{matrix} w \\ z - w \end{matrix}$$

for all i. Then $x \geq x_1 \geq 0$ and $2x \leq w + z - w = z$, completing this case.

Now let $n > 1$, and assume that the lemma holds for lower powers of 2. Since $2^{n-1}(2x_i) \leq z$ for all i, the induction hypothesis provides us with an element $y \in G^+$ such that $2^{n-1}y \leq z$ and $2x_i \leq y$ for all i. By the case above, there exists $x \in G^+$ such that $2x \leq y$ and each $x_i \leq x$. As $2^n x \leq 2^{n-1}y \leq z$, the induction step is complete. $\square$

THEOREM 16.12. *Let (G, u) be a partially ordered abelian group with order-unit, satisfying countable interpolation, and let H be an ideal of G. Then the following conditions are equivalent:*

(a) *G/H is archimedean.*

(b) *G/H is norm-complete.*

(c) *$H = \ker(X)$ for some $X \subseteq S(G, u)$.*

(d) *$H^+ = \ker(X)^+$ for some $X \subseteq S(G, u)$.*

PROOF. Let $\overline{G} = G/H$ and $\bar{u} = u + H$. Then (c) holds if and only if $\ker(Y) = \{0\}$ for some $Y \subseteq S(\overline{G}, \bar{u})$, and (d) holds if and only if $\ker(Y)^+ = \{0\}$ for some $Y \subseteq S(\overline{G}, \bar{u})$. Thus there is no loss of generality in assuming that $H = \{0\}$. As the theorem holds trivially in the case that $G = \{0\}$, we may also assume that G is nonzero.

(a) $\Rightarrow$ (b): Recall from Proposition 1.24 that G is unperforated, whence G is a dimension group. In view of Lemma 7.13, the only element $x \in G$ which satisfies $\|x\| = 0$ is the element $x = 0$. Hence, the norm-topology on G is Hausdorff. Now by Lemma 15.1, we need only show that given any sequence $\{x_1, x_2, \ldots\}$ in G^+ such that $2^n x_n \leq u$ for all n, the series $\sum x_n$ converges in norm in G. Note that $\|x_n\| \leq 1/2^n$ for all n.

For all $k, n \in \mathbb{N}$ with $k > n$, we have

$$\|x_{n+1} + \cdots + x_k\| \leq \|x_{n+1}\| + \cdots + \|x_k\| \leq 1/2^{n+1} + \cdots + 1/2^k < 1/2^n,$$

and hence $2^n(x_{n+1} + \cdots + x_k) \leq u$, by Lemma 7.13. Then for each n, Lemma 16.11 shows that there is an element $y_n \in G^+$ such that $2^n y_n \leq u$ and

$$x_{n+1} + \cdots + x_k \leq y_n$$

for all $k > n$. Note that $\|y_n\| \leq 1/2^n$. For all $k > n$ we have

$$x_1 + \cdots + x_k = x_1 + \cdots + x_n + x_{n+1} + \cdots + x_k \leq x_1 + \cdots + x_n + y_n.$$

In addition, $x_1 + \cdots + x_k \le x_1 + \cdots + x_n \le x_1 + \cdots + x_n + y_n$ for all $k \le n$. By countable interpolation, there exists $x \in G$ such that

$$x_1 + \cdots + x_k \le x \le x_1 + \cdots + x_n + y_n$$

for all $k, n \in \mathbb{N}$. In particular,

$$0 \le x - (x_1 + \cdots + x_n) \le y_n$$

for all n, whence $\|x - (x_1 + \cdots + x_n)\| \le \|y_n\| \le 1/2^n$ for all n. Therefore $\sum x_n$ converges to x.

(b) $\Rightarrow$ (c): Since G is norm-complete, the order-unit metric on G is actually a metric. Hence, any nonzero element $x \in G$ satisfies $\|x\| > 0$, and then Proposition 7.12 shows that $s(x) \ne 0$ for some $s \in S(G, u)$. Thus $\ker(S(G, u)) = \{0\}$.

(c) $\Rightarrow$ (d) a priori.

(d) $\Rightarrow$ (a): We first show that G is 2-unperforated. Because G is directed, it suffices to show that whenever $x, y \in G^+$ and $2x \le 2y$, then $x \le y$. For each $n = 0, 1, 2, \ldots$, we construct elements $a_n, b_n \in G^+$ such that $x = a_n + b_n$ while $a_n \le y$ and $2^n b_n \le x$. To begin, set $a_0 = 0$ and $b_0 = x$.

Now assume that $a_0, \ldots, a_n$ and $b_0, \ldots, b_n$ have been constructed, for some n. Since

$$b_n \le 2b_n = 2(x - a_n) \le 2(y - a_n),$$

we must have $b_n = c + d$ for some $c, d \in G^+$ such that $c \le y - a_n$ and $d \le y - a_n$. Interpolation then provides an element $e \in G^+$ satisfying

$$\begin{matrix} c \\ d \end{matrix} \quad \le \quad e \quad \le \quad \begin{matrix} b_n \\ y - a_n. \end{matrix}$$

Note that $b_n = c + d \le 2e$. Set $a_{n+1} = a_n + e$ and $b_{n+1} = b_n - e$, and observe that $a_{n+1} + b_{n+1} = a_n + b_n = x$. We have $0 \le a_{n+1} \le y$ and $0 \le b_{n+1} \le e$ from the inequalities above. Since

$$2b_{n+1} \le b_{n+1} + e = b_n,$$

we also have $2^{n+1} b_{n+1} \le 2^n b_n \le x$, completing the induction step.

As $a_n \le x$ and $a_n \le y$ for all n, countable interpolation yields an element $z \in G$ such that

$$a_n \quad \le \quad z \quad \le \quad \begin{matrix} x \\ y \end{matrix}$$

for all n. Then $0 \le x - z \le x - a_n = b_n$ and so $2^n(x - z) \le x$ for all n. Consequently,

$$x - z \in \ker(X)^+ = \{0\},$$

and thus $x = z \le y$. Therefore G is 2-unperforated.

To prove that G is archimedean, it suffices to show that whenever $x, y, z \in G^+$ with $2^n x \le 2^n y + z$ for all $n \in \mathbb{N}$, then $x \le y$. For each n, Proposition 2.19

shows that $x = v_n + w_n$ for some $v_n, w_n \in G^+$ such that $v_n \le y$ and $2^n w_n \le z$. By countable interpolation, there exists $v \in G$ such that

$$v_n \quad \le \quad v \quad \le \quad \begin{matrix} x \\ y \end{matrix}$$

for all n. Then $0 \le x - v \le x - v_n = w_n$ and so $2^n(x-v) \le z$ for all n. As a result,

$$x - v \in \ker(X)^+ = \{0\},$$

and thus $x = v \le y$. Therefore G is archimedean. □

COROLLARY 16.13. *Let (G, u) be a partially ordered abelian group with order-unit, satisfying countable interpolation, and let K be the subgroup of G generated by $\ker(S(G,u))^+$. Then K is an ideal of G, and G/K is an archimedean norm-complete dimension group with countable interpolation. Moreover, the map*

$$S(G/K, u+K) \to S(G,u)$$

induced by the quotient map $G \to G/K$ is an affine homeomorphism.

PROOF. Since $\ker(S(G,u))$ is a convex subgroup of G, Proposition 1.9 shows that K is an ideal of G, and we note that $K^+ = \ker(S(G,u))^+$. Then G/K is archimedean and norm-complete by Theorem 16.12, while Proposition 16.4 shows that G/K has countable interpolation. As G/K is unperforated by Proposition 1.24, it is a dimension group.

If $\pi\colon G \to G/K$ is the quotient map, then $S(\pi)$ provides an affine continuous injection of $S(G/K, u+K)$ into $S(G,u)$. As $K \subseteq \ker(S(G,u))$, we see that $S(\pi)$ is surjective. Thus $S(\pi)$ is a continuous bijection and therefore a homeomorphism, by compactness. □

Corollary 16.13 allows most of the results of the previous chapter to be carried over (with minor modifications) to partially ordered abelian groups with order-unit that satisfy countable interpolation. We shall do this explicitly for some of the more important results in the remaining sections of the chapter.

• Functional Representations.

THEOREM 16.14. *Let (G,u) be a nonzero partially ordered abelian group with order-unit, satisfying countable interpolation. Let $\phi\colon G \to \mathrm{Aff}(S(G,u))$ be the natural map, and set*

$$A = \{p \in \mathrm{Aff}(S(G,u)) \mid p(s) \in s(G) \text{ for all discrete } s \in \partial_e S(G,u)\}.$$

Then $\phi(G) = A$ and $\phi(G^+) = A^+$. If $\ker(S(G,u))^+ = \{0\}$, then ϕ provides an isomorphism of (G,u) onto $(A,1)$ (as ordered groups with order-unit).

PROOF. Let K denote the subgroup of G generated by $\ker(S(G,u))^+$, set $\overline{G} = G/K$ and $\bar{u} = u + K$, and let $\pi\colon G \to \overline{G}$ be the quotient map. Then let $\overline{\phi}\colon \overline{G} \to \mathrm{Aff}(S(\overline{G},\bar{u}))$ be the natural map, and set

$$\overline{A} = \{p \in \mathrm{Aff}(S(\overline{G},\bar{u})) \mid p(s) \in s(\overline{G}) \text{ for all discrete } s \in \partial_e S(\overline{G},\bar{u})\}.$$

Since $S(\pi)$ is an affine homeomorphism of $S(\overline{G}, \bar{u})$ onto $S(G, u)$ (Corollary 16.13), the map $\pi^*: (A, 1) \to (\overline{A}, 1)$ induced by $S(\pi)$ is an isomorphism of ordered groups with order-unit. Note that $\pi^*\phi = \overline{\phi}\pi$. Since G is nonzero, $S(G, u)$ is nonempty by Corollary 4.4. Then $S(\overline{G}, \bar{u})$ is nonempty, whence $\overline{G}$ is nonzero.

According to Corollary 16.13, $\overline{G}$ is an archimedean norm-complete dimension group. By Theorem 15.7, $\overline{\phi}$ provides an isomorphism of $(\overline{G}, \bar{u})$ onto $(\overline{A}, 1)$ (as ordered groups with order-unit). In particular, $\overline{\phi}(\overline{G}) = \overline{A}$ and $\overline{\phi}(\overline{G}^+) = \overline{A}^+$, whence $\phi(G) = A$ and $\phi(G^+) = A^+$. $\square$

COROLLARY 16.15. *Let (G, u) be a nonzero monotone σ-complete interpolation group with order-unit, and set*

$$A = \{p \in \mathrm{Aff}(S(G, u)) \mid p(s) \in s(G) \text{ for all discrete } s \in \partial_e S(G, u)\}.$$

Then the natural map $\phi: G \to \mathrm{Aff}(S(G, u))$ provides an isomorphism of (G, u) onto $(A, 1)$ (as ordered groups with order-unit).

PROOF. By Theorem 16.10, G is archimedean and satisfies countable interpolation. Since G is archimedean, $\ker(S(G, u)) = \{0\}$, by Theorem 7.7. Thus Theorem 16.14 applies. $\square$

Recall that the key to the functional representations leading to Theorem 16.14 and Corollary 16.15 was our ability to fit evaluations at elements of an interpolation group (G, u) with order-unit between certain semicontinuous convex and concave functions on the state space $S(G, u)$, as in Theorem 13.5. In general, the strictness may not be removed from the inequalities in that theorem, even when G is an archimedean norm-complete lattice-ordered abelian group, as Example 15.13 shows. Countable interpolation, however, is strong enough to yield a substantial nonstrict version of Theorem 13.5, and we now prepare for this result.

LEMMA 16.16. *Let G be a 2-unperforated directed abelian group with countable interpolation. Let $n \in \mathbb{N}$, and let $z, x_1, x_2, \dots$ be elements of G such that $2^n x_i \le z$ for all i. Then there exists $x \in G$ such that $2^n x \le z$ and each $x_i \le x$.*

PROOF. Choose $u \in G^+$ such that $z + u \ge 0$, and note that $z + 2^n u \ge 0$ as well. For each $i \in \mathbb{N}$, we have

$$\begin{matrix} 2^n 0 \\ 2^n(x_i + u) \end{matrix} \quad \le \quad z + 2^n u.$$

By Corollary 2.22, there exists $y_i \in G$ such that $2^n y_i \le z + 2^n u$ while also $0 \le y_i$ and $x_i + u \le y_i$. As each $y_i \in G^+$, Lemma 16.11 provides us with an element $y \in G^+$ such that $2^n y \le z + 2^n u$ and each $y_i \le y$. Setting $x = y - u$, we conclude that $2^n x \le z$ and that each $x_i \le y_i - u \le x$. $\square$

LEMMA 16.17. *Let (G, u) be a nonzero unperforated partially ordered abelian group with order-unit, satisfying the countable interpolation property. Let*

$\phi: G \to \mathrm{Aff}(S(G,u))$ be the natural map, let $q: S(G,u) \to \mathbb{R}$ be a continuous concave function, and let $x_1, x_2, \ldots \in G$. If $\phi(x_i) \le q$ for all i, there exists $x \in G$ such that $\phi(x) \le q$ and each $x_i \le x$.

PROOF. Set $S = S(G,u)$. We first claim that given any positive real number ε, there exists $y \in G$ such that $\phi(y) \ll q + \varepsilon$ and each $x_i \le y$.

For each $s \in S$, Proposition 11.8 shows that there exists $f_s \in \mathrm{Aff}(S)$ such that $f_s \gg q$ and $f_s(s) < q(s) + \varepsilon$. By compactness, there exist functions $g_1, \ldots, g_k$ in $\mathrm{Aff}(S)$ such that each $g_j \gg q$ and

$$\min\{g_1(s), \ldots, g_k(s)\} < q(s) + \varepsilon$$

for all $s \in S$. Choose a positive real number δ such that each $g_j \gg q + \delta$.

In view of Theorem 7.9, there exist $z_1, \ldots, z_k \in G$ and $n \in \mathbb{N}$ such that

$$g_j - \delta \ll \phi(z_j)/2^n \ll g_j$$

for each j. Then $\phi(x_i) \le q \ll g_j - \delta \ll \phi(z_j)/2^n$ for all i, j, whence $\phi(2^n x_i) \ll \phi(z_j)$ for all i, j. As G is unperforated, Theorem 7.8 shows that $2^n x_i \le z_j$ for all i, j. Using countable interpolation, we obtain $z \in G$ such that $2^n x_i \le z \le z_j$ for all i, j. Then, by Lemma 16.16, there exists $y \in G$ such that $2^n y \le z$ and each $x_i \le y$. For $j = 1, \ldots, k$, we have $2^n y \le z \le z_j$, whence $\phi(y) \le \phi(z_j)/2^n \ll g_j$. Consequently,

$$\phi(y)(s) < \min\{g_1(s), \ldots, g_k(s)\} < q(s) + \varepsilon$$

for all $s \in S$, and thus $\phi(y) \ll q + \varepsilon$, establishing the claim.

Using the claim in the cases $\varepsilon = 1, \frac{1}{2}, \frac{1}{3}, \ldots$, we obtain elements $y_1, y_2, \ldots$ in G such that $\phi(y_j) \ll q + 1/j$ for all j and $x_i \le y_j$ for all i, j. Countable interpolation then provides us with an element $x \in G$ satisfying $x_i \le x \le y_j$ for all i, j. As

$$\phi(x) \le \phi(y_j) \ll q + 1/j$$

for all j, we conclude that $\phi(x) \le q$. $\square$

THEOREM 16.18. *Let (G,u) be a nonzero partially ordered abelian group with order-unit, satisfying countable interpolation. Set $S = S(G,u)$, and let $\phi: G \to \mathrm{Aff}(S)$ be the natural map. Let $p: S \to \mathbb{R}$ be a continuous convex function, and let $q : S \to \mathbb{R}$ be a continuous concave function. Assume that $p \le q$, and that $[p(s), q(s)] \cap s(G)$ is nonempty for all discrete extremal states $s \in S$. Then there exists $x \in G$ such that $p \le \phi(x) \le q$. Moreover, if $q \ge 0$, there exists $x \in G^+$ such that $p \le \phi(x) \le q$.*

PROOF. In view of Corollary 16.13, the problem may be transferred to an unperforated quotient group of G. Thus there is no loss of generality in assuming that G is unperforated.

First assume that $q \ge 0$. Given any positive real number ε, we claim that there exists $y \in G^+$ such that $p \le \phi(y) \le q + \varepsilon$.

For each $i = 1, 2, \ldots$, we have $p - 1/i \ll q + \varepsilon$, and the intersection

$$(p(s) - 1/i, q(s) + \varepsilon) \cap s(G)$$

is nonempty for all discrete $s \in \partial_e S$. Since $q + \varepsilon \gg 0$, Theorem 13.5 says that there exists $x_i \in G^+$ satisfying

$$p - 1/i \ll \phi(x_i) \ll q + \varepsilon.$$

By Lemma 16.17, there exists $y \in G$ such that $\phi(y) \leq q + \varepsilon$ and each $x_i \leq y$. Then

$$p - 1/i \ll \phi(x_i) \leq \phi(y)$$

for all i, and so $p \leq \phi(y)$. As $y \geq x_1 \geq 0$, the claim is established.

Using this claim in the cases $\varepsilon = 1, \frac{1}{2}, \frac{1}{3}, \ldots$, we obtain elements $y_1, y_2, \ldots$ in G^+ such that $p \leq \phi(y_j) \leq q + 1/j$ for all j. Since $\phi(-y_j) \leq -p$ for all j, there exists $z \in G$ such that $\phi(z) \leq -p$ and $-y_j \leq z$ for all j, by Lemma 16.17. Countable interpolation then provides us with an element $x \in G$ such that

$$\begin{matrix} 0 \\ -z \end{matrix} \leq x \leq y_j$$

for all j. Thus $x \in G^+$ and

$$p \leq \phi(-z) \leq \phi(x) \leq \phi(y_j) \leq q + 1/j$$

for all j, whence $p \leq \phi(x) \leq q$.

In the general case, we may choose an integer m such that $q + m \geq 0$. Then $p + m \leq q + m$, and

$$[p(s) + m, q(s) + m] \cap s(G)$$

is nonempty for all discrete $s \in \partial_e S$. By the case just proved, there exists $w \in G^+$ such that $p + m \leq \phi(w) \leq q + m$. Then $x = w - mu$ is an element of G for which $p \leq \phi(x) \leq q$. □

• Compact Sets of Extremal States.

THEOREM 16.19. *Let (G, u) be a partially ordered abelian group with order-unit, satisfying countable interpolation. Let X be a nonempty compact subset of $\partial_e S(G, u)$, and set $H = \ker(X)$.*

(a) *H is an ideal of G.*

(b) *G/H is an archimedean norm-complete lattice-ordered abelian group with countable interpolation.*

(c) $X = \{s \in \partial_e S(G, u) \mid H^+ \subseteq \ker(s)\}$.

(d) *The quotient map $G \to G/H$ induces a homeomorphism of the space $\partial_e S(G/H, u + H)$ onto X.*

PROOF. Because of Corollary 16.13, we may assume that G is an archimedean norm-complete dimension group. Then Corollary 15.15, Proposition 16.4, and Theorem 15.20 provide the desired results. □

THEOREM 16.20. *Let (G,u) be a partially ordered abelian group with order-unit, satisfying countable interpolation. Let X be a nonempty compact subset of $\partial_e S(G,u)$, set $H = \ker(X)$, and set*

$$B = \{p \in C(X,\mathbb{R}) \mid p(s) \in s(G) \text{ for all discrete } s \in X\}.$$

Then the evaluation map $G \to C(X,\mathbb{R})$ induces an isomorphism of $(G/H, u+H)$ onto $(B,1)$ (as ordered groups with order-unit).

PROOF. Corollary 16.13 and Theorem 15.20. □

In general interpolation groups with order-unit, compact sets of extremal states are topologically arbitrary, since any compact Hausdorff space X is homeomorphic to the extreme boundary of the state space of $(C(X,\mathbb{R}),1)$. However, countable interpolation imposes a severe restriction on the topological structure of compact sets of extremal states, as follows.

DEFINITION. A topological space X is called an *F-space* if disjoint open F_σ subsets of X always have disjoint closures.

For example, every basically disconnected compact Hausdorff space is an F-space. On the other hand, there exist compact Hausdorff F-spaces which are connected (for example, $\beta\mathbb{R}^+ - \mathbb{R}^+$, where $\beta\mathbb{R}^+$ denotes the Stone-Čech compactification of $\mathbb{R}^+$), but we shall not digress into the construction of such examples.

LEMMA 16.21. *Let (G,u) be a partially ordered abelian group with order-unit, satisfying countable interpolation, and let V and W be disjoint compact subsets of $\partial_e S(G,u)$. Then there exists $x \in G$ such that $0 \le x \le u$ while $s(x) = 0$ for all $x \in V$ and $t(x) = 1$ for all $t \in W$.*

PROOF. If V is empty, take $x = u$, while if W is empty, take $x = 0$. Thus we may assume that V and W are nonempty.

Set $X = V \cup W$ and $H = \ker(X)$. Let $p \in C(X,\mathbb{R})$ be the characteristic function of W, and note that $p(s) \in \mathbb{Z} \subseteq s(G)$ for all $s \in X$. Also, $0 \le p \le 1$. In view of Theorem 16.20, there exists a coset $y+H$ in G/H such that $0 \le y+H \le u+H$ while $s(y) = 0$ for all $s \in V$ and $t(y) = 1$ for all $t \in W$.

As $y + H \in (G/H)^+$, there is no loss of generality in assuming that $y \in G^+$. Since $y+H \le u+H$, there exists $a \in H^+$ for which $y \le u+a$. Then $y = x+b$ for some $x, b \in G^+$ such that $x \le u$ and $b \le a$. Observing that $b \in H$, we conclude that $s(x) = s(y) = 0$ for all $s \in V$ and that $t(x) = t(y) = 1$ for all $t \in W$. □

THEOREM 16.22. *Let (G,u) be a partially ordered abelian group with order-unit, satisfying countable interpolation. Then any compact subset X of $\partial_e S(G,u)$ is an F-space.*

PROOF. Certainly X may be assumed to be nonempty. After factoring out $\ker(X)$, we may assume, because of Theorem 16.19, that G is archimedean and that $X = \partial_e S(G,u)$.

Let V and W be any disjoint open F_σ subsets of X. Then V is the union of compact subsets $V_1, V_2, \ldots$, and W is the union of compact subsets $W_1, W_2, \ldots$.

For each $i = 1, 2, \ldots$, note that $X - V$ and V_i are disjoint compact subsets of X. By Lemma 16.21, there exists $x_i \in [0, u]$ such that $s(x_i) = 0$ for all $s \in X - V$ and $t(x_i) = 1$ for all $t \in V_i$. Similarly, for each $j = 1, 2, \ldots$, there exists $y_j \in [0, u]$ such that $s(y_j) = 0$ for all $s \in W_j$ and $t(y_j) = 1$ for all $t \in X - W$.

For any $s \in X - V$, we have $s(x_i) = 0 \leq s(y_j)$ for all i, j. On the other hand, any $t \in V$ lies in $X - W$, whence $t(x_i) \leq 1 = t(y_j)$ for all i, j. Thus $s(x_i) \leq s(y_j)$ for all i, j and all $s \in \partial_e S(G, u)$. Since G is archimedean, Theorem 7.7 shows that $x_i \leq y_j$ for all i, j. By countable interpolation, there exists $z \in G$ such that $x_i \leq z \leq y_j$ for all i, j. In particular, $0 \leq x_1 \leq z \leq y_1 \leq u$.

Any $s \in W$ lies in some W_j, whence $0 \leq s(z) \leq s(y_j) = 0$, and so $s(z) = 0$. Similarly, $t(z) = 1$ for all $t \in V$. Thus W and V are contained in the disjoint closed sets

$$\{s \in X \mid s(z) = 0\} \quad \text{and} \quad \{t \in X \mid t(z) = 1\},$$

whence $\overline{W}$ and $\overline{V}$ are disjoint. Therefore X is an F-space. □

In particular, Theorem 16.22 shows that if X is a compact Hausdorff space such that $C(X, \mathbb{R})$ has countable interpolation, then X must be an F-space. Conversely, if X is any compact Hausdorff F-space, it can be proved that $C(X, \mathbb{R})$ has countable interpolation. Assuming that X is also connected, $C(X, \mathbb{R})$ then provides an example of a lattice-ordered abelian group with countable interpolation for which the extreme boundary of the state space is compact and connected.

• Closed Faces.

THEOREM 16.23. *Let (G, u) be a partially ordered abelian group with order-unit, satisfying countable interpolation. Let F be a nonempty closed face of $S(G, u)$, and set $H = \ker(F)$.*

(a) *H is an ideal of G.*

(b) *G/H is an archimedean norm-complete dimension group with countable interpolation.*

(c) *$F = \{s \in S(G, u) \mid H^+ \subseteq \ker(s)\}$.*

(d) *The quotient map $G \to G/H$ induces an affine homeomorphism from $S(G/H, u + H)$ onto F.*

PROOF. Corollaries 16.13 and 15.22, Proposition 16.4, and Theorem 15.23. □

COROLLARY 16.24. *Let (G, u) be a partially ordered abelian group with order-unit, satisfying countable interpolation. Let $\mathcal{F}$ denote the lattice of closed faces of $S(G, u)$, and let $\mathcal{L}$ be the collection of those ideals H of G for which G/H is archimedean, ordered by inclusion. Then the rule $\sigma(F) = \ker(F)$ defines an order anti-isomorphism σ of $\mathcal{F}$ onto $\mathcal{L}$.*

PROOF. That σ defines a map $\mathcal{F} \to \mathcal{L}$ is immediate from Theorem 16.23. Note that σ is inclusion-reversing. Given any $H \in \mathcal{L}$, set

$$\tau(H) = \{s \in S(G, u) \mid H^+ \subseteq \ker(s)\},$$

which by Lemma 6.11 is a closed face of $S(G,u)$. This defines a map $\tau\colon \mathcal{L} \to \mathcal{F}$, and we note that τ is inclusion-reversing.

For any nonempty $F \in \mathcal{F}$, Theorem 16.23 shows that $\tau\sigma(F) = F$. If F is the empty face of $S(G,u)$, then $\sigma(F) = G$ and $\tau\sigma(F) = F$. Thus $\tau\sigma$ is the identity map on $\mathcal{F}$.

For any $H \in \mathcal{L}$, we have $\ker(S(G/H, u+H)) = \{0\}$ by Theorem 7.7 (because G/H is archimedean), whence $\sigma\tau(H) = H$. Thus $\sigma\tau$ is the identity map on $\mathcal{L}$.

Therefore σ and τ are inverse order anti-isomorphisms. □

THEOREM 16.25. *Let (G,u) be a partially ordered abelian group with order-unit, satisfying countable interpolation. Let F be a nonempty closed face of $S(G,u)$, set $H = \ker(F)$, and set*

$$A = \{p \in \mathrm{Aff}(F) \mid p(s) \in s(G) \text{ for all discrete } s \in \partial_e F\}.$$

Then the evaluation map $G \to \mathrm{Aff}(F)$ induces an isomorphism of $(G/H, u+H)$ onto $(A,1)$ (as ordered groups with order-unit).

PROOF. Corollary 16.13 and Theorem 15.23. □

Just as countable interpolation imposes restrictions on compact sets of extremal states (Theorem 16.22), it also imposes restrictions on closed faces of state spaces, as follows.

THEOREM 16.26. *Let (G,u) be a partially ordered abelian group with order-unit, satisfying countable interpolation. If $F_1, F_2, \ldots$ are any closed faces of $S(G,u)$, then the closure of the convex hull of $\bigcup F_n$ is also a face of $S(G,u)$.*

PROOF. Because of Corollary 16.13, there is no loss of generality in assuming that G is an archimedean norm-complete dimension group.

Let C be the convex hull of $\bigcup F_n$, let F be the closure of C, and set

$$F^* = \{s \in S \mid \ker(F)^+ \subseteq \ker(s)\}.$$

By Lemma 6.11, F^* is a closed face of $S(G,u)$, and we note that $F \subseteq F^*$. If $F \neq F^*$, then as F is closed and convex, $\partial_e F^* \not\subseteq F$, by the Krein-Mil′man Theorem. Choose t in $\partial_e F^* - F$, and note that $t \in \partial_e S(G,u)$.

Since t is not in F, it has a neighborhood V which is disjoint from C. In view of Corollary 12.15, we may assume that $V = \{s \in S \mid s(x) < k\}$ for some $x \in G^+$ and some $k \in \mathbb{N}$. Fix $n \in \mathbb{N}$ for a moment. For any $s \in F_n$, we have $s \in C$ and so $s \notin V$, whence $s(x) \geq k = s(ku)$. By Theorem 15.21, there exists $z_n \in G^+$ such that $z_n \leq x$ and $z_n \leq ku$, while also $s(z_n) = k$ for all $s \in F_n$. Countable interpolation now yields an element $z \in G$ such that

$$z_n \quad \leq \quad z \quad \leq \quad \begin{matrix} x \\ ku \end{matrix} \qquad \text{for all } n \in \mathbb{N}.$$

As $z_n \leq z \leq ku$ for all n, we obtain $s(z) = k$ for all $s \in \bigcup F_n$. Since evaluation at z is an affine continuous function, it follows that $s(z) = k$ for all $s \in F$. Consequently, $ku - z$ is an element of $\ker(F)^+$. As $t \in F^*$, the element $ku - z$ must lie in $\ker(t)$, whence $t(z) = k$. But then $t(x) \geq k$, which contradicts the assumption that $t \in V$. Therefore $F = F^*$, and hence F is a face of $S(G, u)$. $\square$

• **Quotient Groups.** Given a partially ordered abelian group (G, u) with order-unit, satisfying countable interpolation, Corollary 16.24 shows that the archimedean quotient groups of G are exactly the quotients $G/\ker(F)$ for closed faces F of $S(G, u)$. Thus any family of archimedean quotient groups of G corresponds to a family of closed faces of $S(G, u)$, which suggests characterizing the families of closed faces corresponding to various families of archimedean quotient groups. We do so for archimedean lattice-ordered quotient groups, for archimedean quotient groups with general comparability, and for Dedekind σ-complete quotient groups.

THEOREM 16.27. *Let (G, u) be a partially ordered abelian group with order-unit, satisfying countable interpolation, and let H be an ideal of G. Then G/H is an archimedean lattice-ordered abelian group if and only if $H = \ker(X)$ for some compact subset X of $\partial_e S(G, u)$, if and only if $H = \ker(F)$ for some closed face F of $S(G, u)$ such that $\partial_e F$ is compact.*

PROOF. We may assume that $H \neq G$. First suppose that G/H is archimedean and lattice-ordered. By Corollary 16.24, $H = \ker(F)$ for some closed face F of $S(G, u)$. Then F is affinely homeomorphic to $S(G/H, u + H)$, by Theorem 16.23, and so $\partial_e F$ is homeomorphic to $\partial_e S(G/H, u + H)$. As G/H is lattice-ordered, Corollary 12.19 shows that $\partial_e F$ is compact.

Next suppose that $H = \ker(F)$ for some closed face F of $S(G, u)$ such that $\partial_e F$ is compact. In this case, $H = \ker(\partial_e F)$ by the Krein-Mil′man Theorem.

Finally, suppose that $H = \ker(X)$ for some compact subset X of $\partial_e S(G, u)$. In view of Theorem 16.12, G/H is an archimedean norm-complete dimension group. By Theorem 16.19, $\partial_e S(G/H, u + H)$ is homeomorphic to X and so is compact. Hence, Corollary 15.10 shows that G/H is lattice-ordered. $\square$

COROLLARY 16.28. *Let (G, u) be a monotone σ-complete interpolation group with order-unit. Then G is lattice-ordered if and only if (G, u) satisfies general comparability, if and only if $\partial_e S(G, u)$ is compact.*

PROOF. If G is lattice-ordered, G is Dedekind σ-complete by Lemma 16.7, and hence Theorem 9.9 shows that (G, u) satisfies general comparability. If (G, u) satisfies general comparability, then $\partial_e S(G, u)$ is compact by Corollary 8.15.

Now assume that $\partial_e S(G, u)$ is compact. By Theorem 16.10, G is archimedean and has countable interpolation. As G is archimedean, the kernel of $\partial_e S(G, u)$ is zero, by Theorem 7.7. Hence, Theorem 16.27 shows that G is lattice-ordered. $\square$

THEOREM 16.29. *Let (G, u) be a partially ordered abelian group with order-unit, satisfying countable interpolation, and let H be an ideal of G. Then G/H is archimedean with general comparability if and only if $H = \ker(X)$ for some compact totally disconnected subset X of $\partial_e S(G, u)$, if and only if $H = \ker(F)$ for some closed face F of $S(G, u)$ such that $\partial_e F$ is compact and totally disconnected.*

PROOF. We may assume that $H \neq G$. First assume G/H is archimedean and $(G/H, u+H)$ satisfies general comparability. By Corollary 16.24 and Theorem 16.23, $H = \ker(F)$ for some closed face F of $S(G,u)$ and $\partial_e F$ is homeomorphic to $\partial_e S(G/H, u+H)$. Then by Corollary 8.15, $\partial_e F$ is compact and totally disconnected.

If $H = \ker(F)$ for some closed face F of $S(G,u)$ such that $\partial_e F$ is compact and totally disconnected, then also $H = \ker(\partial_e F)$.

Finally, assume that $H = \ker(X)$ for some compact totally disconnected subset X of $\partial_e S(G,u)$. By Theorem 16.19, $\partial_e S(G/H, u+H)$ is homeomorphic to X. Hence, after passing to G/H, there is no loss of generality in assuming that $H = \{0\}$ and that $X = \partial_e S(G,u)$. Set

$$B = \{p \in C(X,\mathbb{R}) \mid p(s) \in s(G) \text{ for all discrete } s \in X\}.$$

By Theorem 16.20, the natural map $\psi\colon G \to C(X,\mathbb{R})$ provides an isomorphism of (G,u) onto $(B,1)$ (as ordered groups with order-unit). In particular, G is archimedean.

Now consider any $x, y \in G$, and set

$$V = \{s \in X \mid s(x) < s(y)\} \quad \text{and} \quad W = \{s \in X \mid s(x) > s(y)\}.$$

Then V and W are disjoint open F_σ subsets of X. As X is an F-space (Theorem 16.22), the closures $\overline{V}$ and $\overline{W}$ are disjoint. Since X is totally disconnected, there exists a clopen set $U \subseteq X$ such that $\overline{V} \subseteq U$ and $\overline{W} \subseteq X - U$. Let $q \in C(X,\mathbb{R})$ be the characteristic function of U, and observe that q is a characteristic element of $(B,1)$. Then $e = \psi^{-1}(q)$ is a characteristic element of (G,u).

By definition of W, we see that $\psi(x) \leq \psi(y)$ on $X - W$. As U is disjoint from $\overline{W}$, we obtain $\psi(x) \leq \psi(y)$ on U. Then $p_q\psi(x) \leq p_q\psi(y)$, whence $p_e(x) \leq p_e(y)$. Similarly, $\psi(x) \geq \psi(y)$ on $X - U$, whence $p_{1-q}\psi(x) \geq p_{1-q}\psi(y)$, and so $p_{u-e}(x) \geq p_{u-e}(y)$. Therefore (G,u) satisfies general comparability. □

THEOREM 16.30. *Let (G,u) be a partially ordered abelian group with order-unit, satisfying countable interpolation, and let H be an ideal of G. Then G/H is Dedekind σ-complete if and only if $H = \ker(X)$ for some compact basically disconnected subset X of $\partial_e S(G,u)$, if and only if $H = \ker(F)$ for some closed face F of $S(G,u)$ such that $\partial_e F$ is compact and basically disconnected.*

PROOF. We may assume that $H \neq G$. First assume that G/H is Dedekind σ-complete. As G/H is directed, it must be lattice-ordered, and then Proposition 9.7 shows that G/H is archimedean. By Corollary 16.24 and Theorem 16.23, $H = \ker(F)$ for some closed face F of $S(G,u)$ and $\partial_e F$ is homeomorphic to $\partial_e S(G/H, u+H)$. Hence, Corollary 9.10 shows that $\partial_e F$ is compact and basically disconnected.

If $H = \ker(F)$ for some closed face F of $S(G,u)$ such that $\partial_e F$ is compact and basically disconnected, then also $H = \ker(\partial_e F)$.

Finally, assume that $H = \ker(X)$ for some compact basically disconnected subset X of $\partial_e S(G,u)$. There is no loss of generality in assuming that $H = \{0\}$

and that $X = \partial_e S(G, u)$. Note from Theorem 9.2 that $C(X, \mathbb{R})$ is Dedekind σ-complete. By Corollary 11.20, $\mathrm{Aff}(S(G, u))$ is isomorphic to $C(X, \mathbb{R})$ (as ordered vector spaces), and hence $\mathrm{Aff}(S(G, u))$ is Dedekind σ-complete. Set

$$A = \{p \in \mathrm{Aff}(S(G, u)) \mid p(s) \in s(G) \text{ for all discrete } s \in X\}.$$

By Theorem 16.25, the natural map $\phi: G \to \mathrm{Aff}(S(G, u))$ provides an isomorphism of (G, u) onto $(A, 1)$ (as ordered groups with order-unit). In particular, G is unperforated.

Now consider any countable subset $\{x_1, x_2, \ldots\}$ of G that is bounded above in G. As $\mathrm{Aff}(S(G, u))$ is Dedekind σ-complete, the set $\{\phi(x_1), \phi(x_2), \ldots\}$ has a supremum q in $\mathrm{Aff}(S(G, u))$. By Lemma 16.17, there exists an element $x \in G$ such that $\phi(x) \leq q$ and each $x_i \leq x$. Since each $\phi(x_i) \leq \phi(x)$, we conclude that $\phi(x) = q$, and thus $x = \bigvee x_i$. Therefore G is Dedekind σ-complete. □

• **Notes.** This chapter is mainly due to the author, Handelman, and Lawrence [**60**]. An example with the properties of Examples 16.1 and 16.8 was constructed in [**60**, Example I.1.7]. Also, there exists a Choquet simplex K such that $\mathrm{Aff}(K)$ is monotone σ-complete but not lattice-ordered, and $\mathrm{Aff}(K)$ is "prime" in the sense that given any nonzero functions f_1, f_2 in $\mathrm{Aff}(K)^+$, there exists a nonzero function f in $\mathrm{Aff}(K)^+$ such that each $f_i \geq f$ [**60**, Theorem IV.18.4]. Proposition 16.3 is [**60**, Proposition I.1.6], while Proposition 16.9 was proved by the author. Proposition 16.5 is a slight generalization of [**60**, Proposition I.2.9], and Corollary 16.6 is [**60**, Theorem I.4.8(a)]. Theorem 16.10, under the additional assumption that G is directed, was proved by Handelman, Higgs, and Lawrence [**72**, Theorem 1.3 and Lemma 1.2].

The equivalence of conditions (a), (c), (d) in Theorem 16.12 is [**60**, Corollary I.5.9], while the implication (d) $\Rightarrow$ (b) was proved for the case that $H = \{0\}$ in [**60**, Theorem I.6.6]. Theorem 16.14 is [**60**, Theorem I.9.4]. A version of Theorem 16.18 in which p and q are assumed to be affine continuous functions is given in [**60**, Theorem I.9.2]. Theorem 16.19(c) is [**60**, Corollary I.7.4(a)]. In Theorem 16.20, the results that the evaluation map $G \to C(X, \mathbb{R})$ maps G onto B and G^+ onto B^+ are [**60**, Corollary I.9.6]. That $\beta\mathbb{R}^+ - \mathbb{R}^+$ is an example of a compact connected F-space was proved by Gillman and Henriksen [**46**, Example 2.8]. Theorem 16.22 is [**60**, Theorem I.8.4]. The result that $C(X, \mathbb{R})$, for X a compact Hausdorff space, has countable interpolation if and only if X is an F-space, is due to Seever [**112**, Theorem 1.1].

Parts (a), (c), (d) of Theorem 16.23 are [**60**, Theorem I.7.5(a) and Corollary I.7.4(b)]. In Theorem 16.25, the results that the evaluation map $G \to \mathrm{Aff}(F)$ maps G onto A and G^+ onto A^+ are [**60**, Corollary I.9.5]. Theorem 16.26 is [**60**, Theorem I.7.13]. Theorems 16.27, 16.29, and 16.30 were first proved only in reference to kernels of closed faces (without mention of kernels of compact sets of extremal states) in [**60**, Theorems I.7.11(b), I.8.5, I.8.7]. The second equivalence in Corollary 16.28 is [**60**, Theorem I.8.6].

CHAPTER 17

Extensions of Dimension Groups

Having described and classified the simple dimension groups (Theorem 14.16 and Proposition 14.3), a natural sequel would be to investigate the "simple-by-simple" dimension groups, i.e., dimension groups G possessing ideals H such that H and G/H are both simple. Such a dimension group G may be called an extension of H by G/H and can be viewed as built from H and G/H. Thus the problem arises, given (simple) dimension groups H and K, of finding all dimension groups which can be obtained as extensions of H by K.

This chapter is designed as an introduction to the problems of existence and classification of dimension groups occurring as extensions of a given dimension group H by a given dimension group K. Complete solutions to these problems are far from known, and so we restrict ourselves to a few basic results and some representative examples, to provide a starting point for readers interested in extension questions. For example, dimension group extensions of $\mathbb{Z}$ by any simple dimension group K may easily be classified: aside from the direct sum extension, there are only lexicographic extensions based on the various abelian group extensions of $\mathbb{Z}$ by K. Extensions of H by K in the cases that $K = \mathbb{Z}$ or H is a dense subgroup of $\mathbb{R}$ are also studied in detail.

For consideration of extensions G of H by K such that G possesses an order-unit u, some additional data are introduced. Namely, we assume that an order-unit v in K is given (to correspond to $u+H$) and that a suitable subset D of H^+ is given (to correspond to $H \cap [0,u]$). In this situation, we may say that (G,u) is an extension of (H,D) by (K,v). After some basic properties of such extensions are considered, the case in which H is a dense subgroup of $\mathbb{R}$ and $D = H \cap [0,\alpha)$ for some positive real number α is studied in detail.

In the final section of the chapter, we take up the fundamental question of the existence of dimension group extensions of (H,D) by (K,v). Such extensions do not exist in general, but it is easy to construct an extension of (H,D) by $(\mathbb{Z},1)$. With suitable technical hypotheses on D, in part amounting to divisibility conditions, a general existence theorem for dimension group extensions of (H,D) by (K,v) is derived. In particular, when H is simple and K is nonzero, such extensions always exist.

• Extensions.

DEFINITION. A *short exact sequence of partially ordered abelian groups* is a diagram

$$0 \to H \xrightarrow{\tau} G \xrightarrow{\pi} K \to 0$$

of partially ordered abelian groups and positive homomorphisms such that the underlying sequence of abelian groups and group homomorphisms is exact (that is, τ is injective, π is surjective, and $\tau(H) = \ker(\pi)$), while also $\tau^{-1}(G^+) = H^+$ and $\pi(G^+) = K^+$.

Given a short exact sequence as described above, we note that τ provides an isomorphism of H onto $\tau(H)$ (as ordered groups), that $\tau(H)$ is a convex subgroup of G, and that π induces an isomorphism of the quotient group $G/\tau(H)$ onto K (as ordered groups). Of course, if H is directed, then $\tau(H)$ is an ideal of G.

DEFINITION. Let H and K be partially ordered abelian groups. An *extension of H by K* (in the category of partially ordered abelian groups) is any short exact sequence

$$0 \to H \to G \to K \to 0$$

of partially ordered abelian groups.

The canonical example of an extension is the natural exact sequence

$$0 \to H \xrightarrow{\tau} G \xrightarrow{\pi} G/H \to 0,$$

where H is an ideal in a partially ordered abelian group G, while τ is the inclusion map and π is the quotient map. Given partially ordered abelian groups H and K, we may construct the direct sum extension

$$0 \to H \xrightarrow{\tau} H \oplus K \xrightarrow{\pi} K \to 0,$$

where τ is the natural injection of H into the left-hand factor of $H \oplus K$, and π is the natural projection of $H \oplus K$ onto K.

DEFINITION. Let H and K be partially ordered abelian groups, and let

$$E: \qquad 0 \to H \xrightarrow{\tau} G \xrightarrow{\pi} K \to 0$$

be a short exact sequence of abelian groups. The set

$$G^+ = \tau(H^+) \cup \pi^{-1}(K^+ - \{0\})$$

is a strict cone in G, using which G becomes a partially ordered abelian group. Then E becomes an extension of H by K, called the *lexicographic extension* of H by K based on the given short exact sequence of abelian groups. If $G = H \oplus K$ and τ and π are the natural injection and projection maps, then G with the ordering described above is just the lexicographic direct sum of K with H. In this case, E is called the *lexicographic sum extension* of H by K.

We concentrate attention on extensions of a dimension group H by a dimension group K. Unfortunately, the group appearing in the middle of an extension of H by K need not be a dimension group, as the following example shows.

EXAMPLE 17.1. *There exist simple noncyclic dimension groups H and K and an extension*

$$0 \to H \to G \to K \to 0$$

of H by K such that G is not an interpolation group.

PROOF. Define G and H as in Example 3.2, and set $K = G/H$. Then G is a partially ordered abelian group without interpolation, H is an ideal of G, and H and K are dimension groups. We have the natural extension

$$0 \to H \xrightarrow{\tau} G \xrightarrow{\pi} K \to 0,$$

where τ is the inclusion map and π is the quotient map. As noted in Example 3.2, H is isomorphic to $(\mathbb{R}^2, \ll)$, while K is isomorophic to $\mathbb{R}$. Thus H and K are simple noncyclic dimension groups. □

Due to the existence of extensions with the properties of Example 17.1, we introduce the following terminology.

DEFINITION. Let H and K be dimension groups. A *dimension group extension of H by K* is any extension

$$0 \to H \to G \to K \to 0$$

such that G is a dimension group.

LEMMA 17.2. *Let H and K be dimension groups. Then every lexicographic extension of H by K is a dimension group extension.*

PROOF. Let $0 \to H \xrightarrow{\tau} G \xrightarrow{\pi} K \to 0$ be a lexicographic extension of H by K. Since $\tau(H)$ is isomorphic to H, it is a dimension group.

Consider any $x_1, x_2, y_1, y_2 \in G$ satisfying $x_i \leq y_j$ for all i, j. Since $\pi(x_i) \leq \pi(y_j)$ for all i, j, there exists $a \in K$ such that $\pi(x_i) \leq a \leq \pi(y_j)$ for all i, j. Choose $v \in G$ such that $\pi(v) = a$. We have $x_i - v \leq y_j - v$ for all i, j, and it suffices to find an element $z \in G$ satisfying $x_i - v \leq z \leq y_j - v$ for all i, j. Thus we may assume, without loss of generality, that $\pi(x_i) \leq 0 \leq \pi(y_j)$ for all i, j.

First suppose that each $y_j \notin \tau(H)$. Since each $\pi(x_i) \leq 0$, there exists $z \in \tau(H)$ such that each $x_i \leq z$. As $\pi(z) = 0 < \pi(y_j)$ for each j, we also have $z \leq y_j$ for each j.

Next, suppose that $y_1 \in \tau(H)$ but $y_2 \notin \tau(H)$. Then $\pi(y_1) = 0 < \pi(y_2)$, whence $y_1 \leq y_2$, and so $x_i \leq y_1 \leq y_j$ for all i, j. Similarly, if $y_1 \notin \tau(H)$ but $y_2 \in \tau(H)$, then $x_i \leq y_2 \leq y_j$ for all i, j.

Finally, we may assume that each $y_j \in \tau(H)$. Similarly, we may also assume that each $x_i \in \tau(H)$. As $\tau(H)$ is a dimension group, there exists $z \in \tau(H)$ such that $x_i \leq z \leq y_j$ for all i, j.

Therefore G is an interpolation group.

If $K = \{0\}$, then G is isomorphic to H and so is directed. If K is nonzero, then since it is directed there exists a nonzero element $b \in K^+$. Given any $x_1, x_2 \in G$, choose $c \in K$ such that each $\pi(x_i) \leq c$, and choose $x \in G$ such that $\pi(x) = c + b$. Then each $\pi(x_i) < \pi(x)$, whence each $x_i \leq x$. Thus G is directed.

Finally, Lemma 3.6 shows that G is unperforated. Therefore G is a dimension group. □

• Some Examples.

DEFINITION. Let H and K be partially ordered abelian groups, and for $i = 1, 2$ let

$$E_i: \qquad 0 \longrightarrow H \xrightarrow{\tau_i} G_i \xrightarrow{\pi_i} K \longrightarrow 0$$

be an extension of H by K. The extensions E_1 and E_2 are *equivalent* if and only if there exists an isomorphism (of ordered groups) $\sigma: G_1 \to G_2$ such that $\sigma\tau_1 = \tau_2$ and $\pi_2\sigma = \pi_1$, that is, such that the following diagram commutes:

$$\begin{array}{ccccccccc} 0 & \longrightarrow & H & \xrightarrow{\tau_1} & G_1 & \xrightarrow{\pi_1} & K & \longrightarrow & 0 \\ & & \mathrm{id} \downarrow & & \sigma \downarrow & & \downarrow \mathrm{id} & & \\ 0 & \longrightarrow & H & \xrightarrow{\tau_2} & G_2 & \xrightarrow{\pi_2} & K & \longrightarrow & 0 \end{array}$$

Note that equivalence of extensions does provide an equivalence relation on the class of all extensions of H by K. A general problem is to describe and classify, up to equivalence, all dimension group extensions of a dimension group H by a dimension group K. In this section, we consider a few cases in which information about equivalence classes of dimension group extensions is readily available.

LEMMA 17.3. *Let H and K be simple dimension groups, and let*

$$E: \qquad 0 \to H \xrightarrow{\tau} G \xrightarrow{\pi} K \to 0$$

be a dimension group extension of H by K. Then E is equivalent to the direct sum extension if and only if G contains an ideal different from $\{0\}$, $\tau(H)$, G.

PROOF. If E is equivalent to the direct sum extension, there exists an ideal L in G such that π restricts to an isomorphism of L onto K. As H and K are nonzero, it follows that L is distinct from $\{0\}$, $\tau(H)$, G.

Conversely, assume that G contains an ideal L different from $\{0\}$, $\tau(H)$, G. Recall from Proposition 2.4 that $L \cap \tau(H)$ and $L + \tau(H)$ are ideals of G. Now $L \cap \tau(H)$ is an ideal of $\tau(H)$, whence $L \cap \tau(H)$ equals either $\{0\}$ or $\tau(H)$. Consequently, $L \not\subseteq \tau(H)$. As a result, $L + \tau(H)$ properly contains $\tau(H)$, and so its image in K, namely $\pi(L)$, is a nonzero ideal of K. Then $\pi(L) = K$, whence $L + \tau(H) = G$. Since $L \neq G$, it follows that $L \not\supseteq \tau(H)$, and hence $L \cap \tau(H) = \{0\}$. Thus $G = \tau(H) \oplus L$ as an abelian group.

By Proposition 2.6, $G = \tau(H) \oplus L$ as an ordered group. Consequently, π restricts to an isomorphism of L onto K (as ordered groups). As a result, we obtain an isomorphism (of ordered groups) $\sigma: H \oplus K \to G$ such that $\sigma(x, 0) = \tau(x)$ for all $x \in H$ and $\pi\sigma(0, y) = y$ for all $y \in K$. If E' denotes the direct sum extension

$$0 \longrightarrow H \xrightarrow{\tau'} H \oplus K \xrightarrow{\pi'} K \longrightarrow 0,$$

then $\sigma\tau' = \tau$ and $\pi\sigma = \pi'$. Therefore E' and E are equivalent. □

PROPOSITION 17.4. *Let K be a simple dimension group. Then any dimension group extension of $\mathbb{Z}$ by K is either lexicographic or equivalent to the direct sum extension.*

PROOF. Let $E: 0 \to \mathbb{Z} \xrightarrow{\tau} G \xrightarrow{\pi} K \to 0$ be a dimension group extension of $\mathbb{Z}$ by K. If G contains an ideal different from $\{0\}$, $\tau(\mathbb{Z})$, G, then Lemma 17.3 shows that E is equivalent to the direct sum extension.

Now assume that $\{0\}$, $\tau(\mathbb{Z})$, G are the only ideals of G. Given any element $u \in G^+$, the convex subgroup of G generated by u is an ideal of G. Hence, if $u \notin \tau(\mathbb{Z})$, this convex subgroup equals G and in particular contains $\tau(\mathbb{Z})$.

We first claim that if $x \in G^+ - \tau(\mathbb{Z})$, then $\tau(1) \leq x$. Since the convex subgroup generated by x contains $\tau(\mathbb{Z})$, we obtain $\tau(1) \leq nx$ for some $n \in \mathbb{N}$. Then $\tau(1) = x_1 + \cdots + x_n$ for some $x_i \in G^+$ such that each $x_i \leq x$, and we may assume that $x_1 \neq 0$. Since $0 < x_1 \leq \tau(1)$, we thus have $x_1 \in \tau(\mathbb{Z})^+$. As $\tau(1)$ is an atom in $\tau(\mathbb{Z})^+$, we obtain $x_1 = \tau(1)$. Thus $\tau(1) \leq x$, as claimed.

Next, we claim that if $x \in G^+ - \tau(\mathbb{Z})$, then $\tau(n) \leq x$ for all $n \in \mathbb{Z}$. This is clear if $n \leq 0$, so assume that $n > 0$. By the claim above, $\tau(1) \leq x$. Then $x - \tau(1)$ is an element of $G^+ - \tau(\mathbb{Z})$, and a second application of the claim yields $\tau(1) \leq x - \tau(1)$, whence $\tau(2) \leq x$. Continuing by induction, we obtain $\tau(n) \leq x$, as claimed.

Given any element $y \in \pi^{-1}(K^+ - \{0\})$, we may choose $x \in G^+$ such that $\pi(x) = \pi(y)$. Note that $\pi(x) \neq 0$, whence $x \notin \tau(\mathbb{Z})$. As $x - y \in \tau(\mathbb{Z})$, the last claim shows that $x - y \leq x$, and so $y \geq 0$. Thus

$$\tau(\mathbb{Z}^+) \cup \pi^{-1}(K^+ - \{0\}) \subseteq G^+.$$

The reverse inclusion is clear. Therefore E is a lexicographic extension of $\mathbb{Z}$ by K. □

Given a simple dimension group K, the equivalence classes of lexicographic dimension group extensions of $\mathbb{Z}$ by K are of course in bijection with the set $\operatorname{Ext}_{\mathbb{Z}}(K, \mathbb{Z})$ of equivalence classes of abelian group extensions of $\mathbb{Z}$ by K. In view of Proposition 17.4, there is exactly one other equivalence class of dimension group extensions of $\mathbb{Z}$ by K.

For example, if K is a free abelian group, then all abelian group extensions of $\mathbb{Z}$ by K split. In this case, there are exactly two equivalence classes of dimension group extensions of $\mathbb{Z}$ by K, represented by the direct sum extension and the lexicographic sum extension.

THEOREM 17.5. *Let H and K be nonzero dimension groups, let G be the abelian group $H \oplus K$, and let $\tau: H \to G$ and $\pi: G \to K$ be the natural injection and projection maps. Assume that H has an order-unit v and that there are no discrete states on (H, v). Set $S = S(H, v)$, and let $\phi: H \to \operatorname{Aff}(S)$ be the natural map.*

(a) *Let $f: K \to \operatorname{Aff}(S)$ be a group homomorphism. Then G becomes a dimension group with positive cone*

$$P_f = \tau(H^+) \cup \{(x, y) \in G \mid y \geq 0 \text{ and } \phi(x) + f(y) \gg 0\},$$

and the diagram

$$E_f: \qquad 0 \to H \xrightarrow{\tau} (G, P_f) \xrightarrow{\pi} K \to 0$$

is a dimension group extension of H by K. In case H is simple,

$$P_f = \{(0,0)\} \cup \{(x,y) \in G \mid y \geq 0 \text{ and } \phi(x) + f(y) \gg 0\}.$$

(b) *Let $f, g: K \to \mathrm{Aff}(S)$ be group homomorphisms. Then E_f and E_g are equivalent if and only if there exists a group homomorphism $k: K \to H$ such that $f - g = \phi k$.*

PROOF. (a) It is clear that P_f is a strict cone in G, and so G becomes a partially ordered abelian group with positive cone P_f. It is also clear that G is unperforated. In case H is simple, any nonzero element $x \in H^+$ is an order-unit (Lemma 14.1), whence $\phi(x) \gg 0$. From this observation the alternate description of P_f is immediate.

Consider any elements (x_1, y_1) and (x_2, y_2) in G. As K is directed, there exists $y \in K$ such that each $y_i \leq y$. Since the functions $\phi(x_i) + f(y_i - y)$ are bounded, there is a positive integer n such that

$$n \gg \phi(x_i) + f(y_i - y)$$

for each i. Then $\phi(nv - x_i) + f(y - y_i) \gg 0$ for each i, and consequently each $(x_i, y_i) \leq (nv, y)$. Thus G is directed.

By definition, $H^+ \subseteq \tau^{-1}(P_f)$. If $x \in H$ and $\tau(x) \in \tau(H^+)$, then obviously $x \in H^+$, while if $\phi(x) \gg 0$, then $x \in H^+$ by Theorem 7.8. Thus $\tau^{-1}(P_f) = H^+$. By definition, $\pi(P_f) \subseteq K^+$. Given $y \in K^+$, choose $n \in \mathbb{N}$ such that $n + f(y) \gg 0$. Then (nv, y) is an element of P_f, and $\pi(nv, y) = y$. Thus $\pi(P_f) = K^+$.

Therefore E_f is an extension of H by K. It only remains to show that G has interpolation.

Because of Proposition 2.8, it suffices to show that whenever $a_1, b_1 \in \tau(H)$ and $a_2, b_2 \in G$ with $a_i \leq b_j$ for all i, j, there exists $c \in \tau(H)$ such that $c \leq b_2$ and each $a_i \leq c$. If $b_2 \in \tau(H)$, we may just take $c = b_2$. Hence, we may assume that $b_2 \notin \tau(H)$.

Write $a_1 = (x_1, 0)$ and $b_1 = (v_1, 0)$ for some $x_1, v_1 \in H$. Write $a_2 = (x_2, y_2)$ and $b_2 = (v_2, w_2)$ for some $x_2, v_2 \in H$ and some $y_2, w_2 \in K$ with $w_2 \neq 0$. As $(x_2, y_2) \leq (v_1, 0)$ and $(x_1, 0) \leq (v_2, w_2)$, we see that $y_2 \leq 0$ and $w_2 > 0$. Since $(x_1, 0) \leq (v_2, w_2)$ with $w_2 > 0$, we obtain $\phi(x_1) \ll \phi(v_2) + f(w_2)$. Similarly, since $(x_2, y_2) \leq (v_2, w_2)$ with $y_2 \leq 0 < w_2$, we obtain

$$\phi(x_2) + f(y_2) \ll \phi(v_2) + f(w_2).$$

As there are no discrete states in S, Corollary 13.7 says that $\phi(H)$ is dense in $\mathrm{Aff}(S)$. Consequently, there exists an element $z \in H$ such that

$$\begin{matrix} \phi(x_1) \\ \phi(x_2) + f(y_2) \end{matrix} \quad \ll \quad \phi(z) \quad \ll \quad \phi(v_2) + f(w_2).$$

Then $(z,0)$ is an element of $\tau(H)$ for which

$$a_2 = (x_2, y_2) \leq (z,0) \leq (v_2, w_2) = b_2.$$

Since $\phi(x_1) \ll \phi(z)$, we have $x_1 \leq z$ by Theorem 7.8, whence $a_1 \leq (z,0)$.

Therefore G does have interpolation and hence is a dimension group.

(b) Observe that the group automorphisms σ of G for which $\sigma\tau = \tau$ and $\pi\sigma = \pi$ are exactly the group endomorphisms of G of the form $1 + \tau k \pi$, where k is any group homomorphism from K to H. Thus E_f and E_g are equivalent if and only if there exists a group homomorphism $k: K \to H$ such that $(1+\tau k\pi)(P_f) = P_g$. Hence, it suffices to prove, for any group homomorphism $k: K \to H$, that $(1 + \tau k\pi)(P_f) = P_g$ if and only if $f - g = \phi k$.

First assume that $f - g = \phi k$. Obviously

$$(1 + \tau k\pi)(\tau(H^+)) = \tau(H^+).$$

Given $(x,y) \in G$, observe that $(1 + \tau k\pi)(x,y) = (x + k(y), y)$, and that

$$\phi(x + k(y)) + g(y) = \phi(x) + (\phi k + g)(y) = \phi(x) + f(y).$$

Hence, $\phi(x+k(y))+g(y) \gg 0$ if and only if $\phi(x)+f(y) \gg 0$. Thus $(1+\tau k\pi)(P_f) = P_g$.

Conversely, assume that $(1 + \tau k\pi)(P_f) = P_g$. Since $1 - \tau k\pi$ is the inverse of $1 + \tau k\pi$, we also have $(1 - \tau k\pi)(P_g) = P_f$. Now consider any nonzero element $y \in K^+$. Given any positive real number ε, there exists $x \in H$ satisfying

$$f(y) - \varepsilon \ll \phi(x) \ll f(y),$$

because of Corollary 13.7. Then $(-x, y) \in P_f$, whence the element

$$(k(y) - x, y) = (1 + \tau k\pi)(-x, y)$$

lies in P_g. As $y \neq 0$, we obtain $\phi(k(y) - x) + g(y) \gg 0$, and hence

$$(\phi k + g)(y) \gg \phi(x) \gg f(y) - \varepsilon.$$

Thus $(\phi k+g)(y) \geq f(y)$. By symmetry, $(-\phi k+f)(y) \geq g(y)$, and so $(f-g)(y) = \phi k(y)$. Therefore $f - g$ and ϕk agree on K^+. As K is directed, we conclude that $f - g = \phi k$. □

COROLLARY 17.6. *Let (H,v) be a nonzero dimension group with order-unit, such that there are no discrete states on (H,v). Set $S = S(H,v)$, and let $\phi: H \to \mathrm{Aff}(S)$ be the natural map. Let G be the abelian group $H \oplus \mathbb{Z}$, and let $\tau: H \to G$ and $\pi: G \to \mathbb{Z}$ be the natural injection and projection maps.*

(a) *Let $q \in \mathrm{Aff}(S)$. Then G becomes a dimension group with positive cone*

$$P_q = \tau(H^+) \cup \{(x,n) \in G \mid n \geq 0 \text{ and } \phi(x) + nq \gg 0\},$$

and the diagram

$$E_q: \qquad 0 \to H \xrightarrow{\tau} (G, P_q) \xrightarrow{\pi} \mathbb{Z} \to 0$$

is a dimension group extension of H by $\mathbb{Z}$. In case H is simple,

$$P_q = \{(0,0)\} \cup \{(x,n) \in G \mid n \geq 0 \text{ and } \phi(x) + nq \gg 0\}.$$

(b) *Let* $q, r \in \mathrm{Aff}(S)$. *Then* E_q *and* E_r *are equivalent if and only if* $q - r \in \phi(H)$.

PROOF. Since the group homomorphisms from $\mathbb{Z}$ to an abelian group A are just the maps $n \mapsto na$, for $a \in A$, this corollary is an immediate consequence of Theorem 17.5. □

In the situation of Corollary 17.6, we obtain an injection from the group $\mathrm{Aff}(S)/\phi(H)$ into the family of equivalence classes of dimension group extensions of H by $\mathbb{Z}$. For instance, if H is countable, there are uncountably many equivalence classes of dimension group extensions of H by $\mathbb{Z}$.

COROLLARY 17.7. *Let* H *be a dense subgroup of* $\mathbb{R}$ *such that* $1 \in H$, *and let* K *be a nonzero dimension group. Let* G *be the abelian group* $H \oplus K$, *and let* $\tau: H \to G$ *and* $\pi: G \to K$ *be the natural injection and projection maps.*

(a) *Let* $f: K \to \mathbb{R}$ *be a group homomorphism. Then* G *becomes a dimension group with positive cone*

$$P_f = \{(0,0)\} \cup \{(x,y) \in G \mid y \geq 0 \text{ and } x + f(y) > 0\},$$

and the diagram

$$E_f: \qquad 0 \to H \xrightarrow{\tau} (G, P_f) \xrightarrow{\pi} K \to 0$$

is a dimension group extension of H *by* K.

(b) *Let* $f, g: K \to \mathbb{R}$ *be group homomorphisms. Then* E_f *and* E_g *are equivalent if and only if* $(f - g)(K) \subseteq H$.

(c) *Suppose that* K *is simple and that all abelian group extensions of* H *by* K *split. Then any dimension group extension of* H *by* K *is equivalent to the direct sum extension, the lexicographic sum extension, or one of the extensions constructed in* (a).

PROOF. Observe that H is a simple dimension group with order-unit 1. The inclusion map $s: H \to \mathbb{R}$ is the only state on $(H, 1)$, and s is not discrete. The state space $S = S(G, u)$ is thus a singleton, and $\mathrm{Aff}(S)$ can be identified with $\mathbb{R}$ (as constant functions). With this identification, the natural map $\phi: H \to \mathrm{Aff}(S)$ is just the inclusion map $H \to \mathbb{R}$.

(a) and (b) now follow immediately from Theorem 17.5.

(c) By hypothesis, any abelian group extension of H by K is equivalent to the direct sum extension. Hence, we need only consider dimension group extensions of the form

$$E: \qquad 0 \to H \xrightarrow{\tau} (G, C) \xrightarrow{\pi} K \to 0,$$

where G has been made into a dimension group with positive cone C.

Assume that E is not equivalent to either the direct sum extension or the lexicographic sum extension. By Lemma 17.3, the only ideals in G are $\{0\}$, $\tau(H)$, G. As E is not equivalent to the lexicographic sum extension, there must exist an element $a \in H$ and a nonzero element $b \in K^+$ such that $(a, b) \notin C$. Thus $\tau(-a) \nleq (0, b)$ in G. Note that $\tau(-na) \nleq (0, nb)$ for all $n \in \mathbb{N}$, because G is unperforated.

For each $y \in K^+$, define

$$f(y) = \sup\{x \in H \mid \tau(x) \leq (0, y)\}.$$

Since $\pi(C) = K^+$, there is some $w \in C$ such that $\pi(w) = y$. Then $(0, y) - w = \tau(z)$ for some $z \in H$, whence $\tau(z) \leq (0, y)$, and so $f(y) \geq z > -\infty$. As b is a strictly positive element in the simple dimension group K, it must be an order-unit, and so $y \leq nb$ for some $n \in \mathbb{N}$. Then $\pi(0, y) \leq \pi(0, nb)$, and hence

$$(0, y) \leq (0, nb) + \tau(z')$$

for some $z' \in H$. Since $\tau(-na) \nleq (0, nb)$, we obtain

$$\tau(-na) \nleq (0, y) - \tau(z'),$$

and so $\tau(z' - na) \nleq (0, y)$. Now given any $x \in H$ satisfying $\tau(x) \leq (0, y)$, it follows that $z' - na \nleq x$, whence $x < z' - na$. Consequently,

$$f(y) \leq z' - na < \infty.$$

Thus f is a map from K^+ to $\mathbb{R}$. We next show that f is additive.

Given $y_1, y_2 \in K^+$ and a positive real number ε, there exist $x_1, x_2 \in H$ such that $\tau(x_i) \leq (0, y_i)$ and $x_i > f(y_i) - \varepsilon/2$ for each i. Then $x_1 + x_2$ is an element of H such that $\tau(x_1 + x_2) \leq (0, y_1 + y_2)$, whence

$$f(y_1 + y_2) \geq x_1 + x_2 > f(y_1) + f(y_2) - \varepsilon.$$

Hence, $f(y_1 + y_2) \geq f(y_1) + f(y_2)$.

Choose $u_1, u_2 \in H$ such that $\tau(u_i) \leq (0, y_i)$ for each i. As H is directed, there exists $u \in H$ such that each $u_i \geq u$, whence

$$(0, y_i) \geq \tau(u_i) \geq \tau(u) = (u, 0)$$

for each i. Set $v = -u$, so that $(v, y_i) \geq 0$ for each i. Given any positive real number ε, there exists $x \in H$ such that $\tau(x) \leq (0, y_1 + y_2)$ and $x > f(y_1 + y_2) - \varepsilon$. By increasing v if necessary, we may assume that $x + 2v \geq 0$ as well. Now

$$\tau(x + 2v) = \tau(x) + (2v, 0) \leq (v, y_1) + (v, y_2).$$

By Riesz decomposition, $\tau(x + 2v) = z_1 + z_2$ for some $z_i \in G^+$ such that $z_i \leq (v, y_i)$ for each i. As $0 \leq z_i \leq \tau(x + 2v)$, each $z_i = \tau(x_i)$ for some $x_i \in H$, and we note that $x_1 + x_2 = x + 2v$. Then $\tau(x_i - v) \leq (0, y_i)$ for each i, whence

$$f(y_1) + f(y_2) \geq x_1 - v + x_2 - v = x > f(y_1 + y_2) - \varepsilon.$$

Hence, $f(y_1) + f(y_2) \geq f(y_1 + y_2)$. Thus f is additive, as claimed. Since K is directed, f extends uniquely to a group homomorphism from K to $\mathbb{R}$, which we also denote by f. We claim that $P_f = C$.

First, consider any $(x, y) \in G$ such that $y \geq 0$ and $x + f(y) > 0$. Since $y \in K^+$ and $f(y) > -x$, there exists $z \in H$ such that $\tau(z) \leq (0, y)$ and $z > -x$. Then

$$(0, y) \geq \tau(z) > \tau(-x) = (-x, 0),$$

whence $(x, y) \geq 0$. Thus $P_f \subseteq C$.

Finally, consider any $w \in C$. If $w \in \tau(H)$, then $w \in \tau(H^+)$, and so $w = (x, 0)$ for some $x \in H^+$, whence $w \in P_f$. Now suppose that $w \notin \tau(H)$. The convex subgroup of G generated by w is an ideal of G not contained in $\tau(H)$, and hence this ideal must equal G. As a result, $\tau(1) \leq nw$ for some $n \in \mathbb{N}$. Write $w = (x, y)$ for some $x \in H$ and $y \in K$, and note that $y = \pi(w) \geq 0$. Since $\tau(1) \leq nw = (nx, ny)$, we have $\tau(1 - nx) \leq (0, ny)$, and so

$$nf(y) = f(ny) \geq 1 - nx.$$

Then $n(x + f(y)) \geq 1$, and hence $x + f(y) > 0$, from which we obtain $w \in P_f$. Thus $C \subseteq P_f$.

Therefore $C = P_f$, and consequently $E = E_f$. $\square$

In the situation of Corollary 17.7(c), we obtain a bijection of the group

$$\mathrm{Hom}_{\mathbb{Z}}(K, \mathbb{R}) / \mathrm{Hom}_{\mathbb{Z}}(K, H)$$

onto the family of equivalence classes of those dimension group extensions of H by K not equivalent to the direct sum extension or the lexicographic sum extension. For example, if $H = \mathbb{R}$ and K is any simple dimension group, then all abelian group extensions of H by K split because H is divisible, and we conclude that there are exactly three equivalence classes of dimension group extensions of H by K.

• **Extensions with Order-Units.** We now turn our attention to building dimension group extensions

$$0 \to H \xrightarrow{\tau} G \xrightarrow{\pi} K \to 0$$

in which the middle group G possesses an order-unit u. In this case, the quotient group K has an order-unit, namely $\pi(u)$, and so we shall assume the existence of an order-unit in K as part of our data. The group H, however, need not have an order-unit, but there is a subset of H^+, namely $\tau^{-1}[0, u]$, which can be used as a substitute for an order-unit, in the following manner.

DEFINITION. Let H be a partially ordered abelian group. A *generating interval in* H^+ is any convex upward-directed subset $D \subseteq H^+$ such that every element of H^+ is a sum of elements from D.

Note that any generating interval in H^+ must contain 0. For example, H^+ is a generating interval of itself. For another example, if H is an interpolation group with an order-unit w, then $[0, w]$ is a generating interval in H^+ (the final part of the definition is satisfied because of Riesz decomposition). The following lemma shows that generating intervals appear naturally in dimension group extensions with order-units.

LEMMA 17.8. *Let* $0 \to H \xrightarrow{\tau} G \xrightarrow{\pi} K \to 0$ *be a short exact sequence of directed interpolation groups. If* u *is an order-unit in* G*, then* $\tau^{-1}[0, u]$ *is a generating interval in* H^+.

PROOF. Set $D = \tau^{-1}[0, u]$. It is clear that D is a convex subset of H^+. Given $x_1, x_2 \in D$, we may choose $y \in H$ such that each $x_i \leq y$. Then

$$\begin{matrix} \tau(x_1) \\ \tau(x_2) \end{matrix} \leq \begin{matrix} \tau(y) \\ u, \end{matrix}$$

and hence there exists $z \in G$ such that

$$\begin{matrix} \tau(x_1) \\ \tau(x_2) \end{matrix} \leq z \leq \begin{matrix} \tau(y) \\ u. \end{matrix}$$

As $\tau(H)$ is a convex subgroup of G, we must have $z = \tau(x)$ for some $x \in H$. Then $x \in D$ and each $x_i \leq x$, which proves that D is upward directed.

Given $x \in H^+$, there is some $n \in \mathbb{N}$ such that $\tau(x) \leq nu$. Then

$$\tau(x) = y_1 + \cdots + y_n$$

for some $y_i \in G^+$ such that each $y_i \leq u$. As $0 \leq y_i \leq \tau(x)$, each $y_i = \tau(x_i)$ for some $x_i \in H^+$. Hence, $x_1, \ldots, x_n$ are elements of D whose sum is x. Thus D is a generating interval in H^+. $\square$

In order to build dimension group extensions with order-units, we shall take as data one dimension group with a specified generating interval and one dimension group with a specified order-unit.

DEFINITION. Let H and K be partially ordered abelian groups, let D be a generating interval in H^+, and let v be an order-unit in K. An *extension of* (H, D) *by* (K, v), denoted

$$E: \qquad 0 \to (H, D) \xrightarrow{\tau} (G, u) \xrightarrow{\pi} (K, v) \to 0,$$

consists of an extension

$$E_0: \qquad 0 \to H \xrightarrow{\tau} G \xrightarrow{\pi} K \to 0$$

of H by K together with an order-unit $u \in G$ such that $\tau^{-1}[0, u] = D$ and $\pi(u) = v$. If E_0 is a dimension group extension (that is, H, G, K are all dimension groups), then E is called a *dimension group extension of* (H, D) *by* (K, v).

For example, if $D = [0, w]$ for some order-unit $w \in H$, then the direct sum extension of H by K provides an extension

$$0 \to (H, D) \to (H \oplus K, (w, v)) \to (K, v) \to 0$$

of (H, D) by (K, v). However, if D is not of this form (i.e., if D has no maximal element), there is no direct sum extension of (H, D) by (K, v).

Given a lexicographic dimension group extension

$$0 \to (H, D) \xrightarrow{\tau} (G, u) \xrightarrow{\pi} (K, v) \to 0$$

with K nonzero, observe that $D = H^+$, because $\tau(x) < u$ for all $x \in H$. Under certain conditions, all dimension group extensions of (H, H^+) by (K, v) are lexicographic, as in the following proposition. Example 17.10 shows that this does not hold in general.

PROPOSITION 17.9. *Let H be a dimension group, and let (K, v) be a simple dimension group with order-unit. Then all dimension group extensions of (H, H^+) by (K, v) are lexicographic.*

PROOF. Let E: $0 \to (H, H^+) \xrightarrow{\tau} (G, u) \xrightarrow{\pi} (K, v) \to 0$ be any dimension group extension of (H, H^+) by (K, v).

Consider any element x in $\pi^{-1}(K^+ - \{0\})$. As $\pi(x)$ is a strictly positive element in the simple dimension group K, it is an order-unit, and hence $v \leq \pi(nx)$ for some $n \in \mathbb{N}$. Then $\pi(nx - u) \geq 0$, whence $nx - u = y + \tau(a)$ for some $y \in G^+$ and some $a \in H$. Choose $b \in H^+$ such that $-a \leq b$. Since $\tau^{-1}[0, u] = H^+$, we have $\tau(-a) \leq \tau(b) \leq u$, and so

$$nx = y + u - \tau(-a) \geq y \geq 0,$$

whence $x \geq 0$.

Thus $\pi^{-1}(K^+ - \{0\}) \subseteq G^+$, from which we conclude that E is a lexicographic extension. □

EXAMPLE 17.10. *Let H, A, B be nonzero dimension groups, and assume that A has an order-unit w. Let K be the lexicographic direct product of A with B, and let v be the order-unit $(w, 0)$ in K. Then there exists a dimension group extension of (H, H^+) by (K, v) which is not lexicographic.*

PROOF. Let G be the lexicographic direct product of A with $H \times B$, and let u be the order-unit $(w, (0, 0))$ in G. Note from Proposition 3.3 that G and K are dimension groups.

Define homomorphisms $\tau: H \to G$ and $\pi: G \to K$ according to the rules

$$\tau(y) = (0, (y, 0)) \quad \text{and} \quad \pi(x, (y, z)) = (x, z),$$

and observe that

$$E: \qquad 0 \to (H, H^+) \xrightarrow{\tau} (G, u) \xrightarrow{\pi} (K, v) \to 0$$

is a dimension group extension of (H, H^+) by (K, v). Choose strictly positive elements $p \in H$ and $q \in B$, and set $r = (0, (-p, q))$. Then $\pi(r) > 0$ but $r \notin G^+$. Thus E is not lexicographic. □

• Some Examples.

DEFINITION. Let H and K be partially ordered abelian groups, let D be a generating interval in H^+, and let v be an order-unit in K. For $i = 1, 2$, let

$$E_i: \qquad 0 \longrightarrow (H, D) \xrightarrow{\tau_i} (G_i, u_i) \xrightarrow{\pi_i} (K, v) \longrightarrow 0$$

be an extension of (H, D) by (K, v). Then E_1 and E_2 are *equivalent* if and only if there exists an isomorphism $\sigma: (G_1, u_1) \to (G_2, u_2)$ (of ordered groups with order-unit) such that $\sigma\tau_1 = \tau_2$ and $\pi_2\sigma = \pi_1$.

The following theorems expose two situations in which the dimension group extensions of (H, D) by (K, v) can be classified up to equivalence.

THEOREM 17.11. *Let H be a dimension group, and let (K, v) be a simple dimension group with order-unit. Set*

$$J = \{k(v) \mid k \in \mathrm{Hom}_{\mathbf{Z}}(K, H)\}.$$

Then the family of all equivalence classes of dimension group extensions of (H, H^+) by (K, v) can be arranged in bijection with the group

$$\mathrm{Ext}_{\mathbf{Z}}(K, H) \times (H/J).$$

PROOF. Any abelian group extension

$$0 \to H \xrightarrow{\tau} G \xrightarrow{\pi} K \to 0$$

of H by K can be made into a lexicographic dimension group extension of H by K. Since $v > 0$ (because K is nonzero), any element $u \in G$ satisfying $\pi(u) = v$ becomes an order-unit in G, and

$$0 \to (H, H^+) \xrightarrow{\tau} (G, u) \xrightarrow{\pi} (K, v) \to 0$$

becomes a dimension group extension of (H, H^+) by (K, v).

Thus every abelian group extension of H by K supports at least one dimension group extension of (H, H^+) by (K, v). Hence, it suffices to show that the family of those dimension group extensions of (H, H^+) by (K, v) which are based on a given abelian group extension

$$0 \to H \xrightarrow{\tau} G \xrightarrow{\pi} K \to 0$$

may be arranged in bijection with H/J. As any such dimension group extension is lexicographic (Proposition 17.9), only the choices of order-unit in G distinguish among the extensions.

Assume now that G has been given the lexicographic extension ordering, and choose an element $u \in G$ satisfying $\pi(u) = v$. The elements $u' \in G$ for which $\pi(u') = v$ are exactly the elements $u + \tau(x)$ for $x \in H$, and all such elements are order-units in G. Thus the dimension group extension of (H, H^+) by (K, v) based on the given abelian group extension are exactly the lexicographic extensions

$$E_x: \qquad 0 \to (H, H^+) \xrightarrow{\tau} (G, u + \tau(x)) \xrightarrow{\pi} (K, v) \to 0$$

for $x \in H$. Consequently, we need only show, for any $x, y \in H$, that E_x and E_y are equivalent if and only if $x - y \in J$.

If $x - y \in J$, then $x - y = k(v)$ for some group homomorphism $k: K \to H$. In this case, the map $\sigma = 1 + \tau k \pi$ is an isomorphism (of ordered groups with order-unit) of $(G, u + \tau(y))$ onto $(G, u + \tau(x))$ such that $\sigma\tau = \tau$ and $\pi\sigma = \pi$. Thus E_x and E_y are equivalent.

Conversely, if E_x and E_y are equivalent, there exists an isomorphism (of ordered groups with order-unit)

$$\sigma: (G, u + \tau(y)) \to (G, u + \tau(x))$$

such that $\sigma\tau = \tau$ and $\pi\sigma = \pi$. As $(\sigma - 1)\tau = 0$, we see that $\sigma - 1$ induces a group homomorphism $f: K \to G$ such that $f\pi = \sigma - 1$. Then $\pi f \pi = \pi\sigma - \pi = 0$, whence $f\pi(G) \subseteq \tau(H)$, and so $f(K) \subseteq \tau(H)$. Consequently, there is a group homomorphism $k: K \to H$ such that $f = \tau k$. Thus $\sigma = 1 + \tau k \pi$. Since

$$u + \tau(x) = \sigma(u + \tau(y)) = u + \tau(y) + \tau k(v),$$

we conclude that $x - y = k(v)$, and hence $x - y \in J$, as desired. □

For example, let H be a dimension group which is a rational vector space, and let (K, v) be any simple dimension group with order-unit. Since H is divisible, $\mathrm{Ext}_{\mathbb{Z}}(K, H)$ is zero, and all elements of H can be obtained as $k(v)$ for group homomorphisms $k: K \to H$. Hence, Theorem 17.11 shows that all dimension group extensions of (H, H^+) by (K, v) are equivalent.

For another example, let H be a dimension group which is a free abelian group, and let (K, v) be a simple dimension group with order-unit such that K is a rational vector space. In this case, $\mathrm{Hom}_{\mathbb{Z}}(K, H)$ is zero, and we conclude from Theorem 17.11 that there are infinitely many equivalence classes of dimension group extensions of (H, H^+) by (K, v).

THEOREM 17.12. *Let H be a dense subgroup of $\mathbb{R}$ such that $1 \in H$, let α be a positive real number, and set $D_\alpha = H \cap [0, \alpha)$. Let (K, v) be a nonzero dimension group with order-unit, let G denote the abelian group $H \oplus K$, and let $\tau: H \to G$ and $\pi: G \to K$ be the natural injection and projection maps.*

(a) *Let $f: K \to \mathbb{R}$ be a group homomorphism such that $f(v) \in \alpha + H$. Then G becomes a dimension group with positive cone*

$$P_f = \{(0,0)\} \cup \{(x, y) \in G \mid y \geq 0 \text{ and } x + f(y) > 0\},$$

the element $u_f = (\alpha - f(v), v)$ is an order-unit in (G, P_f), and the diagram

$$E_f: \qquad 0 \to (H, D_\alpha) \xrightarrow{\tau} (G, P_f, u_f) \xrightarrow{\pi} (K, v) \to 0$$

is a dimension group extension of (H, D_α) by (K, v).

(b) *Let $f, g: K \to \mathbb{R}$ be group homomorphisms such that $f(v), g(v) \in \alpha + H$. Then E_f and E_g are equivalent if and only if $(f - g)(K) \subseteq H$.*

(c) *Suppose that K is simple and that all abelian group extensions of H by K split. Then any dimension group extension of (H, D_α) by (K, v) is equivalent to one of the extensions constructed in* (a).

PROOF. Note that since K is nonzero, $v > 0$.

(a) By Corollary 17.7, G becomes a dimension group with positive cone P_f, and

$$0 \to H \xrightarrow{\tau} (G, P_f) \xrightarrow{\pi} K \to 0$$

is a dimension group extension of H by K.

Since $v > 0$ and $\alpha > 0$, we see that $u_f \in P_f$. Now consider any (x, y) in G. As v is an order-unit in K, there exists $n \in \mathbb{N}$ such that $y \leq nv$. By increasing n if necessary, we may also assume that $n\alpha > x + f(y)$. Then $nv - y \geq 0$ and

$$n\alpha - f(nv) - x + f(nv - y) > 0,$$

whence $(x, y) \leq (n\alpha - f(nv), nv) = nu_f$. Thus u_f is an order-unit in (G, P_f).

Obviously $\pi(u_f) = v$. Now consider any $x \in H^+$. As $v \neq 0$, we have $\tau(x) \leq u_f$ if and only if $(x, 0) < (\alpha - f(v), v)$, if and only if

$$\alpha - f(v) - x + f(v) > 0,$$

if and only if $x \in D_\alpha$. Hence, $\tau^{-1}[0, u_f] = D_\alpha$.

Therefore E_f is a dimension group extension of (H, D_α) by (K, v).

(b) If E_f and E_g are equivalent, then the underlying dimension group extensions of H by K are equivalent. In this case, $(f - g)(H) \subseteq K$ by Corollary 17.7.

Conversely, assume that $(f - g)(H) \subseteq K$. By Corollary 17.7, there exists an isomorphism (of ordered groups) $\sigma: (G, P_f) \to (G, P_g)$ such that $\sigma\tau = \tau$ and $\pi\sigma = \pi$. As $\sigma\tau = \tau$, we have $\sigma(a, 0) = (a, 0)$ for all $a \in H$. Since $\pi\sigma = \pi$, we also have $\sigma(u_f) = (x, v)$ for some $x \in H$, and we desire $x = \alpha - g(v)$.

Given any positive real number ε, there exists $b \in H$ such that $\alpha - \varepsilon < b < \alpha$. Then

$$\alpha - f(v) - b + f(v) > 0,$$

whence $u_f - (b, 0) \in P_f$, and so $(x, v) - (b, 0) \in P_g$. Consequently, $x - b + g(v) > 0$, and hence

$$x > b - g(v) > \alpha - g(v) - \varepsilon.$$

Thus $x \geq \alpha - g(v)$.

Given any positive real number ε, there exists $c \in H$ such that

$$x + g(v) - \varepsilon < c < x + g(v).$$

Then $x - c + g(v) > 0$, whence $(x, v) - (c, 0) \in P_g$, and so $u_f - (c, 0) \in P_f$. Consequently, $\alpha - f(v) - c + f(v) > 0$, and hence $x + g(v) < c + \varepsilon < \alpha + \varepsilon$. Thus $x \leq \alpha - g(v)$.

Therefore $x = \alpha - g(v)$, and so $\sigma(u_f) = u_g$. Now σ is an isomorphism of ordered groups with order-unit, of (G, P_f, u_f) onto (G, P_g, u_g), and hence E_f is equivalent to E_g.

(c) Since D_α has no maximal elements, no dimension group extension of (H, D_α) by (K, v) can be based on the direct sum dimension group extension of H by K. As $D_\alpha \neq H^+$, no dimension group extension of (H, D_α) by (K, v) can be lexicographic. Hence, by Corollary 17.7, any dimension group extension of (H, D_α) by (K, v) is equivalent to one of the form

$$E: \qquad 0 \to (H, D_\alpha) \xrightarrow{\tau} (G, P_f, u) \xrightarrow{\pi} (K, v) \to 0,$$

where f is a group homomorphism from K to $\mathbb{R}$ and u is an order-unit in (G, P_f). Since $\pi(u) = v$, we have $u = (x, v)$ for some $x \in H$.

Given any positive real number ε, there exists $a \in D_\alpha$ such that $a > \alpha - \varepsilon$. Then $\tau(a) \leq u$, whence $(x, v) - (a, 0) \in P_f$, and so $x - a + f(v) > 0$. Consequently,

$$x > a - f(v) > \alpha - f(v) - \varepsilon.$$

Thus $x \geq \alpha - f(v)$.

Given any positive real number ε, there exists $b \in H$ such that $\alpha < b < \alpha + \varepsilon$. Then $b \in H^+$ but $b \notin D_\alpha$, whence $f(b) \nleq u$. As a result, $(x, v) - (b, 0) \notin P_f$, and so $x - b + f(v) \leq 0$, whence

$$x \leq b - f(v) < \alpha - f(v) + \varepsilon.$$

Thus $x \leq \alpha - f(v)$.

Therefore $x = \alpha - f(v)$. In particular, $\alpha - f(v) \in H$, and hence $f(v) \in \alpha + H$. Moreover, $u = u_f$, and thus $E = E_f$. □

COROLLARY 17.13. *Let H be a dense subgroup of $\mathbb{R}$ such that $1 \in H$, let α be a positive real number, and set $D_\alpha = H \cap [0, \alpha)$. Let (K, v) be a simple dimension group with order-unit, and assume that all abelian group extensions of H by K split. Then the family of all equivalence classes of dimension group extensions of (H, D_α) by (K, v) can be arranged in bijection with the group*

$$\mathrm{Hom}_{\mathbb{Z}}(K/\mathbb{Z}v, \mathbb{R}/H).$$

PROOF. Let A denote the set of those group homomorphisms $f: K \to \mathbb{R}$ such that $f(v) \in \alpha + H$, and for $f, g \in A$ define $f \sim g$ if and only if $(f - g)(K) \subseteq H$. Then $\sim$ is an equivalence relation on A. Set

$$B = \mathrm{Hom}_{\mathbb{Z}}(K/\mathbb{Z}v, \mathbb{R}/H).$$

In view of Theorem 17.12, it suffices to find a bijection between $A/\sim$ and B.

Since K is a nonzero dimension group, the element v in K has infinite order, and so there exists a group homomorphism $\mathbb{Z}v \to \mathbb{R}$ sending v to α. As $\mathbb{R}$ is divisible, this map extends to a group homomorphism $f_0: K \to \mathbb{R}$ such that $f_0(v) = \alpha$. In particular, $f_0 \in A$.

Given any $f \in A$, note that $(f - f_0)(v) \in H$. As a result, $f - f_0$ induces a homomorphism from $K/\mathbb{Z}v$ to $\mathbb{R}/H$, which we shall call $\phi(f)$. Thus we obtain a map $\phi: A \to B$. For $f, g \in A$, observe that $f \sim g$ if and only if

$$[(f - f_0) - (g - f_0)](K) \subseteq H,$$

if and only if $\phi(f) = \phi(g)$. Thus ϕ induces an injection $\overline{\phi}$ of $A/\sim$ into B.

Now consider any $q \in B$. Composing q with the quotient map $K \to K/\mathbb{Z}v$, we obtain a group homomorphism $q': K \to \mathbb{R}/H$. As all abelian group extensions of H by K split (that is, $\mathrm{Ext}_{\mathbb{Z}}(K, H) = \{0\}$), it follows that q' lifts to a group homomorphism $q'': K \to \mathbb{R}$. Then $q''(v) \in H$, and so $q'' + f_0$ lies in A. Since q'' induces q, we conclude that $\phi(q'' + f_0) = q$. Thus ϕ is surjective.

Therefore $\overline{\phi}$ is a bijection of $A/\sim$ onto B. □

In the situation of Corollary 17.13, the splitting hypothesis is satisfied if either H is divisible or K is free abelian. For instance, if $H = \mathbb{R}$, we conclude that all dimension group extensions of (H, D_α) by (K, v) are equivalent. On the other hand, if $H = \mathbb{Q}$ and $K/\mathbb{Z}v$ is not a torsion group, there are uncountably many equivalence classes of dimension group extensions of (H, D_α) by (K, v).

• **Existence of Extensions with Order-Units.** Perhaps the most fundamental question concerning dimension group extensions is one of existence. Namely, given dimension groups H and K, a generating interval D in H^+, and an order-unit v in K, when can the existence of dimension group extensions of (H, D) by (K, v) be guaranteed? Although such extensions do not always exist, as will be shown in Example 17.15, sufficient conditions for the existence of extensions are not hard to find, and we present several such existence theorems in this section. We begin with a very easy existence result, analogous to the method by which a unit can be adjoined to a ring without one.

PROPOSITION 17.14. *Let H be a dimension group, and let D be a generating interval in H^+. Then there exists a dimension group extension of (H, D) by $(\mathbb{Z}, 1)$.*

PROOF. Let G be the abelian group $H \oplus \mathbb{Z}$, and set

$$C = \{(x, n) \in G \mid n \geq 0 \text{ and } x + na \geq 0 \text{ for some } a \in D\}.$$

We claim that C is a strict cone in G.

Obviously $(0, 0) \in C$. Given (x_1, n_1) and (x_2, n_2) in C, each $n_i \geq 0$, and there exist $a_1, a_2 \in D$ such that $x_i + n_i a_i \geq 0$ for each i. As D is upward directed, there is some $a \in D$ such that each $a_i \leq a$. Then

$$(x_1 + x_2) + (n_1 + n_2)a \geq (x_1 + n_1 a_1) + (x_2 + n_2 a_2) \geq 0,$$

whence $(x_1, n_1) + (x_2, n_2) \in C$, proving that C is closed under addition. Now C is a cone in G. Strictness is clear.

Thus G becomes a partially ordered abelian group with positive cone C. Since H is unperforated, it is clear that G is unperforated.

We next show that the element $u = (0, 1)$ is an order-unit in G. Given $(x, n) \in G$, choose $y \in H^+$ such that $x \leq y$. Then $y = a_1 + \cdots + a_m$ for some elements $a_i \in D$. There exists $a \in D$ such that each $a_i \leq a$, and we note that $x \leq y \leq ma$. Choose $k \in \mathbb{N}$ such that $k \geq n$ and $k \geq n + m$. Then

$$-x + (k - n)a \geq -x + ma \geq 0,$$

whence $(-x, k - n) \geq 0$, and consequently $(x, n) \leq ku$. Then u is an order-unit in G, as claimed. In particular, it follows that G is directed.

Given elements $(x_1, n_1), (x_2, n_2), (y_1, k_1), (y_2, k_2)$ in G such that $(x_i, n_i) \leq (y_j, k_j)$ for all i, j, we have $n_i \leq k_j$ for all i, j. There is some $m \in \mathbb{Z}$ satisfying $n_i \leq m \leq k_j$ for all i, j, and there is no loss of generality in subtracting $(0, m)$ from each (x_i, n_i) and each (y_j, k_j). Hence, we may assume that $n_i \leq 0 \leq k_j$ for all i, j. For any i, j, since $(x_i, n_i) \leq (y_j, k_j)$ we have

$$y_j - x_i + (k_j - n_i)a_{ij} \geq 0$$

for some $a_{ij} \in D$. There exists $a \in D$ such that each $a_{ij} \leq a$. As each $k_j - n_i \geq 0$, we obtain

$$y_j - x_i + (k_j - n_i)a \geq 0$$

for all i, j, and so $x_i + n_i a \leq y_j + k_j a$ for all i, j. Since H has interpolation, there exists $z \in H$ such that

$$x_i + n_i a \leq z \leq y_j + k_j a$$

for all i, j, from which it follows that

$$(x_i, n_i) \leq (z, 0) \leq (y_j, k_j)$$

for all i, j. Thus G has interpolation.

Therefore (G, u) is a dimension group with order-unit.

Let $\tau: H \to G$ and $\pi: G \to \mathbb{Z}$ be the natural injection and projection maps. It is clear that $\tau^{-1}(C) = H^+$ and $\pi(C) = \mathbb{Z}^+$, whence

$$0 \to H \xrightarrow{\tau} G \xrightarrow{\pi} \mathbb{Z} \to 0$$

is a dimension group extension of H by $\mathbb{Z}$. Obviously $\pi(u) = 1$. For any $x \in H^+$, we have $\tau(x) \leq u$ if and only if $(-x, 1) \in C$, if and only if $x \leq a$ for some $a \in D$, if and only if $x \in D$ (because D is convex). Hence, $\tau^{-1}[0, u] = D$. Therefore

$$0 \to (H, D) \xrightarrow{\tau} (G, u) \xrightarrow{\pi} (\mathbb{Z}, 1) \to 0$$

is a dimension group extension of (H, D) by $(\mathbb{Z}, 1)$. □

EXAMPLE 17.15. *There exists a dimension group H with a generating interval D in H^+ such that for any dimension group (K, v) with order-unit and any integer $m \geq 2$, there are no dimension group extensions of (H, D) by (K, mv).*

PROOF. Set $H_n = \mathbb{Z}$ for all $n \in \mathbb{N}$, and let H be the direct sum of the H_n. Let D be the set of those $x \in H$ such that $x_n \in \{0, 1\}$ for all $n \in \mathbb{N}$. It is clear that H is a dimension group and that D is a generating interval in H^+.

Now consider any dimension group (K, v) with order-unit and any integer $m \geq 2$. Suppose that there exists a dimension group extension

$$0 \to (H, D) \xrightarrow{\tau} (G, u) \xrightarrow{\pi} (K, mv) \to 0$$

of (H, D) by (K, mv). Choose $w \in G^+$ such that $\pi(w) = v$. Then $\pi(u - mw) = 0$, and so $u - mw = \tau(p)$ for some $p \in H$. Choose $q \in H^+$ such that $p \leq q$ and $-p \leq q$. Then

$$u \leq mw + \tau(q) \quad \text{and} \quad mw \leq u + \tau(q).$$

Choose $k \in \mathbb{N}$ such that $q_k = 0$, and define $x \in H$ so that $x_k = 1$ while $x_n = 0$ for all $n \neq k$. Then $x \in D$, and we observe that $x \wedge q = mx \wedge q = 0$.

Since $x \in D$, we have $\tau(x) \leq u \leq mw + \tau(q)$. Using Riesz decomposition in G, we infer that

$$x = y_1 + \cdots + y_m + z$$

for some elements $y_1, \ldots, y_m, z \in H^+$ such that $\tau(y_i) \leq w$ for each i and $z \leq q$. Then $z \leq x \wedge q$, and so $z = 0$. Hence, some $y_i \neq 0$, say $y_1 \neq 0$. As x is an atom in H^+, we obtain $y_1 = x$. Thus $\tau(x) \leq w$.

Now $\tau(mx) \leq mw \leq u + \tau(q)$. Using Riesz decomposition again, we find that $mx = s + t$ for some $s, t \in H^+$ such that $\tau(s) \leq u$ and $t \leq q$. Then $t \leq mx \wedge q$, whence $t = 0$, and so $mx = s$. But now $\tau(mx) \leq u$, and hence $mx \in D$, which is impossible (because $m > 1$).

Therefore there are no dimension group extensions of (H, D) by (K, mv). $\square$

In particular, if H and D are as in Example 17.15, there are no dimension group extensions of (H, D) by $(\mathbb{Z}, m)$ when $m \geq 2$. This is essentially due to the fact that no nonzero elements of D are divisible by m. The following proposition shows that if a generating interval contains enough elements divisible by m, then extensions by $(\mathbb{Z}, m)$ can be constructed.

PROPOSITION 17.16. *Let H be a dimension group, let D be a generating interval in H^+, and let $m \in \mathbb{N}$. Assume that for any $a \in D$ there exists $b \in D$ such that $mb \in D$ and $a \leq mb$. Then there exists a dimension group extension of (H, D) by $(\mathbb{Z}, m)$.*

PROOF. We claim that the set $D' = \{a \in H^+ \mid ma \in D\}$ is a generating interval in H^+. It is clear that D' is a convex subset of H^+. Given $a_1, a_2 \in D'$, each $ma_i \in D$, and so there is some $a \in D$ such that each $ma_i \leq a$. By hypothesis, there exists $b \in D'$ such that $a \leq mb$. Then each $ma_i \leq mb$, whence each $a_i \leq b$. Thus D' is upward directed.

Any $x \in H^+$ can be written as $x = a_1 + \cdots + a_k$ for some $a_i \in D$. For each i, there exists $b_i \in D'$ such that $a_i \leq mb_i$. Then $a_i = c_{i1} + \cdots + c_{im}$ for some $c_{ij} \in H^+$ such that $c_{ij} \leq b_i$ for all j. Since D' is convex, it follows that each $c_{ij} \in D'$. Also

$$x = \sum_{i=1}^{k} a_i = \sum_{i=1}^{k} \sum_{j=1}^{m} c_{ij}.$$

Thus D' is a generating interval in H^+, as claimed.

By Proposition 17.14, there exists a dimension group extension

$$0 \to (H, D') \xrightarrow{\tau} (G, u) \xrightarrow{\pi} (\mathbb{Z}, 1) \to 0$$

of (H, D') by $(\mathbb{Z}, 1)$. Note that mu is an order-unit in G and that $\pi(mu) = m$. Given any $a \in D$, there exists $b \in D'$ such that $a \leq mb$. Then $\tau(b) \leq u$, whence $\tau(a) \leq mu$.

Now consider any $a \in H$ such that $0 \leq \tau(a) \leq mu$, and note that $a \in H^+$. Using Riesz decomposition, we infer that $a = a_1 + \cdots + a_m$ for some elements $a_i \in H^+$ such that $\tau(a_i) \leq u$ for each i. Hence, each $a_i \in D'$, and so there exists $b \in D'$ such that each $a_i \leq b$. As $mb \in D$ and $0 \leq a \leq mb$, we conclude that $a \in D$. Thus $\tau^{-1}[0, mu] = D$. Therefore

$$0 \to (H, D) \xrightarrow{\tau} (G, mu) \xrightarrow{\pi} (\mathbb{Z}, m) \to 0$$

is a dimension group extension of (H, D) by $(\mathbb{Z}, m)$. $\square$

More general existence theorems for dimension group extensions may be obtained by imposing conditions on the generating intervals somewhat stronger than those used in Proposition 17.16, as follows.

LEMMA 17.17. *Let H be a dimension group, and let D be a convex subset of H^+ such that $0 \in D$. Assume that for any $a \in D$ there exists an order-unit $x \in H$ such that $a + x \in D$. Then for any $a \in D$ and any $m \in \mathbb{N}$ there exists $b \in D$ such that $mb \in D$ and $a \le mb$.*

PROOF. We may assume that H is nonzero. Since $0 \in D$, our hypotheses imply in particular that H contains an order-unit u. Set $S = S(H, u)$, and let $\phi: H \to \mathrm{Aff}(S)$ be the natural map.

Let $a \in D$ and $m \in \mathbb{N}$. There exist order-units $x_1, x_2, \ldots$ in H such that the elements $a + x_1, a + x_1 + x_2, \ldots$ all lie in D. In particular, the element

$$c = a + x_1 + x_2 + \cdots + x_{2m}$$

lies in D. As x_1 is an order-unit, $\phi(x_1) \gg 0$ and so $\phi(a) \ll \phi(c)$. Consequently, $\phi(a)/m \ll \phi(c)/m$. We shall use Theorem 13.5 to obtain an element $b \in H$ such that $\phi(a)/m \ll \phi(b) \ll \phi(c)/m$.

Thus consider any discrete state $s \in S$. Then $s(H) = (1/n)\mathbb{Z}$ for some $n \in \mathbb{N}$. As each x_i is an order-unit, $s(x_i) > 0$, whence $s(x_i) \ge 1/n$ for each i. Hence,

$$(\phi(c)/m)(s) - (\phi(a)/m)(s) = s(c-a)/m = (s(x_1) + \cdots + s(x_{2m}))/m \ge 2/n.$$

As a result, $((\phi(a)/m)(s), (\phi(c)/m)(s)) \cap s(H)$ is nonempty. Now by Theorem 13.5, there exists $b \in H$ such that

$$\phi(a)/m \ll \phi(b) \ll \phi(c)/m,$$

whence $\phi(a) \ll \phi(mb) \ll \phi(c)$. Then $a \le mb \le c$, by Theorem 7.8. Note in particular that $b \ge 0$. Since $0 \le b \le mb \le c$ and D is convex, we conclude that $b, mb \in D$. □

LEMMA 17.18. *Let H be a dimension group, and let D be a generating interval in H^+. Assume that for any $a \in D$ there exist $b \in D$ and $n \in \mathbb{N}$ such that $(n+1)a \le nb$. Then for any $a \in D$ and any $m \in \mathbb{N}$ there exists $b \in D$ such that $mb \in D$ and $a \le mb$.*

PROOF. Let $a \in D$ and $m \in \mathbb{N}$. Let A be the convex subgroup of H generated by a, and note that A is an ideal of H. Then A is a dimension group, and a is an order-unit in A. Observe that $D \cap A$ is a convex subset of A^+ such that $0 \in D \cap A$. In view of Lemma 17.17, it suffices to prove that for any $c \in D \cap A$ there exists an order-unit $x \in A$ such that $c + x \in D \cap A$.

Since D is upward directed, there exists $d \in D$ such that $a \le d$ and $c \le d$. As $c \in A$, we have $c \le ka$ for some $k \in \mathbb{N}$, and we note that $a \le ka$. By interpolation, there exists $e \in H$ such that

$$\begin{matrix} a \\ c \end{matrix} \le e \le \begin{matrix} d \\ ka. \end{matrix}$$

Since $a \le e \le ka$, we see that $e \in A$ and that e is an order-unit in A. From $a \le e \le d$, we obtain $e \in D$, and hence $e \in D \cap A$. As $c \le e$, it is enough to find an order-unit $x \in A$ such that $e + x \in D \cap A$. Thus there is no loss of generality in assuming that c is an order-unit in A.

By hypothesis, there exist $p \in D$ and $n \in \mathbb{N}$ such that $(n+1)c \leq np$. In particular, $nc \leq np$, whence $c \leq p$. Since $c \leq n(p-c)$, we must have $c = c_1 + \cdots + c_n$ for some elements $c_i \in H^+$ such that each $c_i \leq p - c$. Each $c_i \leq c$ as well, and so there exists $x \in H$ such that

$$c_i \quad \leq \quad x \quad \leq \quad \begin{matrix} p-c \\ c \end{matrix}$$

for all $i = 1, \ldots, n$. As $0 \leq x \leq c$, we find that $x \in A$. Moreover,

$$c = c_1 + \cdots + c_n \leq nx,$$

whence x is an order-unit in A. Since $x \leq p - c$, we conclude that $0 \leq c + x \leq p$, and so $c + x \in D$. Thus $c + x \in D \cap A$, as desired. □

THEOREM 17.19. *Let H be a dimension group, let D be a generating interval in H^+, and let (K, v) be a nonzero dimension group with order-unit. Assume that for any $a \in D$ there exist $b \in D$ and $n \in \mathbb{N}$ such that $(n+1)a \leq nb$. Then there exists a dimension group extension of (H, D) by (K, v).*

PROOF. Let G be the abelian group $H \oplus K$. Since K is nonzero, there exists a state s on (K, v), by Corollary 4.4. Set

$$C = \{(x,y) \in G \mid y \geq 0 \text{ and } nx + ka \geq 0 \text{ for some } a \in D, \\ n \in \mathbb{N},\ k \in \mathbb{Z}^+ \text{ such that } k/n \leq s(y)\}.$$

We claim that C is a strict cone in G.

It is clear that $(0,0) \in C$. Given (x_1, y_1) and (x_2, y_2) in C, each $y_i \geq 0$ and $n_i x_i + k_i a_i \geq 0$ for some $a_i \in D$, some $n_i \in \mathbb{N}$, and some $k_i \in \mathbb{Z}^+$ such that $k_i/n_i \leq s(y_i)$. The integers n_1, k_1, n_2, k_2 may be replaced by $n_2 n_1, n_2 k_1, n_1 n_2$, $n_1 k_2$. Hence, we may assume that $n_1 = n_2$. Set $n = n_1 = n_2$. There exists $a \in D$ such that each $a_i \leq a$, and we note that

$$nx_i + k_i a \geq nx_i + k_i a_i \geq 0$$

for each i. Now $n(x_1 + x_2) + (k_1 + k_2)a \geq 0$ and

$$(k_1 + k_2)/n \leq s(y_1) + s(y_2) = s(y_1 + y_2),$$

whence $(x_1 + x_2, y_1 + y_2) \in C$. Thus C is closed under addition.

Consider any $(x, y) \in C$ such that also $(-x, -y) \in C$. Then $y \geq 0$ and $-y \geq 0$, whence $y = 0$. In addition, $n_1 x + k_1 a_1 \geq 0$ and $-n_2 x + k_2 a_2 \geq 0$ for some $a_1, a_2 \in D$, some $n_1, n_2 \in \mathbb{N}$, and some $k_1, k_2 \in \mathbb{Z}^+$ such that $k_1/n_1 \leq s(y)$ and $k_2/n_2 \leq s(-y)$. Since $y = 0$, it follows that $k_1 = k_2 = 0$, whence $n_1 x \geq 0$ and $-n_2 x \geq 0$. As H is unperforated, we conclude that $x = 0$. Thus $(x, y) = (0, 0)$.

Therefore C is a strict cone in G, as claimed. Consequently, G becomes a partially ordered abelian group with positive cone C.

We next show that the element $u = (0, v)$ is an order-unit in G. Given $(x, y) \in G$, choose $z \in H^+$ such that $x \leq z$. Then $z = a_1 + \cdots + a_k$ for some elements $a_i \in D$. There exists $a \in D$ such that each $a_i \leq a$, and we note that $x \leq z \leq ka$. Choose $m \in \mathbb{N}$ such that $m \geq k + s(y)$ and $y \leq mv$. Then $mv - y \geq 0$ and $(-x) + ka \geq 0$. Since

$$k \leq m - s(y) = s(mv - y),$$

we see that $(-x, mv - y) \in C$, and hence $(x, y) \leq mu$. Thus u is an order-unit in G, as desired. In particular, it follows that G is directed.

Consider any $(x, y) \in G$ such that $m(x, y) \geq 0$ for some $m \in \mathbb{N}$. Then $my \geq 0$, whence $y \geq 0$. In addition, $nmx + ka \geq 0$ for some $a \in D$, some $n \in \mathbb{N}$, and some $k \in \mathbb{Z}^+$ such that $k/n \leq s(my)$. As $k/nm \leq s(y)$, we see that $(x, y) \geq 0$. Thus G is unperforated.

Let $\tau\colon H \to G$ and $\pi\colon G \to K$ be the natural injection and projection maps. Given $x \in H$, we have $\tau(x) \geq 0$ if and only if $nx + ka \geq 0$ for some $a \in D$, some $n \in \mathbb{N}$, and some $k \in \mathbb{Z}^+$ such that $k/n \leq s(0) = 0$, if and only if $nx \geq 0$ for some $n \in \mathbb{N}$, if and only if $x \geq 0$. Thus $\tau^{-1}(C) = H^+$. Obviously $\pi(C) \subseteq K^+$. Given any $y \in K^+$, we observe that $(0, y) \in C$, and of course $\pi(0, y) = y$. Thus $\pi(C) = K^+$. Hence,

$$0 \to H \xrightarrow{\tau} G \xrightarrow{\pi} K \to 0$$

is an extension of H by K.

Obviously $\pi(u) = v$. Given any $a \in D$, note that $1(-a) + 1a = 0$ and $1/1 = s(v)$, whence $(-a, v) \in C$ and so $\tau(a) \leq u$. Conversely, consider any $b \in H$ for which $0 \leq \tau(b) \leq u$, and note that $b \in H^+$. Then $(-b, v) \in C$, whence $n(-b) + kc \geq 0$ for some $c \in D$, some $n \in \mathbb{N}$, and some $k \in \mathbb{Z}^+$ such that $k/n \leq s(v) = 1$. As $k \leq n$, we have $kb \leq nb \leq kc$ and so $b \leq c$, whence $b \in D$. Thus $\tau^{-1}[0, u] = D$. Therefore

$$0 \to (H, D) \xrightarrow{\tau} (G, u) \xrightarrow{\pi} (K, v) \to 0$$

is an extension of (H, D) by (K, v).

It only remains to prove that G has interpolation. Thus consider any (x_1, y_1), $(x_2, y_2), (z_1, w_1), (z_2, w_2)$ in G such that $(x_i, y_i) \leq (z_j, w_j)$ for all i, j. In particular, $y_i \leq w_j$ for all i, j. There exists $y \in K$ such that $y_i \leq y \leq w_j$ for all i, j, and there is no loss of generality in subtracting $(0, y)$ from each (x_i, y_i) and each (z_j, w_j). Hence, we may assume that $y_i \leq 0 \leq w_j$ for all i, j.

For $i, j = 1, 2$, we have $(z_j - x_i, w_j - y_i) \in C$, and so

$$n_{ij}(z_j - x_i) + k_{ij}a_{ij} \geq 0$$

for some $a_{ij} \in D$, some $n_{ij} \in \mathbb{N}$, and some $k_{ij} \in \mathbb{Z}^+$ such that

$$k_{ij}/n_{ij} \leq s(w_j - y_i).$$

Since the fractions k_{ij}/n_{ij} may be rewritten with a common denominator $n \in \mathbb{N}$, there is no loss of generality in assuming that each $n_{ij} = n$. There exists $a \in D$

such that each $a_{ij} \le a$, and we note that

$$n(z_j - x_i) + k_{ij}a \ge n(z_j - x_i) + k_{ij}a_{ij} \ge 0$$

for all i, j.

By hypothesis, there exist $b \in D$ and $p \in \mathbb{N}$ such that $(p+1)a \le pb$. Set $q = p + 1$, and observe that

$$qn(z_j - x_i) + pk_{ij}b \ge qn(z_j - x_i) + qk_{ij}a \ge 0$$

for all i, j. For each i, j, either $k_{ij} = 0$ or else

$$pk_{ij}/qn < k_{ij}/n \le s(w_j - y_i) = s(w_j) + s(-y_i),$$

because $p/q < 1$. As we may replace a, n, k_{ij} by b, qn, pk_{ij}, there is no loss of generality in assuming that for all i, j, either $k_{ij} = 0$ or $k_{ij}/n < s(w_j) + s(-y_i)$.

For each i, j, we have $y_i \le 0 \le w_j$, whence $s(w_j) \ge 0$ and $s(-y_i) \ge 0$. Consequently, we infer that $k_{ij}/n = \alpha_{ij} + \beta_{ij}$ for some $\alpha_{ij}, \beta_{ij} \in \mathbb{Q}^+$ such that $\alpha_{ij} \le s(w_j)$ and $\beta_{ij} \le s(-y_i)$. Choose $m \in \mathbb{N}$ and r_1, r_2, t_1, t_2 in $\mathbb{Z}^+$ such that

$$r_j/m = \max\{\alpha_{1j}, \alpha_{2j}\} \quad \text{and} \quad t_i/m = \max\{\beta_{i1}, \beta_{i2}\}$$

for $j, i = 1, 2$. Then $r_j/m \le s(w_j)$ for each j and $t_i/m \le s(-y_i)$ for each i, while also

$$k_{ij}/n = \alpha_{ij} + \beta_{ij} \le (r_j + t_i)/m$$

for all i, j.

By Lemma 17.18, there exists $c \in D$ such that $mc \in D$ and $a \le mc$. For each $i, j = 1, 2$, we have

$$\begin{aligned} mn(z_j - x_i) + mn(r_j + t_i)c &\ge mn(z_j - x_i) + mk_{ij}mc \\ &\ge mn(z_j - x_i) + mk_{ij}a \ge 0, \end{aligned}$$

whence $z_j - x_i + (r_j + t_i)c \ge 0$, and so $x_i - t_ic \le z_j + r_jc$. There exists $x \in H$ such that

$$x_i - t_ic \le x \le z_j + r_jc$$

for all i, j.

For each $i = 1, 2$, we have $-y_i \ge 0$ and

$$m(x - x_i) + t_i(mc) \ge 0$$

with $mc \in D$ and $t_i/m \le s(-y_i)$, whence $(x - x_i, -y_i) \in C$ and so $(x_i, y_i) \le (x, 0)$. For each $j = 1, 2$, we have $w_j \ge 0$ and

$$m(z_j - x) + r_j(mc) \ge 0$$

with $mc \in D$ and $r_j/m \le s(w_j)$, whence $(z_j - x, w_j) \in C$ and so $(x, 0) \le (z_j, w_j)$. Thus G has interpolation.

Therefore G is a dimension group. $\square$

COROLLARY 17.20. *Let H be a dimension group, let D be a generating interval in H^+, and let (K, v) be a nonzero dimension group with order-unit. Assume that for any $a \in D$, there exists an order-unit $x \in H$ such that $a+x \in D$. Then there exists a dimension group extension of (H, D) by (K, v).*

PROOF. Because of Theorem 17.19, we need only show that given any $a \in D$, there exist $b \in D$ and $n \in \mathbb{N}$ such that $(n+1)a \leq nb$. By hypothesis, there exists an order-unit $x \in H$ such that $a + x \in D$. Set $b = a + x$. Since x is an order-unit, there exists $n \in \mathbb{N}$ such that $a \leq nx$, and we observe that $(n+1)a \leq na + nx = nb$. $\square$

COROLLARY 17.21. *Let H be a simple dimension group, let D be a generating interval in H^+, and let (K, v) be a nonzero dimension group with order-unit. Then there exists a dimension group extension of (H, D) by (K, v).*

PROOF. First assume that D has a maximal element w, and note that because D is upward directed, $D = [0, w]$. Since H is nonzero and D generates H^+, we must have $w > 0$. Hence, w is an order-unit in H, by simplicity. Thus in this case the direct sum extension provides a dimension group extension of (H, D) by (K, v).

Now assume that D has no maximal elements. In view of Corollary 17.20, it suffices to show that given any $a \in D$, there exists an order-unit $x \in H$ such that $a + x \in D$. As D has no maximal elements, there exists $b \in D$ such that $a < b$. Set $x = b - a$. Since $x > 0$ and H is simple, x is an order-unit in H. Thus the conditions of Corollary 17.20 are satisfied. $\square$

• **Notes.** The results of this chapter are due to Handelman and the author [**67, 58**]. (In [**67**], the only extensions considered are extensions $0 \to H \xrightarrow{\tau} G \xrightarrow{\pi} K \to 0$ which are "essential" in the sense that all nonzero ideals of G have nonzero intersection with $\tau(H)$.) Example 17.1 is in [**67**, Section VII] and Lemma 17.2 is [**67**, Lemma II.1], while Proposition 17.4 is equivalent to [**67**, Lemma II.5(b)(i)]. Theorem 17.5 and Corollaries 17.6 and 17.7 are contained in [**58**, Theorems 5.1, 5.2, 5.3]. A special case of Proposition 17.9 appeared in [**67**, Lemma II.5(b)(ii)], while Example 17.10 was constructed by the author. The case of Theorem 17.11 in which H is simple is contained in [**67**, Theorem II.6]. Theorem 17.12 and Corollary 17.13 are contained in [**58**, Theorem 7.15, Corollary 7.16]. Proposition 17.14 is [**58**, Proposition 7.1], while Example 17.15 and Proposition 17.16 are due to the author. Theorem 17.19 is [**58**, Theorem 7.24].

Epilogue: Further K-Theoretic Applications.

We resume the discussion begun in the Prologue, giving brief sketches of other situations in which the theory of interpolation groups can be applied, via K_0, to the study of certain rings and C^*-algebras.

• **Pseudo-Rank Function Spaces and Trace Spaces.** The collection $\mathsf{P}(R)$ of all pseudo-rank functions on a regular ring R (with 1), like the collection of all states on a partially ordered abelian group with order-unit, is naturally a compact convex set. Namely, $\mathsf{P}(R)$ is a compact convex subset of the product space R^R [**49**, Proposition 16.17], and in fact $\mathsf{P}(R)$ is a Choquet simplex [**49**, Theorem 17.5]. Moreover, the canonical bijection between $S(K_0(R), [R])$ and $\mathsf{P}(R)$ is an affine homeomorphism [**49**, Proposition 17.12]. Hence, to realize a given Choquet simplex K as $\mathsf{P}(R)$ for some regular ring R, it suffices to realize K as $S(K_0(R), [R])$. In the metrizable case, this can always be done, because of our ability to realize any countable dimension group as K_0 of an ultramatricial algebra. Namely, if K is metrizable, then K is affinely homeomorphic to $S(G, u)$ for some countable simple dimension group (G, u) with order-unit (Theorem 14.12). Since there exists a simple unital ultramatricial algebra R with $(K_0(R), [R]) \cong (G, u)$, we get $\mathsf{P}(R)$ affinely homeomorphic to K. This proves the author's result that every metrizable Choquet simplex is affinely homeomorphic to P of a simple unit-regular ring [**48**, Theorem 5.1; **49**, Theorem 17.23]. Some nonmetrizable Choquet simplices may be realized via the observation that K_0 of a tensor product of ultramatricial algebras is isomorphic to a corresponding tensor product of the K_0's of the factors [**59**, Proposition 3.4]. As a consequence, the author and Handelman proved that given any nonempty direct product X of compact metric spaces, there exists a tensor product R of simple ultramatricial algebras such that $\mathsf{P}(R)$ is affinely homeomorphic to $M_1^+(X)$ [**59**, Theorem 5.1].

Parallel procedures can be used to realize metrizable Choquet simplices as trace spaces of unital C^*-algebras. Given a unital C^*-algebra A, the set A_{sa} of self-adjoint elements of A becomes a partially ordered real vector space with positive cone

$$A_{\text{sa}}^+ = \{x \in A_{\text{sa}} \mid \text{spectrum}(x) \subseteq \mathsf{R}^+\}$$

and order-unit 1 [**52**, Proposition 6.1]. A *state on* A is any linear functional $A \to \mathbb{C}$ which restricts to a state on $(A_{\mathrm{sa}}, 1)$. Since all states on $(A_{\mathrm{sa}}, 1)$ extend uniquely to states on A, the collection of all states on A may be identified with the state space of $(A_{\mathrm{sa}}, 1)$. A *tracial state* on A is any state t such that $t(xx^*) = t(x^*x)$ for all $x \in A$, and a *normalized finite trace* on A is the restriction to A_{sa}^+ of any tracial state. The *trace space* of A is the collection $T(A)$ of all normalized finite traces on A. We identify $T(A)$ with the collection of all tracial states on A, which is a compact convex subset of $S(A_{\mathrm{sa}}, 1)$. In fact, $T(A)$ is a Choquet simplex [**110**, Theorem 3.1.18].

For a matrix algebra $M_n(\mathbb{C})$, the trace space $T(M_n(\mathbb{C}))$ is a singleton, as is the state space of $(K_0(M_n(\mathbb{C})), [M_n(\mathbb{C})])$. Using the observation that the functors $T(—)$ and $S(K_0(—), [—])$ convert finite products to finite coproducts and convert direct limits to inverse limits, it follows that the trace space of any unital AF C^*-algebra A is affinely homeomorphic to $S(K_0(A), [A])$ [**13**, Corollary 3.2]. Given any metrizable Choquet simplex K, there is a countable simple dimension group (G, u) with order-unit such that $S(G, u)$ is affinely homeomorphic to K (Theorem 14.12). Hence, by choosing a simple unital AF C^*-algebra A for which $(K_0(A), [A]) \cong (G, u)$, we obtain Blackadar's result that any metrizable Choquet simplex K is affinely homeomorphic to $T(A)$ for some simple unital AF C^*-algebra A [**13**, Theorem 3.9].

• **Metrically Complete Regular Rings.** A norm-like function N^* may be defined on any regular ring R (with 1) by setting $N^*(x)$ equal to the supremum of the values $N(x)$ for $N \in \mathbb{P}(R)$ (or zero, if R has no pseudo-rank functions). It is easily checked that the rule $d(x, y) = N^*(x - y)$ then defines a pseudo-metric d on R [**53**, Lemma 1.2]. In case d is a metric and R is complete with respect to d, we say that R is N^*-*complete.* For instance, if there exists a positive integer n such that all nilpotent elements $x \in R$ satisfy $x^n = 0$, then $N^*(y) \geq 1/n$ for all nonzero elements $y \in R$, and so R is N^*-complete [**53**, Theorem 1.3]. This occurs, for instance, if R can be embedded in a direct product of $n \times n$ matrix rings over division rings.

Because of the canonical bijection between $\mathbb{P}(R)$ and $S(K_0(R), [R])$, it follows that $\|[xR]\| = N^*(x)$ for all $x \in R$, and so N^*-completeness of R is related to norm-completeness of $K_0(R)$. Specifically, the author proved that if R is N^*-complete, then $(K_0(R), [R])$ is an archimedean norm-complete interpolation group [**53**, Theorem 2.11]. Since archimedean groups are unperforated, $K_0(R)$ is of course a dimension group in this case. Thus the structure theory for archimedean norm-complete dimension groups developed in Chapter 15 applies to K_0 of any N^*-complete regular ring R.

For example, the lattice anti-isomorphism between the lattice of closed faces of $S(K_0(R), [R])$ and the lattice of norm-closed ideals of $K_0(R)$ obtained from Theorem 15.24 yields a lattice anti-isomorphism between the lattice of closed faces of $\mathbb{P}(R)$ and the lattice of N^*-closed two-sided ideals of R. Under this anti-isomorphism, the singleton closed faces of $\mathbb{P}(R)$ (i.e., the faces $\{N\}$ where N is an extreme point of $\mathbb{P}(R)$) correspond to the maximal two-sided ideals of R (all of which are N^*-closed [**53**, Corollary 1.14]). Because of the Krein-Mil′man Theorem, the only closed face of $\mathbb{P}(R)$ which contains $\partial_e \mathbb{P}(R)$ is $\mathbb{P}(R)$ itself. Hence, applying the lattice anti-isomorphism, the intersection of the maximal two-sided ideals of R must be zero [**53**, Corollary 4.6]. Alternatively, this can be obtained by using Corollary 15.29 to see that the intersection of the maximal ideals of $K_0(R)$ is zero.

The affine continuous function representation for archimedean norm-complete dimension groups (Theorem 15.7) provides an affine continuous function representation for $K_0(R)$. Under the canonical affine homeomorphism between $S(K_0(R), [R])$ and $\mathbb{P}(R)$, a discrete extremal state s on $(K_0(R), [R])$ corresponds to an extremal pseudo-rank function P with a discrete range of values. Specifically, if $s(K_0(R)) = (1/m)\mathbb{Z}$ for some $m \in \mathbb{N}$, then $P(R) = \{0, \frac{1}{m}, \frac{2}{m}, \dots, 1\}$, and this occurs if and only if $R/\ker(P)$ is isomorphic to an $m \times m$ matrix ring over a division ring. Set $A_P = (1/m)\mathbb{Z}$ in this case, and for all other extremal pseudo-rank functions P set $A_P = \mathbb{R}$. Then there is a natural isomorphism of $(K_0(R), [R])$ onto $(A, 1)$, where

$$A = \{q \in \operatorname{Aff}(\mathbb{P}(R)) \mid q(P) \in A_P \text{ for all } P \in \partial_e \mathbb{P}(R)\}$$

[**53**, Theorem 4.11].

To give an easy application of this affine continuous function representation of $K_0(R)$, assume, for some fixed positive integer t, that all simple artinian factor rings of R (if there are any) are $t \times t$ matrix rings (over some other rings, not necessarily division rings). Then if $P \in \partial_e \mathbb{P}(R)$ and $R/\ker(P)$ is isomorphic to an $m \times m$ matrix ring over a division ring, we must have $t|m$, whence $1/t \in A_P$. As a result, the constant function $1/t$ belongs to the group A given above. From the isomorphism of $(K_0(R), [R])$ onto $(A, 1)$, it follows that $[R] = t[C]$ for some $[C] \in K_0(R)^+$. Since R is unit-regular [**53**, Theorem 2.3], the module R is isomorphic to a direct sum of t copies of C, whence the ring R is isomorphic to a $t \times t$ matrix ring (over the endomorphism ring of C) [**53**, Corollary 4.14].

• **Aleph-Nought-Continuous Regular Rings.** In any regular ring R, the collection $L(R_R)$ of principal right ideals forms a lattice, with finite intersections for finite infima and finite sums for finite suprema [**49**, Theorems 1.1, 2.3]. The ring R is said to be $\aleph_0$-*continuous* if the lattice $L(R_R)$ is $\aleph_0$-continuous in the sense that (a) every countable subset of $L(R_R)$ has an infimum and a supremum in $L(R_R)$; (b) whenever $A \in L(R_R)$ and $B_1 \subseteq B_2 \subseteq \cdots$ in $L(R_R)$, then $A \wedge (\bigvee B_i) = \bigvee (A \wedge B_i)$; (c) whenever $A \in L(R_R)$ and $B_1 \supseteq B_2 \supseteq \cdots$ in $L(R_R)$, then $A \vee (\bigwedge B_i) = \bigwedge (A \vee B_i)$. (Since $L(R_R)$ is anti-isomorphic to the lattice of principal left ideals of R [**49**, Theorem 2.5], this definition is left-right symmetric.) Equivalently, R is $\aleph_0$-continuous if and only if given any countably generated right (left) ideal I of R, there exists a principal right (left) ideal $J \supseteq I$

such that every nonzero right (left) ideal contained in J has nonzero intersection with I [**49**, Corollary 14.4].

Handelman proved that every $\aleph_0$-continuous regular ring is unit-regular [**64**, Theorem 3.2; **49**, Theorem 14.24], and the author proved that every $\aleph_0$-continuous regular ring is N^*-complete [**53**, Theorem 1.8]. Hence, the structure theories for archimedean norm-complete dimension groups and N^*-complete regular rings yield a structure theory for $\aleph_0$-continuous regular rings. However, the structure theory for $\aleph_0$-continuous regular rings was first derived from the structure theory for monotone σ-complete interpolation groups, as follows.

Handelman, Higgs, and Lawrence proved that K_0 of any $\aleph_0$-continuous regular ring R is a (directed) monotone σ-complete interpolation group [**72**, Proposition 2.1] and that such groups are archimedean [**72**, Theorem 1.3]. Since $K_0(R)$ is archimedean, they obtained $\ker(\mathsf{P}(R)) = 0$, from which it follows that the intersection of the maximal two-sided ideals of R is zero [**72**, Theorem 2.3]. If M is any maximal two-sided ideal of R, then the existence of a state on $(K_0(R/M), [R/M])$ implies the existence of a pseudo-rank function P on R/M, and $\ker(P) = 0$ because R/M is a simple ring. As a consequence, R/M contains no uncountable direct sums of nonzero principal right or left ideals, and using this countability condition, Handelman proved that R/M is a right and left self-injective ring [**64**, Corollary 3.2]. Thus, since the intersection of the maximal two-sided ideals of R is zero, R is a subdirect product of simple right and left self-injective rings.

The structure theory for monotone σ-complete interpolation groups (Chapter 16) was developed by the author, Handelman, and Lawrence [**60**] and applied to $K_0(R)$. For example, the affine continuous function representation for such groups (Corollary 16.15) led to a complete representation of $K_0(R)$ in terms of affine continuous functions on $\mathsf{P}(R)$ [**60**, Theorem II.15.1]. As a consequence, if all simple factor rings of R are $t \times t$ matrix rings (for some fixed positive integer t), then R is a $t \times t$ matrix ring [**60**, Theorem II.15.3].

Also, the characterizations of various types of quotient groups of monotone σ-complete interpolation groups (Theorems 16.27, 16.29, 16.30) yield characterizations of various types of factor rings of R. For instance, from the characterization of Dedekind σ-complete quotient groups (Theorem 16.30) it follows that a factor ring R/J is $\aleph_0$-continuous with "general comparability" if and only if J equals the kernel of some compact basically disconnected subset of $\partial_e\mathsf{P}(R)$ [**60**, Theorem II.14.13]. (A regular ring S satisfies general comparability if and only if for every $x, y \in S$ there exists a central idempotent $e \in S$ such that exS embeds in eyS while $(1-e)yS$ embeds in $(1-e)xS$. In case S is unit-regular, this holds if and only if $(K_0(S), [S])$ satisfies general comparability [**60**, Proposition II.14.2].)

• **Finite Rickart C^*-Algebras.** A *Rickart C^*-algebra* is a C^*-algebra A within which the right annihilator of any element x (that is, the right ideal

$\{a \in A \mid xa = 0\}$) equals the principal right ideal generated by some projection p (that is, $p = p^* = p^2$). This is a generalization of the concept of an *AW^*-algebra*, which is a C^*-algebra in which the right annihilator of any *subset* is a principal right ideal generated by a projection. In particular, all von Neumann algebras (W^*-algebras) are Rickart C^*-algebras. A *finite* C^*-algebra is a unital C^*-algebra A such that all elements $x \in A$ satisfying $xx^* = 1$ also satisfy $x^*x = 1$.

The K-theory of a finite Rickart C^*-algebra A can be investigated with the aid of an auxiliary $\aleph_0$-continuous regular ring R which is also **-regular*, i.e., there is an involution * on R such that every principal right ideal of R is generated by a projection. Handelman proved that A is a *-subring of an $\aleph_0$-continuous *-regular ring R such that the only projections in R are those in A [**64**, Theorem 2.1]. The ring R is essentially unique (up to a *-ring isomorphism which is the identity on A), and is called *the regular ring of A*. Handelman also proved that the inclusion map $A \to R$ induces an isomorphism of $(K_0(A), [A])$ onto $(K_0(R), [R])$. (A proof for the case that A has no one-dimensional representations is given in [**54**, Theorem 5.2].)

In particular, $K_0(A)$ is a monotone σ-complete interpolation group, and the structure theory for such groups yields a corresponding structure theory for A, in exactly the same manner as for $\aleph_0$-continuous regular rings. (However, this structure theory was first derived from the structure theory for $\aleph_0$-continuous regular rings, via the regular ring of A.) For example, A is a subdirect product of simple AW^*-algebras, and so A can be embedded in a finite AW^*-algebra [**72**, Theorem 3.1]. For another example, if the dimension of every finite-dimensional irreducible representation of A is divisible by a fixed positive integer t, then A is a $t \times t$ matrix ring over some other finite Rickart C^*-algebra [**60**, Theorem III.16.8].

• **Extensions of AF C^*-Algebras.** The classification of AF C^*-algebras in terms of countable dimension groups can also be used to classify various types of AF C^*-algebras. For example, the simple unital AF C^*-algebras are classified by the countable simple dimension groups with order-unit, and these are determined by Proposition 14.3 and Theorem 14.16. A natural next step is to study extensions of one simple AF C^*-algebra by another, as follows.

Let A and C be AF C^*-algebras, with C unital. A *unital (C^*-algebra) extension of A by C* is a diagram

$$0 \to A \xrightarrow{\phi} B \xrightarrow{\psi} C \to 0$$

where B is a unital C^*-algebra, ϕ is a C^*-algebra isomorphism of A onto a closed two-sided ideal of B, and ψ is a C^*-algebra map that induces an isomorphism of $B/\phi(A)$ onto C. Elliott and Brown proved that B is necessarily AF [**36**, Corollary 3.3; **16**, Theorem; **29**, Theorem 9.9], so that B can be classified in terms of $K_0(B)$. To classify the unital extensions of A by C, apply K_0 to the entire extensions. Thus K_0 of an extension as given above yields a diagram

$$0 \longrightarrow (K_0(A), D(A)) \xrightarrow{K_0(\phi)} (K_0(B), [B]) \xrightarrow{K_0(\psi)} (K_0(C), [C]) \longrightarrow 0$$

which is a dimension group extension of $(K_0(A), D(A))$ by $(K_0(C), [C])$ [**67**, Lemma I.5; **58**, p. 84]. Conversely, any dimension group extension of $(K_0(A), D(A))$ by $(K_0(C), [C])$ is equivalent to K_0 of a unital extension of A by C [**67**, Proposition I.7; **58**, p. 84].

Several definitions of equivalence for unital extensions of A by C are possible. For classification via K_0, define unital extensions

$$0 \longrightarrow A \xrightarrow{\phi_1} B_1 \xrightarrow{\psi_1} C \longrightarrow 0 \quad \text{and} \quad 0 \longrightarrow A \xrightarrow{\phi_2} B_2 \xrightarrow{\psi_2} C \longrightarrow 0$$

to be equivalent if and only if there exists a commutative diagram

$$\begin{array}{ccccccccc} 0 & \longrightarrow & A & \xrightarrow{\phi_1} & B_1 & \xrightarrow{\psi_1} & C & \longrightarrow & 0 \\ & & \alpha\downarrow & & \beta\downarrow & & \gamma\downarrow & & \\ 0 & \longrightarrow & A & \xrightarrow{\phi_2} & B_2 & \xrightarrow{\psi_2} & C & \longrightarrow & 0 \end{array}$$

in which β is a C^*-algebra isomorphism of B_1 onto B_2 while α and γ are "approximately inner" C^*-algebra automorphisms of A and C (that is, α and γ are norm-limits of sequences of inner automorphisms). The given unital extensions of A by C are equivalent in this sense if and only if the corresponding dimension group extensions of $(K_0(A), D(A))$ by $(K_0(C), [C])$ are equivalent [**67**, Lemma I.5; **17**, Theorem 1; **58**, Proposition 9.1].

Therefore the problem of classifying all unital extensions of A by C can be translated into the problem of classifying all dimension group extensions of $(K_0(A), D(A))$ by $(K_0(C), [C])$.

For example, the author and Handelman proved some general existence theorems for unital extensions of A by C in this manner. If A has no nonzero unital homomorphic images, [**58**, Proposition 7.5] shows that the generating interval $D(A)$ satisfies the hypotheses of Theorem 17.19, and so there exist dimension group extensions of $(K_0(A), D(A))$ by $(K_0(C), [C])$. Consequently, there exist unital extensions of A by C in this case [**58**, Theorem 9.9]. In particular, this holds if A is simple and unitless.

In case A and C are simple and A is unitless, a unital extension of A by C has only three closed ideals (i.e., there are only three closed ideals in the algebra B in the middle of the extension), namely, 0, the whole algebra, and the image of A. More generally, a unital extension of A by C is *stenotic* provided every ideal of the extension either contains or is contained in the image of A. Stenotic dimension group extensions are defined in the same way, and a unital extension of A by C is stenotic if and only if K_0 of the extension is a stenotic extension of dimension groups.

The author and Handelman classified stenotic dimension group extensions in several cases [**58**, Section VII] and translated the results into classifications of stenotic unital C^*-algebra extensions in the corresponding cases [**58**, Section

IX]. For example, let H be a nonzero dimension group, let D be a generating interval in H^+ satisfying the condition of Theorem 17.19, and let m be a positive integer. If H has an order-unit, the equivalence classes of stenotic dimension group extensions of (H, D) by $(\mathbb{Z}, m)$ can be arranged in bijection with the group H/mH [**58**, Theorem 7.14]. Correspondingly, let A be a nonzero AF C^*-algebra with no nonzero unital homomorphic images, and assume that A contains a projection p such that ApA is dense in A. Then for any positive integer m, the equivalence classes of stenotic unital extensions of A by $M_m(\mathbb{C})$ can be arranged in bijection with the group $K_0(A)/mK_0(A)$ [**58**, Theorem 9.10].

Analogous patterns of course appear for classification of extensions of ultramatricial algebras, as indicated in [**51**].

Open Problems

The most fundamental question relating to the material in this book is a general meta-question: find new situations in which partially ordered Grothendieck groups can be used to classify or provide structural information about some mathematical systems. Such situations will most likely demand new results in the theory of partially ordered abelian groups, particularly if the groups that appear are not interpolation groups. On a more mundane level, we present in this section a list of open problems more directly related to the theory of interpolation groups as developed in this book. Solutions to the more technical of these problems would help to smooth out the development of the theory, but might not be of sufficient general interest to justify publication. In any case, the author would certainly wish to hear of solutions to any of these problems.

• **Perforation.**

1. Remove the "unperforated" hypothesis from as many theorems as possible. In particular, extend results from dimension groups to directed interpolation groups. For instance, does Theorem 13.5 hold for arbitrary interpolation groups? (A positive solution to problem 2 would provide a general technique for use with this problem.)

2. Given a partially ordered abelian group G, define relations $\leq_1$ and $\leq_2$ on G so that elements $x, y \in G$ satisfy $x \leq_1 y$ if and only if $mx \leq my$ for some $m \in \mathbb{N}$, while $x \leq_2 y$ if and only if $2^n x \leq 2^n y$ for some $n \in \mathbb{N}$. If G is an interpolation group, is either $(G, \leq_1)$ or $(G, \leq_2)$ an interpolation group? (In case $\leq_1$ or $\leq_2$ is not antisymmetric, factor an appropriate ideal out of G, as in Proposition 1.1.)

• **Dimension Groups.**

3. Is every dimension group isomorphic to a direct limit of a countable sequence of lattice-ordered abelian groups? (This holds for all countable dimension groups, because of Theorem 3.17. Moreover, every dimension group is isomorphic to a (possibly uncountable) direct limit of lattice-ordered abelian groups, by Theorem 3.21.)

4. Is every (metrizable) Choquet simplex affinely homeomorphic to the state space of a (countable) archimedean simple dimension group with order-unit? (By Corollary 14.9 and Theorem 14.12, every (metrizable) Choquet simplex is affinely homeomorphic to the state space of a (countable) simple dimension group (not necessarily archimedean). Also, any Choquet simplex is affinely homeomorphic to the state space of an archimedean dimension group (not necessarily simple), by Corollary 11.5.)

5. There exists a metrizable Choquet simplex K, called the *Poulsen simplex*, such that $\partial_e K$ is dense in K [**105**]. Moreover, K is unique up to affine homeomorphism, and every metrizable Choquet simplex is affinely homeomorphic to a closed face of K [**92**, Theorems 2.3, 2.5]. Find necessary and sufficient conditions on a countable dimension group (G, u) with order-unit so that $S(G, u)$ is affinely homeomorphic to K. (By Theorem 14.12, there exists at least one countable dimension group with order-unit whose state space is affinely homeomorphic to K.) If possible, extend the conditions to uncountable dimension groups with order-unit and obtain nonmetrizable Choquet simplices with dense extreme boundaries.

6. Given a nonempty set X, we may construct a dimension group (G, u) with order-unit and a map $f: X \to [0, u]$ with the following "semi-universal" mapping property: given any dimension group (H, v) with order-unit and any map $g: X \to [0, v]$, there exists a normalized positive homomorphism (not necessarily unique) $h: (G, u) \to (H, v)$ such that $hf = g$. (Namely, choose an element $u \notin X$, let G_0 be a free abelian group with basis $X \cup \{u\}$, and make G_0 into a partially ordered abelian group with positive cone generated by the set

$$X \cup \{u\} \cup \{u - x \mid x \in X\}.$$

Then construct unperforated partially ordered abelian groups $G_0 \subseteq G_1 \subseteq \cdots$ such that whenever $x_1, x_2, y_1, y_2 \in G_n$ with $x_i \leq y_j$ for all i, j, there exists $z \in G_{n+1}$ such that $x_i \leq z \leq y_j$ for all i, j, and G_{n+1} is generated by G_n and these z's. The union of the G_n is the desired dimension group G, and the inclusion map $X \to [0, u]$ is the desired map f.) If X is countably infinite, then G is countable. Is $S(G, u)$ affinely homeomorphic to the Poulsen simplex? (It follows from the semi-universal mapping property that every countable dimension group with order-unit is isomorphic to a quotient of G. Consequently, because of Theorem 14.12, every metrizable Choquet simplex is affinely homeomorphic to a closed face of $S(G, u)$.) In the uncountable case, is $\partial_e S(G, u)$ always dense in $S(G, u)$?

- **Order-Unit Norms.**

7. Let (G, u) be an interpolation group with order-unit, and let H be an ideal of G. Is $\|x + H\| = \inf\{\|y\| : y \in x + H\}$ for all $x \in G$? (This holds for dimension groups by Proposition 7.15.)

8. Let (G, u) be a partially ordered abelian group with order-unit, let $\overline{G}$ be the norm-completion of G, and let $\psi: G \to \overline{G}$ be the natural map. Is $\overline{G}$ partially ordered? Is $\psi(u)$ an order-unit in $\overline{G}$? If G has interpolation, does $\overline{G}$ have interpolation? Assuming the first two questions have positive answers, is $\overline{G}$ norm-complete? In particular, does the order-unit metric on $\overline{G}$ coincide with the metric obtained from the completion process? (All these questions have positive answers in case G is a dimension group, by Theorems 15.2 and 15.3.)

• **Norm-Completeness.**

9. Derive the structure theory for archimedean norm-complete dimension groups (e.g., Corollaries 15.15, 15.22, Theorems 15.20, 15.23) directly from the affine continuous function representation (Theorem 15.7).

10. Let (G, u) be an archimedean norm-complete dimension group with order-unit. If $\partial_e S(G, u)$ is compact and totally disconnected, does (G, u) have general comparability? If $\partial_e S(G, u)$ is compact and basically disconnected, is G Dedekind complete? If $\partial_e S(G, u)$ is compact and extremally disconnected, is G Dedekind complete? (In case G satisfies countable interpolation, Theorems 16.29 and 16.30 provide positive answers for the first two questions.)

11. Let (G, u) be an archimedean norm-complete dimension group with order-unit, and let H be an ideal of G. If $H^+ = \ker(X)^+$ for some $X \subseteq S(G, u)$, is H norm-closed in G? (This holds if $H = \ker(X)$, by Theorem 15.6, or if G has countable interpolation, by Theorem 16.12.)

12. Let (G, u) be an archimedean norm-complete dimension group with order-unit, and let θ be the continuous bijection of $\partial_e S(G, u)$ onto MaxSpec(G) given in Theorem 15.32. Does θ map all closed subsets of $\partial_e S(G, u)$ onto compact subsets of MaxSpec(G)?

13. Let (G, u) be an archimedean norm-complete dimension group with order-unit, set $S = S(G, u)$, and let $\phi: G \to \mathrm{Aff}(S)$ denote the natural map. Let $p: S \to \{-\infty\} \cup \mathbb{R}$ be an upper semicontinuous convex function, and let $q: S \to \mathbb{R} \cup \{\infty\}$ be a lower semicontinuous concave function. Assume that $p \leq q$ and that $p(s), q(s) \in s(G) \cup \{\pm\infty\}$ for all discrete $s \in \partial_e S$. Does there exist $x \in G$ such that $p \leq \phi(x) \leq q$? If $q \geq 0$, does there exist $x \in G^+$ such that $p \leq \phi(x) \leq q$? (Example 15.13 shows that this need not happen if in place of $p(s), q(s) \in s(G) \cup \{\pm\infty\}$ it is assumed that $[p(s), q(s)] \cap s(G)$ is nonempty. On the other hand, in case G has countable interpolation and p, q are continuous real-valued functions, Theorem 16.18 provides positive answers for both questions.)

14. Let K be a Choquet simplex, and for $x \in \partial_e K$ let A_x be either $\mathbb{R}$ or $(1/m_x)\mathbb{Z}$ for some $m_x \in \mathbb{N}$. Set

$$A = \{f \in \mathrm{Aff}(K) \mid f(x) \in A_x \text{ for all } x \in \partial_e K\}.$$

Then $(A, 1)$ is an archimedean norm-complete partially ordered abelian group with order-unit, and by Theorem 15.7 every archimedean norm-complete dimension group with order-unit is isomorphic to some $(A, 1)$. Find necessary and sufficient conditions on the groups A_x for A to be a dimension group. (Example 15.12 shows that A does not always have interpolation.)

• **Countable Interpolation and Monotone σ-Completeness.**

15. Is every directed abelian group with countable interpolation unperforated?

16. Is every directed abelian group with countable interpolation isomorphic to a quotient group of a monotone σ-complete dimension group?

17. Let (G, u) be a partially ordered abelian group with order-unit, satisfying countable interpolation, and let F be a closed face of $S(G, u)$. Find necessary and sufficient conditions on F for $G/\ker(F)$ to be monotone σ-complete. (In case $\partial_e F$ is compact, $G/\ker(F)$ is monotone σ-complete if and only if $\partial_e F$ is basically disconnected, by Theorems 16.27 and 16.30.)

18. If (G, u) is a monotone σ-complete interpolation group with order-unit, and X is a compact subset of $\partial_e S(G, u)$, then X is an F-space by Theorem 16.22. Is X totally disconnected, or even basically disconnected? Equivalently, does every archimedean lattice-ordered quotient group of G satisfy general comparability, or even Dedekind σ-completeness? (In case $X = \partial_e S(G, u)$, Theorems 16.27 and 16.30 provide positive answers for both questions. On the other hand, monotone σ-completeness may not be replaced by countable interpolation, since $C(X, \mathbb{R})$ has countable interpolation whenever X is a compact Hausdorff F-space [**112**, Theorem 1.1], and there exist compact Hausdorff connected F-spaces [**46**, Example 2.8].)

19. Let (G, u) be a partially ordered abelian group with order-unit, satisfying countable interpolation, set $S = S(G, u)$, and let $\phi: G \to \text{Aff}(S)$ be the natural map. Let $p: S \to \{-\infty\} \cup \mathbb{R}$ be an upper semicontinuous convex function, and let $q: S \to \mathbb{R} \cup \{\infty\}$ be a lower semicontinuous concave function. Assume that $p \leq q$ and that $[p(s), q(s)] \cap s(G)$ is nonempty for all discrete $s \in \partial_e S$. Does there exist $x \in G$ such that $p \leq \phi(x) \leq q$? If $q \geq 0$, does there exist $x \in G^+$ such that $p \leq \phi(x) \leq q$? (Both questions have positive answers in case p and q are continuous and real-valued, by Theorem 16.18.)

20. Characterize those Choquet simplices K for which $\text{Aff}(K)$ has countable interpolation or is monotone σ-complete. For instance, if all compact subsets of $\partial_e K$ are F-spaces, does $\text{Aff}(K)$ have countable interpolation? (In case $\partial_e K$ is compact, it follows from [**112**, Theorem 1.1] that $\text{Aff}(K)$ has countable interpolation if and only if $\partial_e K$ is an F-space. Moreover, $\text{Aff}(K)$ is Dedekind σ-complete

if and only if $\partial_e K$ is compact and basically disconnected, by Corollaries 12.19, 11.20, and 9.3.)

21. Let K be a Choquet simplex, and for $x \in \partial_e K$ let A_x be either $\mathbb{R}$ or $(1/m_x)\mathbb{Z}$ for some $m_x \in \mathbb{N}$. Set

$$A = \{f \in \mathrm{Aff}(K) \mid f(x) \in A_x \text{ for all } x \in \partial_e K\}.$$

Find necessary and sufficient conditions on the groups A_x for A to have countable interpolation, or for A to be monotone σ-complete. (Compare problem 14.)

• **Extensions.**

22. Let H be a dimension group, let D be a generating interval in H^+, and let (K, v) be a dimension group with order-unit. Find necessary and sufficient conditions for the existence of a dimension group extension of (H, D) by (K, v). (See Example 17.15 and Theorem 17.19.) In particular, does such an extension exist if for each $a \in D$ and $m \in \mathbb{N}$ there exists $b \in D$ such that $mb \in D$ and $a \leq mb$?

23. Let H and K be dimension groups. Find data with which to classify all dimension group extensions of H by K. An extension

$$0 \to H \xrightarrow{\tau} G \xrightarrow{\pi} K \to 0$$

is *essential* provided every nonzero ideal of G has nonzero intersection with $\tau(H)$, and it is *stenotic* provided every ideal of G either contains $\tau(H)$ or is contained in $\tau(H)$. Find data with which to classify all essential or stenotic dimension group extensions of H by K. (See [**58**, Theorems 6.5, 6.6, 6.7].) Similarly, given a generating interval D in H^+ and an order-unit $v \in K$, find data with which to classify all, all essential, or all stenotic dimension group extensions of (H, D) by (K, v). (See [**58**, Corollaries 7.10, 7.16, Theorems 7.11, 7.14].)

24. Let H be a dimension group, let D be a generating interval in H^+, and let (K, v) be a dimension group with order-unit. If there exists a stenotic dimension group extension of (H, D) by (K, v), then by [**58**, Lemma 7.6] for any $a \in D$ there exist $b \in D$ and $n \in \mathbb{N}$ such that $(n+1)a \leq nb$. Conversely, if this condition holds, and if H and K are countable, does there exist a stenotic dimension group extension of (H, D) by (K, v)? (See Theorem 17.19 and [**58**, Corollary 7.12, Theorem 7.20]. If K is allowed to be uncountable, there need not exist any stenotic dimension group extensions of (H, D) by (K, v) [**58**, Example 7.17]. A positive answer in the countable case would imply that given any AF C^*-algebras A and C such that C is unital while A has no nonzero unital homomorphic images, there would exist a stenotic unital C^*-algebra extension of A by C, because of [**58**, Proposition 9.5].)

25. Let H be a dimension group, let D be a generating interval in H^+, and let (K, v) be a dimension group with order-unit. For $n \in \mathbb{N}$, let S_n be the family

of all equivalence classes of stenotic dimension group extensions of (H, D) by (K, nv), and set $S = \bigcup S_n$. As shown in [**58**, Section VIII], there are natural addition operations $S_m \times S_n \to S_{m+n}$ for all $m, n \in \mathbb{N}$, using which S becomes an abelian semigroup. In some cases, there exists an abelian group E such that $S \cong E \times \mathbb{N}$ [**58**, Theorems 8.6, 8.7]. Is S always isomorphic to the direct product of an abelian group with $\mathbb{N}$? Does there exist a natural addition operation on S_1 under which S_1 becomes an abelian semigroup or an abelian group?

- **Tensor Products.**

26. Given partially ordered abelian groups $G_1, \ldots, G_n$, their tensor product $G_1 \otimes \cdots \otimes G_n$ can be made into a partially ordered abelian group with positive cone $(G_1 \otimes \cdots \otimes G_n)^+$ equal to the set of all sums of pure tensors of the form $x_1 \otimes \cdots \otimes x_n$ where each $x_i \in G_i^+$ [**59**, Proposition 2.1]. Is every such tensor product of interpolation groups an interpolation group? (By [**59**, Proposition 2.3], every tensor product of dimension groups is a dimension group.)

27. What universal mapping properties are satisfied by tensor products in the category of partially ordered abelian groups? For instance, given partially ordered abelian groups G_1, G_2, G, order-units $u_i \in G_i$, and positive homomorphisms $f_i \colon G_i \to G$, what conditions on f_1 and f_2 ensure that there exists a positive homomorphism $f \colon G_1 \otimes G_2 \to G$ such that $f(x \otimes u_2) = f_1(x)$ for all $x \in G_1$ and $f(u_1 \otimes y) = f_2(y)$ for all $y \in G_2$?

28. The tensor product of an infinite family $\{(G_i, u_i) \mid i \in I\}$ of partially ordered abelian groups with order-unit is defined as the direct limit of the tensor products $(\bigotimes_{i \in A} G_i, \bigotimes_{i \in A} u_i)$ over nonempty finite subsets $A \subseteq I$, using maps

$$\left(\bigotimes_{i \in A} G_i, \bigotimes_{i \in A} u_i\right) \to \left(\bigotimes_{i \in B} G_i, \bigotimes_{i \in B} u_i\right)$$

(for $A \subseteq B$) sending $\bigotimes_{i \in A} x_i$ to $(\bigotimes_{i \in A} x_i) \otimes (\bigotimes_{i \in B - A} u_i)$. Characterize those dimension groups with order-unit which are isomorphic to tensor products of countable dimension groups with order-unit or which are isomorphic to quotient groups of tensor products of countable dimension groups with order-unit. (All such dimension groups appear as K_0 of tensor products of ultramatricial algebras, or as K_0 of factor rings of tensor products of ultramatricial algebras, because of [**59**, Proposition 3.4].)

29. Given an arbitrary (nonmetrizable) Choquet simplex K, does there exist a family $\{K_i \mid i \in I\}$ of metrizable Choquet simplices such that K is affinely homeomorphic to a closed face of the state space of $\bigotimes_{i \in I}(\mathrm{Aff}(K_i), 1)$? (If so, K is affinely homeomorphic to $\mathrm{P}(R)$ for some N^*-complete unit-regular ring R, by the method of [**59**, Theorem 5.2].)

• **Miscellany.**

30. Can every partially ordered abelian group be embedded in a simple interpolation group? (It can be shown that every partially ordered abelian group can be embedded in a (not necessarily simple) interpolation group.) Can every torsion-free (unperforated) partially ordered abelian group be embedded in a torsion-free (unperforated) simple interpolation group?

31. Let G be an interpolation group, let A be a finite group of ordered group automorphisms of G, and let G^A be the fixed subgroup $\{x \in G \mid \alpha(x) = x$ for all $\alpha \in A\}$. Is G^A an interpolation group? (This can be proved for a dimension group G such that whenever $x < z$ in G, there exists $y \in G$ with $x \leq my \leq z$, where $m = \mathrm{card}(A)$. It follows that the result holds for all simple dimension groups.)

32. In the category of convex sets, is every inverse limit of simplices a simplex? (This can be proved for inverse limits of simplices which are affinely isomorphic to bases for Dedekind complete lattice cones.)

33. Find a simpler derivation for Theorem 13.5, or for Corollary 13.6. In particular, can these results be proved without proving the Dedekind complete case first and using metric completions? (For one result that can be done without using metric completions, see the proof of the extremal state criterion for dimension groups discussed following Theorem 12.14.)

Bibliography

1. E. M. Alfsen, *On the geometry of Choquet simplexes*, Math. Scand. **15** (1964), 97–110.

2. ——, *On the decomposition of a Choquet simplex into a direct convex sum of complementary faces*, Math. Scand. **17** (1965), 169–176.

3. ——, *Compact convex sets and boundary integrals*, Ergebnisse der Math. und ihre Grenzgebiete, Band 57, Springer-Verlag, Berlin, 1971.

4. T. Andô, *On fundamental properties of a Banach space with a cone*, Pacific J. Math. **12** (1962), 1163–1169.

5. G. Ascoli, *Sugli spazi lineari metrici e le loro varietà lineari*, Annali di Matematica Pura ed Applicata (4) **10** (1932), 33–81.

6. L. A. Asimow and A. J. Ellis, *Convexity theory and its applications in functional analysis*, Academic Press, New York, 1980.

7. G. Aumann, *Über die Erweiterung von additiven monotonen Funktionen auf regulär geordneten Halbgruppen*, Arch. Math. (Basel) **8** (1957), 422–427.

8. H. Bauer, *Geordnete Gruppen mit Zerlegungseigenschaft*, Sitzungsberichte Bayer. Akad. Wiss. Math.-Nat. Klasse, 1958, pp. 25–36.

9. ——, *Šilovscher Rand und Dirichletsches Problem*, Ann. Inst. Fourier (Grenoble) **11** (1961), 89–136.

10. ——, *Kennzeichnung kompakter Simplexe mit abgeschlossener Extremalpunktmenge*, Arch. Math. (Basel) **14** (1963), 415–421.

11. G. Birkhoff, *Lattice-ordered groups*, Ann. of Math. (2) **43** (1942), 298–331.

12. ——, *Lattice theory*, 3rd ed., Amer. Math. Soc. Colloq. Publ., vol. 25, Amer. Math. Soc., Providence, R. I., 1967.

13. B. E. Blackadar, *Traces on simple* AF C^**-algebras*, J. Funct. Anal. **38** (1980), 156–168.

14. N. Boboc and A. Cornea, *Cônes des fonctions continues sur un espace compact*, C. R. Acad. Sci. Paris **261** (1965), 2564–2567.

15. O. Bratteli, *Inductive limits of finite-dimensional* C^**-algebras*, Trans. Amer. Math. Soc. **171** (1972), 195–234.

16. L. G. Brown, "Extensions of AF algebras: the projection lifting problem," in *Operator Algebras and Applications*. Part 1, edited by R. V. Kadison, Proc. Sympos. Pure Math., vol. 38, Amer. Math. Soc., Providence, R. I., 1982, pp. 175–176.

17. L. G. Brown and G. A. Elliott, *Extensions of* AF*-algebras are determined by* K_0, C. R. Math. Rep. Acad. Sci. Canada **4** (1982), 15–19.

18. W. D. Burgess and D. E. Handelman, *The* N^**-metric completion of regular rings*, Math. Ann. **261** (1982), 235–254.

19. G. Choquet, *Existence et unicité des représentations intégrales au moyen des points extrémaux dans les cônes convexes*, Séminaire Bourbaki (1956/57), Exposé 139, Secrétariat Mathématique, Paris, 1959.

20. ——, *Lectures on analysis*. Vol. II, Benjamin, New York, 1969.

21. A. H. Clifford, *Partially ordered abelian groups*, Ann. of Math. (2) **41** (1940), 465–473.

22. J. Cuntz and W. Krieger, *Topological Markov chains with dicyclic dimension groups*, J. Reine Angew. Math. **320** (1980), 44–51.

23. E. B. Davies and G. F. Vincent-Smith, *Tensor products, infinite products, and projective limits of Choquet simplexes*, Math. Scand. **22** (1968), 145–164.

24. D. A. Edwards, *Séparation des fonctions réelles définies sur un simplexe de Choquet*, C. R. Acad. Sci. Paris **261** (1965), 2798–2800.

25. ——, *The affine continuous functions on a Choquet simplex*, Summer School on Topological Algebra Theory (Bruges, 1966), Brussels, 1967, pp. 233–241 (mimeo.).

26. D. A. Edwards and G. Vincent-Smith, *A Weierstrass-Stone theorem for Choquet simplexes*, Ann. Inst. Fourier (Grenoble) **18** (1968), 261–282.

27. E. G. Effros, *Structure in simplexes*, Acta Math. **117** (1967), 103–121.

28. ——, *Structure in simplexes*. II, J. Funct. Anal. **1** (1967), 379–391.

29. ——, *Dimensions and* C^**-algebras*, CBMS Regional Conf. Series in Math., no. 46, Amer. Math. Soc., Providence, R. I., 1981.

30. E. G. Effros, D. E. Handelman, and C.-L. Shen, *Dimension groups and their affine representations*, Amer. J. Math. **102** (1980), 385–407.

31. E. G. Effros and C.-L. Shen, *Dimension groups and finite difference equations*, J. Operator Theory **2** (1979), 215–231.

32. ——, *Approximately finite* C^**-algebras and continued fractions*, Indiana Univ. Math. J. **29** (1980), 191–204.

33. ——, *The geometry of finite rank dimension groups*, Illinois J. Math. **25** (1981), 27–38.

34. M. Eidelheit, *Zur Theorie der konvexen Mengen in linearen normierten Räumen*, Studia Math. **6** (1936), 104–111.

35. G. A. Elliott, *On the classification of inductive limits of sequences of semisimple finite-dimensional algebras*, J. Algebra **38** (1976), 29–44.

36. ——, *Automorphisms determined by multipliers on ideals of a* C^**-algebra*, J. Funct. Anal. **23** (1976), 1–10.

37. ——, *On totally ordered groups*, Preprint Series no. 10, Copenhagen Univ. Mat. Inst., Copenhagen, 1978.

38. ——, *On totally ordered groups, and K_0*, Ring Theory Waterloo 1978, D. Handelman and J. Lawrence, editors, Lecture Notes in Math., vol. 734, Springer-Verlag, Berlin, 1979, pp. 1–49.

39. A. J. Ellis, *The duality of partially ordered normed linear spaces*, J. London Math. Soc. **39** (1964), 730–744.

40. C. J. Everett and S. Ulam, *On ordered groups*, Trans. Amer. Math. Soc. **57** (1945), 208–216.

41. H. Freudenthal, *Teilweise geordnete Moduln*, Proc. Akad. Wetenschappen Amsterdam **39** (1936), 641–651.

42. L. Fuchs, *Partially ordered algebraic systems*, Pergamon Press, Oxford, 1963.

43. ——, *Riesz groups*, Ann. Scuola Norm. Sup. Pisa Cl. Sci. (4) **19** (1965), 1–34.

44. ——, *Riesz vector spaces and Riesz algebras*, Queens Univ. Papers in Pure and Applied Math., no. 1, Kingston, Ontario, 1966.

45. B. Fuchssteiner and W. Lusky, *Convex cones*, North-Holland, Amsterdam, 1981.

46. L. Gillman and M. Henriksen, *Rings of continuous functions in which every finitely generated ideal is principal*, Trans. Amer. Math. Soc. **82** (1956), 366–391.

47. K. R. Goodearl, *Choquet simplexes and σ-convex faces*, Pacific J. Math. **66** (1976), 119–124.

48. ——, *Algebraic representations of Choquet simplexes*, J. Pure Appl. Algebra **11** (1977), 111–130.

49. ——, *Von Neumann regular rings*, Pitman, London, 1979.

50. ——, *Artinian and noetherian modules over regular rings*, Comm. Algebra **8** (1980), 477–504.

51. ——, *Extensions of simple by simple unit-regular rings*, Ring Theory Antwerp 1980, F. van Oystaeyen, editor, Lecture Notes in Math., vol. 825, Springer-Verlag, Berlin, 1980, pp. 42–58.

52. ——, *Notes on real and complex C^*-algebras*, Shiva, Nantwich (Cheshire), 1982.

53. ——, *Metrically complete regular rings*, Trans. Amer. Math. Soc. **272** (1982), 275–310.

54. ——, "Partially ordered Grothendieck groups," in *Algebra and its applications*, edited by H. L. Manocha and J. B. Srivastava, Dekker, New York, 1984, pp. 71–90.

55. ——, *Partially ordered Grothendieck groups*, Publ. Sec. Mat. Univ. Autònoma Barcelona **29** (1984), 77–103.

56. K. R. Goodearl and D. E. Handelman, *Rank functions and K_0 of regular rings*, J. Pure Appl. Algebra **7** (1976), 195–216.

57. ———, *Metric completions of partially ordered abelian groups*, Indiana Univ. Math. J. **29** (1980), 861–895.

58. ———, *Stenosis in dimension groups and* AF C^**-algebras*, J. Reine Angew. Math. **332** (1982), 1–98.

59. ———, *Tensor products of dimension groups and* K_0 *of unit-regular rings*, Canad. J. Math. (to appear).

60. K. R. Goodearl, D. E. Handelman, and J. W. Lawrence, *Affine representations of Grothendieck groups and applications to Rickart* C^**-algebras and* $\aleph_0$*-continuous regular rings*, Mem. Amer. Math. Soc. No. 234 (1980).

61. K. R. Goodearl and R. B. Warfield, Jr., *State spaces of* K_0 *of noetherian rings*, J. Algebra **71** (1981), 322–378.

62. M. Goullet de Rugy, *Géométrie des simplexes*, Centre de Documentation Universitaire, Paris, 1968.

63. D. Handelman, *Perspectivity and cancellation in regular rings*, J. Algebra **48** (1977), 1–16.

64. ———, "Finite Rickart C^*-algebras and their properties," *Studies in Analysis*, Adv. in Math. Suppl. Studies, vol. 4, Academic Press, 1979, pp. 171–196.

65. ———, *Positive matrices and dimension groups affiliated to* C^**-algebras and topological Markov chains*, J. Operator Theory **6** (1981), 55–74.

66. ———, *Free rank* $n+1$ *dense subgroups of* $\mathbb{R}^n$ *and their endomorphisms*, J. Funct. Anal. **46** (1982), 1–27.

67. ———, *Extensions for* AF C^**-algebras and dimension groups*, Trans. Amer. Math. Soc. **271** (1982), 537–573.

68. ———, *Reducible topological Markov chains via* K_0*-theory and* Ext, Contemp. Math. **10** (1982), 41–76.

69. ———, *Ultrasimplicial dimension groups*, Arch. Math. (Basel) **40** (1983), 109–115.

70. ———, *Positive polynomials and product type actions of compact groups*, Mem. Amer. Math. Soc. No. 320 (1985).

71. ———, *Deciding eventual positivity of polynomials*, Ergodic Theory Dynamical Systems (to appear).

72. D. Handelman, D. Higgs, and J. Lawrence, *Directed abelian groups, countably continuous rings, and Rickart* C^**-algebras*, J. London Math. Soc. (2) **21** (1980), 193–202.

73. D. Handelman and W. Rossmann, *Product type actions of finite and compact groups*, Indiana Univ. Math. J. **33** (1984), 479–509.

74. ———, *Actions of compact groups on* AF *algebras*, Illinois J. Math. **29** (1985), 51–95.

75. F. Jellett, *Homomorphisms and inverse limits of Choquet simplexes*, Math. Z. **103** (1968), 219–226.

76. R. V. Kadison, *A representation theory for commutative topological algebras*, Mem. Amer. Math. Soc. No. 7 (1951).

77. ——, *Unitary invariants for representations of operator algebras*, Ann. of Math. (2) **66** (1957), 304–379.

78. L. V. Kantorovitch, *Sur la théorie générale des opérations dans les espaces semi-ordonnés*, Dokl. Akad. Nauk SSSR **10** (1936), 283–286.

79. ——, *On the moment problem for a finite interval*, Dokl. Akad. Nauk SSSR **14** (1937), 531–537.

80. ——, *Lineare halbgeordnete Räume*, Mat. Sb. **2(44)** (1937), 121–168.

81. ——, *Linear operations in semi-ordered spaces*. I, Mat. Sb. **7(49)** (1940), 209–284.

82. J. L. Kelley, *Note on a theorem of Krein and Milman*, J. Osaka Inst. Sci. Tech., Part 1 Math. Phys. **3** (1951), 1–2.

83. J. L. Kelley and I. Namioka, *Linear topological spaces*, Van Nostrand, Princeton, N. J., 1963.

84. D. G. Kendall, *Simplexes and vector lattices*, J. London Math. Soc. **37** (1962), 365–371.

85. M. Krein, *On positive additive functionals in linear normed spaces*, Zapiski Naukovo-Doslidnovo Inst. Mat. Mech. Kharkiv. Mat. Tovaristva (4) **14** (1937), 227–237. (Ukrainian; French summary)

86. ——, *Sur la décomposition minimale d'une fonctionelle linéaire en composantes positives*, Dokl. Akad. Nauk SSSR **28** (1940), 18–22.

87. M. Krein and D. Mil′man, *On extreme points of regular convex sets*, Studia Math. **9** (1940), 133–137.

88. W. Krieger, *On dimension functions and topological Markov chains*, Invent. Math. **56** (1980), 239–250.

89. J. W. Lawrence, *A simple interpolation group which is not unperforated* (to appear).

90. A. J. Lazar and J. Lindenstrauss, *Banach spaces whose duals are L_1 spaces and their representing matrices*, Acta Math. **126** (1971), 165–193.

91. J. Lindenstrauss, *Extension of compact operators*, Mem. Amer. Math. Soc. No. 48 (1964).

92. J. Lindenstrauss, G. Olsen, and Y. Sternfeld, *The Poulsen simplex*, Ann. Inst. Fourier (Grenoble) **28** (1978), 91–114.

93. S. Mazur, *Über konvexe Mengen in linearen normierten Räumen*, Studia Math. **4** (1933), 70–84.

94. D. Mil′man, *Characteristics of extremal points of regularly convex sets*, Dokl. Akad. Nauk SSSR **57** (1947), 119–122. (Russian)

95. H. Minkowski, "Theorie der konvexen Körper, insbesondere Begründung ihres Oberflächenbegriffs", in *Gesammelte Abhandlungen von Hermann Minkowski*. Vol. II, Teubner, Leipzig, 1911; reprint, Chelsea, New York, 1967, pp. 131–229.

96. H. Nakano, *Über das System aller stetigen Funktionen auf einem topologischen Raum*, Proc. Imp. Acad. Tokyo **17** (1941), 308–310.

97. T. Nakayama, *On Krull's conjecture concerning completely integrally closed integrity domains.* I, Proc. Imp. Acad. Tokyo **18** (1942), 185–187.

98. I. Namioka, *Partially ordered linear topological spaces*, Mem. Amer. Math. Soc. No. 24 (1957).

99. I. Namioka and R. R. Phelps, *Tensor products of compact convex sets*, Pacific J. Math. **31** (1969), 469–480.

100. A. L. Peressini, *Ordered topological vector spaces*, Harper & Row, New York, 1967.

101. R. R. Phelps, *Lectures on Choquet's theorem*, Van Nostrand, Princeton, N. J., 1966.

102. ——, "Integral representations for elements of convex sets," in *Studies in functional analysis*, edited by R. G. Bartle, Studies in Math., vol. 21, Math. Assoc. of Amer., Washington, D. C., 1980, pp. 115–157.

103. M. Pimsner and D. Voiculescu, *Exact sequences for K-groups and* Ext-*groups of certain cross-product C^*-algebras*, J. Operator Theory **4** (1980), 93–118.

104. ——, *Imbedding the irrational rotation C^*-algebra into an* AF-*algebra*, J. Operator Theory **4** (1980), 201–210.

105. E. T. Poulsen, *A simplex with dense extreme points*, Ann. Inst. Fourier (Grenoble) **11** (1961), 83–87.

106. N. Riedel, *A counterexample to the unimodular conjecture on finitely generated dimension groups*, Proc. Amer. Math. Soc. **83** (1981), 11–15.

107. ——, *Dimension groups and traces of simple approximately finite dimensional C^*-algebras*, Preprint TUM-INFO-7910, Tech. Univ. München, 1979.

108. M. A. Rieffel, *C^*-algebras associated with irrational rotations*, Pacific J. Math. **93** (1981), 415–429.

109. F. Riesz, *Sur quelques notions fondamentales dans la théorie générale des opérations linéaires*, Ann. of Math. **41** (1940), 174–206.

110. S. Sakai, *C^*-algebras and W^*-algebras*, Ergebnisse der Math. und ihre Grenzgebiete, Band 60, Springer-Verlag, Berlin, 1971.

111. H. H. Schaefer, *Banach lattices and positive operators*, Springer-Verlag, Berlin, 1974.

112. G. L. Seever, *Measures on F-spaces*, Trans. Amer. Math. Soc. **133** (1968), 267–280.

113. Z. Semadeni, *Free compact convex sets*, Bull. Acad. Polon. Sci. Sér. Sci. Math. Astr. Phys. **13** (1965), 141–146.

114. ——, *Banach spaces of continuous functions.* Vol. I, PWN, Warsaw, 1971.

115. C.-L. Shen, *On the classification of the ordered groups associated with the approximately finite-dimensional C^*-algebras*, Duke Math. J. **46** (1979), 613–633.

116. M. H. Stone, *A general theory of spectra.* II, Proc. Nat. Acad. Sci. U.S.A. **27** (1941), 83–87.

117. ——, *Pseudo-norms and partial orderings in abelian groups*, Ann. of Math. **48** (1947), 851–856.

118. ——, *Boundedness properties in function-lattices*, Canad. J. Math. **1** (1949), 176–186.

119. J. R. Teller, *On partially ordered groups satisfying the Riesz interpolation property*, Proc. Amer. Math. Soc. **16** (1965), 1392–1400.

120. B. Vulich, *Une définition du produit dans les espaces semi-ordonnés linéaires*, Dokl. Akad. Nauk SSSR **26** (1940), 850–854.

Index

absolutely continuous positive homomorphism, 197
AF C^*-algebra, xviii
Aff(−), 113
affine
  combination, 74
  homeomorphism, 75
  isomorphism, 74
  map, 74
  span, 74
  subspace, 74
affinely
  dependent, 74
  homeomorphic, 75
  independent, 74
  isomorphic, 74
$\aleph_0$-continuous regular ring, 311
antitone map, xxii
approximately finite-dimensional C^*-algebra, xviii
archimedean, 20
atom, 48
AW^*-algebra, 313

$B(-,-)$, 129
base for convex cone, 154
basically disconnected space, 141
boolean algebra, xxii
bounded homomorphism, 123

$C(-,-)$, 5
category
  of compact convex sets, 75
  of convex sets, 74
  of partially ordered abelian groups, 12
    with order-unit, 12
characteristic element, 127
Choquet simplex, 163
classical simplex, 153
closed interval, xxi
complementary face, 161
complete lattice, xxi
completion with respect to positive homomorphism, 190
concave function, 173
conditionally
  complete, 7
  σ-complete, 8
cone, 3, 154
continuous extension of positive homomorphism to completion, 197
convex
  combination, 73
  cone, 154
  function, 173
  hull, 74
  set, 73
  subgroup, 8
    generated by (−), 8
  subset of partially ordered set, 8

coproduct
of compact convex sets, 77
of convex sets, 76
of partially ordered abelian groups, 13, 17
countable interpolation property, 263
C^*-algebra extension, 313

$D(-)$, xv
DCC, 50
Dedekind
complete, 7
complete lattice cone, 162
σ-complete, 8
descending chain condition, 50
dimension group, xvii, 44
extension, 287, 295
dimension
of affine subspace, 74
of convex set, 74
direct convex sum, 75
directed, xxi, 4
abelian group, 4
subgroup, 4
direct limit of partially ordered abelian groups, 13, 14
directly finite, xv
discrete ordering, 3
discrete state, 70
downward directed, xxi
dual, xxi
ordering, xxi

$\varepsilon_{(-)}$, 89
equivalent
extensions, 288, 296, 314
unital C^*-algebra extensions, 314
essential dimension group extension, 321
extension, 28, 286, 295, 313
with order-unit, 295
extreme
boundary, 78
point, 78

face, 79
generated by $(-)$, 79
finite C^*-algebra, 313
F-space, 279

general comparability, 131, 312
generating interval, 294
Grot$(-)$, xiv
Grot$(-)^+$, xiv
Grothendieck group, xiv

ideal, xviii, 8
interpolation, 23
group, 23
inverse limit
of compact convex sets, 77
of convex sets, 75
of partially ordered abelian groups, 14, 15
irrational rotation algebra, xvi
isotone map, xxii

ker$(-)$, 97
kernel of set of states, 97
Krein-Mil′man Theorem, 85
$K^0(-)$, xiv
$K_0(-)$, xiv, xv

$L(-)$, 311
lattice, xxi
cone, 156
homomorphism, xxii
isomorphism, xxii
of principal right ideals, 311
lattice-ordered abelian group, 5
lexicographic
direct product, 18
direct sum, 18
extension, 286
ordering, 18, 286
sum extension, 286
$\varinjlim$, 14
$\varprojlim$, 14, 15
locally convex, 84
Hausdorff space, 84

locally convex space, 84
lower semicontinuous function, 172

$M_1^+(-)$, 87
matricial algebra, xvi
Max $B(-,-)$, 131
maximal ideal, 256
MaxSpec$(-)$, 258
metric derived from positive
homomorphism, 190
Minkowski functional, 82
monotone σ-complete, 269

$\mathbb{N}$, xxii
natural
affine homeomorphism of $(-)$ onto
$S(\mathrm{Aff}(-), 1)$, 115
extension of positive
homomorphism to
completion, 192
homeomorphism of $(-)$ onto
$\partial_e M_1^+(-)$, 90
map from $(-)$
to $\mathrm{Aff}(S(-,-))$, 117
to $C(\partial_e S(-,-), \mathbb{R})$, 117
normalized
finite trace, 310
positive homomorphism, 12
norm-closed, 122
norm-complete, 237
norm-completion, 238
norm of bounded
homomorphism, 123
norm-topology, 122
N^*, 310
N^*-complete regular ring, 310

open interval, xxi
order
anti-automorphism, xxii
anti-isomorphism, xxii
order-automorphism, xxii
order-convex subset, 8
order-isomorphism, xxii
order-preserving map, xxii
order-reversing map, xxii
order-unit, xv, 4
metric, 237
norm, 120

$p_{(-)}$, 128
$\mathbb{P}(-)$, 309
partial order, xxi
partially ordered
abelian group, xv, 1
real Banach space, 113
set, xxi
vector space, 1
perforated, 19
permutation automorphism, 228
point mass, 89
pointwise ordering, xxi, 4
poset, xxi
positive
cone, 3
convex combination, 73
element, 2
homomorphism, 12
σ-convex combination, 97, 117
Poulsen simplex, 318
pre-order, 1
pre-ordered abelian group, xv, 1
probability measure, 87
product
of compact convex sets, 77
of convex sets, 75
ordering, 12
projection, 313
base, 129
pseudo-rank function, xix
pullback, 15
pushout, 15, 16, 17

$\mathbb{Q}^+$, 3
quotient ordering, 10

$\mathbb{R}^+$, 3
rational convex combination, 111
regular ring, xviii
of finite Rickart C^*-algebra, 313
relatively bounded map, 39

restricted directed product, 13
reverse
  lexicographic direct sum, 18
  lexicographic ordering, 18
Rickart C^*-algebra, 312
Riesz
  decomposition properties, 23
  interpolation property, xvii, 23

$S(-)$, 101
$S(-,-)$, 95
semicontinuous function, 172
short exact sequence of partially ordered abelian groups, 286
σ-convex
  combination, 97, 117
  hull, 117
  subset, 117
simple partially ordered abelian group, xviii, 218
simplex, 153, 156, 163
simplicial
  basis, 47
  group, 47
stably isomorphic, xiv
standard finite-dimensional simplex, 153
*-regular ring, 313
state, xix, 60, 310
  space, 95
  space functor, 101
stenotic
  dimension group extension, 314, 321
  unital C^*-algebra extension, 314
strict
  cone, 3, 154
  convex cone, 154
  interpolation, 31
  ordering, xxi, 5, 12
strictly positive element, 2
subadditive map, 38
sublinear functional, 81
$T(-)$, 310
tensor product of partially ordered abelian groups, 322
  with order-unit, 322
torsion-free rank of abelian group, 230
totally ordered, xxi
  abelian group, 1
total order, xxi
trace space, 310
tracial state, 310
translation-invariant, 1

ultramatricial algebra, xvi
unital C^*-algebra extension, 313
unitification, xv
unit-regular ring, xv
unperforated, xvii, 19
upper semicontinuous function, 172
upward directed, xxi

von Neumann regular ring, xviii

$\mathbb{Z}^+$, xxii, 3

$[-]$, xiv
$\ll$, xxi, 5, 12
$\leqslant$, xxi, 5, 12
$\leq_C$, 3
$\leq^+$, 38
$\wedge$, xxii
$\vee$, xxii
$(-)^+$, 3
$\partial_e(-)$, 78
$\|\cdot\|$, 120, 123
$\|\cdot\|_{(-)}$, 120, 123
$|\cdot|_{(-)}$, 189

ABCDEFGHIJ — 89876